Symmetries in Science XIV

Journal of Physics: Conference Series Volume 237

Bregenz, Austria
19-24 July 2009

ISBN: 978-1-61738-747-0
ISSN: 1742-6588

Printed from e-media with permission by:

Curran Associates, Inc.
57 Morehouse Lane
Red Hook, NY 12571

Printed by Curran Associates, Inc. (2010)

For permission requests, please contact the Institute of Physics at the address below.

Institute of Physics
Dirac House, Temple Back
Bristol BS1 6BE UK

Phone: 44 1 17 929 7481
Fax: 44 1 17 920 0979

techtracking@iop.org

Additional copies of this publication are available from:

Curran Associates, Inc.
57 Morehouse Lane
Red Hook, NY 12571 USA
Phone: 845-758-0400
Fax: 845-758-2634
Email: curran@proceedings.com
Web: www.proceedings.com

TABLE OF CONTENTS

Symmetries in Science XIV

July 19–24, 2009

Kloster Mehrerau, Bregenz, Austria

This volume of the proceedings "Symmetries in Science XIV" is dedicated to the memory of our colleagues and dear friends **Marcos Moshinsky** and **Yuriĭ Smirnov** who regularly participated in these Symposia and were a great inspiration to many. We shall miss them.
Dieter Schuch and Michael Ramek

The international symposium "Symmetries in Science XIV" held at Collegium Mehrerau in Bregenz, Austria from July 19-24, 2009, attended by 32 scientists from 11 countries, was an experiment, performed by theoreticians. Aim of this experiment was to find out if the desire to revive or even continue this conference series was stronger than the very restricted pecuniary boundary conditions. It obviously was!

After its establishment by Bruno Gruber in 1979, the biennial series settled in the very stimulating atmosphere of the monastery Mehrerau, which provided the ideal environment for a limited number of invited participants to exchange ideas, without parallel sessions, and pursue deeper discussions (at the latest in the evening at "Gasthof Lamm").

When the conference series terminated in 2003, former participants were quite disappointed. Meeting again at several (larger) conferences in subsequent years, there were repeated expressions of "the lack of a Bregenz-type meeting in our field nowadays" and the question of a possible "revitalization", even without external funding. After some hesitation, but also driven by our own desire to reinstate the series, we consulted Bruno who not only approved wholeheartedly but also offered his full support. It all finally led to the symposium in July 2009. The atmosphere was really like in the "good old days" and the interesting and thought-provoking presentations culminated in the publication of these Proceedings. We are grateful to Carl Bender for establishing contact with IOP making it possible for us to publish these Proceedings in the *Journal of Physics Conference Series.*

A majority of the participants contributed to these Proceedings though some were unable to do so as their results were either previously submitted or published elsewhere. In order to make these contributions also available to readers of the Proceedings, we now list the respective references as follow:

- Albertsson C, Hlavatý L and Šnobl L 2008 On the Poisson-Lie T-plurality of boundary conditions *J. Math. Phys.* **49** 032301-23, arXiv:0706.0820;
- Bender C M, Brody D C and Hook D W 2008 Quantum effects in classical systems having complex energy *J. Phys. A: Math. Theor.* **41** 352003 (15 pp);
- Kanellopoulos V, Kleber M and Kramer T 2009 Use of Lambert's theorem for the n-dimensional Coulomb problem *Phys. Rev. A* **80** 012101;
- Kramer P 2010 Platonic topology and CMB fluctuations: homotopy anisotropy and multipole selection rules *Class. Quantum Grav.* **27** 095013 (27 pp), arXiv:0909.2758v1;
- Quesne C 2009 Solvable rational potentials and exceptional orthogonal polynomials in supersymmetric quantum mechanics *SIGMA* **5** 084 (24 pp).

On the other hand, some colleagues planning on attending the meeting had to forfeit their participation due to unforeseeable circumstances. As they had prepared their contributions particularly for our symposium, we have also allowed their input. The articles of Kibler, Leach, Reinisch and Wolf have therefore been included though the authors were absent.

Marcos Moshinsky and Yuriĭ Smirnov, two long-standing colleagues and friends who attended the symposium on multiple occasions passed away in the year before the conference. A memorial service was held at the chapel of the Mehrerau during the meeting. The commemorative words from Father Vinzenz were based on information from our own experience, from an article by Alejandro Frank in Letras Libres Junio 2009, p.54 (a monthly literary magazine published by Editorial Vuelta, in Spanish) and articles from "TESTIMONIOS, Marcos Moshinsky: 80 años de vida y 60 años de trabajo científico" 2001 Frank A and Wolf K B eds. Two contributions dedicated especially in their memory are also included in these Proceedings.

It was especially rewarding and greatly appreciated that symposium-founder Bruno Gruber attended all the sessions and that Dr. Hubert Regner, a distinguished official of the provincial administration and ardent supporter of the symposia for over twenty years, honoured us with a visit and an encouraging address to the participants.

We wish to express our sincere gratitude to the local community, particularly the Schenk Family and the staff of Collegium Mehrerau for the selfless friendship, generosity and kind hospitality they offered our gathering. It made a lasting impression on participants and guests alike and provided an excellent basis for fruitful scientific discussions and personal interactions.

This and the positive resonance from participants have encouraged us to take the experiment a step further to "Symmetries in Science 2011"!

Thanks also to Yvette for continuous and reliable support. The conference and proceedings would probably not have materialized without her.

Frankfurt am Main and Graz, June 2010
Dieter Schuch
Michael Ramek

An interesting modular structure associated to Landau levels

S. Twareque Ali

Department of Mathematics and Statistics, Concordia University, Montréal, Québec, CANADA H3G 1M8

E-mail: `stali@mathstat.concordia.ca`

Abstract. We analyze here, from a von Neumann algebraic point of view, the well known problem of a charged particle, moving on the infinite two-dimensional plane under the influence of a constant magnetic field, perpendicular to this plane. The energy levels of this system, known in the literature as Landau levels, are those of the harmonic oscillator, with each level being infinitely degenerate. There is an interesting algebraic property associated with this problem, that we discuss here. It arises from the fact that there are two mutually commuting von Neumann algebras associated to this system, which correspond to the two possible directions of the constant magnetic field. This commuting pair has a modular structure, in the sense of the Tomita-Takesaki theory, leading to the existence of KMS states for the system with physically interesting properties. An example of such a state is the Gibbs state related to the interaction part of the Landau Hamiltonian, which in this case is invariant under the modular automorphism group, generated by this Hamiltonian. Also associated to the problem is a family of orthogonal polynomials, the complex Hermite polynomials, which form a basis of the underlying Hilbert space.

1. Introduction

The quantum mechanical Hamiltonian of an electron moving in a constant electromagnetic field has eigenvalues identical to those of the harmonic oscillator, except that each level is infinitely degenerate. These are the well-known Landau levels [1], their degeneracy being explained by the fact that quantum mechanically, the centre of the (classical circular) orbit of the electron cannot be precisely determined. If the direction of the magnetic field is reversed, one obtains a second Hamiltonian, similar to the first, but commuting with it. Being oscillator type Hamiltonians, these can be written in terms of two sets of creation and annihilation operators, which mutually commute. In fact, the von Neumann algebras generated by this commuting pair of sets are mutual commutants, leading to the existence of a modular structure, in the sense of the Tomita-Takesaki theory [12]. The KMS state predicted by the theory is given essentially by the Gibbs state for a harmonic oscillator. The KMS state is then invariant under the action of the time evolution generated by the associated modular operator and obtainable in terms of the interaction Hamiltonian. The modular conjugation operator simply interchanges the two possible orientations of the magnetic field and mathematically, in an equivalent complex representation, amounts to a complex conjugation of the relevant quantities. We also look at some related complex analytic properties of the system, touching on complex Hermite polynomials which generate a basis for the underlying Hilbert space of the problem, in this particular complex analytic representation.

Some earlier results along this direction were reported in [3, 4, 5], so that much of this paper is of a review nature. The general mathematical theory of the modular structure associated to the Hilbert space of Hilbert Schmidt operators has also been described in detail in [6] (Chapter 8) and [7] (Section II). Indeed, a structure similar to that reviewed here appears is a large number of quantum optical systems (see [7] and references to the earlier literature therein). The modular structure associated to the canonical commutation relations on $L^2(\mathbb{R}^2)$ was already observed in [8] and further elaborated in [5]. In [9] it was shown how a similar structure is present in the set of discrete representations of a locally compact group. To emphasize again, in this report we draw attention the existence of a modular structure in a rather well known physical problem, *viz* that of the Landau levels. As just indicated, the interesting point here is that the two commuting algebras correspond to the two possible directions of the magnetic field, giving them a specific physical interpretation. Furthermore, the modular automorphism group is mediated by the interaction Hamiltonian.

2. The physical problem

Consider a single electron of unit charge, placed in the xy-plane and subjected to a constant magnetic field, pointing along the *positive z-direction*. Choosing the magnetic vector potential to be $\vec{A}^{\uparrow} := \vec{A} = \frac{1}{2}(-y, x, 0)$ (so that the magnetic field, $\vec{B} = \nabla \times \vec{A}^{\uparrow} = (0, 0, 1)$), the classical Hamiltonian of the system, in some rationalized units, can be written as

$$H_{\text{elec}} = \frac{1}{2}(\vec{p} - \vec{A})^2 = \frac{1}{2}\left(p_x + \frac{y}{2}\right)^2 + \frac{1}{2}\left(p_y - \frac{x}{2}\right)^2 , \tag{1}$$

On the Hilbert space $\tilde{\mathfrak{H}} = L^2(\mathbb{R}^2, dxdy)$ of our problem, we make the replacements

$$p_x + \frac{y}{2} \longrightarrow Q_- = -i\frac{\partial}{\partial x} + \frac{y}{2}, \qquad p_y - \frac{x}{2} \longrightarrow P_- = -i\frac{\partial}{\partial y} - \frac{x}{2} . \tag{2}$$

Then, $[Q_-, P_-] = iI_{\tilde{\mathfrak{H}}}$ and the quantum Hamiltonian corresponding to H_{elec} becomes

$$H^{\uparrow} = \frac{1}{2}\left(P_-^2 + Q_-^2\right) . \tag{3}$$

The eigenvalues of this Hamiltonian, the so-called *Landau levels*, are $E_\ell = (\ell + \frac{1}{2})$, $\ell = 0, 1, 2, \ldots \infty$, each level being infinitely degenerate, and we will denote the corresponding normalized eigenvectors by $\Psi_{n\ell}$, with $\ell = 0, 1, 2, \ldots, \infty$, indexing the energy level and $n = 0, 1, 2, \ldots, \infty$, the degeneracy at each level.

The degeneracy of the eigenvalues can be understood in the following way. If the magnetic field were aligned along the *negative z-axis* (with $\vec{A}^{\downarrow} = \frac{1}{2}(y, -x, 0)$ and $\vec{B} = \nabla \times \vec{A}^{\downarrow} = (0, 0, -1)$), the corresponding quantum Hamiltonian would be

$$H^{\downarrow} = \frac{1}{2}\left(P_+^2 + Q_+^2\right) . \tag{4}$$

with

$$Q_+ = -i\frac{\partial}{\partial y} + \frac{x}{2}, \qquad P_+ = -i\frac{\partial}{\partial x} - \frac{y}{2}, \tag{5}$$

and $[Q_+, P_+] = iI_{\tilde{\mathfrak{H}}}$. The two sets of operators $\{Q_\pm, P_\pm\}$, mutually commute:

$$[Q_+, Q_-] = [P_+, Q_-] = [Q_+, P_-] = [P_+, P_-] = 0 . \tag{6}$$

Thus, $[H^\downarrow, H^\uparrow] = 0$ and the eigenvectors $\Psi_{n\ell}$ of $H^\uparrow$ can be chosen so that they are also the eigenvectors of $H^\downarrow$ in the manner

$$H^\downarrow \Psi_{n\ell} = \left(n + \frac{1}{2}\right)\Psi_{n\ell}\,, \qquad H^\uparrow \Psi_{n\ell} = \left(\ell + \frac{1}{2}\right)\Psi_{n\ell}\,, \tag{7}$$

so that $H^\downarrow$ lifts the degeneracy of $H^\uparrow$ and vice versa. It ought to be pointed out, however, that this lifting of the degeneracy is not just a consequence of the commutativity of the two Hamiltonians $H^{\uparrow,\downarrow}$, but more so of the existence of the two commuting algebras generated by the $\{Q_\pm, P_\pm\}$, or equivalently, by the $\{A_\pm, A^*_\pm\}$, in (9) below.

We also see that the quantized Hamiltonian may be split into a free part H_0 and an interaction or angular momentum part, $H^\uparrow_{\text{int}}$:

$$\begin{cases} H^\uparrow = H_0 + H^\uparrow_{\text{int}}, \\ H_0 = H_{0,x} + H_{0,y} = \frac{1}{2}\left(\widehat{p}_x^2 + \dfrac{\widehat{x}^2}{4}\right) + \frac{1}{2}\left(\widehat{p}_y^2 + \dfrac{\widehat{y}^2}{4}\right), \\ H^\uparrow_{\text{int}} = -\dfrac{1}{2}(\widehat{x}\widehat{p}_y - \widehat{y}\widehat{p}_x) = -\widehat{l}_z\ . \end{cases} \tag{8}$$

with the usual definitions of $\widehat{x}, \widehat{p}_x$, etc. Of course, $[\widehat{x}, \widehat{p}_x] = [\widehat{y}, \widehat{p}_y] = iI_{\tilde{\mathfrak{H}}}$, while all the other commutators are zero.

Let us next define

$$\begin{aligned} A_+ &= \frac{1}{\sqrt{2}}(Q_+ + iP_+) \qquad & A^*_+ &= \frac{1}{\sqrt{2}}(Q_+ - iP_+) \\ A_- &= \frac{1}{\sqrt{2}}(iQ_- - P_-) \qquad & A^*_- &= \frac{1}{\sqrt{2}}(-iQ_- - P_-). \end{aligned} \tag{9}$$

These satisfy the commutation relations,

$$[A_\pm, A^*_\pm] = 1\ , \tag{10}$$

with all other commutators being zero. In terms of these, we may write the two Hamiltonians as (see (3) and (4)),

$$H^\uparrow = N_- + \frac{1}{2}I_{\tilde{\mathfrak{H}}}\,, \quad H^\downarrow = N_+ + \frac{1}{2}I_{\tilde{\mathfrak{H}}}\,, \quad \text{with} \quad N_\pm = A^*_\pm A_\pm\ . \tag{11}$$

Furthermore,

$$H_0 = \frac{1}{2}(N_+ + N_- + 1) \quad \text{and} \quad H^\uparrow_{\text{int}} = -\frac{1}{2}(N_+ - N_-)\,, \quad H^\downarrow_{\text{int}} = \frac{1}{2}(N_+ - N_-)\ . \tag{12}$$

The joint eigenstates of $H^\uparrow$ and $H^\downarrow$ are now easily written down. Let Ψ_{00} be such that $A_-\Psi_{00} = A_+\Psi_{00} = 0$. Then,

$$\Psi_{n\ell} := \frac{1}{\sqrt{n!\ell!}}\left(A^*_+\right)^n \left(A^*_-\right)^\ell \Psi_{00}, \tag{13}$$

where $n, \ell = 0, 1, 2, \ldots$ and we have, $N_+\Psi_{n\ell} = n\Psi_{n\ell}\,,\ \ N_-\Psi_{n\ell} = \ell\Psi_{n\ell}\,,\ \ H_0\Psi_{n\ell} = \frac{1}{2}(n+\ell+1)\Psi_{n\ell}$ and $H^\uparrow_{\text{int}}\Psi_{n\ell} = \frac{1}{2}(n-\ell)\Psi_{n\ell}$. A simple argument then shows that the vectors $\Psi_{n\ell}$ span the Hilbert space $\mathfrak{H} = L^2(\mathbb{R}^2, dx\ dy)$.

We now change over to a complex representation, introducing the complex variable $z = \frac{1}{\sqrt{2}}(x - iy)$ and its associated derivative $\partial_z = \frac{1}{\sqrt{2}}(\partial_x + i\partial_y)$. In terms of theses

$$A_- = \frac{1}{2}\overline{z} + \partial_z\ , \quad A_+ = \frac{1}{2}z + \partial_{\overline{z}} \qquad \text{and} \qquad A_-^* = \frac{1}{2}z - \partial_{\overline{z}}\ , \quad A_+^* = \frac{1}{2}\overline{z} - \partial_z\ , \tag{14}$$

In this representation, the ground state $\Psi_{00}(\overline{z}, z)$ is the solution of the equations $A_+\Psi_{00}(\overline{z}, z) = A_-\Psi_{00}(\overline{z}, z) = 0$, so that, $\Psi_{00}(\overline{z}, z) = \sqrt{\frac{1}{2\pi}}\, e^{-\frac{1}{2}|z|^2}$. Thus, using (13) we get

$$\Psi_{n\ell}(\overline{z}, z) = \frac{1}{\sqrt{n!\ell!}}\left(\frac{1}{2}\overline{z} - \partial_z\right)^n \left(\frac{1}{2}z - \partial_{\overline{z}}\right)^\ell \Psi_{00}(z, \overline{z})\ . \tag{15}$$

When using this representation, we shall write our Hilbert space as $\widetilde{\mathfrak{H}} = L^2(\mathbb{C}, \frac{d\overline{z}\wedge dz}{i})$ and then it is useful to make the further unitary transformation

$$\mathcal{V} : L^2(\mathbb{C}, \frac{d\overline{z}\wedge dz}{i}) \longrightarrow L^2(\mathbb{C}, d\nu(\overline{z}, z)) \quad \text{where} \quad d\nu(\overline{z}, z) = \frac{e^{-|z|^2}}{2\pi}\, \frac{d\overline{z}\wedge dz}{i}\ , \tag{16}$$

to the more convenient Hilbert space $L^2(\mathbb{C}, d\nu(\overline{z}, z))$, using the mapping

$$(\mathcal{V}\Psi)(\overline{z}, z) = \sqrt{2\pi}\, e^{\frac{|z|^2}{2}}\ \Psi(\overline{z}, z)\ , \tag{17}$$

and to rewrite all the operators in question on this new space. Note that this space contains the two subspaces $\mathfrak{H}_{\text{hol}}$ and $\mathfrak{H}_{\text{a-hol}}$, of holomorphic and antiholomorphic functions, respectively. Both these subspaces contain the constant unit vector, $H_{00}(\overline{z}, z) = 1\ , \forall(\overline{z}, z)$. Apart from this one vector, all other vectors in the complementary subspaces of $\mathfrak{H}_{\text{hol}}$ and $\mathfrak{H}_{\text{a-hol}}$ are mutually orthogonal. Then,

$$\begin{aligned} \mathcal{A}_- &:= \mathcal{V}A_-\mathcal{V}^{-1} = \partial_z\ , & \mathcal{A}_+ &:= \mathcal{V}A_+\mathcal{V}^{-1} = \partial_{\overline{z}}\ , \\ \mathcal{A}_-^* &:= \mathcal{V}A_-^*\mathcal{V}^{-1} = z - \partial_{\overline{z}}\ , & \mathcal{A}_+^* &:= \mathcal{V}A_+^*\mathcal{V}^{-1} = \overline{z} - \partial_z\ , \end{aligned} \tag{18}$$

and

$$\begin{aligned} \mathcal{N}_+ &:= \mathcal{V}N_+\mathcal{V}^{-1} = \mathcal{A}_+^*\mathcal{A}_+ = -\partial_z\partial_{\overline{z}} + \overline{z}\partial_{\overline{z}} \\ \mathcal{N}_- &:= \mathcal{V}N_-\mathcal{V}^{-1} = \mathcal{A}_-^*\mathcal{A}_- = -\partial_z\partial_{\overline{z}} + z\partial_z\ . \end{aligned} \tag{19}$$

Writing $H_{n\ell} = \mathcal{V}\Psi_{n\ell}$, for the transformed basis vectors (15), we have,

$$\begin{aligned} H_{n\ell}(\overline{z}, z) &= \frac{1}{\sqrt{n!\ell!}}\left(\overline{z} - \partial_z\right)^n \left(z - \partial_{\overline{z}}\right)^\ell H_{00}(\overline{z}, z) = \frac{1}{\sqrt{n!\ell!}}\left(\mathcal{A}_+^*\right)^n \left(\mathcal{A}_-^*\right)^\ell H_{00}(\overline{z}, z)\ , \\ &= \frac{1}{\sqrt{n!\ell!}}\left(\overline{z} - \partial_z\right)^n (z^\ell) = \frac{1}{\sqrt{n!\ell!}}\left(z - \partial_{\overline{z}}\right)^\ell (\overline{z})^n\ . \end{aligned} \tag{20}$$

Also,

$$H_{n0}(\overline{z}, z) = \frac{\overline{z}^n}{\sqrt{n!}} \quad \text{and} \quad H_{0\ell}(\overline{z}, z) = \frac{z^\ell}{\sqrt{\ell!}}\ , \tag{21}$$

so that $\mathfrak{H}_{\text{a-hol}}$ is spanned by the vectors $H_{n0}\ , n = 0, 1, 2, \ldots$, and the space $\mathfrak{H}_{\text{hol}}$ by the vectors $H_{0\ell}\ ,\ \ell = 0, 1, 2, \ldots$.

The vectors $H_{n\ell}$ are joint eigenstates of the number operators:

$$\mathcal{N}_+ H_{n\ell} = \ell H_{n\ell}\,, \qquad \mathcal{N}_- H_{n\ell} = n H_{n\ell}\,. \tag{22}$$

Moreover, writing

$$\begin{aligned} \mathcal{H}^{\uparrow} &= \mathcal{V} H^{\uparrow} \mathcal{V}^{-1}\,, & \mathcal{H}^{\downarrow} &= \mathcal{V} H^{\downarrow} \mathcal{V}^{-1}\,, \\ \mathcal{H}_0 &= \mathcal{V} H_0 \mathcal{V}^{-1}\,, & \mathcal{H}_{\text{int}}^{\uparrow,\downarrow} &= \mathcal{V} H_{\text{int}}^{\uparrow,\downarrow} \mathcal{V}^{-1}\,, \end{aligned} \tag{23}$$

for the two Hamiltonians, we clearly have,

$$\mathcal{H}^{\uparrow} H_{n\ell} = \left(\ell + \frac{1}{2}\right) H_{n\ell}\,, \qquad \mathcal{H}^{\downarrow} H_{n\ell} = \left(n + \frac{1}{2}\right) H_{n\ell}\,. \tag{24}$$

The functions $h_{n,k}(\overline{z}, z) = \sqrt{n!k!}\; H_{nk}(\overline{z}, z)$ are just the *complex Hermite polynomials* [10, 11], also obtainable as:

$$h_{n,k}(\overline{z}, z) = (-1)^{n+k}\, e^{|z|^2} \partial_z^n \partial_{\overline{z}}^k\, e^{-|z|^2}\,. \tag{25}$$

Explicitly, the $h_{n,k}$ are given by

$$h_{n,k}(\overline{z}, z) = n!\, k! \sum_{j=0}^{n \curlyvee k} \frac{(-1)^j}{j!} \frac{(\overline{z})^{n-j}}{(n-j)!} \frac{z^{k-j}}{(k-j)!}\,, \tag{26}$$

where $n \curlyvee k$ denotes the smaller of the two numbers n and k. In particular,

$$h_{0,k}(\overline{z}, z) = z^k \quad \text{and} \quad h_{n,0}(\overline{z}, z) = \overline{z}^n\,. \tag{27}$$

One also has the useful series expansion,

$$\sum_{n=0}^{\infty} \sum_{k=0}^{\infty} \frac{v^n \overline{u}^k}{n!k!} h_{n,k}(\overline{z}, z) = \sum_{n=0}^{\infty} \sum_{k=0}^{\infty} \frac{v^n \overline{u}^k}{\sqrt{n!k!}} H_{nk}(\overline{z}, z) = e^{\overline{u}z + v\overline{z} - \overline{u}v}\,. \tag{28}$$

Furthermore,

$$h_{n,k}(\overline{z}, z) = h_{k,n}(z, \overline{z}) \quad \text{and} \quad h_{n,k}(\overline{z}, z) = ((\mathcal{A}_+^*)^n h_{0,k})(\overline{z}, z)\,. \tag{29}$$

They also satisfy the recursion relations,

$$\begin{aligned} h_{n+1,k}(\overline{z}, z) &= \overline{z}\, h_{n,k}(\overline{z}, z) - k\, h_{n,k-1}(\overline{z}, z) \\ h_{n,k+1}(\overline{z}, z) &= z\, h_{n,k}(\overline{z}, z) - n\, h_{n-1,k}(\overline{z}, z)\,. \end{aligned} \tag{30}$$

If we formally take z to be real in (25), the complex Hermite polynomials $h_{n,k}(\overline{z}, z)$ become the well-known real Hermite polynomials $h_{n+k}(x)$:

$$h_n(x) = (-1)^n\, e^{x^2} \partial_x^n\, e^{-x^2}\,.$$

The above complex representation of the problem is

3. Summary of the Tomita-Takesaki theory

In the next two sections we shall make a connection between the Landau levels obtained above and the Tomita-Takesaki theory of modular Hilbert algebras [12]. But first, we give in this section a highly condensed summary of this theory.

Let $\mathfrak{A}$ be a von Neumann algebra on a Hilbert space $\mathfrak{H}$ and $\mathfrak{A}'$ its commutant. Let $\mathbf{\Phi} \in \mathfrak{H}$ be a unit vector which is *cyclic and separating* for $\mathfrak{A}$. Then the corresponding state φ on the algebra, $\langle \varphi \,;\, A\rangle = \langle \mathbf{\Phi} \mid A\mathbf{\Phi}\rangle$, $A \in \mathfrak{A}$, is faithful and normal. Consider the antilinear map,

$$S : \mathfrak{H} \longmapsto \mathfrak{H}\,, \qquad S A\mathbf{\Phi} = A^*\mathbf{\Phi}\,,\ \forall A \in \mathfrak{A}\,. \tag{31}$$

Since $\mathbf{\Phi}$ is cyclic, this map is densely defined and in fact it can be shown that it is closable. We denote its closure again by S and write its polar decomposition as

$$S = J\Delta^{\frac{1}{2}} = \Delta^{-\frac{1}{2}}J\,, \quad \text{with} \quad \Delta = S^*S\,. \tag{32}$$

The operator Δ, called the *modular operator*, is positive and self-adjoint. The operator J, called the *modular conjugation operator*, is antiunitary and satisfies $J = J^*$, $J^2 = I_{\mathfrak{H}}$. The antiunitarity of J is to be understood as: $\langle J\phi \mid J\psi\rangle = \langle \psi \mid \phi\rangle$, $\forall \phi, \psi \in \mathfrak{H}$.

Since Δ is self-adjoint, using its spectral representation, we see that for $t \in \mathbb{R}$, the family of operators $\Delta^{-\frac{it}{\beta}}$, for some fixed $\beta > 0$, defines a unitary family of automorphisms of the algebra $\mathfrak{A}$. Denoting these automorphisms by $\alpha_\varphi(t)$, we may write,

$$\alpha_\varphi(t)[A] = \Delta^{\frac{it}{\beta}} A \Delta^{-\frac{it}{\beta}}\,, \quad \forall A \in \mathfrak{A}\,. \tag{33}$$

Thus, they constitute a strongly continuous one-parameter group of automorphisms, called the *modular automorphism group*. Denoting the generator of this one-parameter group by $\mathbf{H}_\varphi$, we get

$$\Delta^{-\frac{it}{\beta}} = e^{it\mathbf{H}_\varphi} \quad \text{and} \quad \Delta = e^{-\beta\mathbf{H}_\varphi}\,. \tag{34}$$

It can then be shown that the state φ is invariant under this automorphism group:

$$e^{-\beta\mathbf{H}_\varphi}\mathbf{\Phi} = \mathbf{\Phi}\,, \qquad \Delta^{\frac{it}{\beta}}\,\mathfrak{A}\,\Delta^{-\frac{it}{\beta}} = \mathfrak{A}\,, \tag{35}$$

and the antilinear map J interchanges $\mathfrak{A}$ with its commutant $\mathfrak{A}'$:

$$J\mathfrak{A}J = \mathfrak{A}'\,. \tag{36}$$

Finally, the state φ can be shown to satisfy the *KMS (Kubo-Martin-Schwinger) condition* (see for example [13]), with respect to the automorphism group $\alpha_\varphi(t)$, $t \in \mathbf{R}$, in the following sense: for any two $A, B \in \mathfrak{A}$, the function

$$F_{A,B}(t) = \langle \varphi \,;\, A\,\alpha_\varphi(t)[B]\rangle\,, \tag{37}$$

has an extension to the strip $\{z = t + iy \mid t \in \mathbb{R},\ y \in [0, \beta]\} \subset \mathbb{C}$ such that $F_{A,B}(z)$ is analytic in the open strip $(0, \beta)$ and continuous on its boundaries. Moreover, it also satisfies the boundary condition (*at inverse temperature* β),

$$\langle \varphi \,;\, A\,\alpha_\varphi(t + i\beta)[B]\rangle = \langle \varphi \,;\, \alpha_\varphi(t)[B]\,A\rangle\,, \quad t \in \mathbb{R}\,. \tag{38}$$

4. The case of Hilbert-Schmidt operators

We now proceed to demonstrate how a simple example of the Tomita-Takesaki theory is provided by two naturally commuting von-Neumann algebras carried by a Hilbert space of Hilbert-Schmidt operators. (A much fuller discussion is given, e.g., in [6, 7]). In the next section we shall transfer this structure to the Hilbert space of the Landau problem and the two commuting von-Neumann algebras generated by the operators $\mathcal{A}_\pm, \mathcal{A}_\pm^*$ and the identity operator, using a specific unitary map.

Let $\mathfrak{H}$ be a (complex, separable) Hilbert space of dimension N (finite or infinite) and $\{\zeta_i\}_{i=1}^N$ an orthonormal basis of it ($\langle \zeta_i \mid \zeta_j \rangle = \delta_{ij}$). We denote by $\mathcal{B}_2(\mathfrak{H}) \simeq \mathfrak{H} \otimes \overline{\mathfrak{H}}$ the space of all Hilbert-Schmidt operators on $\mathfrak{H}$. This is a Hilbert space with scalar product

$$\langle X \mid Y \rangle_2 = \mathrm{Tr}[X^* Y] .$$

The vectors,

$$\{X_{ij} = |\zeta_i\rangle\langle\zeta_j| \mid i, j = 1, 2, \ldots, N\} , \tag{39}$$

form an orthonormal basis of $\mathcal{B}_2(\mathfrak{H})$,

$$\langle X_{ij} \mid X_{k\ell} \rangle_2 = \delta_{ik}\delta_{\ell j} .$$

In particular, the vectors,

$$\mathbb{P}_i = X_{ii} = |\zeta_i\rangle\langle\zeta_i| , \tag{40}$$

are one dimensional projection operators on $\mathfrak{H}$. In what follows I will denote the identity operator on $\mathfrak{H}$ and I_2 that on $\mathcal{B}_2(\mathfrak{H})$.

We identify a special class of linear operators on $\mathcal{B}_2(\mathfrak{H})$, denoted by $A \vee B,\ A, B \in \mathcal{L}(\mathfrak{H})$, which act on a vector $X \in \mathcal{B}_2(\mathfrak{H})$ in the manner:

$$(A \vee B)(X) = AXB^* .$$

It is easy to see see that

$$(A \vee B)^* = A^* \vee B^* \quad \text{and} \quad (A_1 \vee B_1)(A_2 \vee B_2) = (A_1 A_2) \vee (B_1 B_2) . \tag{41}$$

Next we specify the two special von Neumann algebras,

$$\mathfrak{A}_\ell = \{A_\ell = A \vee I \mid A \in \mathcal{L}(\mathfrak{H})\} , \qquad \mathfrak{A}_\mathrm{r} = \{A_\mathrm{r} = I \vee A \mid A \in \mathcal{L}(\mathfrak{H})\} , \tag{42}$$

which turn out to be both factors and mutual commutants:

$$(\mathfrak{A}_\ell)' = \mathfrak{A}_\mathrm{r} , \qquad (\mathfrak{A}_\mathrm{r})' = \mathfrak{A}_\ell , \qquad \mathfrak{A}_\ell \cap \mathfrak{A}_\mathrm{r} = \mathbb{C} I_2 . \tag{43}$$

The operator $J : \mathcal{B}_2(\mathfrak{H}) \longrightarrow \mathcal{B}_2(\mathfrak{H})$, acting on the basis vectors X_{ij} in (39) as

$$J X_{ij} = X_{ji} \Longrightarrow J^2 = I_2 \quad \text{and} \quad J(|\phi\rangle\langle\psi|) = |\psi\rangle\langle\phi| , \quad \forall \phi, \psi \in \mathfrak{H} , \tag{44}$$

is antiunitary, and in fact,

$$J \mathfrak{A}_\ell J = \mathfrak{A}_\mathrm{r} . \tag{45}$$

4.1. A KMS state

We take $N = \infty$ and for some $\beta > 0$, consider the sequence of positive numbers

$$\alpha_n = (1 - e^{-\beta})e^{-(n-1)\beta} , \qquad \sum_{n=1}^{\infty} \alpha_n = 1 . \tag{46}$$

Let

$$\mathbf{\Phi} = \sum_{n=1}^{\infty} \alpha_n^{\frac{1}{2}} \mathbb{P}_n = \sum_{n=1}^{\infty} \alpha_n^{\frac{1}{2}} X_{nn} \in \mathcal{B}_2(\mathfrak{H}) . \tag{47}$$

We note the following properties of $\mathbf{\Phi}$:

- $\mathbf{\Phi}$ defines a *vector state* φ on the von Neumann algebra $\mathfrak{A}_\ell$. This follows from the fact that for any $A \vee I \in \mathfrak{A}_\ell$, we may define the state φ on $\mathfrak{A}_\ell$ by

$$\langle \varphi \, ; \, A \vee I \rangle = \langle \mathbf{\Phi} \mid (A \vee I)(\mathbf{\Phi}) \rangle_2 = \text{Tr}[\mathbf{\Phi}^* A \mathbf{\Phi}] = \text{Tr}[\rho_\varphi A] , \quad \text{with} \quad \rho_\varphi = \sum_{i=1}^{\infty} \alpha_i \mathbb{P}_i . \tag{48}$$

  Explicitly, $\mathbf{\Phi}$ represents the mixed state ρ_φ, defined on the algebra $\mathcal{L}(\mathfrak{H})$, as a vector state defined on an algebra which is isomorphic to the initial algebra, but acting on the enlarged Hilbert space of Hilbert-Schmidt operators on $\mathfrak{H}$ (this is a concrete and explicit example of the well known GNS construction).
- The state φ is faithful and normal. Normality follows from the last equality in (48) and the fact that ρ_φ is a density matrix. To check for faithfulness, note that for any $A \vee I \in \mathfrak{A}_\ell$,

$$\langle \varphi \, ; \, (A \vee I)^*(A \vee I) \rangle = \text{Tr}[\rho_\varphi A^* A] = \sum_{i=1}^{\infty} \alpha_i \|A\zeta_i\|^2 ,$$

  from which it follows that $\langle \varphi \, ; \, (A \vee I)^*(A \vee I) \rangle = 0$ if and only if $A = 0$ (since the ζ_i are an orthonormal basis set and the $\alpha_i > 0$), hence if and only if $A \vee I = 0$.
- The vector $\mathbf{\Phi}$ is cyclic and separating for $\mathfrak{A}_\ell$. Indeed, cyclicity follows from the fact that if $X \in \mathcal{B}_2(\mathfrak{H})$ is orthogonal to all $(A \vee I)\mathbf{\Phi}$, $A \in \mathcal{L}(\mathfrak{H})$, then

$$\text{Tr}[X^* A \mathbf{\Phi}] = \sum_{i=1}^{\infty} \alpha_i^{\frac{1}{2}} \langle \zeta_i \mid X^* A \zeta_i \rangle = 0, \qquad \forall A \in \mathcal{L}(\mathfrak{H}) .$$

  Taking $A = X_{k\ell}$, we easily get from the above equality, $\langle \zeta_\ell \mid X^* \zeta_k \rangle = 0$ and since this holds for all k, ℓ, we get $X = 0$. In the same way, $\mathbf{\Phi}$ is also cyclic for $\mathfrak{A}_{\text{r}}$, hence separating for $\mathfrak{A}_\ell$, i.e., $(A \vee I)\mathbf{\Phi} = (B \vee I)\mathbf{\Phi} \Longleftrightarrow A \vee I = B \vee I$.

4.2. Time evolution and modular automorphism

It will turn out (see below) that the state φ constructed above is indeed KMS for the time evolution $\alpha_\varphi(t)$, $t \in \mathbb{R}$, that we now define on the algebra $\mathfrak{A}_\ell$. With respect to this evolution, this state will be shown to satisfy,

$$\langle \varphi \, ; \, A_\ell \, \alpha_\varphi(t + i\beta)[B_\ell] \rangle = \langle \varphi \, ; \, \alpha_\varphi(t)[B_\ell] \, A_\ell \rangle , \quad \forall A_\ell, B_\ell \in \mathfrak{A}_\ell . \tag{49}$$

Furthermore, the function,

$$F_{A_\ell, B_\ell}(z) = \langle \varphi \, ; \, A_\ell \, \alpha_\varphi(z)[B_\ell] \rangle , \tag{50}$$

is analytic in the strip $\{\Im(z) \in (0, \beta)\}$ and continuous on its boundaries.

Define the operators,

$$\mathbf{P}_{ij} = \mathbb{P}_i \vee \mathbb{P}_j , \qquad i, j = 1, 2, \ldots, \infty \tag{51}$$

the $\mathbb{P}_i$ being the projection operators on $\mathfrak{H}$ defined in (40). Then, the $\mathbf{P}_{ij}$ are projection operators on the Hilbert space $\mathcal{B}_2(\mathfrak{H})$. Next, using ρ_φ in (48), define the *Hamiltonian* H_φ as:

$$\rho_\varphi = e^{-\beta H_\varphi} \Longrightarrow H_\varphi = -\frac{1}{\beta}\sum_{i=1}^{\infty} (\ln \alpha_i)\, \mathbb{P}_i , \tag{52}$$

and using it define the operators:

$$H_\varphi^\ell = H_\varphi \vee I , \qquad H_\varphi^{\mathrm{r}} = I \vee H_\varphi , \qquad \mathbf{H}_\varphi = H_\varphi^\ell - H_\varphi^{\mathrm{r}} . \tag{53}$$

A straightforward computation then establishes the spectral form of $\mathbf{H}_\varphi$:

$$\mathbf{H}_\varphi = -\frac{1}{\beta}\sum_{i,j=1}^{\infty} \ln\left[\frac{\alpha_i}{\alpha_j}\right] \mathbf{P}_{ij} = \sum_{i,j=1}^{\infty} (i-j)\mathbf{P}_{ij} . \tag{54}$$

Writing

$$\Delta_\varphi := \sum_{i,j=1}^{\infty} \left[\frac{\alpha_i}{\alpha_j}\right] \mathbf{P}_{ij} = e^{-\beta \mathbf{H}_\varphi} , \tag{55}$$

we define a time evolution operator on $\mathcal{B}_2(\mathfrak{H})$:

$$e^{i\mathbf{H}_\varphi t} = [\Delta_\varphi]^{-\frac{it}{\beta}} , \qquad t \in \mathbb{R} , \tag{56}$$

and we note that, for any $X \in \mathcal{B}_2(\mathfrak{H})$,

$$\begin{aligned} e^{i\mathbf{H}_\varphi t}(X) &= \sum_{i,j=1}^{\infty} \left[\frac{\alpha_i}{\alpha_j}\right]^{-\frac{it}{\beta}} \mathbf{P}_{ij}(X) = \left[\sum_{i=1}^{\infty} (\alpha_i)^{-\frac{it}{\beta}} \mathbb{P}_i\right] \vee \left[\sum_{j=1}^{\infty} (\alpha_j)^{-\frac{it}{\beta}} \mathbb{P}_j\right](X) \\ &= e^{iH_\varphi t}(X) e^{-iH_\varphi t} , \end{aligned}$$

so that

$$e^{i\mathbf{H}_\varphi t} = e^{iH_\varphi t} \vee e^{iH_\varphi t} , \tag{57}$$

where H_φ is the operator defined in (52). From the definition of the vector $\mathbf{\Phi}$ in (47), it is clear that it commutes with H_φ and hence that it is invariant under this time evolution:

$$e^{i\mathbf{H}_\varphi t}(\mathbf{\Phi}) = e^{iH_\varphi t}\, \mathbf{\Phi}\, e^{-iH_\varphi t} = \mathbf{\Phi} . \tag{58}$$

Finally, using $e^{i\mathbf{H}_\varphi t}$ we define the time evolution α_φ on the algebra $\mathfrak{A}_\ell$, in the manner (see (33)):

$$\alpha_\varphi(t)[A_\ell] = e^{i\mathbf{H}_\varphi t}\, A_\ell\, e^{-i\mathbf{H}_\varphi t} \qquad \forall A_\ell \in \mathfrak{A}_\ell . \tag{59}$$

Writing $A_\ell = A \vee I$, $A \in \mathcal{L}(\mathfrak{H})$, and using the composition law (41), we see that

$$e^{i\mathbf{H}_\varphi t}\, A_\ell\, e^{-i\mathbf{H}_\varphi t} = \left[e^{iH_\varphi t}\, A\, e^{-iH_\varphi t}\right] \vee I , \tag{60}$$

so that by virtue of (48),

$$\langle \varphi \,;\, \alpha_\varphi(t)[A_\ell]\rangle = \mathrm{Tr}\left[\rho_\varphi\, e^{iH_\varphi t}\, A\, e^{-iH_\varphi t}\right] = \langle \varphi \,;\, A_\ell \rangle , \tag{61}$$

since ρ_φ and H_φ commute. Thus, the state φ is invariant under the time evolution α_φ.

To obtain the KMS condition (49), combining (59) and (60), we first note that, with $A_\ell = A \vee I$, $B_\ell = B \vee I$,

$$A_\ell \alpha_\varphi(t)[B_\ell] = \left[A e^{iH_\varphi t}\, B\, e^{-iH_\varphi t}\right] \vee I\ .$$

Hence, again using (48),

$$F_{A_\ell, B_\ell}(t) = \langle \varphi\, ;\, A_\ell\, \alpha_\varphi(t)[B_\ell]\rangle = \mathrm{Tr}\left[\rho_\varphi A e^{iH_\varphi t}\, B\, e^{-iH_\varphi t}\right] = \mathrm{Tr}\left[\rho_\varphi e^{-iH_\varphi t}\, A\, e^{iH_\varphi t} B\right]\ ,$$

the last equality following from the commutativity of ρ_φ and H_φ. Thus, since $\rho_\varphi = e^{-\beta H_\varphi}$,

$$F_{A_\ell, B_\ell}(t + i\beta) = \mathrm{Tr}\left[\rho_\varphi e^{-iH_\varphi t} e^{\beta H_\varphi}\, A\, e^{iH_\varphi t} e^{-\beta H_\varphi} B\right] = \mathrm{Tr}\left[e^{-iH_\varphi t}\, A\, e^{iH_\varphi t} \rho_\varphi B\right]\ ,$$

so that

$$\langle \varphi\, ;\, A_\ell\, \alpha_\varphi(t + i\beta)[B_\ell]\rangle = \mathrm{Tr}\left[\rho_\varphi e^{iH_\varphi t}\, B\, e^{-iH_\varphi t} A\right] = \langle \varphi\, ;\, \alpha_\varphi(t)[B_\ell]\, A_\ell\rangle\ ,$$

which is the KMS condition.

Finally note that from the definition in (52), it follows that H_φ is the quantum Hamiltonian (up to an additive constant) of a one-dimensional harmonic oscillator, so that the state ρ_φ is just the Gibbs state for the harmonic oscillator.

4.3. The antilinear operator S_φ

We now analyze the antilinear operator $S_\varphi : \mathcal{B}_2(\mathfrak{H}) \longrightarrow \mathcal{B}_2(\mathfrak{H})$, which acts as (see (31))

$$S_\varphi(A_\ell \mathbf{\Phi}) = A_\ell^* \mathbf{\Phi}\ , \qquad \forall A_\ell \in \mathfrak{A}_\ell\ . \tag{62}$$

Taking $A_\ell = A \vee I$,

$$S_\varphi(A_\ell \mathbf{\Phi}) = A_\ell^* \mathbf{\Phi}\ , \quad \forall A_\ell \in \mathfrak{A}_\ell \quad \Longleftrightarrow \quad S_\varphi(A\mathbf{\Phi}) = A^* \mathbf{\Phi}\ , \quad \forall A \in \mathcal{L}(\mathfrak{H})\ .$$

Using (47) we may write,

$$S_\varphi(A\mathbf{\Phi}) = A^*\mathbf{\Phi} \quad \Longrightarrow \quad \sum_{i=1}^{\infty} \alpha_i^{\frac{1}{2}} S_\varphi(A\mathbb{P}_i) = \sum_{i=1}^{\infty} \alpha_i^{\frac{1}{2}} A^* \mathbb{P}_i\ .$$

Taking $A = X_{k\ell}$ (see (39)) and using $X_{k\ell}\mathbb{P}_i = \delta_{\ell i} X_{ki}$, we then get

$$\alpha_\ell^{\frac{1}{2}} S_\varphi(X_{k\ell}) = \alpha_k^{\frac{1}{2}} S X_{\ell k} \quad \Longrightarrow \quad S_\varphi(X_{k\ell}) = \left[\frac{\alpha_k}{\alpha_\ell}\right]^{\frac{1}{2}} X_{\ell k}\ . \tag{63}$$

Since any $A \in \mathcal{L}(\mathfrak{H})$ can be written as $A = \sum_{i,j=1}^{\infty} a_{ij} X_{ij}$, where $a_{ij} = \langle \zeta_i \mid A\zeta_j\rangle$, and furthermore, since $\mathbf{P}_{ij}(X_{k\ell}) = X_{ij}\delta_{ik}\delta_{j\ell}$, we obtain using (44) and (55),

$$S_\varphi = J[\Delta_\varphi]^{\frac{1}{2}}\ , \tag{64}$$

which in fact, also gives the polar decomposition of S_φ.

In this way, within the framework of the Hilbert space of Hilbert-Schmidt operators, we have obtained all the ingredients of a modular structure, as outlined in Section 3. We now proceed to transfer this entire structure to the Hilbert space $L^2(\mathbb{C}, d\nu)$ (see (16)) incorporating the Landau levels.

5. Modular structure of the Landau level problem

Going back to the Hilbert space $\mathfrak{H}$ (this time taken to be infinite-dimensional) with orthonormal basis $\{\zeta_n\}_{n=0}^{\infty}$, we consider the one-dimensional oscillator Hamiltonian,

$$H_{\text{osc}} = \sum_{n=0}^{\infty} \left(n + \frac{1}{2}\right) \mathbb{P}_n = \frac{1}{2}(P^2 + Q^2) , \qquad \mathbb{P}_n = |\zeta_n\rangle\langle\zeta_n| , \tag{65}$$

Q, P being the usual position and momentum operators satisfying $[Q, P] = iI_{\mathfrak{H}}$.

On $\mathfrak{H}$ we construct the operator

$$V(z, \overline{z}) = 2\pi e^{|z|^2} e^{-i(za^\dagger + \overline{z}a)} , \qquad \text{with} \quad a = \frac{1}{\sqrt{2}}(Q + iP) , \quad a^\dagger = \frac{1}{\sqrt{2}}(Q - iP) . \tag{66}$$

for each $z \in \mathbb{C}$. Consider now the map $\mathcal{U} : \mathcal{B}_2(\mathfrak{H}) \longrightarrow L^2(\mathbb{C}, d\nu)$, with

$$(\mathcal{U}X)(x, y) = \frac{1}{(2\pi)^{\frac{1}{2}}} \text{Tr}[V(z, \overline{z})^* X] . \tag{67}$$

This map is known to be unitary (see, for example, [6]). Furthermore, as shown in [5] it induces the remarkable transformations,

$$\mathcal{U}\begin{pmatrix} a \vee I_{\mathfrak{H}} \\ a^\dagger \vee I_{\mathfrak{H}} \end{pmatrix} \mathcal{U}^{-1} = \begin{pmatrix} \mathcal{A}_+ \\ \mathcal{A}_+^* \end{pmatrix} , \qquad \mathcal{U}\begin{pmatrix} I_{\mathfrak{H}} \vee a \\ I_{\mathfrak{H}} \vee a^\dagger \end{pmatrix} \mathcal{U}^{-1} = \begin{pmatrix} \mathcal{A}_- \\ \mathcal{A}_-^* \end{pmatrix} , \tag{68}$$

and thus,

$$\mathcal{U}\begin{pmatrix} H_{\text{osc}} \vee I_{\mathfrak{H}} \\ I_{\mathfrak{H}} \vee H_{\text{osc}} \end{pmatrix} \mathcal{U}^{-1} = \begin{pmatrix} \mathcal{H}^\downarrow \\ \mathcal{H}^\uparrow \end{pmatrix} , \qquad \mathcal{U}X_{n\ell} = H_{n-1\ell-1}, \tag{69}$$

where the $X_{n\ell}$ are the basis vectors defined in (39) and the $H_{n\ell}$ are the normalized complex Hermite polynomials defined in (20). The two sets of operators, $\{\mathcal{A}_+, \mathcal{A}_+^*\}$ and $\{\mathcal{A}_-, \mathcal{A}_-^*\}$, generate the two von Neumann algebras $\mathfrak{U}_+$ and $\mathfrak{U}_-$, respectively, with $\mathcal{U}\mathfrak{A}_\ell\mathcal{U}^{-1} = \mathfrak{U}_+$ and $\mathcal{U}\mathfrak{A}_r\mathcal{U}^{-1} = \mathfrak{U}_-$. Thus physically, the two commuting algebras, which are also mutual commutants and are both factors, in view of (43), correspond to the two directions of the magnetic field. The KMS state $\boldsymbol{\mathcal{X}} = \mathcal{U}\boldsymbol{\Phi}$, with $\boldsymbol{\Phi}$ given by (47) is just the *Gibbs equilibrium state* for this physical system, i.e., for the *interaction Hamiltonian* $\boldsymbol{\mathcal{H}}_\varphi$ (see (73) below). Note also, that the existence of the unitary map $\mathcal{U}$ is also a consequence of the well known theorem of von Neumann on representations of the canonical commutation relations. Here we simply have an explicit example of it.

Let us consider some further aspects of the modular structure, now in the context of the physical problem of the electron in a constant magnetic field. The map $\mathcal{U}$ transforms the modular conjugation map J in (44) to $\mathcal{J} = \mathcal{U}J\mathcal{U}^{-1}$ which basically acts by complex conjugation (see (18) and (29)):

$$(\mathcal{J}H_{n\ell})(\overline{z}, z) = H_{\ell n}(\overline{z}, z) = H_{n\ell}(z, \overline{z}) \quad \text{and} \quad \mathcal{J}\mathcal{A}_\pm\mathcal{J}^{-1} = \mathcal{A}_\mp = \overline{\mathcal{A}_\pm} , \tag{70}$$

etc. (The overline indicates complex conjugation of the variables appearing in the definitions of the operators). Furtheremore, with $\mathcal{H}_0, \mathcal{H}^\uparrow$, etc., as in (23),

$$\mathcal{J}\mathcal{H}_0\mathcal{J} = \mathcal{H}_0, \qquad \mathcal{J}\mathcal{H}_{\text{int}}^\uparrow\mathcal{J} = \mathcal{H}_{\text{int}}^\downarrow = -\mathcal{H}_{\text{int}}^\uparrow \quad \Rightarrow \quad \mathcal{J}\mathcal{H}^\uparrow\mathcal{J} = \mathcal{H}^\downarrow , \tag{71}$$

meaning that the modular conjugation map interchanges the two physical setups, corresponding to the two directions of the magnetic field.

The KMS state on the algebra $\mathfrak{U}_+$, which is a vector state, is given by the vector (see (47) and (46)),

$$\boldsymbol{\mathcal{X}} = \mathcal{U}\boldsymbol{\Phi} = (1 - e^{-\beta})^{\frac{1}{2}} \sum_{n=0}^{\infty} e^{-\frac{n\beta}{2}} H_{nn} \in L^2(\mathbb{C}, d\nu(\overline{z}, z)) , \qquad \mathcal{J}\boldsymbol{\mathcal{X}} = \boldsymbol{\mathcal{X}} . \tag{72}$$

The Hamiltonian

$$\boldsymbol{\mathcal{H}}_\varphi = \mathcal{H}^\uparrow - \mathcal{H}^\downarrow = 2\mathcal{H}^\uparrow_{\text{int}} = -2(\mathcal{N}_+ - \mathcal{N}_-) , \tag{73}$$

then gives the modular operator,

$$\Delta_\varphi = \exp[-\beta \boldsymbol{\mathcal{H}}_\varphi] = \sum_{n,\ell=0}^{\infty} e^{-\beta(n-\ell)} |H_{n\ell}\rangle\langle H_{n\ell}| , \tag{74}$$

and the one-parameter automorphism group,

$$\Delta_\varphi^{-\frac{it}{\beta}} = \exp[i\boldsymbol{\mathcal{H}}_\varphi t] = \exp[2i\mathcal{H}^\uparrow_{\text{int}} t] , \quad t \in \mathbb{R} , \qquad \Delta_\varphi^{-\frac{it}{\beta}} \boldsymbol{\mathcal{X}} = \boldsymbol{\mathcal{X}} , \tag{75}$$

which stabilizes $\boldsymbol{\mathcal{X}}$. In other words, the modular automorphism is basically the time evolution generated by the interaction Hamiltonian. One also verifies that

$$\Delta_\varphi^{-\frac{it}{\beta}} \mathfrak{U}_+ \Delta_\varphi^{\frac{it}{\beta}} = \mathfrak{U}_+ . \tag{76}$$

6. Discussion

As set out in the Introduction, our aim in this paper was to make a link between a somewhat arcane (at least from a physicist's point of view) mathematical theory of modular structures in von Neumann algebras and the rather well-known physical problem of quantum mechanically describing an electron moving in a constant magnetic field and the associated Landau levels. We have shown here how the two experimental setups, corresponding to the two possible directions of the magnetic field are mathematically reflected in the appearance of two commuting von Neumann algebras. The effect of the modular conjugation operator of the algebraic theory is translated, in the experimental context, into a change of direction of the magnetic field. The equilibrium KMS state turns out to be the Gibbs state obtained from the interacting Hamiltonian and, being invariant under the modular conjugation, is independent of the choice of the direction of the magnetic field and thus reflects a symmetry of the problem. What is intriguing is that this symmetry is essentially the symmetry under complex conjugation. Let us just point out again that the complex analytic representation introduced here is unitarily equivalent to the initial representation on $L^2(\mathbb{R}^2)$. Its interest lies more in the identification of the modular conjugation operator with a simple complex conjugation. Additionally, as shown in [3], there are interesting families of coherent states that can be readily obtained from this representation. There is also the possibility that, starting from general families of coherent states on other Hilbert spaces of holomorphic functions, one could arrive at other pairs of commuting von Neumann algebras and thus, other modular structures.

Acknowledgements

The author would like to thank the organizers of the Symmetries in Science Symposium (2009) in Bregenz, in particular, Profs. D. A. Schuch and M. Ramek for inviting him to the meeting and giving him the opportunity to present a talk on which this paper is based.

References

[1] Cohen-Tannoudji , Diu B and Laloe F 1977 *Quantum Mechanics, Vol. I* (Paris: Hermann)
[2] Takesaki M 2002 *Theory of Operator Algebras I*, (Berlin: Springer)
[3] Ali S T, Bagarello F and Honnouvo G 2010 Modular structures on trace class operators and applications to Landau levels *J. Phys.* A **43** 105202 1-17 (online version)
[4] Ali S T and Bagarello F 2005 Some physical appearances of vector coherent states and coherent states related to degenerate Hamiltonians *J. Math. Phys.* **46** 053518 1 - 18 (online version)
[5] Ali S T and Emch G G 1986 Geometric quantization: Modular reduction theory and coherent states *J. Math. Phys.* **27** 2936 - 43
[6] Ali S T , Antoine J P and Gazeau J P 2000 *Coherent States, Wavelets and Their Generalizations* (New York: Springer)
[7] Bach V, Fröhlich J and Sigal I M 2000 Return to equilibrium *J. Math. Phys.* **41** 3985 - 4060
[8] Emch G G 1981 Prequantization and KMS structures *International J. Theor. Phys.* **20** 891 - 904
[9] Ali S T, El Gradechi A M and Emch G G 1994 Modular algebras in geometric quantization *J. Math. Phys.* **35** 8237 - 43
[10] Abdelkader Intissar and Ahmed Intissar 2006 Spectral properties of the Cauchy transform on $L_2(\mathbb{C}, e^{-|z|^2}\lambda(z)$ *J. Math. Anal. Appl.* **313** 400 - 18
[11] Ghanmi A 2008 A class of generalized complex Hermite polynomials *J. Math. Anal. Appl.* **340**, 1395 - 406
[12] Takesaki M 1970 *Tomita's Theory of Modular Hilbert Algebras and its Applications* (New York: Springer).
[13] Haag R, Hugenholtz N M and Winnink M 1967 On the equilibrium states in quantum statistical mechanics *Commun. Math. Phys.* **5** 215 - 36

Modified Noether theorem and arrow of time in quantum mechanics

V V Asadov[1], **O V Kechkin**[1,2]

[1] Neur OK–III, Lomonosov Moscow State University, Vorob'jovy Gory, 119899 Moscow, Russia
[2] Skobeltsyn Institute of Nuclear Physics, Lomonosov Moscow State University, Vorob'jovy Gory, 119899 Moscow, Russia

E-mail: asadov@neurok.com,kechkin@gmail.com

Abstract. Relativistic quantum mechanics is presented with modified Noether theorem. It was shown that Noether charges are related with thermodynamic potentials in such scheme. Broken symmetries generated by thermodynamic mode lead to gravity appearance as effective quantum field.

1. Introduction

Symmetries play a very important role in modern theoretical and mathematical physics. Of course, it is possible to study abstract dynamical systems, which do not possess any symmetry. However, from phenomenology we know that all dynamical systems which describe the real world are characterized by some group of explicit and hidden symmetries definitely [1]–[4]. It can be said, that all realistic field theories, as well as all string theory models are some specific realizations of corresponding symmetry groups. Moreover, quantization procedure of such kind of dynamical systems is understood as some non-trivial map of the starting classical theory which conserves its basic symmetry properties. Actually, in the standard approach to quantization all anomalies (the results of original symmetry breaking) must vanish identically if the correct quantization procedure was really found. Thus, symmetries are considered as native property of the physical world; they characterize the realistic dynamical systems and help us establish a correct formalism for dynamics description. This explains the role of symmetries in scientific activity from the theoretical physics point of view.

In mathematical physics context, symmetries of dynamical equations for a given physical system are important because they can be used for integration of these equations – i.e. for the solution of dynamical problem. The corresponding approach to the solution of ordinary and partial differential equations was introduced by Sofus Lee. Now this activity includes the study of the so-called integrable systems – the systems, which can be understood as the most symmetric ones in the class of dynamical systems of the given type. More precise, integrable systems are characterized by their symmetry groups completely: knowledge of symmetry groups allows one to solve corresponding dynamical equations for arbitrary initial data in case of integrable systems. Of course, this possibility is considered as the most optimistic for these systems with highly non-linear dynamics [5]–[9].

In our modification of quantum and classical dynamics, symmetries also play a crucial role. We develop our approach using some hidden symmetry of original standard dynamical system in such way, when the final system also possesses this hidden symmetry, but in fact satisfies the dynamics of a more general type. Namely, we start from the standard reversible dynamics and, finally, obtain the dynamics with an arrow of time, – i.e. the irreversible dynamical system, which can be considered as a thermodynamical modification of the original one. Our main goal is the unification of the fundamental laws of quantum and classical dynamics with the Second Law of thermodynamics – one of the main goals of all theoretical and mathematical physics of our days [10]–[13]. Also the problem of violation of time reversal invariance was considered in publications [14]–[17].

2. Symmetries and conservation laws in standard quantum and classical dynamics

In this section we give a review of well-known facts related to the symmetries in the standard classical and quantum dynamics. Here and below we restrict our consideration by the class of dynamical systems which possess Hamilton's formalism in their description. More precisely, our programm is based on some 'minimal modification' of such kind of dynamical systems to the form which is not completely the standard Hamiltonian one. For example, we lead to classical limit of the modified quantum systems which satisfies the improved system of Hamilton's equations. The modification found, generates irreversible effects and gives base for the thermodynamical formalism.

2.1. Symmetry definition

Let us consider a dynamical system with coordinates q^a and Hamiltonian H; the dynamical temporal variable ('Hamilton's time') we denote as t. For the quantum variant of the theory we will use the wave-function Ψ in the Schrödinger's picture; for the corresponding classical limit of this system in its Hamilton's form we must add to the set of canonical variables the momentums p^a. We will understand the symmetry of system of the given type as a transformation in the unified space of independent and dependent dynamical variables (including all derivatives of independent variables with respect to dependent ones) which preserve dynamical equations describing this system. The unified space mentioned, consists of t, q^a, Ψ, $\Psi_{,t}$, $\Psi_{,q^a}$ and derivatives of the wave-function Ψ of the highest order with respect to t and q^a in the quantum case. For the classical variant of the theory, this space includes t, q^a, p^a, $q^a_{,t}$, $p^a_{,t}$ and also the derivatives of q^a and p^a, with respect to t and q^a, again. Of course, we consider t and q^a as independent dynamical variables. Here q^a are understood as coordinates of the corresponding representation of the quantum realization of the theory. A solution of the corresponding dynamical problem is the set of functions $q^a = q^a(t)$ and $p^a = p^a(t)$ in the classical case, and the wave-function $\Psi = \Psi(t,\, q^a)$ in the quantum variant of dynamics.

It is clear that symmetries can be defined alternatively, as such transformations of the discussed type that map an arbitrary solution of dynamical equations describing the system to a corresponding solution of the same system. It is easy to see that the symmetry transformations form a group with respect to the operation of their successive application. This group includes the discrete and continuous maps; all these symmetries are important for the study of dynamical properties of the system under consideration.

2.2. Discrete symmetries

In theoretical physics, the most important discrete symmetries include C (charge), P (spatial parity) and T (time) inversions. The first of them means the change of sign of all the charges (not only electric charges) to the opposite ones. The second transformation maps the left orientation to the right one, and vice versa. The third symmetry is understood as the change of given 'time direction' to the opposite direction (the transformation which includes the map $t \to -t$) without any consequences for the system dynamics. Also, it can be said that T-transformation maps the initial and final data of the dynamics to each other. For further discussion, we call the situation where the broken T-symmetry is present in the dynamics of the system the 'arrow of time'. It is important to note that for the wide class of theories, the rigid symmetry with respect to the combined CPT transformation (the CPT-theorem) was established. In our modification of quantum and classical dynamics presented below, an appearance of the arrow of time without the destruction of fundamental principles of quantum and the corresponding classical theory seems the most important result.

2.3. Noether symmetries

Continuous transformations are related to the transformation parameters that vary continuously on some subspace of the corresponding (multidimensional) parameter set. The most famous and well-known continuous transformations are the contact and Noether symmetries. They are important especially in the study of specific realistic systems that arise in the framework of the conventional theoretical physics. For each Noether symmetry, there exists a corresponding Noether current, which yields the null-divergence condition.

2.4. Noether charges and their conservation

It is a well known fact that any Noether symmetry is accompanied by a corresponding Noether charge – the integral quantity (over the coordinates q^a), which does not depend on the Hamilton's parameter of evolution t. This conservation law follows from the null-divergent condition mentioned; this is known as the famous Noether theorem. For our purposes it is important to stress that Noether charges are numbers (not functions of some variables), which can be expressed in terms of initial conditions of the problem. Also, it must be noted that independent Noether symmetries generate the independent sets of Noether charges in the standard quantum or classical system.

In contrast to this conventional situation, in the modified dynamics presented below Noether symmetries do not lead to any conservation laws. Using other words, it can be said that in modified quantum and classical dynamics, developed below, Noether charges become some functions on the set of extra parameters of the theory. If one considers some dependence of extra parameters on the evolutionary parameter t, one obtains possibility to study corresponding dynamics of the Noether charges on this Hamilton's time. In fact, we relate such kind of dependence with thermodynamical mode fixing. Finally, for the general thermodynamic mode the relation between Noether symmetries and conservation laws becomes broken. Actually, in our form of modified dynamics, Noether charges are some specific functions of time, not constant quantities. However, in a special case of thermodynamic mode with, for example, conserved energy, one has the energy integral conservation up to construction. To clarify and explain all these unusual properties of modified quantum and classical systems let us consider them in detail below.

3. Modified quantum dynamics, time dependence of Noether charges and arrow of time

In this section, a general theoretical scheme for the modified quantum dynamics is presented and discussed. This scheme can be considered as a 'totally complexified' standard quantum theory. Performing that modification allows one to unify quantum and thermodynamical principles in a consistent form. The thermodynamic behavior of the system is related to dependence of its quantum evolution on temperature regime, which appears in the framework of the modification performed. The most interesting prediction of our extension of the standard quantum theory is the presence of the arrow of time for quantum evolution with physically well-motivated temperature regimes. Such kind of time irreversibility can be used for microscopic explanation of the Second Law of thermodynamics. We give some basics for the unified quantum and thermodynamical principles and show that the entropy production actually takes place in the modified dynamics of the type presented. The results of this section are based on the publication [10].

3.1. Modified quantum dynamics

We base modified quantum theory on the standard Schrödinger's equation and on analytic condition

$$i\hbar\partial_\tau \Psi = \hat{H}\Psi, \qquad \partial_{\tau^*}\Psi = 0, \tag{1}$$

– i.e. on the system of two linear differential dynamical relations with respect to some evolutionary parameter τ. Our operator of evolution is considered as a non-Hermitian on ($\hat{H} \neq \hat{H}^+$) in a general case. We preserve the name 'Hamiltonian' for the operator $\hat{H}$ because it is conjugated to the evolutionary parameter τ through the Schrödinger's equation in the conventional form. It can be said that the theory under consideration is 'totally complexified' in view of the general complex nature of the wave-function Ψ, evolutionary parameter τ and operator of evolution $\hat{H}$.

Then, the theory is considered as totally analytical: in addition to the second relation in Eq. (1), we restrict the system by the conditions of analyticity

$$\partial_\tau \hat{H} = \partial_{\tau^*} H = 0 \tag{2}$$

with respect to its Hamiltonian. Let us now give the interpretation for all quantities in the theory. For the extended objects introduced above we use the following decompositions:

$$\tau = t - \frac{i\hbar}{2}\beta, \qquad \hat{H} = \hat{E} - \frac{i\hbar}{2}\hat{\Gamma}. \tag{3}$$

Here, parameters t and β are understood as 'usual' time and inverse temperature, respectively (the temperature is measured in energy units). Then, for the Hermitian operators $\hat{E}$ and $\hat{\Gamma}$ we use the interpretation of operators of energy and decay. In the special case of trivial decay operator (when $\hat{\Gamma} = 0$) and infinite temperature (when $\beta = 0$) one deals with the standard quantum dynamics. In this sense the theory under discussion includes the conventional quantum theory, i.e. the correspondence principle is satisfied.

Using this principle, it is possible to identify the energy operator $\hat{E}$ with the Hermitian Hamiltonian of the standard form of quantum system under modification. Then, the proposed modification, is related to the complex extension of the evolutionary parameter (i.e. to the map $t \to \tau$), and to the introduction of the decay operator Γ in accordance with Eq. (3). Of course, both operators $\hat{E}$ and $\hat{\Gamma}$ must be taken as independent on t and β: this follows from the generalized stationary condition (2). It is clear that the presented theoretical scheme must be specified with respect to the choice of the decay operator $\hat{\Gamma}$. We propose to use, for this specification, the commutativity condition

$$\left[\hat{E}, \hat{\Gamma}\right] = 0, \tag{4}$$

which simplifies the extremely analytical structure of this theory and also has an evident physical basis.

Namely, let us assume that for the general element Ψ of solution space of dynamical system (1)–(4) one has:

$$\Psi = \sum_n C_n \Psi_n, \tag{5}$$

where C_n = const are the arbitrary complex constants and the eigen-vector Ψ_n satisfies the eigen-problem

$$\hat{H}\Psi_n = \left(E_n - \frac{i\hbar}{2}\Gamma_n\right)\Psi_n. \tag{6}$$

Here E_n and Γ_n are the eigen-values for operators $\hat{E}$ and $\hat{\Gamma}$, respectively; the condition (4) guarantees the correctness of Eqs. (5)–(6). Then, using the standard quantum formalism (for the non-normalized states), for probability P_n to find the state Ψ, given by Eq. (5), in the eigen-state Ψ_n, one obtains:

$$P_n = \frac{w_n}{Z}, \tag{7}$$

where $Z = \sum w_n$, and

$$w_n = r_n \exp\left[-\left(E_n\beta + \Gamma_n t\right)\right] \tag{8}$$

with $r_n = |C_n|^2$. It is seen that the quantity Z is the norm of the state Ψ, the constant parameter r_n is defined by initial data of the problem and plays the role of a weight coefficient in this probability distribution. Also, it is clear that the function Z can be considered as the statistical sum for this system. In the special case of $\beta = 0$, one immediately interprets the dynamical parameter β as inverse temperature (in the energy units) if the eigen-values E_n are identified with conventional energy levels of the system.

For interpretation of the set of constants Γ_n, let us consider a special state with only two nontrivial probabilities P_0 and P_N, where P_N is the small perturbation over the 'ground' probability P_0:

$$P_0 + P_N = 1, \qquad P_0 \geq P_N. \tag{9}$$

Using Eqs. (7)–(8), it is not difficult to show that for such distribution of probabilities

$$P_N \sim \exp\left(-\Gamma_N t\right) \tag{10}$$

in case of β = const (i..e., for the isothermal evolution). Thus, the parameter $t_N = 1/\Gamma_N$ plays the role of time of life of the state Ψ in the eigen-state Ψ_N in the conventional sense of physics of resonances. Of course, for the general solution (5) of the problem, dynamics of probabilities P_n is more complicated. Let us note that condition (4) guarantees the possibility of simultaneous measurement of the energy levels and times of life for the corresponding states of the system. It is possible to say that in the theory under consideration, times of life are not less important parameters than the physical energies; exactly such situation takes place in the resonance phenomena. Our hypothesis is related with this evident fact: we suppose that the complex analytic extension (1)–(4) of the standard quantum theory describes adequately the world of nonstable particles at least. Below we present the relativistic models, based on the general scheme (1)–(4) and on the explicitly realized conception of the multi-universe world.

3.2. Quantum correlations in modified quantum dynamics

Correlations between different physical variables play a crucial role in the theory under discussion. Here, we give all necessary definitions and discuss the physical sense of the introduced quantities.

First of all, for expectation value of operator $\hat{A}$ in the state $|\Psi>$ we use the conventional Dirac's formula

$$\bar{A} = \frac{<\Psi|\hat{A}|\Psi>}{<\Psi|\Psi>}. \tag{11}$$

Of course, the relation

$$Z =< \Psi|\Psi > \tag{12}$$

holds, and some explicit rule for the calculation of all matrix elements and scalar products in the theory is supposed. We define correlation $A \circ B$ of the quantum observables $\hat{A}$ and $\hat{B}$ as a natural binary generalization

$$A \circ B = \overline{AB} + \overline{BA} - 2\bar{A}\bar{B} \tag{13}$$

of conventional formula $D_A^2 = \overline{A^2} - \bar{A}^2$ for dispersion of quantity $\hat{A}$. It is seen that from Eq. (13) it follows the relation

$$A \circ A = 2D_A^2 \tag{14}$$

in the special case of $\hat{B} = \hat{A}$. Let us stress that we preserve the definition of dispersion, so that if operators $\hat{A}$ and $\hat{B}$ satisfy the commutation relation $[\hat{A}, \hat{B}] = i\hbar$, than the uncertainty relationship $D_A^2 D_B^2 \geq \hbar^2/4$ holds. In the relativistic realization of the theory under consideration we will study the coherent states; for these states, which are coherent with respect to the observables $\hat{A}$ and $\hat{B}$ one has the equality

$$D_A^2 D_B^2 = \frac{\hbar^2}{4}. \tag{15}$$

3.3. Symmetries in modified quantum dynamics

Symmetries also play extremely important role in the modified quantum theory (1)–(4). Let us now formalize a symmetry conception which was given in the section 2. Following the standard mathematical definition, we name operator $\hat{S}$ the symmetry operator if transformed quantum state $\hat{S}\Psi$ satisfies the dynamical equations (1) together with the original quantum state Ψ. It is easy to see that any symmetry operator $\hat{S}$ must commute with the Hamiltonian of the system:

$$\left[\hat{S}, \hat{H}\right] = 0; \tag{16}$$

this follows from linearity of Eq. (1). From Eq. (4) it follows immediately that the decay operator $\hat{\Gamma}$ is one of the symmetry operators of the system. From the pure mathematical point of view, the formal scheme (1)–(4) is defined completely as an extension of the standard quantum theory if one takes the decay operator $\hat{\Gamma}$ in the form of some function of the energy operator $\hat{E}$, i. e. $\hat{\Gamma} = f(\hat{E})$. In this special case the relation (4) becomes an equality. However, for relativistic realizations of the modified quantum dynamics under consideration, such kind of extensions of the standard theory seems not realistic. It is possible to show that for the nonlinear function $f(\hat{E})$ the resulting modified field theory contains (infinite) set of terms with highest derivatives with respect to coordinates of the physical space-time. Below we develop an alternative approach to the decay operator construction. It is based on the use of the multi-universe conception in quantum theory and on the possibility of its explicit realization in the framework of the general scheme (1)–(4).

3.4. Thermodynamic modes and quantum observer

In the previous section, isothermal evolution appeared – i.e. the special regime of consideration of the theory where quantities are specified by the restriction

$$\beta = \text{const}. \tag{17}$$

Of course, this restriction does not follow from the general theoretical scheme (1)–(4) for the modified quantum dynamics, because in this scheme the dynamical parameters t and β are completely independent. Thus, the restriction (17) must be understood as some additional condition, which actually specifies all physical quantities calculated on the base of the solution Ψ of Eq. (1). Another natural extra condition imposed on the quantum evolution is related with the conservation of expectation value of the system energy:

$$\bar{E} = \text{const.} \tag{18}$$

This special mode we name as an 'adiabatic regime'; it is clear that this mode corresponds to some special relation between the dynamical parameters t and β.

Starting with the examples discussed above, let us consider some observable which is described by the operator $\hat{A}$. Let us consider the expectation value of this observable, i.e. the quantity $\bar{A}$. It is clear that $\bar{A}$ is a specific function of the dynamical parameters t and β, $\bar{A} = \bar{A}\,(t,\,\beta)$. Actually, all the coordinates q^a of the quantum representation were integrated out through the process of the expectation value construction. Thus, the special evolution mode with $\bar{A}\,(t,\,\beta) = \text{const}$ defines (in not explicit form) the 'temperature regime' (or the 'temperature curve')

$$\beta = \beta\,(t) \tag{19}$$

such that the relation $A\,(t,\,\beta\,(t)) = \text{const}$ becomes the equality for any given value of the Hamilton's time t. Finally, we postulate that the specified temperature regime (19) completes the general scheme (1)–(4); the resulting theory can be studied and compared with phenomenology. Let us note that imposing the temperature regime can be interpreted as introducing the macroscopic observer into the general quantum theory scheme. Actually, fixing some special temperature regime (19), this observer defines the macroscopic experiment conditions. These conditions may have really different nature – this evident fact is reflected by extremely wide class of the temperature curves (19) which can be considered within the general quantum microscopic dynamics (1)–(4).

3.5. Time dependence of expectation values of symmetry operators: arrow of time appearance

Now we are ready to state that the modified quantum dynamics given by the general scheme (1)–(4) and the temperature regime (19) fixed, agree with the Second Law of thermodynamics. More precisely, we state that for the wide class of the temperature regimes (19), the modified quantum dynamics (1)–(4) possesses an arrow of time. In this subsection we prove this statement for the physically well-motivated isothermal and adiabatic regimes (17) and (18). In our proof we will operate with the most important expectation values of the theory, defined by its Hamiltonian:

$$\bar{E} = \sum_n P_n E_n, \qquad \bar{\Gamma} = \sum_n P_n \Gamma_n. \tag{20}$$

For the isothermal evolution (17), a straightforward calculation leads to the following remarkable result:

$$\frac{d\bar{\Gamma}}{dt} = -D_{\Gamma}^2, \tag{21}$$

where D_{Γ}^2 denotes the squared dispersion of the decay operator $\hat{\Gamma}$. Thus, for theories with not negative probabilities P_n the inequality

$$\frac{d\bar{\Gamma}}{dt} \le 0 \tag{22}$$

holds for the arbitrary value of time parameter t. Note that this inequality holds for any initial data which specifies the solution Ψ of the dynamical equations (1). Thus, we deal with the irreversible in time evolution, or with the 'arrow of time' in the isothermal quantum evolution.

Then, repeating the calculations in the case of adiabatic evolution (18), one obtains the following result:

$$\frac{d\bar{\Gamma}}{dt} = -D_{\Gamma}^2 \left[1 - \frac{\left(\overline{E\Gamma} - \bar{E}\bar{\Gamma}\right)^2}{D_E^2 D_{\Gamma}^2} \right], \tag{23}$$

where D_E^2 is the squared dispersion for the energy operator. Of course, Eq. (23) is much more complicated than the previous one given by Eq. (21). However, using the Cauchy-Bunyakovsky inequality (for the theory with non-negative probability set), one concludes that the therm in the square brackets in Eq. (23) is not negative. From this fact it follows that the inequality (22) remains true for the case of adiabatic evolution. Thus, the adiabatic quantum evolution is also irreversible – i.e., it also possesses the arrow of time. The presence of the arrow of time, or time irreversibility in the quantum system evolution, means that this evolution can agree with the Second Law of thermodynamics.

To clarify the situation, let us study the dynamics of the entropy function Σ, defined by the relation

$$\Sigma = \sum_n P_n \log P_n = \overline{\log r} - \bar{E}\beta - \bar{\Gamma}t - \log Z, \tag{24}$$

where $\overline{\log r} = \sum P_n \log r_n$. Using the straightforward calculations, it is possible to prove that the following remarkable inequality holds for any $t \geq 0$ in the isothermal and adiabatic regimes of evolution:

$$\frac{d\Sigma_{\rm phys}}{dt} = -t\frac{d\bar{\Gamma}}{dt} \geq 0. \tag{25}$$

Here we have used Eq. (22), and the quantity $\Sigma_{\rm phys}$ is introduced as the renormalized entropy (24):

$$\Sigma_{\rm phys} = \Sigma - \overline{\log r}. \tag{26}$$

We name this new entropy function as the 'physical entropy'. It is necessary to stress that for the special statistics with equal values of the weight coefficients r_n one has $\overline{\log r} = \text{const}$, and from Eq. (25) it follows the same relation for the original entropy function Σ. Note that exactly such kind of statistics is realized in nature through the microcanonial Gibbs distribution. Thus, the arrow of time detected by the inequality (22) actually corresponds to the Second Law of thermodynamics.

Also, our conclusion shows that for the consistent explanation of the Second Law of thermodynamics in the microscopic framework, one must introduce the non-trivial decay operator into the fundamental quantum system description. Thus, we propose to see the decay processes as the reason for the irreversibility of dynamics of physical systems in the real world.

4. Quasi-classical limit: modified classical dynamics, time dependence of Noether charges and arrow of time

Classical limit of quantum theory must describe the 'worlds around us', i.e. the macroscopic reality. Usually, in constructing the classical limit of the specific quantum dynamical system, one neglects the terms, which include the Plank constant. Thus, this limit is related to the calculation of all necessary theory quantities and all corresponding dynamical relations in the regime with $\hbar \to 0$. However, taking into account the really small value of the Plank constant is not the single condition which can be imposed on the fundamental quantum theory. Actually, the crucial property of the quantum theory – the non-localized nature of the dynamical quantities – does not break at the limit of $\hbar = 0$. Thus, it is possible to develop the theory with non-trivial distribution of probability density and vanishing value of the Plank constant. This type of generalized dynamics, being classical, differs essentially from the dynamics with a completely localized probability density which is usually identified with the adequate candidate to the macroscopical world description. In this section we construct both types of classical limit for the generalized quantum dynamics developed above. It is shown that explanation of irreversible properties in evolution of macroscopic systems can be obtained only in the classical limit of their quantum dynamics with non-locality saved. Thus, the Second Law of thermodynamics provides the true choice between the two variants of classical limits for real systems: one must choose as correct the dynamics with non-trivial (not localized completely) distributions of the classical probability density. The results of this section are based on the publications [10]–[11].

4.1. Definition of modified classical dynamics

Let us start our consideration with the one-component wave- function Ψ (the more complicated quantum systems are discussed below). Such wave function can be represented in the exponential form

$$\Psi = \exp\left(\frac{iS}{\hbar}\right), \tag{27}$$

where the complex phase S can be decomposed to its real and imaginary parts S_1 and S_2 in the following way:

$$S = S_1 + \frac{i\hbar}{2} S_2. \tag{28}$$

Our goal is to establish the dynamical relations which follow from the quantum equations (1) at the limit $\hbar \to 0$ (we will use the same notation S_1 and S_2 for the limit values of the phase components).

To solve the corresponding mathematical problem, let us use the coordinate representation of the original quantum theory; then for the momenta, one has the standard relation $\hat{p}_k = i\hbar\partial_{x^k}$. Study of the general quantum scheme (1)–(4) with $\hat{H}$ realized as some function of the total set of the canonical variables q_k and $\hat{p}_k$ in the limit $\hbar = 0$ leads to the following result for the real part S_1 of the complex quantum phase:

$$S_{1,t} = -E, \qquad S_{1,\beta} = 0. \tag{29}$$

It is seen that the function S_1 does not depend on the temperature parameter β and satisfies the standard Hamilton-Jacoby equation. Note that the classical function E in Eq. (29) is constructed from the quantum operator $\hat{E} = E(q_k, \hat{p}_k)$ using the standard substitution

$$\hat{p}_k \to S_{1,q^k} \tag{30}$$

of the analytical mechanics. To simplify the form of the remaining relations, let us introduce the following new quantity S_e:

$$S_e = S_2 - \beta E, \tag{31}$$

which is explicitly entropy-like. Then, for the remaining dynamical relations we obtain:

$$S_{e,t} + E_{,p^T} S_{e,q} = \Gamma_\star, \qquad S_{e,\beta} = 0, \tag{32}$$

i.e. the function S_e is also β-independent at the limit $\hbar \to 0$. Here the quantity $\Gamma_\star$ means the shifted decay dynamical variable; it is defined as

$$\Gamma_\star = \Gamma + \mathrm{tr}\left(E_{,pp^T} S_{1,qq^T}\right), \tag{33}$$

whereas the notation $p = S_{1,q}$ corresponds to Eq. (30) and is understood in its natural matrix form. Of course, the substitution (30) is performed in all relations, which need this action for their existence. Finally: Eqs. (29) and (32) play the role of the fundamental quantum dynamical relations in the limit $\hbar \to 0$. They generalize the conventional Hamilton-Jacoby equation to this case of modified classical dynamics.

Now let us consider the dynamics of soliton-like excitations of probability density ρ of the theory. This quantity is defined as

$$\rho = \frac{\exp(-S_2)}{Z}, \tag{34}$$

where, for the norm of the solution, the following explicit formula can be used:

$$Z = \int dq \, \exp(-S_2). \tag{35}$$

It is seen that the probability density is normalized to the unit, because $\int dq\,\rho \equiv 1$. Our goal is to study the dynamics of maxima of this function: we will identify classical positions $q = q_{\text{classic}}$ of these solitons with the set of maxima using natural arguments. Thus, for definition of the classical position we have:

$$\max_q \rho(q, t, \beta) = \rho\left(q_{\text{classic}}(t, \beta), t, \beta\right). \tag{36}$$

It is clear that such kind of description of this soliton-like dynamic is incomplete under construction and does not take into account the processes of the soliton deformation. Below we will detect the kinematical terms which describe correctly the solitonic formation and decreasing. Now our goal is to derive the dynamical relations for the quantities $q = q_{\text{classic}} = q_{\text{classic}}(t)$ and $p = p_{\text{classic}} \equiv S_{1,q}(t, q_{\text{classic}}(t))$. Using the necessary differential conditions for the extremal principle (36), after its differentiation with respect to t and β, we obtain:

$$\begin{aligned} q_{,t} &= E_{,p} - A_2^{-1}\Gamma_{\star;q}, & q_{,\beta} &= -A_2^{-1}E_{;q}, \\ p_{,t} &= -E_{,q} - A_1 A_2^{-1}\Gamma_{\star;q}, & p_{,\beta} &= -A_1 A_2^{-1}E_{;q}. \end{aligned} \tag{37}$$

Eqs. (37) generalize Hamilton's equations to the case of β-dependence of the canonical variables. Also, these equations contain extra terms related to the matrices A_1 and A_2, which are defined as

$$A_1 = S_{1,qq^T}, \qquad A_2 = S_{2,qq^T}; \tag{38}$$

their kinematical sense is clarified in the next subsection. It is shown that these terms describe the non-locality of the soliton's distribution in the phase space of the theory. In Eqs. (37), the total derivative with respect to q is defined by the relation

$$_{;q} = \frac{\partial}{\partial q} + A_1 \frac{\partial}{\partial p}. \tag{39}$$

Finally, to calculate the total t-derivative of the set of canonical variables, one must use Eqs. (37) and the standard chain rule

$$\frac{d}{dt} = \frac{\partial}{\partial t} + \dot{\beta}\frac{\partial}{\partial \beta}, \tag{40}$$

where $\dot{\beta} = d\beta/dt$. Finally, Eqs. (29), (32) give the generalization of the Hamilton-Jacoby equations, whereas the system of equations (37) modify the Hamilton's system. Together, these two systems of dynamical relations describe the modified classical dynamics which is the limit of the fundamental quantum dynamics (1)–(4) at the limit $\hbar \to 0$.

4.2. Correlations and double quasi-classical limit

The limit $\hbar \to 0$ leads to the introduction of the dynamical variables like the quantity $A = A(q, p) = A(q, S_{1,q})$ which arise from the initial quantum operator $\hat{A} = A(q, \hat{p})$. In the limit under consideration, the expectation value $\bar{A}$ is given by the conventional relation

$$\bar{A} = \int dq\, \rho\, A, \tag{41}$$

where the probability density is defined by Eqs. (34)–(35) with the limit value of the function S_2. The main statement of this subsection is that the relations

$$q \circ q^T \approx A_2^{-1}, \qquad p \circ q^T \approx A_1 A_2^{-1}, \qquad p \circ p^T \approx A_1 A_2^{-1} A_1 \tag{42}$$

hold in the kinematics of the theory if one considers the regime with

$$A_2^{-1} \approx 0. \tag{43}$$

Note that the system (37) transforms into conventional Hamilton's equations in the limit $A_2^{-1} \to 0$. Actually, all the additional terms there have a correlation nature, whereas this limit is exactly the limit of infinitesimal correlations. It is interesting to note that not only the coordinate-coordinate correlations affect the system dynamics, but also the coordinate-momentum and momentum-momentum ones. Thus, the dynamics 'feels' the total set of correlations of the canonical variables of the theory. It is seen that conventional Hamilton's dynamics arises in the double limit with $\hbar \to 0$ and $A_2^{-1} \to 0$. We name this double limit as the 'hard' one, whereas the 'soft' limit will refer with the limit procedure $\hbar \to 0$ only. Let us summarize these

$$\begin{array}{lll} \text{soft classical limit}: & \hbar \to 0 & \text{correlations} \sim \mathrm{A}_2^{-1} \neq 0, \\ \text{hard classical limit}: & \hbar \to 0 & \text{correlations} \sim \mathrm{A}_2^{-1} \to 0. \end{array} \tag{44}$$

There is another way to detect the correlation nature of the quantity A_2^{-1}, which is crucial for the hard classical limit construction. Namely, let us consider the decomposition of the function S_2 in vicinity of the classical position q_{classic} at the given values of the time t and inverse temperature β. First of all, for the first differential of this function we have $dS_2 = 0$ here in view of the necessary condition corresponding to the maximum principle given by Eq. (36). Then, the sufficient condition can be guaranteed by the second differential d^2S_2 positive definition. It is easy to see that $d^2S_2 = dq^T A_2 dq$ in view of the second relation in Eq. (38). Thus, d^2S_2 becomes the infinite positive term at the limit $A_2^{-1} \to +0$. Finally, using the relations (34)–(35) between functions S_2 and ρ, it is possible to show that

$$\lim_{A_2^{-1} \to 0} \rho = \delta\left(q - q_{\text{classic}}\right). \tag{45}$$

Thus, the limit procedure with $A_2^{-1} \to 0$ corresponds to the localization of the soliton-like probability density distribution and arising of the delta-functional localization at the limit $A_2^{-1} = 0$.

4.3. Modified Newtonian force: MOND and General Relativity effects out of correlations
In this subsection we study the one special system to demonstrate the role of correlations in the soft variant of classical dynamics. We would like to show that the correlation terms modify the dynamics and can explain, for example, the classical effects in the General Relativity. Thus, it is possible to say that the non-locality generates additional forces which can lead to arising of effective gravity.

Actually, let us study the Second Law of classical mechanics in the soft limit of the modified quantum theory. It is given by the standard relation

$$\frac{dp}{dt} = F_m, \tag{46}$$

where the total force F_m can be obtained from Eqs. (37) for the momentum p, and the relation (40) for the total t-derivative must be taken into account. The result reads:

$$F_m = -E_{,q} - A_1 A_2^{-1}\left(\Gamma_{\star;q} + \frac{d\beta}{dt}E_{;q}\right); \tag{47}$$

from this it is seen that the modified force value F_m differs from the conventional one $F = -E_{,q}$ exactly by the terms of the correlation type (compare Eq. (47) with Eqs. (42)). Also, it is clear that in the hard type of the classical limit these terms vanish, and we recover the standard Newton's relation for the potential force. This situation takes place in a general case of the theory under consideration.

Now let us specify the theory and take its energy function as the Kepler's one,

$$E = \sum_{k=1}^{3} \frac{p_k^2}{2m} - \frac{GMm}{r} \tag{48}$$

whereas for the decay function we will take its trivial value, i.e. we will study the system with $\Gamma = 0$. Then, for the shifted decay variable (33) we will obtain the relation

$$\Gamma_\star = \mathrm{tr}\left(E_{,pp^T} S_{1,qq^T}\right) = \frac{1}{m}\Delta S_1, \tag{49}$$

which defines the force F_m completely. The explicit form for its k-th component reads:

$$F_{mk} = F_{0k} - \frac{d\left[\left(A_2^{-1}\right)_{kl}\Delta S_{1,x_l}\right]}{dt} - \frac{1}{m}\left(A_1 A_2^{-1}\right)_{kl}\Delta S_{1,x_l}, \tag{50}$$

where

$$F_{0k} = -\frac{GMm}{r^2}\frac{x_k}{r} \tag{51}$$

is the k-th component of the classical Newtonian gravity force. Our goal is to calculate the extra terms in Eq. (50) for the case of some natural distribution of probability density, and to study the asymptotic behavior of these force corrections. For this system, the modified Hamilton-Jacoby equation (29), (32) leads to the equations

$$\Sigma_{,t} + \frac{1}{2}\sum_{k=1}^{3}\Sigma_{,x_k}^2 - \frac{GM}{r} = 0, \quad S_{e,t} + \sum_{k=1}^{3}\Sigma_{,x_k} S_{e,x_k} = \Delta\Sigma, \tag{52}$$

where the new function $S_1 = m\Sigma$ has been introduced. Here and below we consider the isothermal evolution ($\beta = \mathrm{const}$), which is natural in the framework of planet dynamics. Possible influence of the temperature regime to the gravity effects will be discussed below – here we are interested in the correlation ones only.

To establish the correlation modification of the Newtonian gravitational force, let us find the solution of Eq. (52) of the spherical type. Using separation of variables, we explore for the corresponding ansatz the following relation:

$$\Sigma = \Sigma(t, r) = -\frac{\alpha t}{2} + f(r), \tag{53}$$

where $\alpha = \text{const}$ (below we consider the case of $\alpha > 0$ for simplicity). From the first equation of (52) one obtains that

$$f^{-1}(r) = \left(\alpha + \frac{2GM}{r}\right)^{\frac{1}{2}}. \tag{54}$$

We would like to calculate the total force F_m which acts on the soliton-like distribution of the probability density. Of course, we interpret this localized object as a planet with the mass m, which moves under the action of the static mass M – i.e. of the Sun.

To perform the calculation, let us suppose, for simplicity that the spatial distribution of probability density is spherically symmetric with the center in the classical position of the planet at the time under consideration. We will not study the deformation of this soliton shape during the dynamics; it is defined by Eq. (52) completely. Taking into account Eqs. (42), for the spherically symmetric soliton we put:

$$\left(A_2^{-1}\right)_{kl} = \epsilon\delta_{kl}, \tag{55}$$

where ϵ is some positive parameter. We will study the highly localized probability density case, where

$$\epsilon \to +0, \tag{56}$$

and the soliton almost coincides with the classical object position. In other words, we consider the weak correlation case which is not exactly the dynamics of a single point mass. Using Eqs. (54) and (55), for the force F_m components up to $\sim \epsilon$ terms one obtains the following result:

$$F_{mk} = F_{0k} - \frac{4\epsilon m}{r^3} \frac{\left(\alpha + \frac{3GM}{r}\right)^2}{\alpha + \frac{2GM}{r}} \frac{x_k}{r}. \tag{57}$$

Eq. (57) gives the modified force acting on the moving mass m in the weak and spherically symmetric correlation case.

It is important to note that the force (57) has the asymptotics

$$F_{mk} = F_{0k} - \frac{4\epsilon m\alpha}{r^3} \frac{x_k}{r} \tag{58}$$

at the spatial infinity $r \to +\infty$. It is seen that the additional term is proportional to $1/r^3$ in this asymptotics – exactly as the Einstein's term in the General Relativity. From this follows the possibility to base the explanation of classical effects of General Relativity on the pure correlation conception. For another asymptotical regime with $0 \ll r \ll 2GM/\alpha$, which can be realized in the case of $\alpha \ll 2GM$, one obtains the following result:

$$F_{mk} = F_{0k} - \frac{18\epsilon GMm}{r^4} \frac{x_k}{r}. \tag{59}$$

Thus, the force (57) is proportional to the squared acceleration $a_0 = F_0/m$ for the large, but not infinite distances. Finally, the force (57) can be considered as the fundamental candidate to the role of the Modified Newtonian Dynamics (MOND) force. Actually, the MOND force possesses exactly the discussed asymptotical properties [20]–[22]. It is important to stress that the modified Newtonian force (57) was obtained using the fundamental quantum principles and consistent procedure of constructing the (soft) classical limit.

4.4. Symmetries in modified classical dynamics

Now let us come back to the general formalism of the theory to study the symmetries and irreversible effects in the soft classical limit of the modified quantum dynamics.

Let us start with the system symmetries. The statement is that the quantum symmetry relation (16) with commutation of the symmetry operator $\hat{S}$ and the Hamiltonian $\hat{H}$ transforms to condition of vanishing of the Poisson bracket

$$\{S,\, H\} = 0 \tag{60}$$

for the symmetry variable $S = S(q,\, p)$ and the Hamilton's function $H = H(q,\, p)$, calculated on the soft limit with $\hbar \to 0$. In particular, we obtain the relation of this type for the decay function Γ. From this, it follows also the zero value of Poisson bracket for the decay and energy dynamical variables:

$$\{\Gamma,\, E\} = 0. \tag{61}$$

Of course, in Eq. (61) the classical quantity $\Gamma = \Gamma(q,\, p) = \Gamma(q,\, S_{1,q})$ is obtained from the corresponding quantum one $\hat{\Gamma} = \Gamma(\hat{q},\, \hat{p})$ using the straightforward calculation at the soft limit $\hbar \to 0$ of the theory (the same situation takes place for the energy function E and all other observables). Thus, all quantum symmetries transform to the corresponding classical ones through the consistent procedure of constructing the classical limit.

4.5. Arrow of time in modified classical dynamics: phase space evolution

The same situation holds for the irreversible properties of the dynamics of classical systems which arise in the original quantum theory (1)–(4) in its soft classical limit. Namely, all conclusions of the subsection 3.5 are preserved when the soft classical limit is taken. The only differences are related to the definition of the expectation values of quantities, which must be used for detecting the arrow of time. For example, the limit of Eq. (20) at $\hbar \to 0$ reads:

$$\bar{E} = \int dq\, \rho\, E \qquad \bar{\Gamma} = \int dq\, \rho\, \Gamma; \tag{62}$$

i.e. the expectation values of classical variables have the conventional definition in terms of the probability density ρ. Then, performing the same calculations, as in the subsection 3.5., we conclude that $d\bar{\Gamma}(t)/dt \le 0$ for arbitrary initial data in the isothermal and adiabatic regimes of evolution. This conclusion is based on Eqs. (21) and (23), which preserve their form in the soft classical dynamics. Moreover, after the introduction of physical entropy

$$\Sigma_{\text{phys}} = \int dq\, \rho \log \rho - \overline{\log r} \tag{63}$$

we obtain the non-decreasing entropy law $d\Sigma_{\text{phys}}/dt \ge 0$ for any value of time $t \ge 0$. The corresponding results for the quantum dynamics are given by Eqs. (25)–(26), which remain true in the classical case.

Finally, we see that the soft classical dynamics preserves, exactly, the irreversible properties of the quantum one. In both quantum and classical regimes, the arrow of time effects are defined by the correlation structures, see Eqs (21) and (23). It is possible to say that the arrow of time for macro-world is the direct consequence of the micro-world arrow of time in the modified theory under discussion. We would like to note that the phenomenology related to the macroscopic physics is based on the Second Law of thermodynamics, i.e. the macroscopic arrow of time exists definitely. Thus, if our theory is realistic, the microscopic dynamics must also possess the irreversible properties, i.e. it must be based on the Hamiltonian with the non-trivial decay operator $\hat{\Gamma}$.

4.6. Arrow of time in modified classical dynamics: soliton trajectory evolution
At the end of this section we would like to show the appearance of the arrow of time in the classical trajectory description. This important dynamical property cannot be related to behavior of expectation values of the dynamical variables, because the trajectory does not coincide with the total probability stream. However, irreversibility effects can be detected surprisingly in the system dynamics using this incomplete dynamical data too.

Namely, let us consider the system with the energy function

$$E = \frac{p^2}{2m} + \alpha r^{\delta}, \tag{64}$$

where α and δ are constant parameters. Let us study dynamics of the classical quantity

$$\Gamma_{\star} = \Gamma + \mathrm{tr}\left(E_{,pp^T} A_1\right). \tag{65}$$

The statement is that the inequality

$$\frac{d\Gamma_{\star}(t)}{dt} \leq 0 \tag{66}$$

holds for the systems with the parameters α and δ satisfying the restriction

$$\alpha(\delta + 1) \geq 0 \tag{67}$$

for arbitrary initial data. Of course, Eqs. (65) and (66) are considered on the classical trajectory with $q = q_{\text{classic}}(t)$ and $p = p_{\text{classic}}$. Note that the special Coulomb case, which is characterized by $\delta = -1$, satisfies the restriction (67) identically. Note that this special case is actually important for different applications.

5. Modified relativistic dynamics, time structure and arrow of time

For application of the modified quantum theory to fundamental physics, it is necessary to develop the relativistic realizations of the general scheme (1)–(4). In this section we formulate a concept of constructing the specific relativistic quantum models of the proposed type and give a universal precept for introducing a decay operator for the system. This precept is related to the famous multi-world concept in quantum theory. In fact, we realize this concept in the framework of general quantum formalism and show that the resulting relativistic theory possesses the arrow of time. The results of this section are based on the publications [10]– [11].

5.1. *Extra time dimension conception*

We propose to construct the relativistic models which satisfy the relations (1)–(4) in the explicitly relativistic form. Namely, our idea is to work with the wave functions

$$\Psi = \Psi\left(\tau,\, x^{\mu}\right), \tag{68}$$

which realizes Poincare group representation, and depends on the complex Hamilton's time τ and the four-dimensional coordinates x^{μ} (where $\mu = 0,\, 1,\, 2,\, 3$) as on the kinematically independent variables. We propose to consider that

$$\tau = \mathrm{inv}, \tag{69}$$

i.e. that Hamilton's time is the Poincare-invariant parameter of evolution. In the framework of this general approach, the Hamiltonian $\hat{H}$ must act on the state vector Ψ in such a way that the vector $\hat{H}\Psi$ transforms exactly as Ψ. Actually, in this case, the dynamical equations (1) become Poincare-invariant relations identically.

We would like to stress that the real part t of the complex evolutionary parameter τ cannot be related to the Minkowskian time x^0 on the kinematical level, i.e.

$$x^0 \neq ct. \tag{70}$$

In this sense, our realization of the modified quantum theory and its classical limit are some dynamical models with double time. Below we study how it is possible to reduce the extra time parameter in the theory of such type. Briefly, we postulate, additionally, a new extremal principle: all extra variables must be fixed on their values which realize maximal value for the probability density for any given value of the space-time coordinates x^{μ}. Here it is important to note that the final theoretical scheme is free of the 'second time' t effectively – it is a conventional four-dimensional quantum or classical theory.

5.2. *Double universe case and arrow of time*

It is clear that the main problem in construction of some specific relativistic realization of the modified quantum theory (1)–(4) is related to the search of the decay operator $\hat{\Gamma}$, which commutes with the energy $\hat{E}$, does not destroy the formal structure of the theory, and possesses a clear physical sense. For example, the non-linear functions $f(\hat{E})$ cannot be taken as realizations of $\hat{\Gamma}$, because the resulting theory will be related to the higher order derivatives with respect to the four-dimensional coordinates x^{μ}, than the starting theory with trivial $\hat{\Gamma}$.

To solve this problem in the universal form, we propose the following procedure of the system 'multiplication'. Let us consider two identical quantum subsystems with the wave functions Ψ_1 and Ψ_2, and the coinciding Hamiltonians $\hat{H}_1 = \hat{H}_2 = \hat{E}$. We unify these subsystems into a total one using the following embedding relations:

$$\Psi = \begin{pmatrix} \Psi_1 \\ \Psi_2 \end{pmatrix}, \qquad \hat{\mathcal{H}} = \hat{\mathcal{E}} - \frac{i\hbar}{2}\,\hat{\Im}, \tag{71}$$

where

$$\hat{\mathcal{E}} = \begin{pmatrix} \hat{E} & 0 \\ 0 & \hat{E} \end{pmatrix}, \qquad \hat{\Im} = \begin{pmatrix} \gamma_1 & 0 \\ 0 & \gamma_2 \end{pmatrix} \tag{72}$$

and

$$\gamma_1 < \gamma_2 = \mathrm{const}. \tag{73}$$

Thus, we suppose to unify the two identical subsystems using the direct sum formalism. This approach leads to arising of the two fundamental constants γ_1 and γ_2 in the theory (in fact, it can be shown that only their difference $\gamma_2 - \gamma_1$ has the fundamental sense).

The main statement related to the realization (71)–(73) of the general scheme (1)–(4) is that for the resulting theory the fundamental inequality

$$\frac{d\bar{\Im}(t)}{dt} \leq 0 \tag{74}$$

actually holds. Thus, this theory possesses the arrow of time and is due to the fundamental thermodynamical framework.

5.3. Example: parity symmetry violation

Let us illustrate the results of the previous subsection by the following explicit example. First of all, let us take

$$-\gamma_1 = \gamma_2 = \gamma > 0; \tag{75}$$

then

$$\hat{\Im} = \gamma\hat{\Pi}, \tag{76}$$

where

$$\hat{\Pi} = \begin{pmatrix} 1 & 0 \\ 0 & -1 \end{pmatrix}. \tag{77}$$

In fact, the restriction (75) is not sufficient – this can be obtained using some simple 'renormalization' of the theory quantities. It is easy to see that $\Pi^2 = 1$; thus, this theory contains the decay operator of the parity type. Let us use here the notation $\Psi_1 = \Psi_+$ and $\Psi_2 = \Psi_-$ for the odd and even eigenstates of the parity operator, respectively. Then the eigenvalue problem in the theory under consideration will take the following simple form:

$$\hat{E}\Psi_{n\pm} = E_n\Psi_{n\pm}, \qquad \Pi\Psi_{n\pm} = \pm\Psi_{n\pm}. \tag{78}$$

Of course, the relation (74) takes place in this theory. We state, however, that the additional remarkable identity

$$\lim_{t\to\pm\infty} \bar{\Im}(t) = \mp\gamma \tag{79}$$

is satisfied for this system for arbitrary initial data and for any temperature mode $\beta = \beta(t)$ which allows the limit temperature values $\beta_\pm = \lim_{t\to\pm\infty}\beta(t)$. Thus, the global evolution of this system (from $t \to -\infty$ to $t \to +\infty$) has the 'kink nature'. Actually, this evolution leads to the map of the subsystem with $-\gamma$-value of the decay operator to the subsystem with its $+\gamma$-value. In fact, it is possible to find this kink in its explicit form here, using special initial data with $P_{n+} = P_{n-}$ (and, thus, with $Z_+ = Z_-$) at $t = 0$. In this special case, one obtains the following relation for the expectation of the decay operator:

$$\bar{\Im}(t) = -\gamma\tanh(\gamma t). \tag{80}$$

It is seen that Eq. (80) gives kink exactly; this kink relates to the subspaces $\Psi_\pm$ as attractor states at $t \to \pm\infty$. Thus, the global evolution of the system has the sense of transformation of the 'even world' to the 'odd' one (or vise versa).

5.4. Multi-universe conception

It is clear that the approach of introducing a decay operator into the theory proposed in the previous subsection, can be generalized naturally to the case of an arbitrary number of identical realizations of the given quantum system. Namely, let us consider the theory with the wave function

$$\Psi = \begin{pmatrix} \Psi_1 \\ \Psi_2 \\ \Psi_3 \\ . \\ . \end{pmatrix}, \tag{81}$$

and the energy and decay operators given by the following matrices:

$$\hat{\mathcal{E}} = \begin{pmatrix} \hat{E} & 0 & 0 & . & . & . \\ 0 & \hat{E} & 0 & . & . & . \\ 0 & 0 & \hat{E} & . & . & . \\ . & . & . & . & & \\ . & . & . & & . & \\ . & . & . & & & . \end{pmatrix}, \quad \hat{\Im} = \begin{pmatrix} \gamma_1 & 0 & 0 & . & . & . \\ 0 & \gamma_2 & 0 & . & . & . \\ 0 & 0 & \gamma_3 & . & . & . \\ . & . & . & . & & \\ . & . & . & & . & \\ . & . & . & & & . \end{pmatrix}, \tag{82}$$

where

$$\gamma_1 < \gamma_2 < \gamma_3 < \cdots = \mathrm{const}. \tag{83}$$

The statement is, that the fundamental inequality (74) holds in this theory for the well physically-motivated temperature modes again. For example, it is easy to check Eq. (74) for the isothermal regime $\beta = \mathrm{const}$.

Thus, the proposed scheme allows one to transform the reversible theory with the trivial decay operator $\hat{\Gamma} = 0$ and some fixed energy operator $\hat{E}$ to the irreversible in time quantum system with the energy and decay operators, given by Eq. (82). This new quantum 'super system' describes the dynamics of the quantum reality between 'instant worlds' represented by the subsystems

$$\begin{pmatrix} \Psi_1 \\ 0 \\ 0 \\ . \\ . \end{pmatrix}, \quad \begin{pmatrix} 0 \\ \Psi_2 \\ 0 \\ . \\ . \end{pmatrix}, \quad \begin{pmatrix} 0 \\ 0 \\ \Psi_3 \\ . \\ . \end{pmatrix}, \ldots. \tag{84}$$

The arrow of time is 'directed' to some attractor in this 'multi-world' quantum system; for the isothermal evolution, the final 'instant world' is characterized by the minimal operator $\hat{\Im}$ eigenvalue. Thus, for this simplest type of a temperature mode, the final state is the first one in the states row (84). In fact, Eqs. (81)–(84) give realization of the multi-world conception in the quantum theory in its explicit form. Any evolution here is related to some 'migration' of this system through its 'equivalent realizations' (84) from the probability point of view.

6. Modified Dirac's system: relation between different Noether symmetries

In this section we present one specific relativistic quantum theory of the modified type presented above. This theory describes the spinor field in four dimensions with the additional complex parameter of evolution. It is shown that spectrum of this model consists of particles, antiparticles and tachyons. We propose to use the tachyon modes as the source of dark mass in this system, and calculate the tachyon input to the mass expectation value for the given quantum state. We study the Noether structures of the theory, and establish some relation between the gauge charge and mass expectation value. Also, the spinor model with analytic dependence of wave function on the full set of complex coordinates is introduced. This totally complexified theory possesses additional relations between the gauge and energy-momentum charges, which makes it especially attractive for the following applications. The results of this section are based on the publications [12]–[13]. Also, the integrals of motion for systems with nonstationary conditions were discussed in papers [18]–[19].

6.1. Definition of modified Dirac's system

First of all, let us note that the conventional Dirac's equation for spinor field Ψ with the mass m can be written in the form of eigenvalue problem

$$\hat{E}\Psi = mc^2\,\Psi \tag{85}$$

with respect to the energy operator

$$\hat{E} = -c\left(\gamma^4\right)^{-1}\gamma^\mu\hat{p}_\mu. \tag{86}$$

Here the matrix γ^4 can be taken in one of the two following forms:

$$\gamma^4 = \left\{ \begin{array}{c} -I \\ \gamma^5 \end{array} \right\}, \tag{87}$$

which gives rise to two possible branches of the theory – for normal and abnormal spinors, respectively, whereas $\hat{p}_\mu = i\hbar\partial_{x^\mu}$ is the standard momentum operator. From Eq. (85) it follows that the observable

$$\hat{m} = \frac{1}{c^2}\hat{E} \tag{88}$$

plays the role of the mass operator in the theory. It is not difficult to prove that the equality

$$\hat{m}^2c^2 = \hat{p}^2 = \eta^{\mu\nu}\hat{p}_\mu\hat{p}_\nu, \tag{89}$$

where $\eta^{\mu\nu} = \mathrm{diag}(1, -1, -1, -1)$ is the inverse Minkowskian metrics, holds for the both realizations (87) of the matrix γ^4. This equality has exactly the form of the famous relation in special relativity between the squared four-dimensional momentum and the mass of the particle. Thus, we have an additional argument to interpret the operator $\hat{m}$, given by Eq. (88), as the mass operator for the spinor system.

Now we are ready to define the modified Dirac's system, in complete accordance to the general scheme (1)–(4), and conception of the relativistic models construction developed in the previous section. Namely, let us define the theory of this type as the system with the energy operator $\hat{E}$ given by Eq. (86), and with the decay operator $\Gamma = 0$. Then, the dynamical relations (1) take the following form:

$$i\hbar\Psi_{,\tau} = -c\left(\gamma^4\right)^{-1}\gamma^\mu\hat{p}_\mu\Psi, \qquad \Psi_{,\tau^*} = 0. \tag{90}$$

Let us note that this theory does not contain any mass parameter fixed – this is a massless five-dimensional dynamical model with one complex and four real coordinates. From the general point of view, the problems (85) and (90) give the stationary and non-stationary Schrödinger's equations for the quantum theory under consideration. From this, it follows that the generalized system (90) contains superposition of the total set of special systems like the system (85), i.e. we deal with some 'complete spinor world' in a framework of the modified spinor theory (90).

For the operator $\hat{A}$ in this theory, the matrix element between the states $|\Psi_1>$ and $|\Psi_2>$ is defined by the relation

$$< \Psi_1|\hat{A}|\Psi_2 >= \int d^4x\,\bar{\Psi}_1\hat{A}\gamma^4\Psi_2. \tag{91}$$

Taking $\hat{A} = I$ and $|\Psi_1>=|\Psi_2>=|\Psi>$ in Eq. (91), we obtain the definition of the statistical sum Z (and the norm of the state $|\Psi>$) in this theory:

$$Z =< \Psi|\Psi >= \int d^4x\,\bar{\Psi}\gamma^4\Psi. \tag{92}$$

For the probability density ρ we obtain immediately the following relation:

$$\rho = \frac{\bar{\Psi}\gamma^4\Psi}{Z}; \tag{93}$$

this result will be discussed additionally in the last subsection of this section. It will be shown that the definitions (91)–(93) are consistent with the fundamental formalism of Noether symmetries in this theory. Here we would like to note that the rule (91) for calculating matrix elements and the mass operator definition given by Eqs. (86) and (88) guarantee the Hermitian property of this mass operator. Namely, one can prove using straightforward calculation that the equality

$$< \Psi_1|\hat{m}|\Psi_2 >=< \Psi_2|\hat{m}|\Psi_1 >^* \tag{94}$$

holds in this theory. This means that $\hat{m} = \hat{m}^+$. This fact gives another argument to the mathematical correctness of the general scheme presented in addition to the one related to Eq. (89). In the following subsections we study the generalized quantum system (90) and its possible applications in details.

6.2. Mass generation mechanism

To perform the Fourier analysis of the theory (90), let us change the coordinate representation to the momentum one. Then, for the wave function in the momentum representation, we have the superposition formula

$$\Psi(\tau, p) = \sum_{k=1}^{4} C_{pk}\Psi_k(\tau, p), \tag{95}$$

where $C_{pk} = \text{const}$, and the Fourier modes $\Psi_k(\tau, p)$ are defined as

$$\Psi_k(\tau, p) = \exp\left(-\frac{ic\epsilon_{pk}\tau}{\hbar}\right)\psi_{pk}. \tag{96}$$

Here the wave functions $\psi_{p,k}$, which are stationary with respect to the complex evolutionary parameter τ, are solutions of the eigenvalue problem

$$\hat{m}\,\psi_{pk} = c^{-1}\epsilon_{pk}\psi_{pk} \tag{97}$$

with $\epsilon_{pk}^2 = p^2$. In fact, Eq. (97) coincides with Eq. (85) up to the notations change; thus, we have decomposed the solution Ψ of the dynamical system (90) to the superposition of the standard type Dirac's modes. This conclusion can be done immediately for the modes with $p^2 > 0$ and $p^0 > 0$, which describe the Dirac particles in the theory, and also for anti-particles with $p^2 > 0$ and $p^0 < 0$. Then, for the massless objects with $p^2 = 0$ this statement holds again – all these three special cases are described by the original Dirac's theory (85)–(86). It is interesting to note that the tachyon modes with $p^2 < 0$ are included into the spectrum of the modified spinor theory too. Thus, the theory under consideration describes the quantum superposition of all the spinor particles, antiparticles and tachyons.

These three types of the spinor objects split the momentum space into three subspaces: the cone of future V_{I} with $p^2 > 0$ and $p^0 > 0$, the cone of past V_{II} with $p^2 > 0$ and $p^0 < 0$, and the the rest part of the momentum space V_{III} with $p^2 < 0$. Using Eqs. (92) and (95)–(97), for the statistical sum Z one obtains the following explicit form:

$$\begin{aligned} Z &= \int_{V_{\mathrm{I}}} d^4p\left[-\left(\left|C_{p1}^{\mathrm{I}}\right|^2+\left|C_{p2}^{\mathrm{I}}\right|^2\right)\exp\left(-c\epsilon_p\beta\right)+\left(\left|C_{p3}^{\mathrm{I}}\right|^2+\left|C_{p4}^{\mathrm{I}}\right|^2\right)\exp\left(c\epsilon_p\beta\right)\right]+ \\ &+ \int_{V_{\mathrm{II}}} d^4p\left[-\left(\left|C_{p1}^{\mathrm{II}}\right|^2+\left|C_{p2}^{\mathrm{II}}\right|^2\right)\exp\left(c\epsilon_p\beta\right)+\left(\left|C_{p3}^{\mathrm{II}}\right|^2+\left|C_{p4}^{\mathrm{II}}\right|^2\right)\exp\left(-c\epsilon_p\beta\right)\right]- \\ &- i\int_{V_{\mathrm{III}}} d^4p\left[\left(C_{p1}^{\mathrm{III}*}C_{p3}^{\mathrm{III}}-C_{p2}^{\mathrm{III}*}C_{p4}^{\mathrm{III}}\right)\exp\left(ic\epsilon_p\beta\right)-\mathrm{c.\,c.}\right], \end{aligned} \tag{98}$$

where the quantity $\epsilon_p = |p^2|^{1/2}$ was introduced, and we have taken the γ^5-variant of the theory for definiteness (see Eq. (87)). It is seen that all the spinor modes generate their input to the statistical sum of the theory. From this fact and using the formula

$$\bar{m} = \frac{<\Psi|\hat{m}|\Psi>}{<\Psi|\Psi>} = c^2\left(\log|Z|\right)_{,\beta}, \tag{99}$$

which can be proved directly, one concludes that all the spinor modes give their input to the expectation value $\bar{m}$ of the mass of this quantum system. Together, Eqs. (98) and (99) represent the mass generation mechanism in the theory (90), which does not contain any mass parameter fixed. Of course, this mass generation mechanism is based on the presence of an extra dimension τ of special type in the theory. We will illustrate this mechanism by one interesting specific example in the next subsection.

6.3. Tachyons in theory: invisibility and dark matter source

It is important to note that

$$<\psi_k(p)|\psi_k(p)>=0 \tag{100}$$

if $p^2 < 0$; this relation follows from Eq. (97) for the case of imaginary value of ϵ_{pk}. Eq. (100) means that every tachyonic mode itself has no effect on the mass expectation value $\bar{m}$ and it is possible to speak about tachyons as about 'invisible matter'. However, from Eq. (98) it follows that the nontrivial combinations of tachyons with different polarizations generate the non-vanishing term in the function Z. Actually, this term is related to the integration over the subspace V_{III} of the momentum space. Thus, we have a possibility to identify the tachyons with invisible matter with non-trivial mass (see Eq. (99) which relates $\bar{m}$ with Z). Finally, the theory can be considered as describing the 'spinor world' (or the 'spinor universe') with the sector of visible matter which consists of particles and antiparticles ($p^2 \geq 0$), and invisible sector constructed of tachyons ($p^2 < 0$). It is possible to introduce the 'defect of mass' $\Delta\bar{m}$,

$$\Delta\bar{m} = \bar{m} - \bar{m}_0, \tag{101}$$

which can be identified with the dark mass in this system. Here the quantity $\bar{m}_0$ is understood as the mass expectation value, calculated over the visible sector of the theory states. Of course, tachyons are the reason of the dark matter in the modified spinor dynamics predictions.

Let us present one exact tachyonic solution with the explicit mass formula. This solution is defined by the parameters, which satisfy the relation

$$-i\left(C_{p1}^{\mathrm{III}\,*}C_{p3}^{\mathrm{III}} - C_{p2}^{\mathrm{III}\,*}C_{p4}^{\mathrm{III}}\right) = r\exp(-i\alpha)\delta(p-p_0), \tag{102}$$

where r and α are the arbitrary real constants, and $\epsilon_0 = \sqrt{|p_0^2|}$. Performing the calculations, one obtains the following remarkable result:

$$\bar{m}_{\mathrm{tach}} = \frac{\epsilon_0}{c}\tan\left(c\epsilon_0\beta + \alpha\right). \tag{103}$$

Thus, for this simplest nontrivial tachyonic state, expectation value for the mass depends on the temperature parameter β, i.e. $\bar{m}_{\mathrm{tach}} = \bar{m}_{\mathrm{tach}}(\beta)$. This is the illustration of the general fact that the dark mass which is generated by the tachyons has temperature dependence, as it follows from Eqs. (98)–(99). Note, that the standard Dirac's modes with $p^2 \geq 0$ has not such kind of thermodynamic properties, because for such pure states $\bar{m} = \epsilon_0 = \mathrm{const}$. Thus, our theory gives an important prediction related to the dark mass problem: the dark mass of the system must depend on the temperature of this system. Alternative approaches to the dark matter problem can be found in [23]–[25].

6.4. Relations between Noether currents in totally analytic theory

In this subsection we come back to the study of analytical properties of the modified spinor dynamics. First of all: Eqs. (90) can be rewritten in the following equivalent form:

$$\gamma^M \Psi_{,x^M} = 0, \qquad \Psi_{,\beta} = -\frac{i\hbar}{2}\Psi_{,x^4}, \tag{104}$$

where the new coordinate notation $x^4 = ct$ was introduced. Then, the first relation in (104) can be derived as the Euler-Lagrange equation from the five-dimensional action

$$S_5 = \int d^5x\, L_5, \tag{105}$$

where the five-dimensional Lagrangian is given by the relation

$$L_5 = \frac{i}{2}\left(\bar{\Psi}\gamma^M\Psi_{,x^M} - \bar{\Psi}_{,x^M}\gamma^M\Psi\right). \tag{106}$$

Thus, we have established the Lagrangian structure in the theory under consideration, which can be used for study of canonical structures and relations between them.

Of course, only the coordinate x^4 possesses a 'partner' (the temperature parameter β) in this system; this fact gives rise to the second relation in Eq. (104). Together with another fact that $t = x^4/c$ is the real part of complex Hamilton's parameter of evolution, we will define the set of Noether charges with respect to split of the (formal) five-dimensional space-time x^M to the original four-dimensional space-time x^μ and the evolutionary variable x^4. Namely, we define the Noether charges as quantities, which do no depend explicitly on the time coordinate x^4, and which are constructed as integrals over the coordinates x^μ (with the differential volume element $d^4x = dx^0dx^1dx^2dx^3$). Performing the standard analysis, we first obtain the gauge charge Q,

$$Q = \int d^4x\, J^4, \tag{107}$$

which is related via integration to the 4-th component of the gauge current four-dimensional vector field

$$J^N = \bar{\Psi}\gamma^N\Psi. \tag{108}$$

The statement is, that the identities

$$\partial_{x^N} J^N = 0, \qquad \partial_{x^4} Q = 0 \tag{109}$$

hold (and that the first relation in Eq. (109) is the consequence of the second one there). Then, the second Noether charge is the energy-momentum five-dimensional vector P_M, which is defined by the relation

$$P_M = \int d^4x\, T^4_M, \tag{110}$$

where T^4_M is the 4-th component of the five-dimensional tensor of energy-momentum:

$$T^N_M = \frac{i}{2}\left(\bar{\Psi}\gamma^N\Psi_{,x^M} - \bar{\Psi}_{,x^M}\gamma^N\Psi\right). \tag{111}$$

Again, the statement is that the equalities

$$\partial_{x^N} T^N_M = 0, \qquad \partial_{x^4} P_M = 0 \tag{112}$$

are satisfied on-shell, and that the second relation follows from the first one in Eq. (112). These properties of the modified spinor dynamics are common with the standard Dirac's theory. The unusual facts are related to the 'hidden' β-dependence of all quantities, which lead to surprising relations between some of them here. Namely, it is easy to prove that the relations

$$T^N_4 = -\frac{1}{\hbar c}\,\partial_\beta J^N, \qquad P_4 = -\frac{1}{\hbar c}\,\partial_\beta Q \tag{113}$$

are satisfied in this theory. For example, the second of them allows one to calculate the conserving quantity P_4 using the gauge charge Q through its β-dependence. Note that Eq. (112) can be obtained from Eq. (109) using the differentiation over the temperature parameter β as direct consequence. Finally, it can be said, that in this theory the symmetry with respect to the x^4-direction follows from the gauge symmetry. In standard theory, such kind of symmetries are completely independent.

At the end of this section let us generalize the modified spinor dynamics to the totally analytic case. It will be shown that this generalization is supported by the symmetry structure of the theory. Namely, let us complexify the coordinates x^μ. Then, the set of five-dimensional coordinates can be represented as

$$X^M = x^M - \frac{i\hbar}{2}\beta^M, \tag{114}$$

and the theory wave-function $\Psi = \Psi(X^M)$ will be defined by the dynamical relation

$$\gamma^M\Psi_{,X^M} = 0. \tag{115}$$

Performing the Noether analysis will lead to the same set of the Noether charges. The remarkable fact is, that, now the relation between translation and gauge symmetries has the covariant form:

$$T^N_M = -\frac{1}{\hbar c}\,\partial_{\beta^M} J^N, \qquad P_M = -\frac{1}{\hbar c}\,\partial_{\beta^M} Q. \tag{116}$$

Thus, the gauge symmetry generates all the translation shifts (in the directions x^M), and the total set of corresponding conserving charges P_N can be calculated using the β^N-dependence of the single gauge charge Q.

7. Perspectives: dynamical reduction mechanism and effective gravity appearing

Double time structure provides the evident problems with the interpretation of the theory. Actually, in the conventional physics, time is a one-dimensional real parameter, which is identified with the Mincowaskian temporal coordinate x^0. In fact, the standard theoretical framework admits only the four-dimensional coordinates x^μ in the role of independent space-time parameters. This means that all possible extra (Kaluza-Klein type) dimensions must be compactified; the resulting physical theory must be expressed in terms of the four-dimensional Minkowskian coordinates only. In accordance with this 'hard' paradigm, the extra times (as well as all other extra or hidden parameters) must be fixed as functions of the coordinates x^μ using well-motivated compactification principle. Finally, the reduced four-dimensional theory will contain a set of the additional four-dimensional fields, which are the fields of compactification under discussion.

These compactification fields can describe different physical interactions in the theory, which can be taken as originally free ones. In this section we propose to relate the fields, generated by extra parameters of the theory, with gravity in the reduced four-dimensional system. The gravity arisen is not fundamental, but the effective field only. The effective metrics is expressed in terms of the solution Ψ of the quantum dynamical relations of the theory. We start our consideration with the modified theory of the scalar field, and continue our study to the more interesting case of the totally analytic quantum theory of the spinor field. In fact, we present two approaches for constructing the effective gravity in the system. In the framework of the first one, gravity effects can be calculated on the soliton trajectory only. In the second approach, we construct gravity as the physical field on the full four-dimensional physical space-time. The results of this section are based on the publications [10]–[11].

7.1. Modified Klein Gordon system and metrics generation

Let us start our effective gravity construction from the study of the case of a modified scalar theory. This simplest four-dimensional relativistic theory arises, if one takes

$$\hat{E} = \eta^{\mu\nu}\hat{p}_\mu\hat{p}_\nu/2m, \tag{117}$$

where $m = \text{const}$ is some fixed parameter of the mass type, and $\hat{\Gamma} = 0$. Then, for the modified Hamilton's equations (37) one obtains:

$$\begin{aligned}
\frac{dx^\mu}{dt} &= \frac{1}{m}\left\{\eta^{\mu\nu}p_\nu - \eta^{\nu\lambda}\left[\left(A_2^{-1}\right)^{\mu\sigma} S_{1,\sigma\nu\lambda} + \dot{\beta}\left(A_2^{-1}A_1\right)^{\mu}_{\ \nu} p_\lambda\right]\right\},\\
\frac{dp_\mu}{dt} &= -\frac{1}{m}\eta^{\nu\lambda}\left[\left(A_1A_2^{-1}\right)_\mu^{\ \sigma} S_{1,\sigma\nu\lambda} + \dot{\beta}\left(A_1A_2^{-1}A_1\right)_{\mu\nu} p_\lambda\right].
\end{aligned} \tag{118}$$

It is easy to see that the trajectory, which corresponds to Eq. (118), becomes the straight line in the case of vanishing correlations. Moreover, in this hard classical limit, the relation $p_\mu = \eta_{\mu\nu}\, m\, dx^\nu/dt$ (which follows from the first relation in Eq. (118)), allows one to identify the Hamilton's time t with the proper time of the moving point mass m in this theory.

To establish the effective gravity in the system, let us compare the relations (118) with conventional equations of motion of the point mass m propagating in the gravitational field with the metrics $g_{\mu\nu}$:

$$\frac{dx^\mu}{dt} = \frac{1}{m}g^{\mu\nu}p_\nu, \qquad \frac{dp_\mu}{dt} = -\frac{1}{2m}g^{\nu\lambda}{}_{,\mu}\,p_\nu\,p_\lambda. \tag{119}$$

We will study the case of weak correlations, i.e. the dynamics of the well-localized soliton in the framework of Eq. (118). It is clear that we must compare it with the regime of the weak gravity in Eq. (119), where it is convenient to put

$$g_{\mu\nu} = \eta_{\mu\nu} + \xi_{\mu\nu} \tag{120}$$

with $g_{(1)\mu\nu} \to 0$). We state that it is possible to identify Eq. (118) at $A_2^{-1} \to 0$ with the system (119)–(120), using some highly non-trivial temperature mode $\beta = \beta(t)$. The result has the especially simple form in the reference frame, which coincides with initial rest frame of the mass. For the infinitesimal potential ξ_{00} the result of this identification reads:

$$\xi_{00} = \left(A_2^{-1}\right)^{0\mu} \left[\frac{1}{mc}\,\eta^{\nu\lambda}\,S_{1,\,\mu\nu\lambda} + \dot{\beta}\,(A_1)_{\mu 0}\right].$$

Note that this potential defines the non-relativistic limit of the General Relativity and provides the special interest for different application.

Thus, uncertainty in the position and momentum value of the point mass can be interpreted by physical observer as a presence of some non-trivial gravity in the system. Note that this gravity is calculated on the classical trajectory only, i.e. on the trajectory of maximums of the probability density. Thus, this approach cannot give information about the effective gravity as about the field defined over the whole four-dimensional physical space-time.

7.2. Reduction of extra parameters and source of quantum gravity: basic conception

Now let us come back to the modified spinor dynamics with the motion equations (90). Our goal is the reduction principle, which allows one to extract the additional variables x^4 and β^M from all observables of the theory. Thus, we are interested in establishing the relations

$$x^4 = x^4\left(x^\mu\right), \qquad \beta^M = \beta^M\left(x^\mu\right), \tag{121}$$

with some specified right sides. The reduction postulate, which we propose reads: the functions (121) must realize a maximum of the probability density at a given space time point x^μ:

$$\max_{x^4,\,\beta^M} \rho\left(x^\mu,\, x^4,\, \beta^M\right) = \rho\left(x^\mu,\, x^4(x^\mu),\, \beta^M(x^\mu)\right). \tag{122}$$

Thus, this principle fixes all extra parameters in the framework of the maximal probability arguments. Exactly the same situation takes place in the Hamilton equations construction, see Eq. (36). Performing calculations, like the ones leading to Eq. (37), we obtain the following defining relations for the functions (121):

$$x^4_{,x^\mu} = -\frac{1}{\rho_{,x^4x^4}}\left[\rho_{,x^4x^\mu} - \left(Q^{-1}\right)^{MN}\rho_{,x^4\beta^M}\left(\rho_{,\beta^Nx^\mu} - \frac{\rho_{,\beta^Nx^4}\rho_{,x^4x^\mu}}{\rho_{,x^4x^4}}\right)\right],$$

$$\beta^M_{,x^\mu} = -\left(Q^{-1}\right)^{MN}\left(\rho_{,\beta^Nx^\mu} - \frac{\rho_{,\beta^Nx^4}\rho_{,x^4x^\mu}}{\rho_{,x^4x^4}}\right), \tag{123}$$

where the 5×5-matrix Q is defined in its components as

$$Q_{MN} = \rho_{,\beta^M\beta^N} - \frac{\rho_{,\beta^Mx^4}\rho_{,\beta^Nx^4}}{\rho_{,x^4x^4}}. \tag{124}$$

Note, that the parameters β^M describe the anisotropy in the probability density distribution related to the given solution (as imaginary parts of the complex coordinates, see Eq. (114)). Thus, the fields (121) specified by Eqs. (123)–(124) give corresponding localized anisotropy fields in the theory. In the standard approach, the fundamental anisotropy properties of the system can be related to gravity. Actually, metric coefficients $g_{\mu\nu} = g_{\mu\nu}(x^\lambda)$ can be used for the adequate anisotropy of space-time description. Our main idea is to try to calculate an effective metrics which exactly corresponds to the anisotropy functions defined by Eq. (124). Such gravity will describe the isotropy breaking in the theory. The relations (121) can be also interpreted as thermodynamic mode of the generalized type, see Eq. (19) for comparison.

7.3. Relationship between thermodynamic mode and metrics
The simplest way to detect the effective gravity in a system is related to the study of the squared mass operator behavior under the reduction process (we present a simplified approach here). First of all, let us define the reduced spinor field ψ as

$$\psi\left(x^{\mu}\right)=\Psi\left[X^{N}\left(x^{\mu}\right)\right]. \tag{125}$$

Let us also introduce the following convenient quantities, which are defined by the reduction relations (124) completely:

$$X^{\lambda}_{,x^{\mu}}=\delta^{\lambda}_{\mu}-\frac{i\hbar c}{2}\beta^{\lambda}_{,x^{\mu}}\equiv A^{\lambda}_{\mu},\qquad X^{4}_{,x^{\mu}}=x^{4}_{,x^{\mu}}-\frac{i\hbar c}{2}\beta^{4}_{,x^{\mu}}\equiv B_{\mu}. \tag{126}$$

Then, performing straightforward calculations, one obtains the following important relation:

$$\psi_{,x^{\mu}}=D^{\lambda}_{\mu}\Psi_{,x^{\lambda}}, \tag{127}$$

where

$$D^{\lambda}_{\mu}=A^{\lambda}_{\mu}-\left(\gamma^{4}\right)^{-1}\gamma^{\lambda}B_{\mu}. \tag{128}$$

Using this, it is possible to establish the reduction formula for the squared operator of momentum of the system (or its squared mass operator). The result reads:

$$\eta^{\sigma\lambda}\hat{p}_{\sigma}\hat{p}_{\lambda}\Psi=\eta^{\sigma\lambda}\left(D^{-1}\right)^{\nu}_{\sigma}\left(D^{-1}\right)^{\mu}_{\lambda}\hat{p}_{\mu}\hat{p}_{\nu}\psi+i\hbar\eta^{\sigma\lambda}\left(D^{-1}\right)^{\nu}_{\sigma}\left(D^{-1}\right)^{\mu}_{\lambda,x^{\nu}}\hat{p}_{\mu}\psi. \tag{129}$$

Our introduction of the effective metrics $g_{\mu\nu}$ in this system is related to appropriate interpretation of Eq. (129). Namely, its left side gives distribution of the squared mass operator of the 'whole' system:

$$\eta^{\sigma\lambda}\hat{p}_{\sigma}\hat{p}_{\lambda}\Psi=\hat{m}^{2}c^{2}\Psi; \tag{130}$$

we will name this operator as the 'total' one here. Then, the first term on the right side of Eq. (129) can be rewritten as

$$\widehat{g^{\mu\nu}}\hat{p}_{\mu}\hat{p}_{\nu}\psi=\hat{m}^{2}_{\mathrm{red}}c^{2}\,\psi, \tag{131}$$

where the metric-type operator $\widehat{g^{\mu\nu}}$ can be expressed in terms of the matrices inverse with respect to the ones given by Eq. (128):

$$\widehat{g^{\mu\nu}}=\frac{1}{2}\eta^{\lambda\sigma}\left\{\left(D^{-1}\right)^{\nu}_{\sigma},\left(D^{-1}\right)^{\mu}_{\lambda}\right\}. \tag{132}$$

We identify $\hat{m}^{2}_{\mathrm{red}}$ with the squared mass operator for the reduced wave function ψ, given by Eq. (125).

In this context, the operator (132) becomes identified with the inverse metric operator in the system under consideration. Its expectation value $\overline{g^{\mu\nu}}$ can be calculated using the general formalism presented in subsection 6.1. Finally, the third term in Eq. (129) can be identified with additional squared mass term, related to the compactification process. We will not discuss it here leaving the corresponding discussion to other works.

7.4. Cosmological solution

We relate the spatially homogeneous and isotropic solutions of the Eq. (90) with cosmological models in effective gravity approach presented in the previous subsections. Here we study these solutions and discuss their remarkable properties. First of all, let us describe them. We will work with subspace of solution space of the theory with

$$\Psi_{,x^i} = 0 \tag{133}$$

(where $i = 1, 2, 3$), which realizes homogeneity and anisotropy in the sense of three-dimensional space. Thus, the corresponding wave-function Ψ will depend on the 'time-like' parameters x^0 and τ only. Let us consider the γ^5-variant of the theory; then, the general solution of Eq. (90) reads:

$$\Psi = \Psi_+ + \Psi_-, \tag{134}$$

where

$$\Psi_\pm = \frac{1}{\sqrt{2}} \begin{pmatrix} f_\pm \\ \mp i f_\pm \end{pmatrix}, \tag{135}$$

and the 2-columns $f_\pm$ are defined as

$$f_\pm = f_\pm\left(x^0 \mp c\tau\right). \tag{136}$$

It is clear that the special solutions $\Psi_\pm$ represent the solitons running forward and backward in respect to Minkowskian time x^0, when the Hamilton's time x^4 increases. Such solutions contain particles and antiparticles, respectively, and can be named as 'universe' and 'anti-universe' in the framework of natural speculation. The total Universe is a superposition of universe and anti-universe in this picture. Then, the observer living in the universe (i.e., constructed of the particles) will not see the anti-universe in his observations, if the anti-universe is described by the wave function of the highly-localized probability density far from the time of the universe and anti-universe solitons 'collision'. This speculation gives explanation to the particle/antiparticle asymmetry in the real Universe.

Then, it is very important to stress that the cosmological solution (134)–(136) of the quantum motion equations (90) possesses coherent states. These states are described by the special solitons with

$$f_\pm = c_\pm \exp\left[-\frac{\left(x^0 \mp c\tau\right)^2}{4\mathcal{D}^2_{x^0}}\right], \tag{137}$$

where $c_\pm$ are the complex constant 2-columns, and $\mathcal{D}_{x^0}$ is the real positive constant parameter. It is possible to prove that Eq. (137) gives the Gauss wave-packet with dispersion $\mathcal{D}_{x^0}$ of the probability density near its maximal value at the average value $\bar{x}^0 = \pm x^4$ of the Minkowskian time x^0. Thus, $\pm x^4$ corresponds to the center of this distribution; i.e. the Hamilton's time x^4 can be interpreted as the 'averaged' Minkowskian time up to the sign here.

Now let us consider the Universe with vanishing anti-universe mode, i.e. let us take $\Psi = \Psi_+$. Performing the calculations, one obtains the following results:

$$\bar{m} = \left(\frac{\hbar}{2\mathcal{D}_{x^0}}\right)^2 \beta \tag{138}$$

– the explicit mass formula, and the uncertainty relation

$$\mathcal{D}_{mc^2}\mathcal{D}_{x^0/c} = \frac{\hbar}{2} \tag{139}$$

with equality realized. Thus, the family (125) of the Gauss soliton solutions actually describes the coherent states of the theory. These states minimize the uncertainty relation between the mass and Minkowskian time. Then, Eq. (138) shows that the established mass formula $\bar{m} = \bar{m}(\beta)$ is completely defined by the nontrivial $\sim \beta$ term, which is inverse proportional to the squared dispersion $\mathcal{D}^2_{x^0}$ of the probability distribution. It is possible to say that the mass of the state is generated by the nontrivial time dispersion of the probability density to find the Dirac particle at a given moment of the physical time. It is seen that $\bar{m}(0) = 0$, so the mass expectation value vanishes at the infinite temperature for the solitons under study. This seems natural in the framework of the light dominated period of the Universe cosmological evolution scenario.

8. Conclusions

Let us summarize the main results of this publication.

(i) Quantum Mechanics modification is done via the non-Hermitian Hamiltonian and complex parameter of evolution;

(ii) Decay operator is taken as the symmetry operator, this maintains the arrow of time appearing at the quantum level;

(iii) Double limit defines the transition to the classical dynamics with the Planck constant and correlations as infinitesimal parameters. Particles can be presented as solitons in the theory;

(iv) Correlations in the soft limit generate the General Relativity type of correction for Newtonian gravity force;

(v) Soft limit with the non-zero correlations keep the arrow of time in the classical dynamics;

(vi) Relativistic extension based on the Poincare group includes the invariant complex parameters of evolution and relativistic Hamiltonian as the Poincare group generator;

(vii) Wave function multiplication approach leads to appearance of the arrow of time in the relativistic version of the theory. Dynamic parity violation is one of the consequences of the arrow of time appearance;

(viii) Dirac system can be modified as well. No mass introduction required in this theory - the system generates the mass itself;

(ix) Gravity can be constructed as an effective field in the modified relativistic theory. To construct the metrics, one must compactify all extra parameters of the theory.

Acknowledgments

Authors would like to thank Profs. B. Ishkhanov, Yu. Rybakov, G. Costa and M. Matone for fruitful discussions while this investigation was carried out. We are really grateful to the organizers of the International Symposium 'Symmetries in Science XIV' for the provided possibility to present and discuss these results of our activity.

References

[1] Sudbery E 1986 *Quantum mechanics and particles of Nature* (Cambridge: Cambridge University Press)
[2] Bogoliubov N and Shirkov D 1983 *Quantum Fields* (London: Benjamin Cummings)
[3] Frampton P H 2000 *Gauge Field Theories* (New York: John Wiley and Sons)
[4] Itzykson C and Zuber J-B 1980 *Quantum Field Theory* (New York: McGraw-Hill)
[5] Arnold V I 1997 *Mathematical Methods of Classical Mechanics* (Berlin: Springer Verlag)
[6] Faddeev L D and Takhtajan L A 1987 *Hamiltonian Methods in the Theory of Solitons* (Berlin: Springer Verlag)
[7] Goldstein H 1980 *Classical Mechanics* (Toronto: Addison-Wesley)
[8] Harnad J, Winternitz P and Sabidussi G, eds 2000 *Integrable Systems: From Classical to Quantum* (Providence, RI: American Mathematical Society)

[9] Korepin V E, Bogoliubov N M and Izergin A G 1997 *Quantum Inverse Scattering Method and Correlation Functions* (Cambridge: Cambridge University Press)
[10] Asadov V V and Kechkin O V 2008 *Moscow University Physics Bulletin* **2** 105
[11] Asadov V V and Kechkin O V 2009 *Bulletin of the Peoples Friendship University of Russia (Mathematics. Informatics. Physics)* **4** 80
[12] Asadov V V and Kechkin O V 2009 *Gravit. Cosmol.* **15** 295
[13] Asadov V V and Kechkin O V 2009 *Bulletin of the Peoples Friendship University of Russia (Mathematics. Informatics. Physics)* **3** 99
[14] Dekker H 1981 *Phys. Rep.* **80** 1
[15] Tarasov V E 2001 *Phys. Lett.* A **288** 173
[16] Celeghini E, Rasetti M and Vitiello G 1992 *Annals Phys.* **215** 156
[17] Feshbach H and Tikochinsky Y 1977 *Trans. New York Acad. Sci.* **38** 44
[18] Castanos O, Lopez-Pena R and Manko V I 1994 *Mod. Phys. Lett.* A **9** 1785
[19] Castanos O, Lopez-Pena R and Manko V I 1994 *J. Phys.* A **27** 1751
[20] Milgrom M 1986 *Astrophys. J.* **302** 617
[21] Milgrom M 2002 *Astrophys. J.* **571** 617
[22] Bekenstein J D 2004 *Phys. Rev.* A **70** 083509
[23] Bertone G, Hooper D and Silk J 2005 *Phys. Rept.* **405** 279
[24] Jungman G, Kamionkowski M and Griest K 1996 *Phys. Rept.* **267** 195
[25] Hooper D and Profumo S 2007 *Phys. Rept.* **267** 195

Deformation of semi-classical symmetries: Poisson-Hopf limit of quantum algebras

A. Ballesteros[1], E. Celeghini[2] and M.A. del Olmo[3]

[1] Departamento de Física, Universidad de Burgos, E-09006, Burgos, Spain
[2] Departimento di Fisica, Università di Firenze and INFN–Sezione di Firenze, I50019 Sesto Fiorentino, Firenze, Italy
[3] Departamento de Física Teórica, Universidad de Valladolid, E-47005, Valladolid, Spain.

E-mail: angelb@ubu.es, celeghini@fi.infn.it, olmo@fta.uva.es

Abstract. We present a method to construct the Poisson-Hopf structure associated to a quantum algebra $U_z(g)$. The procedure is based on the construction of a $\hbar$-family of quantum algebras $U_{z,\hbar}(g)$ and taking the "semiclassical" limit $\hbar \to 0$.

1. Introduction

Quantum groups appeared in relation with the quantization of classical integrable systems by the method of quantum inverse scattering. In this context, Poisson algebras of functions defined on Poisson manifolds that are Lie groups, G, are involved. These structures, called Poisson-Lie groups, are determined by certain compatibility conditions [1] between the Poisson manifold and the Lie group. The quantization of such (commutative) spaces of functions, $\mathrm{Fun}(G)$, following the Moyal spirit [2] (i.e, by constructing a $\star$-product) leads to the quantum groups $\mathrm{Fun}_q(G)$. In these quantum structures there is a deformation parameter $q = e^z$, where z is the constant that governs the noncommutativity of the algebra of functions

$$[f_1, f_2] := f_1 \star f_2 - f_2 \star f_1 = z\{f_1, f_2\} + \mathcal{O}(z^2),$$

being $\{f_1, f_2\}$ the Poisson bracket in $\mathrm{Fun}(G)$ that has to be the "semiclassical limit" (or Poisson limit) of $[f_1, f_2]$, i.e.

$$\{f_1, f_2\} = \lim_{z\to 0} \frac{1}{z}\, [f_1, f_2].$$

In a physical context where the transition from classical to quantum physical models is studied, the deformation parameter z has been interpreted as the Planck constant $\hbar$. But, in other contexts z is a parameter whose meaning has to be elucidated for each particular case. Indeed from the beginning quantum groups and quantum algebras (the Hopf algebras dual of quantum groups [1,3–6]) has been also considered in an abstract way as Hopf algebras, that could describe new symmetries of physically relevant systems where the deformation parameter z (or q) is completely independent with respect to the Planck constant. Thus, for instance, we can mention the $su_q(2)$ invariance of the Heisenberg spin XXZ chain [7,8], where z is the anisotropy of the chain. Also lattice systems related to deformations of inhomogeneous algebras, where z is connected with the lattice length [9, 10], deformations of kinematical symmetries in which the deformation parameter is a fundamental scale [11–15], effective models in nuclear physics [16,17], etc.

Thus, in this paper we consider a mathematical framework for quantum algebras in which two independent parameters, the deformation parameter $z = \log q$ and the Planck constant $\hbar$, are simultaneously considered. From a technical point of view we construct a one-parameter family of equivalent quantizations $U_{z,\hbar}(g)$ depending explicitly on both parameters for any quantum algebra $U_z(g)$. Performing the semiclassical limit $\hbar \to 0$ we obtain the related q-Poisson-Hopf algebra. In this way we can get classical systems as limit of quantum ones endowed with a quantum algebra symmetry. This approach can be interesting to construct new (classical) integrable systems with a parameter z that could control their dynamical behaviour. The method is illustrated by applying it to the standard quantum deformation of $su(2)$. In this elementary case the q-Poisson-Hopf algebra obtained in the $\hbar \to 0$ limit looks, in despite of the diferences of the two structures, to be formally identical to the original quantum algebras, but this is not the case for more complex algebras like $su(3)$ [18].

2. Poisson limit of quantum algebras

A Lie bialgebra [1,6] (g, δ) is a Lie algebra g,

$$[X_i, X_j] = f_{ij}^k X_k, \tag{1}$$

equipped with a cocommutator map $\delta : g \to g \otimes g$, i.e, a skew-symmetric linear map, given by

$$\delta(X_i) = c_i^{jk} X_j \otimes X_k \tag{2}$$

such that the structure constants c_i^{jk} fulfill the following compatibility conditions together with the structure constants of g, f_{ij}^k,

$$c_r^{p,q} f_{s,t}^r = c_s^{p,r} f_{r,t}^q + c_s^{r,q} f_{r,t}^p + c_t^{p,r} f_{s,r}^q + c_t^{r,q} f_{s,r}^p. \tag{3}$$

It is worthy to note that the two sets of structure constants play symmetric roles: (g^*, δ^*) is the Lie bialgebra dual of (g, δ) determined by

$$[x^p, x^q] = c_r^{p,q} x^r, \qquad \delta^*(x^p) = f_{q,r}^p x^q \otimes x^r, \tag{4}$$

where g^* is the dual space of g generated by the set of elements $\{x^i\}$ such that $x^i(X_j) = \delta_j^i$.

The Lie bialgebras play a fundamental role in order to construct explicitly the deformation of a Lie algebra g. Effectively, given an arbitrary Lie bialgebra (g, δ) the quantum algebra $(U_z(g), \Delta_z)$ is the Hopf algebra deformation of the universal enveloping algebra $U(g)$, compatible with the deformed coproduct $\Delta_z(X)$ whose leading order terms are

$$\Delta_z(X) = \Delta_0(X) + z\,\delta(X) + \mathcal{O}[z^2],$$

where Δ_0 is the primitive coproduct (i.e. $\Delta_0(X) = 1 \otimes X + X \otimes 1$). The deformed coproduct can be obtained using the analytic procedure described in [19,20].

Let M be a smooth manifold of (finite) dimension m, and $\mathcal{C}^\infty(M) \equiv \mathrm{Fun}(M)$ the algebra of smooth real-valued functions on M. A *Poisson structure* on M map is a $\mathbb{R}$-bilinear map (Poisson bracket),

$$\{\ ,\ \} : \mathcal{C}^\infty(M) \times \mathcal{C}^\infty(M) \longrightarrow \mathcal{C}^\infty(M)$$

satisfying the following properties ($\forall f_1, f_2, f_3 \in \mathcal{C}^\infty(M)$):

(i) $\{f_1, f_2\} = -\{f_2, f_1\}$ (skew-symmetry),

(ii) $\{f_1, \{f_2, f_3\}\} + \{f_2, \{f_3, f_1\}\} + \{f_3, \{f_1, f_2\}\} = 0$ (Jacobi identity),

(iii) $\{f_1 f_2, f_3\} = f_1\{f_2, f_3\} + \{f_1, f_3\} f_2$ (Leibniz rule).

A map between Poisson manifolds preserving the Poisson brackets of both manifolds is called *Poisson map*.

A *Poisson-Lie group* is a Lie group and a Poisson manifold such that both structures are compatible. In more detail, a Poisson-Lie group is a Lie group G with a Poisson structure $\{\ ,\ \}$ (defined on $\mathcal{C}^\infty(G)$) such that the multiplication on G, $m : G \times G \longrightarrow G$ $(m(g_1, g_2) = g_1 \cdot g_2)$, is a Poisson map.

A *homomorphism of Poisson-Lie groups* will be a homomorphism of Lie groups and also a Poisson map.

A *Poisson algebra* over the field $\mathbb{K}$ is a commutative algebra A over $\mathbb{K}$ equipped with a Poisson bracket (i.e, a skew-symmetric $\mathbb{K}$-module) verifying the Jacobi identity and the Leibniz rule.

A *homomorphism between two Poisson algebras* $(A, \{\ ,\ \}_A)$ and $(B, \{\ ,\ \}_B)$ is a linear map, $f : A \to B$, such that $(\forall a_1, a_2 \in A)$:

(i) $f(a_1 a_2) = f(a_1) f(a_2)$,

(ii) $f(\{a_1, a_2\}_A) = \{f(a_1), f(a_2)\}_B$.

A *Poisson-Hopf algebra A* is a Poisson algebra $(A, \{\ ,\ \}_A)$ which is also a Hopf algebra (A, Δ_A) over $\mathbb{K}$ if Δ_A is a homomorphism of Poisson algebras between $(A, \{\ ,\ \}_A)$ and $(A \otimes A, \{\ ,\ \}_{A\otimes A})$. Note that the Poisson bracket in $A \otimes A$ is defined by

$$\{a_1 \otimes a_1', a_2 \otimes a_2'\}_{A\otimes A} := \{a_1, a_2\}_A \otimes a_1' a_2' + a_1 a_2 \otimes \{a_1', a_2'\}_A.$$

Let us remember that quantization in physics tries to associate to a commutative algebra of physical observables, Fun(M), of a classical mechanical system (with phase space M) a non-commutative algebra of linear operators, $Op(\mathcal{H})$, on the suitable complex Hilbert space, $\mathcal{H}$, of the quantum system. More precisely, the idea of quantization is to find a linear bijective map $\mathbb{Q} : \text{Fun}(M) \to Op(\mathcal{H})$ such that for all $f \in \text{Fun}(M)$

$$\mathbb{Q}(\{f_1, f_2\}) = \frac{\mathbf{i}}{\hbar}[\mathbb{Q}(f_1), \mathbb{Q}(f_2)]. \tag{5}$$

This condition comes from the fact that if f_2 is the classical Hamiltonian of the system and $\mathbb{Q}(f_2)$ its quantum counterpart then $\{f_1, f_2\}$ and $[\mathbb{Q}(f_1), \mathbb{Q}(f_2)]$ give the time evolution of the physical obsevable f_1 in the classical and in the quantum case, respectively. However such a map is not known in general, even for very simple cases.

Let us consider the case of $M = \mathbb{R}^{2n}$ and $\mathcal{H} = L^2(\mathbb{R}^n)$. The fundamental Poisson brackets associated to the canonical variables $(q_1, q_2, \ldots, q_n, p_1, p_2, \ldots, p_n)$ are

$$\{q_i, p_j\} = \delta_{ij}. \tag{6}$$

The canonical quantization [21] allows us to replace the Poisson brackets (6) by the commutators

$$[Q_i, P_j] = \mathbf{i}\hbar\,\delta_{ij}\, I, \tag{7}$$

where Q_i and P_j are the linear operators associated to q_i and p_j respectively (i.e. Q_i and P_j are the quantum counterparts of the classical observables q_i and p_j: the operator Q_i corresponds to multiplication by q_i and P_j to $-\mathbf{i}\hbar\frac{\partial}{\partial q_j}$). In both cases (6) and (7) reflect that the Heisenberg algebra $\mathfrak{h}$ is involved. Now the condition (5) is verified by the canonical coordinates. This observation leads to the idea, pointed out also by Dirac, that the correspondence between classical and quantum theories has to be based on a certain analogy among their mathematical structures, in other words, a homomorphism must exist between the Lie algebras of the classical and quantum observables.

When there is a set of quantum observables X_i that close a Lie algebra

$$[X_i, X_j] = f_{ij}^k X_k$$

in the (classical) Poisson manifold a corresponding algebra can be defined by

$$\{\mathbb{Q}^{-1}(X_i), \mathbb{Q}^{-1}(X_j)\} := \lim_{\hbar\to 0} \frac{\mathrm{i}}{\hbar}[X_i, X_j]. \tag{8}$$

On the other hand, the dual space of g, g^*, can be equipped with a Poisson structure (Kostant-Kirillov-Souriau structure) whose fundamental brackets are

$$\{x^i, x^j\} = f_k^{ij} x^k, \tag{9}$$

where the x^i can be considered as local coordinates on g^* and $f_k^{ij} \equiv f_{ij}^k$. For any two functions on Fun(g^*) their Poisson bracket will be given by

$$\{f_1, f_2\} = \frac{\partial f_1}{\partial x^i}\frac{\partial f_2}{\partial x^j} f_k^{ij}\, x^k.$$

Also we have a new bialgebra (g^*, δ) determined by (9) and (2).

Keeping in mind all these ideas we see that easily we can quantize the Kostant-Kirillov-Souriau structure as follows

$$\{x^i, x^j\} = f_k^{ij} x^k \quad \Longrightarrow \quad [X_i, X_j] = \hbar\, f_{ij}^k\, X_k.$$

So, we have obtained a "new" Lie algebra $g_\hbar$ isomorphic to g^* and also to g. In this way we have built up a one-parameter family of equivalent Lie bialgebras $(g_\hbar, \delta)$ determined by

$$[X_i, X_j] = \hbar\, f_{ij}^k\, X_k \tag{10}$$

and (2). Obviously, the mathematical structure of the original bialgebra (g, δ) is recovered when $\hbar = 1$.

We can quantize this one-parameter family of Lie bialgebras $(g_\hbar, \delta)$ by using the method of the analytical deformation described in [20], and a two-parameter quantum algebra $(U_{z,\hbar}(g), \Delta_{z,\hbar})$, depending explicitly on $\hbar$, is thus obtained.

The Poisson limit (i.e., $\hbar \to 0$) of $(U_{z,\hbar}(g), \Delta_{z,\hbar})$ gives, in a unique way, the Poisson-Hopf algebra $(\text{Fun}(g_z), \Delta_z^{\mathcal{P}})$ which is just a Poisson-Lie structure on the group g_z whose Lie algebra is characterized by δ^* (4), i.e. by the structure tensor f_i^{jk}.

The Poisson bracket on Fun(g_z) is given by

$$\{\mathcal{X}_i, \mathcal{X}_j\} := \lim_{\hbar\to 0} \frac{[X_i, X_j]}{\hbar}, \tag{11}$$

and the coproduct map by

$$\Delta_z^{\mathcal{P}}(\mathcal{X}_i) := \lim_{\hbar\to 0} \Delta_{z,\hbar}(X_i), \tag{12}$$

where $\mathcal{X}_i = \mathbb{Q}^{-1}(X_i)$ can be seen as the local coordinates on the manifold g_z, so the Poisson brackets (11) are the so-called fundamental brackets. The coproduct (12) is a Poisson algebra homomorphism between Fun(g_z) and Fun$(g_z) \otimes$ Fun(g_z). Effectively, for non-deformed Lie algebras, its coproduct is the primitive one and the commutation rules are linear: thus the Poisson limits (11) and (12) lead to the same formal structure where commutation rules have been just replaced by Poisson brackets and the coproduct remains primitive. However, in quantum algebras nonlinear functions of the generators appear at the level of the commutation rules and of the coproduct. This implies that the q-Poisson structure given by the limit $\hbar \to 0$ can be formally different from the original quantum algebra structure. In fact, the Poisson limit allows us to remove contributions in the deformation that arise as reordering terms.

In the next section we will illustrate this general approach constructing the standard q-Poisson algebra associated to $su_q(2)$. In [18] the reader can find, to best of our knowledge, the first example of a Hopf algebra deformation of the Poisson $su(3)$ algebra.

3. The basic example: q-Poisson-Hopf algebra of $su(2)$

In the $su(2)$ case, as it is well-known, two quantum deformations that do exist: the standard one [22, 23], that we consider here, and the non-standard (or Jordanian) deformation [24].

The $su(2)$ algebra is generated by (J_3, J_+, J_-) with commutation rules

$$[J_+, J_-] = 2\, J_3, \qquad [J_3, J_+] = J_+, \qquad [J_3, J_-] = -J_-. \tag{13}$$

The standard $su(2)$ Lie bialgebra is characterized by the cocommutator

$$\delta(J_3) = 0, \qquad \delta(J_+) = J_3 \wedge J_+, \qquad \delta(J_-) = J_3 \wedge J_-. \tag{14}$$

The quantum algebra deformation of (13) and (14) is given, respectively, by

$$[J_+, J_-] = \frac{\sinh(2\, z\, J_3)}{z}, \qquad [J_3, J_+] = J_+, \qquad [J_3, J_-] = -\, J_- \tag{15}$$

and

$$\begin{aligned} \Delta_z(J_3) &= J_3 \otimes 1 + 1 \otimes J_3, \\ \Delta_z(J_+) &= e^{z\, J_3} \otimes J_+ + J_+ \otimes e^{-z\, J_3}, \\ \Delta_z(J_-) &= e^{z\, J_3} \otimes J_- + J_- \otimes e^{-z\, J_3}. \end{aligned} \tag{16}$$

Let us introduce the one-parameter $\hbar$-family of Lie algebras $(su(2)_\hbar, \delta)$ (10) determined by the commutators

$$[J_+, J_-] = 2\, \hbar\, J_3, \qquad [J_3, J_+] = \hbar\, J_+. \qquad [J_3, J_-] = -\hbar\, J_- \tag{17}$$

combined with the cocommutator (14). Since the cocommutator of $su(2)_\hbar$ is the same as of $su(2)$ we would obtain a quantization in which the coproduct (16) does not formally change, however we get

$$[J_+, J_-] = \hbar\, \frac{\sinh(2\, z\, J_3)}{z}, \qquad [J_3, J_+] = \hbar\, J_+, \qquad [J_3, J_-] = -\hbar\, J_-. \tag{18}$$

The $(\hbar, z)$-deformed Casimir operator is now

$$C_{z,\hbar} = \frac{\sinh(zJ_3)}{z} \frac{\sinh z(J_3 - \hbar)}{z} + J_+\, J_-. \tag{19}$$

It is worthy noticing that the following substitution

$$J_3 \to J_3/\hbar, \qquad J_+ \to J_+/\hbar, \qquad J_- \to J_-/\hbar. \qquad z \to \hbar\, z, \tag{20}$$

allows us to reabsorb the parameter $\hbar$ and, hence, to recover the standard expressions (15) and (16) depending only on the parameter z. In other words, $U_z(su(2))$ depends on only one essential parameter instead on two. But, in the limit $\hbar \to 0$ this mapping (20) is singular, so the parameter $\hbar$ is essential in order to define properly the q-Poisson algebra.

The q-Poisson-Hopf algebra $su_q^{\mathcal{P}}(2) \equiv (\mathrm{Fun}(g_z), \Delta_z^{\mathcal{P}})$ is obtained as the commutative algebra of functions defined on g_z endowed with the Poisson bracket and coproduct map obtained by performing, respectively, the limits (11) and (12):

$$\{\mathcal{J}_+, \mathcal{J}_-\} = \frac{\sinh(2\, z\, \mathcal{J}_3)}{z}, \qquad \{\mathcal{J}_3, \mathcal{J}_+\} = \mathcal{J}_+, \qquad \{\mathcal{J}_3, \mathcal{J}_-\} = -\mathcal{J}_-;$$

$$\begin{aligned}
\Delta_z^{\mathcal{P}}(\mathcal{J}_3) &= \mathcal{J}_3 \otimes 1 + 1 \otimes \mathcal{J}_3, \\
\Delta_z^{\mathcal{P}}(J_+) &= e^{z\, \mathcal{J}_3} \otimes \mathcal{J}_+ + \mathcal{J}_+ \otimes e^{-z\, \mathcal{J}_3}, \\
\Delta_z^{\mathcal{P}}(\mathcal{J}_-) &= e^{z\, \mathcal{J}_3} \otimes \mathcal{J}_- + \mathcal{J}_- \otimes e^{-z\, \mathcal{J}_3}.
\end{aligned}$$

The q-deformed Casimir function obtained from (19) performing the limit $\hbar \to 0$ is:

$$C_z^{\mathcal{P}} = \lim_{\hbar \to 0} C_{q,\hbar} = \frac{1}{z^2} \sinh^2(z\, \mathcal{J}_3) + \mathcal{J}_+ \mathcal{J}_-.$$

The non-standard q-Poisson algebra $su_{ns}^{\mathcal{P}}(2)$ and the q-Poisson $su_q^{\mathcal{P}}(3)$ algebra have been also introduced in [18], where the quantum algebras $U_{z,\hbar}^{ns}(su(n))$ and $U_{z,\hbar}(su(3))$ were obtained by making use of the quantization approach described in [20]. In this case $su_q^{\mathcal{P}}(3)$ and $U_{z,\hbar}(su(3))$ are quite different unlike the $su(2)$ cases, where the corresponding q-Poisson-Hopf algebras are formally equivalent to the quantum algebras from which they have been obtained as one can see on simple inspection of the corresponding expressions of the commutators and the coproduct.

4. General framework of $\hbar$-quantization and z-deformation

In the canonical quantization when the phase space is $M = \mathbb{R}^{2n}$ and $\mathcal{H} = L^2(\mathbb{R}^n)$ it is well-known that appears the Heisenberg algebra $\mathfrak{h}$ (6) or $\mathfrak{h}_\hbar$ (7). In some sense, the idea of this quantization would be to find a map between the universal enveloping algebra of $\mathfrak{h}$, $U(\mathfrak{h}_\hbar)$, and $\mathrm{Fun}(\mathfrak{h})$ (in reality, power series in the canonical variables). This map is found constructing a relation between the classical observables (x^i, p^j) and their quantum counterparts(X_i, P_j). We can summarized this quantization in the following diagram

$$\begin{array}{ccc}
\langle X_i, P_j\rangle \equiv \mathfrak{h}_\hbar & \longrightarrow & U(\mathfrak{h}_\hbar) \\
\big\downarrow{\scriptstyle \hbar\to 0} & & \big\downarrow{\scriptstyle \hbar\to 0} \\
\langle x^i, p^j\rangle \equiv \mathfrak{h} & \longrightarrow & \mathrm{Fun}(\mathfrak{h})
\end{array} \tag{21}$$

We can add the procedure of q-deformation to the previous diagram obtaining the following one including the q-deformation together with $\hbar$-quantization

$$\begin{array}{ccc}
\langle X_i, P_j\rangle_z \equiv \mathfrak{H}_z & \longrightarrow & U_z(\mathfrak{h}_\hbar) \\
\big\downarrow{\scriptstyle z\to 0} & & \big\downarrow{\scriptstyle z\to 0} \\
\langle X_i, P_j\rangle \equiv \mathfrak{h}_\hbar & \longrightarrow & U(\mathfrak{h}_\hbar) \\
\big\downarrow{\scriptstyle \hbar\to 0} & & \big\downarrow{\scriptstyle \hbar\to 0} \\
\langle x^i, p^j\rangle \equiv \mathfrak{h} & \longrightarrow & \mathrm{Fun}(\mathfrak{h})
\end{array}$$

The Heisenberg algebra $\mathfrak{h}_\hbar$ is q-deformed obtaining the q-algebra $\mathfrak{H}_z$ [19, 20], that is related with the usual q-algebra $U_z(\mathfrak{h}_\hbar)$, which is the q-deformation of $U(\mathfrak{h}_\hbar)$, making use of a generalized version of the well-known Friedrichs theorem for Lie algebras [25].

The Friedrichs theorem allows us to obtain a Lie algebra basic set of elements $\{X_i\}$ with linear commutation relations and primitive coproducts that generate the pair (g, Δ_0) from an arbitrary basic set of $U(g)$. Its generalized version [20] can be applied to obtain similarly (g_z, Δ) from an arbitrary basic set of $U_z(g)$

$$\begin{array}{ccc}
(g, \Delta_0) & \xleftarrow{\text{Friedrichs theorem}} & U(g) \\
\uparrow{\scriptstyle z\to 0} & & \uparrow{\scriptstyle z\to 0} \\
(g_z, \Delta) & \xleftarrow{\text{Generalized Friedrichs th.}} & U_z(g)
\end{array}$$

In this paper we have considered the $\hbar$-quantization for q-deformed objects. The following diagram, similar to that corresponding to the canonical quantization (21), summarizes these ideas

$$\begin{array}{ccc}
\langle X_i\rangle_z \equiv \mathfrak{g}_{z,\hbar} & \longrightarrow & U_{z,\hbar}(g) \\
\downarrow{\scriptstyle \hbar\to 0} & & \downarrow{\scriptstyle \hbar\to 0} \\
\langle \mathcal{X}^i\rangle_z \equiv g_z & \longrightarrow & \mathrm{Fun}(g_z)
\end{array}$$

Here, we have put in relation $U_{z,\hbar}(g)$ and $Fun(g_z)$ via the semiclassical limit (11).

The results of this paper show the existence of a hierarchy of complexity of Hopf algebras summarized in the following commutative diagram:

$$\begin{array}{ccccc}
(U(g), \Delta_0) & \xleftarrow{\hbar=1} & (U_\hbar(g), \Delta_0) & \xrightarrow{\hbar\to 0} & (\mathrm{Fun}(g), \Delta_0^{\mathcal{P}}) \\
\uparrow{\scriptstyle z\to 0} & & \uparrow{\scriptstyle z\to 0} & & \uparrow{\scriptstyle z\to 0} \\
(U_z(g), \Delta_z) & \xleftarrow{\hbar=1} & (U_{z,\hbar}(g), \Delta_{z,\hbar}) & \xrightarrow{\hbar\to 0} & (\mathrm{Fun}(g_z), \Delta_z^{\mathcal{P}})
\end{array}$$

Thus, we see that quantum algebras $U_{z,\hbar}(g)$ are the richest and most complex structures (they are non-commutative because $\hbar \neq 0$ and non-cocommutative since $z \neq 0$), the q-Poisson algebras $Fun(g_z)$ are intermediate objects (being commutative and non-cocommutative) and, finally, the Lie algebras (and the universal envelopping algebras $U(g)$) are, in some sense, the simplest ones (non-commutative and cocommutative).

The procedure to construct a q-Poisson algebra from a given quantum algebra is thus perfectly defined. First, one goes from $U_z(g)$ to $U_{z,\hbar}(g)$ by constructing analytically the deformation from the very beginning, i.e. starting from the bialgebra $(g_\hbar, \delta)$, and by taking into account explicitly both parameters $(\hbar, z)$. This two-parameter deformation seems to be irrelevant since for any finite value of $\hbar$, by using the Lie algebra automorphism $X \to X/\hbar$ and by rescaling the deformation parameter as $z \to \hbar\, z$, the quantum algebra $(U_{z,\hbar}(g), \Delta_{z,\hbar})$ can be converted into $(U_z(g), \Delta_z)$. But the situation changes when the Poisson limit $(\hbar \to 0)$ is performed since it does not commute with the rescaling automorphism and the result thus obtained is not trivial.

We have presented only the Poisson limit for the commutators and the coproduct, however the extension to counity and antipode is trivial. In both cases the Poisson limit would be similar to the coproduct limit (12).

Commutation rules instead q-commutators have been used in order to make explicit the quantum algebra structure since, in general, the Poisson limit of the q-commutators is not well-defined. This is the reason because q-Poisson structures for algebras of rank greater than one have not been constructed so far. However, the analytical bases approach presented in [19, 20] provides a quantization framework based on pure commutators, thus leading to a well-defined Poisson limit for arbitrary quantum algebras. Hence, analytical bases looks also to have a privileged connection with the semiclassical limit.

In [18] we have shown, moreover, that it is possible to consider the corresponding Poisson limit and to obtain, what could be called, the q-Poisson-Serre relations. This result is based on the fact that the q-Serre relations can be rewritten in terms of commutators [26, 27].

A potential application for this kind of q-Poisson-Hopf algebras could be their use for constructing integrable deformations of integrable systems. In particular $su_q^{\mathcal{P}}(3)$ could be related with higher rank classical Gaudin models [28–31] by means of the approach introduced in [32], or to study the semiclassical limit of the $U_z(su(3))$ dynamics from the viewpoint of [33].

Acknowledgments
This work was partially supported by the Ministerio de Educación y Ciencia of Spain (Projects FIS2009-09002 and MTM2007-67389, all of them with EU-FEDER support), by the Junta de Castilla y León (Project GR224), and by INFN-MICINN (Italy-Spain).

References
[1] Chari V and Pressley A 1994 *A Guide to Quantum Groups* (Cambridge Univ. Press, Cambridge)
[2] Moyal J E 1949 *Proc.Cambridge Phil. Soc. Inst.* **45** 99
[3] Kulish P P and Reshetikhin N 1981 *Proc. Steklov Math. Inst.* **101** 101
[4] Takhtajan L A 1990 *Introduction to quantum group and integrable massive models of Quantum Field Theory*, Mo-Lin Ge and Bao-Heng Zao (eds.), (World Scientific, Singapore) p. 69
[5] Reshetikhin N Y, Takhtadzhyan L A and Faddeev L D 1990 *Leningrad Math. J.* **1** 193
[6] Majid S 1995 *Foundations of Quantum Group Theory* (Cambridge Univ. Press, Cambridge)
[7] Pasquier V and Saleur H 1990 *Nucl. Phys.* **B330** 523
[8] Kulish P P and Sklyanin E K 1991 *J. Phys. A: Math. Gen.* **24** L435
[9] Bonechi F, Celeghini E, Giachetti R, Sorace E and Tarlini M 1992 *Phys. Rev. Lett.* **68** 3718
[10] Bonechi F, Celeghini E, Giachetti R, Sorace E and Tarlini M 1992 *Phys. Rev.* **B46** 5727
[11] Lukierski J, Nowicky A and Ruegg H 1992 *Phys. Lett.* **B293**, 344
[12] Giller S, Koshinski P, Majewski M, Maslanka P and Kunz J 1992 *Phys. Lett.* **B286**, 57
[13] Ruegg H 1993 in *Integrable Systems, Quantum Groups and Quantum Field Theories* NATO ASI Series, L.A. Ibort and M.A. Rodríguez, eds (Kluwer, Netherlands) p. 45
[14] Dobrev V K 1993 *J. Phys. A: Math. Gen.* **26** 1317
[15] Ballesteros A, Herranz F J , del Olmo M A and Santander M 1994 *J. Math. Phys.* **35**, 4928
[16] Celeghini E, Giachetti R, Sorace E and Tarlini M 1992 *Phys. Lett.* **B280** 180
[17] Bonatsos D and Daskaloyannis C 1999 *Prog. Part. Nucl. Phys.* **43** 537
[18] Ballesteros A, Celeghini E and del Olmo M A 2009 *J. Phys. A: Math. Theor.* **42** 275202.
[19] Ballesteros A, Celeghini E and del Olmo M A 2008 *J. Phys. C. S.* **128** 012049
[20] Celeghini E, Ballesteros A and del Olmo M A 2008 *J. Phys. A: Math. Theor.* **41** 304038
[21] Dirac P A M 1930 *The principles of quantum mechanics* (Oxford univ. Press, Oxford)
[22] Drinfeld V G 1987 "Quantum Groups" in *Proceed. of the Inter. Congress of Mathematicians*, Berkeley, 1986, A.M. Gleason (ed.) (AMS, Providence)
[23] Jimbo M 1985 *Lett. Math. Phys.* **10** 63
[24] Ohn C 1992 *Lett. Math. Phys.* **25** 85
[25] Jacobson N 1979 *Lie algebras* (Dover, New York)
[26] Ballesteros A, Celeghini E and del Olmo M A 2006 *J. Phys. A: Math. Gen.* **39** 9161
[27] Ballesteros A, Celeghini E and del Olmo M A 2007 *J. Phys. A: Math. Theor.* **40** 2013
[28] Gaudin M 1998 *J. de Physique* **37** 1087
[29] Sklyanin E K 1992 *Commun. Math. Phys.* **150** 181
[30] Falqui G and Musso F 2003 *J. Phys. A: Math. Gen.* **46** 11655
[31] Lerma H S and Errea B F 2003 *J. Phys. A: Math. Theor.* **2007** 4125
[32] Ballesteros A and Ragnisco O 1998 *J. Phys. A: Math. Gen.* **31** 3791
[33] Da Providencia J 1993 *J. Phys. A: Math. Gen.* **26** 5845.

Maximum entropy production rate in quantum thermodynamics

Gian Paolo Beretta

Università di Brescia, via Branze 38, 25123 Brescia, Italy

E-mail: `beretta@ing.unibs.it`

Abstract. In the framework of the recent quest for well-behaved nonlinear extensions of the traditional Schrödinger–von Neumann unitary dynamics that could provide fundamental explanations of recent experimental evidence of loss of quantum coherence at the microscopic level, a recent paper [Gheorghiu-Svirschevski 2001 *Phys. Rev.* A **63** 054102] reproposes the nonlinear equation of motion proposed by the present author [see Beretta G P 1987 *Found. Phys.* **17** 365 and references therein] for quantum (thermo)dynamics of a single isolated indivisible constituent system, such as a single particle, qubit, qudit, spin or atomic system, or a Bose-Einstein or Fermi-Dirac field. As already proved, such nonlinear dynamics entails a fundamental unifying microscopic proof and extension of Onsager's reciprocity and Callen's fluctuation–dissipation relations to all nonequilibrium states, close and far from thermodynamic equilibrium. In this paper we propose a brief but self-contained review of the main results already proved, including the explicit geometrical construction of the equation of motion from the steepest-entropy-ascent ansatz and its exact mathematical and conceptual equivalence with the maximal-entropy-generation variational-principle formulation presented in Gheorghiu-Svirschevski S 2001 *Phys. Rev.* A **63** 022105. Moreover, we show how it can be extended to the case of a composite system to obtain the general form of the equation of motion, consistent with the demanding requirements of strong separability and of compatibility with general thermodynamics principles. The irreversible term in the equation of motion describes the spontaneous attraction of the state operator in the direction of steepest entropy ascent, thus implementing the maximum entropy production principle in quantum theory. The time rate at which the path of steepest entropy ascent is followed has so far been left unspecified. As a step towards the identification of such rate, here we propose a possible, well-behaved and intriguing, general closure of the dynamics, compatible with the nontrivial requirements of strong separability. Based on the time–energy Heisenberg uncertainty relation, we derive a lower bound to the internal-relaxation-time functionals that determine the rate of entropy generation. This bound entails an upper bound to the rate of entropy generation. By this extreme maximal-entropy-generation-rate ansatz, each indivisible subsystem follows the direction of steepest locally perceived entropy ascent at the highest rate compatible with the time–energy uncertainty principle.

PACS numbers: 03.65.Ta,11.10.Lm,04.60.-m,05.45.-a

1. Introduction

Recent discussions [1, 2, 3] on possible fundamental tests of standard unitary quantum mechanics, related to the existence of "spontaneous decoherence" at the microscopic level, are relevant to understanding and predicting decoherence in important future applications involving nanometric devices, fast switching times, clock synchronization, superdense coding, quantum computation, teleportation, quantum cryptography, etc. where entanglement structure and dynamics play a key role [4]. It has been suggested [1] that long-baseline neutrino oscillation experiments may provide means of testing the existence of spontaneous decoherence and, therefore, the validity of linear and nonlinear extensions of standard unitary QM. In this paper, we restrict our attention to extensions of standard QM that assume a "broader quantum kinematics", i.e., an augmented set of true quantum states described by state operators ρ without the restriction $\rho^2 = \rho$, as recently reproposed in this context [3, 5] and in other contexts [6].

To our knowledge, the first instance where this broader quantum kinematics has been considered is in the pioneering work by Hatsopoulos and Gyftopoulos [7], where an extension of standard QM was proposed so as to provide a non-statistical non-information-theoretic microscopic unification of mechanics and thermodynamics, based on the (revolutionary [8]) ansatz that, for any system, even if strictly isolated and uncorrelated: (1) the "true" quantum state (in the sense analogous to that of the wave function of standard quantum mechanics) is represented by a state operator ρ — a unit-trace, nonnegative-definite, hermitian operator on the Hilbert space $\mathcal{H}$ associated with the system according to standard QM — belonging to a "broader quantum kinematics" that includes pure states ($\rho^2 = \rho$) as well as non-idempotent states ($\rho^2 \neq \rho$); and (2) the "physical" entropy (as opposed to a statistical or information-theoretic entropy related to incoherent stochastic mixtures of true states) is represented by the state functional $-k_B\mathrm{Tr}(\rho \ln \rho)$. References [7, 9] give proofs that only this functional can represent the physical entropy in such context.

The present author, in [10, 11, 12, 13, 14, 15, 16, 17, 18] (see also [5, 19, 20, 21]), addressed the problem of deriving a well-behaved extension of the Schrödinger–von Neumann unitary dynamics meeting the very demanding set of strict requirements that appear to be necessary if one is willing to accept the Hatsopoulos–Gyftopoulos broader kinematics ansatz.

The equation of motion that we proposed and postulated in [10, 11] for the state operator ρ is rederived in the present paper by means of a the explicit geometrical construction also detailed in [12] that clarifies the steepest-entropy-ascent, i.e., maximum entropy production, feature already recognized in [13, 14, 15, 16, 17, 18]. Here, several new interesting additional features related to separability and correlations are proved. As a result, we also show how the equivalent variational formulation in [3] can be extended to the composite system case.

The nonlinear extension of the Schrödinger–von Neumann equation of motion is derived, together with a full discussion of the necessary notation and definitions, in section 3 for a single indivisible system, and in sections 10 and 11 for a general system consisting of M distinguishable indivisible subsystems. However, to support some preliminary and motivating discussion, and to single it out from the many equations in the paper, we anticipate here the display of the proposed equation of motion.

For a single indivisible system, the rate of change of the state operator is given by the sum of the usual unitary Hamiltonian term and a nonlinear, entropy-generating term

$$\frac{\mathrm{d}\rho}{\mathrm{d}t} = -\frac{i}{\hbar}[H,\rho] + \hat{\mathbf{D}}_1(\rho, H, G_i) \ , \tag{1a}$$

$$\hat{\mathbf{D}}_1(\rho, H, G_i) = -\frac{1}{2\tau(\rho)}\left[\sqrt{\rho}D + D^\dagger\sqrt{\rho}\right] \ , \tag{1b}$$

$$D = [\sqrt{\rho}(B\ln\rho)]_{\perp\mathcal{L}\{\sqrt{\rho}I,\sqrt{\rho}H,[\sqrt{\rho}G_i]\}} \ , \tag{1c}$$

where (details in section 3) $[\sqrt{\rho}(B\ln\rho)]_{\perp\mathcal{L}\{\sqrt{\rho}R_i\}}$ denotes the component of operator $\sqrt{\rho}(B\ln\rho)$ orthogonal to the linear manifold $\mathcal{L}\{\sqrt{\rho}R_i\} = \mathcal{L}\{\sqrt{\rho}I,\sqrt{\rho}H,[\sqrt{\rho}G_i]\}$ spanned by all real linear combinations of the set of operators $\{\sqrt{\rho}I,\sqrt{\rho}H,[\sqrt{\rho}G_i]\}$ where I is the identity operator on the Hilbert space $\mathcal{H}$ of the system, $B\ln\rho$ the operator obtained from ρ

by substituting in its spectral expansion each nonzero eigenvalue p_i with $\ln p_i$, H the usual Hamiltonian operator associated with the system in standard QM, the G_i's are additional (not always necessary) hermitian operators commuting with H that we call the non-Hamiltonian generators of the motion and that depend on the structure of the system, such as the i-th-type particle-number operators N_i for a Bose-Einstein or Fermi-Dirac field (in which case $\mathcal{H}$ is a Fock space), or the total momentum component operators P_i of a free particle (in which case Gheorghiu-Svirschevski [3] proved Galileian invariance).

The Hamiltonian term in the equation of motion, $-i[H,\rho]/\hbar$, drives the state operator towards a unitary isoentropic evolution. The entropy-generating but energy-conserving nonunitary term $\hat{\mathbf{D}}_1(\rho, H, G_i)$, instead, drives the state operator towards the local direction of steepest entropy ascent (which is orthogonal to the direction of unitary evolution) compatible with mean values of the non-Hamiltonian time-invariants. Indeed, for a strictly isolated system, both terms maintain independently invariant the values of the trace, $\mathrm{Tr}(\rho I) = 1$, the energy, $\mathrm{Tr}(\rho H) = e(\rho)$, and the mean values of each non-Hamiltonian generator, $\mathrm{Tr}(\rho G_i) = g_i(\rho)$, if any. In other words, the irreversible term pulls the state operator in the direction of the projection of the gradient of the entropy functional $-k_{\rm B}\mathrm{Tr}(\rho \ln \rho)$ onto the hyperplane of constant $\mathrm{Tr}(\rho I)$, $\mathrm{Tr}(\rho H)$, and $\mathrm{Tr}(\rho G_i)$'s. Because the system is isolated, the entropy ceases to increase only when a local entropy maximum is reached, i.e., either an equilibrium state or a limit cycle.

For a general composite of indivisible subsystems, instead, the equation of motion takes the following form. The rate of change of the state operator is given by the sum of the usual unitary Hamiltonian term and a sum of nonlinear terms accounting for "irreversible internal redistribution along the path of locally maximal entropy generation" within each indivisible subsystem,

$$\frac{\mathrm{d}\rho}{\mathrm{d}t} = -\frac{i}{\hbar}[H,\rho] + \hat{\mathbf{D}}_M(\rho, H, G_i) \;, \tag{2a}$$

$$\hat{\mathbf{D}}_M(\rho, H, G_i) = -\sum_{J=1}^{M} \frac{1}{2\tau_J(\rho)} \left[\sqrt{\rho_J} D_J + D_J^\dagger \sqrt{\rho_J}\right] \otimes \rho_{\overline{J}} \;, \tag{2b}$$

$$D_J = \left[\sqrt{\rho_J}(B\ln\rho)^J\right]_{\perp \mathcal{L}\{\sqrt{\rho_J} I_J, \sqrt{\rho_J}(H)^J, [\sqrt{\rho_J}(G_i)^J]\}} \;. \tag{2c}$$

where $(F)^J = \mathrm{Tr}_{\overline{J}}[(I_J \otimes \rho_{\overline{J}})F]$, for $F = B\ln\rho$, H, and G_i, and the other details are defined in sections 10 and 11 where we prove that all the necessary requirements (listed in Appendix A) for a self-consistent and well-defined extension of the Schrödinger–von Neumann equation of motion compatible with thermodynamic requirements are satisfied, together with a set of very intriguing mathematical properties and consistent physical consequences.

In [16], we realized that the internal-relaxation times $\tau_J(\rho)$ need not be constants (as initially assumed in [11]) but can be positive functionals of ρ. In sections 10 and 11, we note that these functionals must satisfy conditions of strong separability conceptually equivalent to those discussed by Czachor [2].

In section 7, we derive a lower bound for $\tau_J(\rho)$ which corresponds to the highest entropy generation rate compatible with the time–energy Heisenberg uncertainty principle. Taking the internal-relaxation times $\tau_J(\rho)$ exactly equal to these nontrivial lower bounds turns out to be compatible with the separability condition, and would complete the nonlinear dynamics in a general way with no need to assume the existence of new physical constants. Not only, then, the state would evolve along the path of steepest entropy ascent, but it would do so at the highest rate compatible with the time–energy uncertainty princple. However, other less extreme options are equally compatible.

We emphasize that the nonlinearity of the dynamics is such that one should not expect that the form of the equation valid for a single elementary (indivisible) constituent of matter be valid also for a system with an internal structure, because the equation of motion must reflect such structure not only through the Hamiltonian operator. However, of course, the superoperator $\hat{\mathbf{D}}_M$ reduces to $\hat{\mathbf{D}}_1$ [see (108) below] when the system has only one indivisible subsystem.

To make the paper self-contained, in section 6 we briefly review the fundamental unifying extensions to all nonequilibrium (dissipative) states of the Onsager reciprocity relations [22] and Callen's fluctuation-dissipation theorem [23] that derive from (1) [16] as well as, in general, from (2) [18].

In Appendix A and Appendix B, we discuss a set of criteria and the definition of stability of equilibrium that a fundamental dynamics must satisfy in order to be compatible with second law and other thermodynamic requirements. In the other appendixes, we discuss some extensions of the mathematics of (1) and (2).

2. The augmented state domain ansatz

The fundamental ansatz that the postulates of quantum mechanics can be successfully supplemented by the first and second principles of thermodynamics by assuming a broader state domain that includes not only $\rho^2 = \rho$ but also $\rho^2 \neq \rho$ state operators, provided that the functional $-k_B \mathrm{Tr}(\rho \ln \rho)$ is taken for the physical entropy, was first proposed (without a dynamical law) by Hatsopoulos and Gyftopoulos [7].

Thirty years ago, the hypothesis of a state domain augmented with respect to that of traditional QM was perceived as adventurous [19] and countercurrent to the prevailing approaches to dissipative quantum dynamics within the frameworks of statistical, stochastic, phenomenological, information-theoretic, chaotic-behavior and bifurcation theories. For this reason, the broader quantum kinematics ansatz [7] and, five years later, the new nonlinear equation of motion have been considered initially as unphysical and substantially ignored (except for a few exceptions [19, 20, 21]), mainly because their motivation appeared to be derived from theoretical reasoning only (for example as recently summarized in [24]).

In search for direct experimental evidence, we derived explicit solutions for a two-level system and computed the effects of the irreversible atomic relaxation implied by the nonlinear equation of motion onto some basic quantum-electrodynamics results on absorption, stimulated emission, and resonance fluorescence from a single two-level atom [14, 15]. The results were obtained in the near-equilibrium linear limit and, of course, in terms of the yet undetermined internal-relaxation-time functional $\tau(\rho)$ that is part of the equation of motion. To our knowledge no one has yet attempted to verify these results experimentally and estimate $\tau(\rho)$.

The recent experimental evidence of loss of quantum coherence [2, 4, 25, 26] and the impressive effort devoted to study nonlinear modifications of the standard Schrödinger equation in the last fourty years [27], finally seem to make more acceptable, if not require, the $\rho^2 \neq \rho$ augmented state domain Hatsopoulos–Gyftopoulos ansatz.

Once the $\rho^2 \neq \rho$ ansatz is accepted, the nonlinear equation of motion we proposed completes the dynamics and holds the promise to provide a microscopic-level explanation of the recent experimental evidence of loss of quantum coherence. It is with this motivation that Gheorghiu-Svirschevski [3] has "rediscovered" Eq. (1) together with many of its known features. [3] contributes to confirm the mathematical validity of this equation, including existence and uniqueness of solutions, and elegantly derives useful expansions and other results in the near-equilibrium linear limit. However, also in [3] the question of defining the form of the relaxation-time functional is left unresolved. A related development is also found in [28].

3. The steepest-entropy-ascent ansatz for an indivisible system

In [13, 16, 17], we emphasized that (1) and (2) have an important geometric interpretation. The Hamiltonian term, $-i[H,\rho]/\hbar$, and the irreversible term, $\hat{\mathbf{D}}_M(\rho, H, G_i)$ [or $\hat{\mathbf{D}}_1(\rho, H, G_i)$], compete with each other in the sense that the first drives the state operator towards a unitary motion tangent to the local constant entropy surface, whereas the second drives it towards the local direction of steepest entropy ascent along the surface with constant mean values of the generators of the motion. In this section, we show more explicitly than in the original papers how (1) can be constructed directly from the steepest-entropy-ascent ansatz.

Let $\mathcal{H}$ ($\dim\mathcal{H} \leq \infty$) be the Hilbert space and H the Hamiltonian operator that are associated with the given (indivisible) elementary constituent system in standard QM. For simplicity, we first consider a system composed of a single elementary constituent. The nontrivial generalization to M constituents is given in section 10.

We assume that the (true) quantum states are one-to-one with the linear hermitian operators ρ on $\mathcal{H}$ with $\mathrm{Tr}(\rho) = 1$ and $\rho \geq 0$. As done in [11], we introduce the square-root state operator $\sqrt{\rho}$ obtained from the spectral expansion of ρ by substituting its eigenvalues with their positive square roots [30].

As a first step to force positivity and hermiticity of the state operator ρ we assume that the equation of motion may be written as

$$\frac{\mathrm{d}\rho}{\mathrm{d}t} = \sqrt{\rho}E + E^\dagger\sqrt{\rho} \tag{3}$$

where the operator E, in general non-hermitian, is defined below after introducing some necessary notation.

We consider the space $\mathcal{L}(\mathcal{H})$ of linear operators on $\mathcal{H}$ equipped with the real scalar product

$$(F|G) = \frac{1}{2}\mathrm{Tr}(F^\dagger G + G^\dagger F) \ . \tag{4}$$

so that for any (time-independent) hermitian R in $\mathcal{L}(\mathcal{H})$ the corresponding mean-value state functional, its local gradient operator with respect to $\sqrt{\rho}$, and its rate of change are

$$r(\rho) = \mathrm{Tr}(\rho R) = (\sqrt{\rho}|\sqrt{\rho}R) \ , \tag{5}$$

$$\nabla_{\sqrt{\rho}}\, r(\rho) = \frac{\partial r(\rho)}{\partial\sqrt{\rho}} = \sqrt{\rho}R + R\sqrt{\rho} \ , \tag{6}$$

$$\frac{\mathrm{d}r(\rho)}{\mathrm{d}t} = \mathrm{Tr}(\frac{\mathrm{d}\rho}{\mathrm{d}t}R) = 2\,(E|\,\sqrt{\rho}R) \ . \tag{7}$$

Moreover, the entropy state functional, its gradient operator with respect to $\sqrt{\rho}$, and its rate of change are

$$s(\rho) = -k_{\mathrm{B}}\mathrm{Tr}(\rho\ln\rho) = -k_{\mathrm{B}}(\sqrt{\rho}|\sqrt{\rho}\ln\rho) \ , \tag{8}$$

$$\nabla_{\sqrt{\rho}}\, s(\rho) = \frac{\partial s(\rho)}{\partial\sqrt{\rho}} = -2k_{\mathrm{B}}\left[\sqrt{\rho} + \sqrt{\rho}\ln\rho\right] \ , \tag{9}$$

$$\frac{\mathrm{d}s(\rho)}{\mathrm{d}t} = -k_{\mathrm{B}}\left[\mathrm{Tr}(\frac{\mathrm{d}\rho}{\mathrm{d}t}) + \mathrm{Tr}(\frac{\mathrm{d}\rho}{\mathrm{d}t}\ln\rho)\right] = \left(E\,\middle|\,\frac{\partial s(\rho)}{\partial\sqrt{\rho}}\right) = \left(\frac{\partial s(\rho)}{\partial\sqrt{\rho}}\,\middle|\,E\right) . \tag{10}$$

Now it is easy to see from (7) that the values of the mean functionals $r_i(\rho)$ are time invariant if and only if E is orthogonal to $\sqrt{\rho}R_i$, for all i, i.e., if it is orthogonal to the linear manifold $\mathcal{L}\{\sqrt{\rho}R_i\}$ spanned by the set of operators $\{\sqrt{\rho}R_i\}$ in which we always have $\sqrt{\rho}R_0 = \sqrt{\rho}I$ (to preserve $\mathrm{Tr}\rho = 1$) and $\sqrt{\rho}R_1 = \sqrt{\rho}H$ (to conserve energy), plus the additional non-Hamiltonian generators of the motion as already discussed.

It is noteworthy that

$$E_H = \frac{i}{\hbar}\sqrt{\rho}\,[H + c(\rho)I] \ , \tag{11}$$

with $c(\rho)$ any real functional of ρ, yields [through (3)] the Hamiltonian part of the equation of motion, and is orthogonal to $\mathcal{L}\{\sqrt{\rho}R_i\}$ as well as to the entropy gradient $\partial s(\rho)/\partial\sqrt{\rho}$.

Introducing the notation [16]

$$\Delta F = F - \mathrm{Tr}(\rho F)I \ , \tag{12}$$

$$\langle\Delta F\Delta G\rangle = \langle\Delta G\Delta F\rangle = (\sqrt{\rho}\Delta F|\sqrt{\rho}\Delta G) = \frac{1}{2}\mathrm{Tr}(\rho\{\Delta F, \Delta G\}) \ , \tag{13}$$

for F and G hermitian in $\mathcal{L}(\mathcal{H})$ and $\{\cdot,\cdot\}$ the usual anticommutator, and defining the shortest characteristic time associated with the Hamiltonian part of the equation of motion by the relation (see Appendix C)

$$\frac{1}{\tau_H(\rho)^2} = 4\,(E_H|\,E_H) \ , \tag{14}$$

we find, from (11) and $(\sqrt{\rho}|\sqrt{\rho}) = 1$,

$$\tau_H(\rho)^2\langle\Delta H\Delta H\rangle = \hbar^2/4 - [c(\rho) + \mathrm{Tr}(\rho H)]^2\tau_H(\rho)^2 \ , \tag{15}$$

so that only the choice $c = -e(\rho) = -\mathrm{Tr}(\rho H)$ is compatible with the Heisenberg time–energy uncertainty relation $\tau_H(\rho)^2\langle\Delta H\Delta H\rangle \geq \hbar^2/4$ which is then satisfied with strict equality. Therefore,

$$E_H = \frac{i}{\hbar}\sqrt{\rho}\,\Delta H \ , \tag{16}$$

and

$$\tau_H(\rho)^2\langle\Delta H\Delta H\rangle = \hbar^2/4 \ . \tag{17}$$

We may note that the foregoing discussion on the choice of $c(\rho)$ is somewhat artificial because any $c(\rho)$ cancels out in (3). However, the line of thought becomes important when we apply it later to the irreversible part of the dynamics.

For the general rate of change of the state operator, we let

$$E = E_H + E_D \ , \tag{18}$$

so that (3) may be rewritten as

$$\frac{\mathrm{D}\rho}{\mathrm{D}t} = \frac{\mathrm{d}\rho}{\mathrm{d}t} + \frac{i}{\hbar}[H,\rho] = \sqrt{\rho}E_D + E_D^\dagger\sqrt{\rho} \ , \tag{19}$$

and we assume E_D orthogonal to $\mathcal{L}\{\sqrt{\rho}R_i\}$ but in the direction of the entropy gradient operator. We cannot take E_D directly proportional to $\partial s(\rho)/\partial\sqrt{\rho}$ as such, because the entropy gradient in general has a component along $\mathcal{L}\{\sqrt{\rho}R_i\}$ which would not preserve the mean values of the generators of the motion. We must take E_D proportional to the component of the entropy gradient orthogonal to $\mathcal{L}\{\sqrt{\rho}R_i\}$, namely, for the indivisible system,

$$E_D = \frac{1}{4k_\mathrm{B}\tau(\rho)}\left[\frac{\partial s(\rho)}{\partial\sqrt{\rho}}\right]_{\perp\mathcal{L}\{\sqrt{\rho}R_i\}} \tag{20a}$$

$$= \frac{1}{4k_\mathrm{B}\tau(\rho)}\left(\frac{\partial s(\rho)}{\partial\sqrt{\rho}} - \left[\frac{\partial s(\rho)}{\partial\sqrt{\rho}}\right]_{\mathcal{L}\{\sqrt{\rho}R_i\}}\right) \tag{20b}$$

$$= -\frac{1}{2\tau(\rho)}\left(\sqrt{\rho}\ln\rho - [\sqrt{\rho}\ln\rho]_{\mathcal{L}\{\sqrt{\rho}R_i\}}\right) \tag{20c}$$

$$= -\frac{1}{2\tau(\rho)}D \tag{20d}$$

where $[\partial s(\rho)/\partial\sqrt{\rho}]_{\mathcal{L}\{\sqrt{\rho}R_i\}}$ denotes the projection of $\partial s(\rho)/\partial\sqrt{\rho}$ onto $\mathcal{L}\{\sqrt{\rho}R_i\}$ and

$$D = \sqrt{\rho}\ln\rho - [\sqrt{\rho}\ln\rho]_{\mathcal{L}\{\sqrt{\rho}R_i\}} \tag{21a}$$

$$= [\sqrt{\rho}\ln\rho]_{\perp\mathcal{L}\{\sqrt{\rho}R_i\}} \tag{21b}$$

$$= \frac{\begin{vmatrix} \sqrt{\rho}\ln\rho & \sqrt{\rho}R_0 & \cdots & \sqrt{\rho}R_i & \cdots \\ (\sqrt{\rho}\ln\rho|\sqrt{\rho}R_0) & (\sqrt{\rho}R_0|\sqrt{\rho}R_0) & \cdots & (\sqrt{\rho}R_i|\sqrt{\rho}R_0) & \cdots \\ \vdots & \vdots & \ddots & \vdots & \ddots \\ (\sqrt{\rho}\ln\rho|\sqrt{\rho}R_i) & (\sqrt{\rho}R_0|\sqrt{\rho}R_i) & \cdots & (\sqrt{\rho}R_i|\sqrt{\rho}R_i) & \cdots \\ \vdots & \vdots & \ddots & \vdots & \ddots \end{vmatrix}}{\Gamma(\{\sqrt{\rho}R_i\})} \tag{21c}$$

$$= -\frac{1}{k_B} \frac{\begin{vmatrix} \sqrt{\rho}\Delta S & \sqrt{\rho}\Delta R_1 & \cdots & \sqrt{\rho}\Delta R_i & \cdots \\ \langle \Delta S \Delta R_1 \rangle & \langle \Delta R_1 \Delta R_1 \rangle & \cdots & \langle \Delta R_i \Delta R_1 \rangle & \cdots \\ \vdots & \vdots & \ddots & \vdots & \ddots \\ \langle \Delta S \Delta R_i \rangle & \langle \Delta R_1 \Delta R_i \rangle & \cdots & \langle \Delta R_i \Delta R_i \rangle & \cdots \\ \vdots & \vdots & \ddots & \vdots & \ddots \end{vmatrix}}{\Gamma(\{\sqrt{\rho}R_i\})} \tag{21d}$$

where in writing the last equation, which follows [29] from the well-known properties of determinants together with (12) and (13) and the fact that $R_0 = I$, we defined for convenience the entropy operator

$$S = -k_B B \ln \rho = -k_B \ln(\rho + I - B) \tag{22a}$$

with B obtained from ρ by substituting in its spectral expansion each nonzero eigenvalue with unity, i.e., more formally, $B = B^2 = P_{\mathrm{Ran}\,\rho} = I - P_{\mathrm{Ker}\,\rho} = P_{\perp \mathrm{Ker}\,\rho}$, so that

$$B = B^2 \ , \quad [B, \rho] = 0 \ , \quad B\rho = \rho \ . \tag{22b}$$

Because operator B is the projector onto the range of ρ (and $I - B$ the projector onto the kernel of ρ) it is easy to show that S is well defined for any ρ, even if singular. In particular, when $\rho^2 = \rho$, S is the null operator.

The Gram determinant at the denominator,

$$\Gamma(\{\sqrt{\rho}R_i\}) = \det[\{(\sqrt{\rho}R_i|\sqrt{\rho}R_j)\}] \tag{23a}$$

$$= \begin{vmatrix} (\sqrt{\rho}R_0|\sqrt{\rho}R_0) & \cdots & (\sqrt{\rho}R_i|\sqrt{\rho}R_0) & \cdots \\ \vdots & \ddots & \vdots & \ddots \\ (\sqrt{\rho}R_0|\sqrt{\rho}R_i) & \cdots & (\sqrt{\rho}R_i|\sqrt{\rho}R_i) & \cdots \\ \vdots & \ddots & \vdots & \ddots \end{vmatrix} \tag{23b}$$

$$= \det[\{\langle \Delta R_{i>0} \Delta R_{j>0} \rangle\}] \tag{23c}$$

$$= \begin{vmatrix} \langle \Delta R_1 \Delta R_1 \rangle & \cdots & \langle \Delta R_i \Delta R_1 \rangle & \cdots \\ \vdots & \ddots & \vdots & \ddots \\ \langle \Delta R_1 \Delta R_i \rangle & \cdots & \langle \Delta R_i \Delta R_i \rangle & \cdots \\ \vdots & \ddots & \vdots & \ddots \end{vmatrix} , \tag{23d}$$

is always strictly positive by virtue of the linear independence of the operators in the set $\{\sqrt{\rho}R_i\}$.

By the well-known properties of determininats, it is easy to verify that E_H and E_D are orthogonal, i.e., $(E_D|E_H) = 0$, and that the mean value of each R_i is conserved, i.e., $(E_D|\sqrt{\rho}R_i) = 0$, therefore, $\mathrm{Tr}(\rho I)$, $\mathrm{Tr}(\rho H)$, and the $\mathrm{Tr}(\rho G_i)$'s are constants of the motion.

A further, compact expression can be written if we choose a set of operators $\{\sqrt{\rho}A_i\}$ which like the set $\{\sqrt{\rho}R_i\}$ spans $\mathcal{L}\{\sqrt{\rho}I, \sqrt{\rho}H, [\sqrt{\rho}G_i]\}$ but, in addition, forms an orthonormal set, e.g., obtained from $\sqrt{\rho}I$, $\sqrt{\rho}H$, $\sqrt{\rho}G_i$ by a Gram-Schmidt orthogonalization procedure followed by normalization. For example,

$$A_1 = I \ , \tag{24a}$$

$$A_2 = \frac{\Delta H}{\sqrt{\langle \Delta H \Delta H \rangle}} \ , \tag{24b}$$

$$A_3 = \frac{\langle \Delta H \Delta H \rangle \Delta G_1 - \langle \Delta H \Delta G_1 \rangle \Delta H}{\sqrt{\langle \Delta H \Delta H \rangle [\langle \Delta H \Delta H \rangle \langle \Delta G_1 \Delta G_1 \rangle - \langle \Delta H \Delta G_1 \rangle^2]}} \ , \tag{24c}$$

$$\cdots \ .$$

Then,

$$(\sqrt{\rho}A_i|\sqrt{\rho}A_j) = \delta_{ij} \ , \tag{25a}$$

$$a_i(\rho) = (\sqrt{\rho}A_i|\sqrt{\rho}A_1) = \delta_{i1} \,, \tag{25b}$$

$$D = \sqrt{\rho}\ln\rho - \sum_{i=1}^{a}(\sqrt{\rho}\ln\rho|\sqrt{\rho}A_i)\sqrt{\rho}A_i \,, \tag{25c}$$

$$\Gamma(\{\sqrt{\rho}A_i\}) = 1 \,. \tag{25d}$$

It is noteworthy that operators A_i are nonlinear functions of ρ and, therefore, in general, vary with ρ as it evolves with time.

We finally obtain the following equivalent expressions for the rate of entropy change

$$\frac{\mathrm{d}s(\rho)}{\mathrm{d}t} = \frac{1}{4k_{\mathrm{B}}\tau(\rho)}\left(\left[\frac{\partial s(\rho)}{\partial\sqrt{\rho}}\right]_{\perp\mathcal{L}\{\sqrt{\rho}R_i\}}\middle|\left[\frac{\partial s(\rho)}{\partial\sqrt{\rho}}\right]_{\perp\mathcal{L}\{\sqrt{\rho}R_i\}}\right) \tag{26a}$$

$$= 4k_{\mathrm{B}}\tau(\rho)\,(E\,|E\,) \tag{26b}$$

$$= 4k_{\mathrm{B}}\tau(\rho)\,(E_D\,|E_D\,) \tag{26c}$$

$$= \frac{1}{k_{\mathrm{B}}\tau(\rho)}\left[\langle\Delta S\Delta S\rangle - \sum_{i=1}^{a}(\langle\Delta S\Delta A_i\rangle)^2\right] \tag{26d}$$

$$= \frac{k_{\mathrm{B}}}{\tau(\rho)}(D|D) \tag{26e}$$

$$= \frac{k_{\mathrm{B}}}{\tau(\rho)}\frac{\Gamma(\sqrt{\rho}\ln\rho,\{\sqrt{\rho}R_i\})}{\Gamma(\{\sqrt{\rho}R_i\})} \tag{26f}$$

$$= \frac{1}{k_{\mathrm{B}}\tau(\rho)}\frac{\Gamma(\sqrt{\rho}S,\{\sqrt{\rho}R_i\})}{\Gamma(\{\sqrt{\rho}R_i\})} \tag{26g}$$

where the Gram determinant

$$\Gamma(\sqrt{\rho}S,\{\sqrt{\rho}R_i\}) = \begin{vmatrix} (\sqrt{\rho}S|\sqrt{\rho}S\,) & (\sqrt{\rho}R_0|\sqrt{\rho}S\,) & \cdots & (\sqrt{\rho}R_i|\sqrt{\rho}S\,) & \cdots \\ (\sqrt{\rho}S|\sqrt{\rho}R_0) & (\sqrt{\rho}R_0|\sqrt{\rho}R_0) & \cdots & (\sqrt{\rho}R_i|\sqrt{\rho}R_0) & \cdots \\ \vdots & \vdots & \ddots & \vdots & \ddots \\ (\sqrt{\rho}S|\sqrt{\rho}R_i) & (\sqrt{\rho}R_0|\sqrt{\rho}R_i) & \cdots & (\sqrt{\rho}R_i|\sqrt{\rho}R_i) & \cdots \\ \vdots & \vdots & \ddots & \vdots & \ddots \end{vmatrix}$$
$$= \begin{vmatrix} \langle\Delta S\Delta S\rangle & \langle\Delta R_1\Delta S\rangle & \cdots & \langle\Delta R_i\Delta S\rangle & \cdots \\ \langle\Delta S\Delta R_1\rangle & \langle\Delta R_1\Delta R_1\rangle & \cdots & \langle\Delta R_i\Delta R_1\rangle & \cdots \\ \vdots & \vdots & \ddots & \vdots & \ddots \\ \langle\Delta S\Delta R_i\rangle & \langle\Delta R_1\Delta R_i\rangle & \cdots & \langle\Delta R_i\Delta R_i\rangle & \cdots \\ \vdots & \vdots & \ddots & \vdots & \ddots \end{vmatrix} \tag{27}$$

is also strictly positive [except when ρ satisfies (28), see below].

Whereas the unitary Hamiltonian term in (1) maintains unchanged all the eigenvalues of ρ, the irreversible term maintains zero the initially zero eigenvalues of ρ and, therefore, conserves the cardinality of the set of zero eigenvalues, $\dim\mathrm{Ker}(\rho)$ (for proofs see [3, 11]). In other words, the equation of motion preserves the rank and nullity of ρ. If the isolated system is prepared in a state that does not 'occupy' the eigenvector $|\psi_\ell\rangle$ of H (and the G_i's), i.e., if $\rho(0)|\psi_\ell\rangle = 0$ (so that $|\psi_\ell\rangle$ is also an eigenvector of ρ corresponding to a zero eigenvalue), then such energy eigenvector remains 'unoccupied' at all times, i.e., $\rho(t)|\psi_\ell\rangle = 0$, the null space of ρ is an invariant of the motion.

This condition preserves an important feature that in standard QM allows remarkable model simplifications: the dynamics is fully equivalent to that of a model system with Hilbert space $\mathcal{H}'$ (a subspace of $\mathcal{H}$) defined by the linear span of all the $|\psi_\ell\rangle$'s such that $\rho(t)|\psi_\ell\rangle \neq 0$ at some time t (and, hence, by our condition, at all times). The relevant operators X'

on $\mathcal{H}'$ (ρ', H', G'_i, ...) are defined from the original X on $\mathcal{H}$ (ρ, H, G_i, ...) so that $\langle\alpha_k|X'|\alpha_\ell\rangle = \langle\alpha_k|X|\alpha_\ell\rangle$ with $|\alpha_k\rangle$ any basis of $\mathcal{H}'$.

It is also consistent with recent experimental tests [31] that rule out, for pure states, deviations from linear and unitary dynamics and confirm that initially unoccupied eigenstates cannot spontaneously become occupied. This fundamental intrinsic feature of our extended microscopic dynamics adds nontrivial experimental and conceptual difficulty to the problem of designing a fundamental test of QM, but preserves within the extended theory the exact validity of all the remarkable successes of QM.

4. Nondissipative states and limit cycles: the Schrödinger–von Neumann limit

Eq. (1) reduces to the Schrödinger–von Neumann equation of motion, $i\hbar\dot\rho = [H,\rho]$, when $\rho^2 = \rho$ (thus entailing the usual unitary hamiltonian dynamics of standard QM), and also when and only when

$$\rho_{\rm nd} = \frac{B\exp(C)}{{\rm Tr}[B\exp(C)]}\;, \quad B^2 = B\;, \quad [B,C] = 0\;, \tag{28}$$

or, equivalently,

$$\sqrt{\rho_{\rm nd}}\ln\rho_{\rm nd} = \sqrt{\rho_{\rm nd}}C - \sqrt{\rho_{\rm nd}}\ln{\rm Tr}[B\exp(C)]\;, \tag{29}$$

where

$$C = -\beta H + \sum_i \nu_i G_i\;, \tag{30}$$

in which case we say that the state is nondissipative, and the solution of the equation of motion is

$$\rho_{\rm nd}(t) = B(t)\exp(C)/{\rm Tr}[B(t)\exp(C)]\;, \tag{31a}$$
$$B(t) = U(t)B(0)U^{-1}(t)\;, \quad U(t) = \exp(-itH/\hbar)\;, \tag{31b}$$

which includes the usual $\rho^2 = \rho$ Schrödinger dynamics when C is the null operator [$\beta = \nu_i = 0$ in (28)] and, therefore, ${\rm Tr}(B) = 1$.

Nondissipative states (28) have a thermal-like distribution (with positive or negative temperatures) over a finite number, ${\rm Tr}(B)$, of "occupied" eigenvectors. Because entropy cannot decrease and $s(\rho)$ is an S-function [32], they are conditionally locally stable equilibrium states or limit cycles. They constitute the "target" highest-entropy states compatible with the mean values of the invariant functionals and the invariant null subspace of unoccupied eigenvectors.

For an initial state ρ commuting with H, i.e., for $[B(0),H] = 0$, by interpreting the entropy $s(\rho)$ as a measure of how "well" the energy is distributed within the isolated system among the available energy levels, the nonlinear dynamics describes a spontaneous internal redistribution of the energy along the path of maximal entropy increase leading towards an "optimally" distributed (highest entropy) state compatible with the condition of maintaining unoccupied the initially unoccupied energy levels.

When $[B(0),H] = 0$ the nondissipative state is an equilibrium state [or stationary state, if $H = H(t)$]. When $[B(0),H] \neq 0$, instead, the unitary evolution in (31) can be regarded as a (constant entropy) limit cycle (or "ridge" as termed in [3]), which coincides with the usual periodic solutions of the Schrödinger equation when B is a one-dimensional projector, $B = |\psi\rangle\langle\psi|$ [${\rm Tr}(B) = 1$].

The equation of motion attracts the state operator towards such highest entropy state or limit cycle by rearranging the nonzero eigenvalues of ρ along the direction of steepest entropy ascent at a rate that depends on the value of internal-relaxation-time functional $\tau(\rho)$. A minor perturbation of the state that changes an initially zero eigenvalue to an arbitrarily small nonzero value would cause an irreversible departure of the state towards a different equilibrium state or limit cycle. Hence, by the definition of stability according to Lyapunov, as long as there are zero eigenvalues of ρ, i.e., unless $B = I$, all equilibrium states and limit cycles are unstable. Without this conclusion, we could not claim that the equation of motion entails the second law of thermodynamics (see Appendix A and Appendix B, and [7, 11, 13, 32], for further discussions of this important point). The question of whether the equation could admit of metastable equilibrium states has not yet been investigated.

The Hamiltonian operator may be set to vary with time in order to model (adiabatic) energy exchange with the surroundings. Then the unitary part of the equation of motion, $-i[H(t),\rho]/\hbar$, may model energy change, but not entanglement nor entropy exchange with the surroundings. The irreversible term, instead, describes internal relaxation only and does not contribute to energy change in any case.

5. Highest-entropy equilibrium states

The only globally stable equilibrium states of the dynamics generated by (1) are

$$\rho_e = \frac{\exp(C)}{\mathrm{Tr}\exp(C)} = \frac{\exp(-\beta H + \sum_i \nu_i G_i)}{\mathrm{Tr}\exp(-\beta H + \sum_i \nu_i G_i)} \; . \tag{32}$$

The definition of global stability, which is stronger than Lyapunov or local stability and is required by the second law [32, 33], is given and discussed in Appendix B. The proof was discussed in [11, 14, 32].

For a system with generators of the motion I, H, N_i, (32) represents in general the grand canonical thermodynamic equilibrium states. It reduces to the canonical equilibrium states if $N_i = \tilde{n}_i I$ for all i's (to the microcanonical if also $H = \tilde{e} I$).

As is well known, states given by (32) are solutions of the constrained maximization problem

$$\max\ s(\rho) \quad \text{subject to} \quad r_i(\rho) = \tilde{r}_i \text{ and } \rho \geq 0 \; , \tag{33}$$

where $s(\rho) = -k_{\mathrm{B}}\mathrm{Tr}(\rho\ln\rho)$, $r_1(\rho) = \mathrm{Tr}(\rho I)$, $r_2(\rho) = \mathrm{Tr}(\rho H)$, $r_i(\rho) = \mathrm{Tr}(\rho G_i)$, $\tilde{r}_1 = 1$, $\tilde{r}_2$ and $\tilde{r}_i$'s given.

The inequality constraint $\rho \geq 0$ can be eliminated recasting the problem in terms of $\sqrt{\rho}$,

$$\max\ s(\sqrt{\rho}) \quad \text{subject to} \quad r_i(\sqrt{\rho}) = \tilde{r}_i \; , \tag{34}$$

where $s(\sqrt{\rho}) = -k_{\mathrm{B}}\mathrm{Tr}[(\sqrt{\rho})^2\ln(\sqrt{\rho})^2] = s(\rho)$ and $r_i(\sqrt{\rho}) = \mathrm{Tr}[(\sqrt{\rho})^2 R_i] = r_i(\rho)$. The method of Lagrange multipliers then gives the condition (necessary but not sufficient)

$$\frac{\partial s(\rho)}{\partial\sqrt{\rho}} - \sum_i \lambda_i \frac{\partial r_i(\rho)}{\partial\sqrt{\rho}} = 0 \; , \tag{35}$$

which, using (6) and (9), becomes

$$-2k_{\mathrm{B}}\sqrt{\rho}\ln\rho - 2k_{\mathrm{B}}\sqrt{\rho} - \sum_i \lambda_i(\sqrt{\rho}R_i + R_i\sqrt{\rho}) = 0 \; . \tag{36}$$

It is noteworthy that (36) is satisfied with obvious identification of multipliers by (29), where $\sum_i \lambda_i\sqrt{\rho}R_i$ commutes with $\sqrt{\rho}\ln\rho$, i.e., with ρ. Therefore, each nondissipative state satisfies the necessary condition (36) although it is not a solution of the maximization problem (34) unless $B = I$.

6. Onsager's reciprocity and Callen's fluctuation-dissipation nonequilibrium relations

First, we introduce a general representation of state operators particularly useful for representing nonequilibrium states. Any ρ can be written as [16]

$$\rho = \frac{B\exp(-\sum_j f_j X_j/k_{\mathrm{B}})}{\mathrm{Tr}B\exp(-\sum_j f_j X_j/k_{\mathrm{B}})} \; , \tag{37}$$

where the hermitian operators in the set $\{I, X_j\}$ span the real space $\mathcal{L}_h(\mathcal{H})$ of linear hermitian operators on $\mathcal{H}$, and B is the usual idempotent operator defined in (22b). Indeed, the operator $B \ln \rho$ is always well-defined and belongs to $\mathcal{L}_h(\mathcal{H})$, so that

$$S = -k_{\rm B} B \ln \rho = f_0 I + \sum_j f_j X_j = f_0 B I + \sum_j f_j B X_j \tag{38}$$

where the second equality follows from $B^2 = B$ and f_0 is a generalized log-partition function

$$f_0 = k_{\rm B} \ln {\rm Tr} B \exp(-\sum_j f_j X_j / k_{\rm B}) \ . \tag{39}$$

Therefore,

$$-\, k_{\rm B} \sqrt{\rho} \ln \rho = f_0 \sqrt{\rho} + \sum_j f_j \sqrt{\rho} X_j \ , \tag{40}$$

$$x_j(\rho) = {\rm Tr}(\rho X_j) \ , \tag{41}$$

$$\sqrt{\rho} \Delta S = \sum_i f_i \sqrt{\rho} \Delta X_i \ , \tag{42}$$

$$\langle \Delta S \Delta S \rangle = \sum_i \sum_j f_i f_j \langle \Delta X_i \Delta X_j \rangle \ , \tag{43}$$

$$s(\rho) = f_0 + \sum_j f_j \, x_j(\rho) \ , \tag{44}$$

$$k_{\rm B} D = -\sum_j f_j \left[\sqrt{\rho} X_j - \sum_{k=1}^{a} (\sqrt{\rho} X_j | \sqrt{\rho} A_k) \sqrt{\rho} A_k \right] \tag{45}$$

where, for simplicity, we use the expression (25c) for D in terms of the orthonomal set $\{\sqrt{\rho} A_i\}$ [an equivalent expression in terms of Gram determinants is readily obtained using (40) in (21c)].

By taking partial derivatives of a symmetrized expansion of f_0, we may also obtain the relations $x_j(\rho) = \partial f_0 / \partial f_j$ and, equating second order partial derivatives, the generalized Maxwell relations

$$\left. \frac{\partial x_i(\rho)}{\partial f_j} \right|_{f_{k \neq j}} = -\frac{1}{k_{\rm B}} \langle \Delta X_i \Delta X_j \rangle = \left. \frac{\partial x_j(\rho)}{\partial f_i} \right|_{f_{k \neq i}} \ . \tag{46}$$

It is noteworthy that state operator (37) can be viewed as the solution of the constrained maximization problem

$$\max \ s(\rho) \tag{47a}$$

$$\text{subject to } x_j(\rho) = {\rm Tr}(\rho X_j) = \tilde{x}_j \ , \tag{47b}$$

$${\rm Tr}(\rho) = 1 \ , \ \rho \geq 0 \ , \text{and } \rho = B \rho B \tag{47c}$$

or, equivalently,

$$\max \ s(\sqrt{\rho}) \tag{48a}$$

$$\text{subject to } x_j(\sqrt{\rho}) = (\sqrt{\rho} | \sqrt{\rho} X_j) = \tilde{x}_j \ , \tag{48b}$$

$$(\sqrt{\rho} | \sqrt{\rho}) = 1 \ , \ \text{and } \sqrt{\rho} = B \sqrt{\rho} B \tag{48c}$$

for a given idempotent operator $B = B^2$ and given mean values $\tilde{x}_j$ of the operators in the set $\{X_j\}$. As a result, the Lagrange multipliers f_j can be viewed as functions of the $\tilde{x}_j$'s, defined implicitly by the relations obtained by substitution of (37) into the constraints. Thus, the expression for the entropy can be written as

$$s(\tilde{\mathbf{x}}) = f_0(\tilde{\mathbf{x}}) + \sum_j f_j(\tilde{\mathbf{x}}) \, \tilde{x}_j \tag{49}$$

from which we readily obtain that

$$f_j = \left. \frac{\partial s(\tilde{\mathbf{x}})}{\partial \tilde{x}_j} \right|_{\tilde{x}_{k \neq j}} \tag{50}$$

may be interpreted as the generalized affinity or force representing the entropy change that corresponds to an independent change in the mean value of the linear observable X_j. Moreover, by equating second order partial derivatives of $s(\tilde{\mathbf{x}})$, we obtain the generalized Maxwell relations

$$\left.\frac{\partial f_i(\tilde{\mathbf{x}})}{\partial \tilde{x}_j}\right|_{\tilde{x}_{k\neq j}} = \frac{\partial^2 s(\tilde{\mathbf{x}})}{\partial \tilde{x}_i \partial \tilde{x}_j} = \left.\frac{\partial f_j(\tilde{\mathbf{x}})}{\partial \tilde{x}_i}\right|_{\tilde{x}_{k\neq i}} . \tag{51}$$

Next, we define the dissipative rate of change of the linear mean-value functional associated with operator X_i,

$$\frac{\mathrm{D}x_i(\rho)}{\mathrm{D}t} = \mathrm{Tr}[\hat{\mathbf{D}}_1(\rho, H, G_i)X_i] = 2(E_D|\sqrt{\rho}X_i) . \tag{52}$$

We use the term dissipative because the irreversible term in the equation of motion conserves but downgrades the energy, i.e., it describes the irreversible conversion (dissipation) of mechanical energy into thermal energy. Using $E_D = -D/2\tau(\rho)$ [(20d)] and (45) yields the linear interrelations between dissipative rates and generalized affinities,

$$\frac{\mathrm{D}x_i(\rho)}{\mathrm{D}t} = \sum_j f_j\, L_{ij}(\rho) , \tag{53}$$

where the coefficients $L_{ij}(\rho)$, which may be interpreted as generalized (Onsager) dissipative conductivities, are the nonlinear functionals of ρ given by the relations

$$L_{ij}(\rho) = L_{ji}(\rho) = \frac{1}{k_{\mathrm{B}}\tau(\rho)}\Bigg[(\sqrt{\rho}X_i|\sqrt{\rho}X_j) - \sum_{k=1}^{a}(\sqrt{\rho}X_i|\sqrt{\rho}A_k)(\sqrt{\rho}A_k|\sqrt{\rho}X_j)\Bigg] \tag{54a}$$

$$= \frac{1}{k_{\mathrm{B}}\tau(\rho)}\left[\langle\Delta X_i \Delta X_j\rangle - \sum_{k=1}^{a}\langle\Delta X_i \Delta A_k\rangle\langle\Delta A_k \Delta X_j\rangle\right] \tag{54b}$$

$$= \frac{1}{k_{\mathrm{B}}\tau(\rho)}\left([\sqrt{\rho}X_i]_{\perp\mathcal{L}\{\sqrt{\rho}R_i\}}\big|\,[\sqrt{\rho}X_j]_{\perp\mathcal{L}\{\sqrt{\rho}R_i\}}\right) \tag{54c}$$

$$= \frac{\begin{vmatrix} \langle\Delta X_i\Delta X_j\rangle & \langle\Delta R_1\Delta X_j\rangle & \cdots & \langle\Delta R_k\Delta X_j\rangle & \cdots \\ \langle\Delta X_i\Delta R_1\rangle & \langle\Delta R_1\Delta R_1\rangle & \cdots & \langle\Delta R_k\Delta R_1\rangle & \cdots \\ \vdots & \vdots & \ddots & \vdots & \ddots \\ \langle\Delta X_i\Delta R_k\rangle & \langle\Delta R_1\Delta R_k\rangle & \cdots & \langle\Delta R_k\Delta R_k\rangle & \cdots \\ \vdots & \vdots & \ddots & \vdots & \ddots \end{vmatrix}}{k_{\mathrm{B}}\tau(\rho)\,\Gamma(\{\sqrt{\rho}R_k\})} . \tag{54d}$$

These relations are at the same time a proof and a generalization of Onsager's reciprocity relations [22] extended to all nonequilibrium (dissipative) states. Moreover, they give explicit expressions for the nonlinear dependence of the dissipative conductivities $L_{ij}(\rho)$ on the state operator ρ, the generators of the motion and the internal-relaxation-time $\tau(\rho)$. Of course, at any stable equilibrium, ρ_{e}, and at any nondissipative state or limit cycle, ρ_{nd}, all dissipative rates are zero and, therefore,

$$\sum_j f_{j\ \mathrm{e/nd}}\, L_{ij}(\rho_{\mathrm{e/nd}}) = 0 . \tag{55}$$

Using (26) and (40), the rate of entropy change may be rewritten as a quadratic form in the generalized affinities,

$$\frac{\mathrm{d}s(\rho)}{\mathrm{d}t} = \sum_i f_i \frac{\mathrm{D}x_i(\rho)}{\mathrm{D}t} = \sum_i\sum_j f_i f_j L_{ij}(\rho) . \tag{56}$$

The symmetric matrix $[\{L_{ij}(\rho)\}]$ is a Gram matrix [see (54c)] and, as such, it is nonegative definite, i.e., its determinant

$$\det[\{L_{ij}(\rho)\}] \geq 0 \ , \tag{57}$$

with strict positivity only if all operators

$$[\sqrt{\rho}X_i]_{\perp \mathcal{L}\{\sqrt{\rho}R_i\}} \tag{58}$$

are linearly independent, in which case (53) may be solved to yield

$$f_j = \sum_i L_{ij}^{-1}(\rho) \frac{\mathrm{D}x_i(\rho)}{\mathrm{D}t} \tag{59}$$

and the rate of entropy change can be written as a quadratic form in the dissipative rates

$$\frac{\mathrm{d}s(\rho)}{\mathrm{d}t} = \sum_i \sum_j L_{ij}^{-1}(\rho) \frac{\mathrm{D}x_i(\rho)}{\mathrm{D}t} \frac{\mathrm{D}x_j(\rho)}{\mathrm{D}t} \ . \tag{60}$$

Eqs. (54) are also a proof and generalization of Callen's fluctuation-dissipation theorem [23] extended to all nonequilibrium (dissipative) states. Indeed, we interpret $\langle \Delta X_i \Delta X_j \rangle$ as the codispersion (covariance) of measurement results of observables X_i and X_j when the system is in state ρ and $\langle \Delta X_i \Delta X_i \rangle$ as the dispersion (or fluctuations) of measurement results of observable X_i. As in Callen's fluctuation-dissipation theorem, the expressions in (54) relate codispersions with generalized conductivities.

A judicious choice of the set $\{\sqrt{\rho}X_j\}$ may greatly simplify these relations. In particular, if $\{\sqrt{\rho}X_j\}$ is any orthogonal extension of the orthonormal subset $\{\sqrt{\rho}A_i\}$, in the sense that $(\sqrt{\rho}X_j|\sqrt{\rho}A_i) = 0$ for all A_i's and $j > a$, then for all i and $j > a$,

$$L_{ij}(\rho) = \frac{1}{k_{\mathrm{B}}\tau(\rho)} \langle \Delta X_i \Delta X_j \rangle \ , \tag{61}$$

which relates directly [through the internal-relaxation time $\tau(\rho)$] covariances and fluctuations in the observables X_i with their associated dissipative conductivities.

Onsager's [22] and Callen's [23] theorems are keystones of our understanding of irreversibility. Indeed, the proof we proposed emerges not from the analysis of the Hamiltonian term of the equation of motion supplemented with the so-called assumption of microscopic reversibility, but from the irreversible term of the equation of motion, i.e., the only term responsible for irreversibility, with no additional assumptions.

Onsager's result [22] was obtained from empirical observations on nonequilibrium phenomena very close to stable thermodynamic equilibrium, so that the list of X_i's was indeed very short, and the result valid only for a limited class of states. Our result generalizes the validity of Onsager's reciprocity relations to all nonequilibrium states, close and far from stable thermodynamic equilibrium. Of course, the price we have to pay to describe nonequilibrium states far from stable equilibrium is that we must use a much larger, possibly infinite list of X_i's in (37).

7. Internal relaxation time vs time–energy Heisenberg uncertainty: maximal-entropy-generation-rate ansatz

By analogy with what was done for the Hamiltonian characteristic time $\tau_H(\rho)$, we define the shortest characteristic time of the irreversible part of the rate of change of the state operator by the relation (see Appendix C)

$$\frac{1}{\tau_D(\rho)^2} = 4(E_D|\,E_D) \ . \tag{62}$$

From (20a) and (21) and $(\sqrt{\rho}|\sqrt{\rho}) = 1$ [or, directly, from (26)], we find

$$\tau_D(\rho)^2 = k_{\mathrm{B}}^2 \tau(\rho)^2 \frac{\Gamma(\{\sqrt{\rho}R_i\})}{\Gamma(\sqrt{\rho}S, \{\sqrt{\rho}R_i\})} = \frac{\tau(\rho)^2}{(D|D)} \ . \tag{63}$$

From this relation we may extract a possible ansatz on the functional form or at least a lower bound for the internal-relaxation-time functional $\tau(\rho)$ in (1) by assuming that also $\tau_D(\rho)$, like $\tau_H(\rho)$, should satisfy the Heisenberg uncertainty relation

$$\tau_D(\rho)^2\langle\Delta H\Delta H\rangle \geq \hbar^2/4 \ . \tag{64}$$

This implies

$$\tau(\rho)^2 \geq \frac{\hbar^2}{4k_B^2}\frac{\Gamma(\sqrt{\rho}S,\{\sqrt{\rho}R_i\})}{\langle\Delta H\Delta H\rangle\,\Gamma(\{\sqrt{\rho}R_i\})} = \frac{\hbar^2(D|D)}{4\langle\Delta H\Delta H\rangle} \tag{65}$$

or, using (26) and (56),

$$\tau(\rho) \geq \frac{\hbar^2}{4k_B\langle\Delta H\Delta H\rangle}\sum_i\sum_j f_i f_j L_{ij}(\rho) \ . \tag{66}$$

Relation (65) implies an upper bound to the rate of entropy generation,

$$\frac{\mathrm{d}s(\rho)}{\mathrm{d}t} \leq \frac{2}{\hbar}\sqrt{\frac{\langle\Delta H\Delta H\rangle\,\Gamma(\sqrt{\rho}S,\{\sqrt{\rho}R_i\})}{\Gamma(\{\sqrt{\rho}R_i\})}} \ . \tag{67}$$

By analogy with the Hamiltonian time, a possible ansatz that is worthy of careful consideration is that also $\tau_D(\rho)$ and, therefore, $\tau(\rho)$ be equal exactly to the lower bound [i.e., strict equality in (65), (66) and (67)], corresponding to the maximal entropy generation rate compatible with the time–energy uncertainty relation. This corresponds to a truly maximal entropy generation dynamics. The square-root state operator $\sqrt{\rho}$ not only moves in the direction of steepest entropy ascent, but it does so at the highest rate compatible with the uncertainty principle.

This ansatz is very intriguing, conceptually appealing and fraught with far-reaching physical consequences. But, like any other hypotheses on the functional form of $\tau(\rho)$, it should be verified against known behavior in worked-out specific cases, such as the examples considered in [3, 15, 20] and the variety of physical problems for which nonlinear modifications of the Schrödinger equation have been deemed necessary but have not yet been resolved [27].

If proved valid, this ansatz would complete the dynamics without need of additional physical constants and would provide a fundamental microscopic foundation of the macroscopic observation that Nature always evolves at the fastest possible rate along the most direct path towards maximum entropy compatible with the system's structure and external constraints. Interpreted in this way, our nonlinear dynamics would provide a unifying fundamental microscopic foundation of all phenomenological theories of irreversible processes advanced in the last seventy years after the pioneering work of Onsager.

8. Variational principle formulation

Following Gheorghiu-Svirschevski [3], the direction of steepest-entropy-ascent can also be found by considering the constrained maximization problem

$$\max \frac{\mathrm{d}s}{\mathrm{d}t}(E_D) \tag{68a}$$

$$\text{subject to } \frac{\mathrm{d}r_i}{\mathrm{d}t}(E_D) = \dot{r}_i \text{ and } (E_D|E_D) = \frac{1}{4\tau_D^2} \ , \tag{68b}$$

where $\dot{r}_i$ and τ_D^2 are given real functionals of ρ. The necessary solving condition in terms of Lagrange multipliers is

$$\frac{\partial}{\partial E_D}\frac{\mathrm{d}s}{\mathrm{d}t} - \sum_i \lambda_i \frac{\partial}{\partial E_D}\frac{\mathrm{d}r_i}{\mathrm{d}t} - \lambda_0\frac{\partial}{\partial E_D}(E_D|E_D) = 0 \ . \tag{69}$$

Using (7) and (10), it becomes

$$\frac{\partial s(\rho)}{\partial\sqrt{\rho}} - 2\sum_i \lambda_i\sqrt{\rho}R_i - \lambda_0 E_D = 0 \ , \tag{70}$$

or, with (9),

$$-2k_{\rm B}\left[\sqrt{\rho}+\sqrt{\rho}\ln\rho\right]-2\sum_i \lambda_i\sqrt{\rho}R_i-\lambda_0 E_D=0\ , \tag{71}$$

where the multipliers must be determined by substitution in the system of constraint equations.

It is easy to verify that our expression for $E_D = -D/2\tau(\rho)$ in (20*a*) and (21) is an explicit solution of (70) in the case $\dot{r}_i = 0$ and τ_D given by the explicit expression in (63). See Appendix E for $\dot{r}_i \neq 0$.

9. Bloch near-equilibrium linear limit

In [11] we have shown that in the neighbourhood of each equilibrium state $\rho_{\rm e}$ {given by (28) with the additional condition $[B,H]=0$}, if we assume $\tau(\rho)$ constant (e.g., $\tau_{\rm e}=\tau(\rho_{\rm e})\neq 0$), the initial state operators $\rho(0)$ in the subset with $B(0)=B_{\rm e}$ and $\mathrm{Tr}(\rho(0)R_i)=\mathrm{Tr}(\rho_{\rm e}R_i)$ (i.e., with the same nullity, rank and mean values of the generators as the equilibrium state) obey a linearized form of the equation of motion that has the form of a Bloch relaxation equation,

$$\frac{\mathrm{d}\rho}{\mathrm{d}t}\rightarrow -\frac{i}{\hbar}[H,\rho]-\frac{\rho-\rho_{\rm e}}{\tau_{\rm e}}\ , \tag{72}$$

so that the solution of the equation of motion is

$$\begin{aligned}\rho(t)\rightarrow\ &\exp(-t/\tau_{\rm e})U(t)\rho(0)U^{-1}(t)\\ +\ &\left[1-\exp(-t/\tau_{\rm e})\right]\rho_{\rm e}\ ,\end{aligned} \tag{73}$$

with $U(t)$ given by (31*b*) {note that if there are no non-Hamiltonian generators the condition $[B,C]=0$ implies $[\rho,H]=0$ and, therefore, $U(t)\rho(0)U^{-1}(t)=\rho(0)$}.

Gheorghiu-Svirschevski [3], derived a more general linearized form for the case with no non-Hamiltonian generators valid also when $[\rho,H]\neq 0$ (but assuming $\tau\rightarrow\tau_{\rm e}\neq 0$), and showed that it yields a generalized Fokker-Planck equation for a free particle, and a Langevin equation for a harmonic oscillator.

It is however noteworthy that such linearized limit behavior should be revised if we take $\tau(\rho)$ equal to the lower bound imposed by the time–energy uncertainty relation [(65) with strict equality], because then, if $\langle\Delta H\Delta H\rangle\neq 0$, $\tau(\rho)\rightarrow 0$ as $\rho\rightarrow\rho_{\rm e}$.

10. Divisible composite systems dynamics: locally perceived overall energy and entropy operators

The composition of the system is embedded in the structure of the Hilbert space as a direct product of the subspaces associated with the individual elementary constituent subsystems, as well as in the form of the Hamiltonian operator.

In this section, we consider a system composed of M distinguishable and indivisible elementary constituent subsystems. For example, each subsystem may be a different elementary particle or a Fermi-Dirac or Bose-Einstein field (in which case the corresponding $\mathcal{H}^J$ is a Fock space). The Hilbert space is

$$\mathcal{H}=\mathcal{H}^1{\otimes}\mathcal{H}^2{\otimes}\cdots{\otimes}\mathcal{H}^M\ , \tag{74}$$

and the Hamiltonian operator

$$H=\sum_{J=1}^{M}H_J{\otimes}I_{\overline{J}}+V\ , \tag{75}$$

where H_J is the Hamiltonian operator on $\mathcal{H}^J$ associated with the J-th subsystem when isolated and V (on $\mathcal{H}$) the interaction Hamiltonian among the M subsystems.

For convenience, we denote by $\mathcal{H}^{\overline{J}}$ the direct product of the Hilbert spaces of all subsystems except the J-th one, so that the Hilbert space of the overall system is

$$\mathcal{H}=\mathcal{H}^J{\otimes}\mathcal{H}^{\overline{J}} \tag{76}$$

and the identity operator $I = I_J \otimes I_{\overline{J}}$.

The subdivision into elementary constituents, each considered as indivisible, is reflected by the structure of the Hilbert space $\mathcal{H}$ as a direct product of subspaces, and is particularly important because it defines the level of description of the system and specifies its elementary structure. This determines also the structure of the nonlinear dynamical law (2) we proposed in [10, 11] and rederive in the next section. In a sense, this is a price we have to pay in order to free ourselves from the assumption of linearity of the law of motion.

In other words, the form of superoperators $\hat{\mathbf{D}}_M$ and $\hat{\mathbf{D}}_1$ are different depending on whether the system is or is not subdivisible into indivisible subsystems, i.e., whether or not it has an internal structure. For example, consistently with the conditions listed in Appendix A, we request that superoperator $\hat{\mathbf{D}}_M$ satisfies the strong separability conditions [2, 20]

$$\begin{aligned}&\hat{\mathbf{D}}_M(\rho_A\otimes\rho_B, H_A\otimes I_B+I_A\otimes H_B, G_{iA}\otimes I_B+I_A\otimes G_{iB})\\ &= \hat{\mathbf{D}}_{M_A}(\rho_A,H_A,G_{iA})\otimes\rho_B+\rho_A\otimes\hat{\mathbf{D}}_{M_B}(\rho_B,H_B,G_{iB}) \text{ for any } \rho_A, \rho_B, \qquad (77a)\\ &\mathrm{Tr}_B[\hat{\mathbf{D}}_M(\rho, H_A\otimes I_B+I_A\otimes H_B, G_{iA}\otimes I_B+I_A\otimes G_{iB})]\\ &= \hat{\mathbf{F}}_{M_A}(\rho,H_A,G_{iA}) \qquad \text{for any } \rho, \qquad (77b)\\ &\mathrm{Tr}_A[\hat{\mathbf{D}}_M(\rho, H_A\otimes I_B+I_A\otimes H_B, G_{iA}\otimes I_B+I_A\otimes G_{iB})]\\ &= \hat{\mathbf{F}}_{M_B}(\rho,H_B,G_{iB}) \qquad \text{for any } \rho, \qquad (77c)\end{aligned}$$

where subsystems A and B are obtained by partitioning the set of constituents $1, 2, \ldots, M$ into two disjoint subsets of M_A and M_B constituents, respectively ($M_A + M_B = M$). Note that, of course, if $\rho = \rho_A\otimes\rho_B$ then we must have $\hat{\mathbf{F}}_{M_A}(\rho,H_A,G_{iA}) = \hat{\mathbf{D}}_{M_A}(\rho_A,H_A,G_{iA})$ and $\hat{\mathbf{F}}_{M_B}(\rho,H_B,G_{iB}) = \hat{\mathbf{D}}_{M_B}(\rho_B,H_B,G_{iB})$.

It is noteworthy that trying to apply the same conditions to superoperator $\hat{\mathbf{D}}_1$ would be physically meaningless. For a subdivisible system, $\hat{\mathbf{D}}_1$ cannot be the irreversible evolution superoperator because it would entail for example exchange of energy between noninteracting subsystems. Conditions (77*b*) and (77*c*) instead, prevent non-locality problems by guaranteeing that changes of the Hamiltonian (or the other generators of the motion) in one of two noninteracting subsystems cannot affect the mean values of local observables of the other subsystem. For example, assume that subsystems A and B are correlated but not interacting. We may switch on a measurement apparatus within B and therefore alter the Hamiltonian H_B. By virtue of (77*b*) and (77*c*), the rate of change of the reduced state operator $\rho_A = \mathrm{Tr}_B(\rho)$ does not depend on H_B and, therefore, all the functionals of ρ_A (local observables) remain unaffected by the change in B, i.e., no faster-than-light communication can occur between B and A (of course, if we exclude the projection postulate [34]).

In addition, we must consider the following additional non-trivial conditions of separate energy conservation of noninteracting (possibly correlated) subsystems (see Appendix A, Condition 6)

$$\mathrm{Tr}[(H_A\otimes I_B)\,\hat{\mathbf{D}}_M(\rho, H_A\otimes I_B+I_A\otimes H_B, G_i)] = 0\ , \qquad (78a)$$

$$\mathrm{Tr}[(I_A\otimes H_B)\,\hat{\mathbf{D}}_M(\rho, H_A\otimes I_B+I_A\otimes H_B, G_i)] = 0\ , \qquad (78b)$$

for any ρ, and of separate entropy nondecrease for uncorrelated (possibly interacting) subsystems (see Appendix A, Condition 7)

$$\mathrm{Tr}[(S_A\otimes I_B)\,\hat{\mathbf{D}}_M(\rho_A\otimes\rho_B, H, G_i)] \geq 0\ , \qquad (79a)$$

$$\mathrm{Tr}[(I_A\otimes S_B)\,\hat{\mathbf{D}}_M(\rho_A\otimes\rho_B, H, G_i)] \geq 0\ , \qquad (79b)$$

for any ρ_A and ρ_B, where $S_A = -k_{\mathrm{B}} B_A \ln\rho_A$, $S_B = -k_{\mathrm{B}} B_B \ln\rho_B$.

For each type of particle in the system, we may write without loss of generality the number-of-particles-of-i-th-type operator associated with the system as

$$N_i = \sum_{J=1}^{M} N_{iJ}\otimes I_{\overline{J}} \qquad \text{for } i = 1, 2, \ldots, r\ , \qquad (80)$$

where N_{iJ} denotes the number-of-particles-of-i-th-type operator associated with the J-th subsystem [35]. In general, we assume that the set of linear hermitian operators I, H, G_i

on $\mathcal{H}$, always including I and H, are the generators of the motion of the composite system. For example, the list of non-Hamiltonian generators G_i may coincide with that of the number operators defined in (80).

Next, we introduce a useful notation that allows the definition of important local observables. For each subsystem J we denote by $\mathcal{L}(\mathcal{H}^J)$ the space of linear operators on $\mathcal{H}^J$ equipped with the real scalar product

$$(F_J|G_J) = \frac{1}{2}\mathrm{Tr}\left(F_J^\dagger G_J + G_J^\dagger F_J\right) . \tag{81}$$

For a given state operator ρ on $\mathcal{H}$, given linear hermitian F and G on $\mathcal{H}$, and each subsystem J, in addition to that already defined in (4), (12) and (13), we define the following convenient notation [11, 16]

$$\rho_J = \mathrm{Tr}_{\overline{J}}(\rho) , \tag{82}$$

$$\rho_{\overline{J}} = \mathrm{Tr}_J(\rho) , \tag{83}$$

$$(F)^J = \mathrm{Tr}_{\overline{J}}[(I_J \otimes \rho_{\overline{J}})F] , \tag{84}$$

$$\begin{aligned}\langle \Delta F \Delta G\rangle^J &= (\sqrt{\rho_J}(\Delta F)^J | \sqrt{\rho_J}(\Delta G)^J) \\ &= \frac{1}{2}\mathrm{Tr}(\rho_J\{(\Delta F)^J, (\Delta G)^J\}) = \langle(\Delta F)^J(\Delta G)^J\rangle ,\end{aligned} \tag{85}$$

where $\mathrm{Tr}_{\overline{J}}$ denotes the partial trace over $\mathcal{H}^{\overline{J}}$, Tr_J over $\mathcal{H}^J$, ρ_J is the reduced state operator of elementary subsystem J and $\rho_{\overline{J}}$ that of the composite of all other subsystems.

In view of the special role they play in the equation of motion, we interpret the operators $(H)^J$ and $(S)^J$,

$$(H)^J = \mathrm{Tr}_{\overline{J}}[(I_J \otimes \rho_{\overline{J}})H] , \tag{86}$$

$$(S)^J = \mathrm{Tr}_{\overline{J}}[(I_J \otimes \rho_{\overline{J}})S] , \tag{87}$$

as "locally perceived overall-system energy" and "locally perceived overall-system entropy" operators, respectively, associated with a measure of how the overall-system energy and entropy operators, H and $S = -k_\mathrm{B} B \ln\rho$, are "felt" locally within the J-th constituent subsystem.

For a given ρ_J, we further define the operators

$$\sqrt{\rho_J} , \tag{88}$$

$$B_J , \tag{89}$$

$$S_J = -k_\mathrm{B} B_J \ln\rho_J , \tag{90}$$

obtained from ρ_J by substituting in its spectral expansion each nonzero eigenvalue p_i with its positive square root $\sqrt{p_i}$, unity and $-k_\mathrm{B}\ln p_i$, respectively. Note that, of course, $S_J \neq (S)^J$. Moreover, for given linear hermitian F_J and G_J on $\mathcal{H}^J$, we define the notation

$$\langle F_J\rangle = \mathrm{Tr}[\rho\,(F_J \otimes I_{\overline{J}})] = \mathrm{Tr}(\rho_J F_J) , \tag{91}$$

$$\langle \Delta F_J \Delta G_J\rangle = (\sqrt{\rho_J}\Delta F_J | \sqrt{\rho_J}\Delta G_J) = \frac{1}{2}\mathrm{Tr}(\rho_J\{\Delta F_J, \Delta G_J\}) . \tag{92}$$

S_J may be interpreted as the subsystem entropy operator only if subsystem J is not correlated with the other subsystems, i.e., if ρ can be written as

$$\rho = \rho_J \otimes \rho_{\overline{J}} ; \tag{93}$$

then the subsystem entropy is defined and given by the nonlinear state functional of the reduced state operator,

$$\begin{aligned}s_J(\rho_J) &= \langle S_J\rangle = \mathrm{Tr}(\rho_J S_J) = -k_\mathrm{B}\mathrm{Tr}(\rho_J \ln\rho_J) \\ &= \mathrm{Tr}[\rho(S_J \otimes I_{\overline{J}})] = (\sqrt{\rho_J}|\sqrt{\rho_J}S_J) ,\end{aligned} \tag{94}$$

$$S = S_J \otimes I_{\overline{J}} + I_J \otimes S_{\overline{J}} , \tag{95}$$

$$s(\rho) = s_J(\rho_J) + s_{\overline{J}}(\rho_{\overline{J}}) , \tag{96}$$

and we also have $(\Delta S)^J = \Delta S_J$.

When subsystem J is correlated, instead, its entropy is not defined; however, the functional

$$s^J(\rho) = \langle (S)^J \rangle = \mathrm{Tr}[\rho_J (S)^J] = (\sqrt{\rho_J}|\sqrt{\rho_J}(S)^J) \tag{97}$$

may be interpreted as the subsystem's local perception of the overall-system entropy. Only when J is uncorrelated, $s^J(\rho) = s(\rho) = s_J(\rho_J) + s_{\overline{J}}(\rho_{\overline{J}})$.

The energy is defined for a subsystem J only if it is not interacting with the other subsystems, i.e., if H can be written as

$$H = H_J \otimes I_{\overline{J}} + I_J \otimes H_{\overline{J}} ; \tag{98}$$

then it is given by the mean-value state functional

$$e_J(\rho_J) = \langle H_J \rangle = \mathrm{Tr}(\rho_J H_J) = (\sqrt{\rho_J}|\sqrt{\rho_J} H_J) , \tag{99}$$

and we also have $(\Delta H)^J = \Delta H_J$.

When subsystem J is interacting, instead, its energy is not defined; however the functional

$$e^J(\rho) = \langle (H)^J \rangle = \mathrm{Tr}[\rho_J (H)^J] = (\sqrt{\rho_J}|\sqrt{\rho_J}(H)^J) \tag{100}$$

may be interpreted as the subsystem's local perception of the overall-system energy. Only when J is non-interacting, $e^J(\rho) = e(\rho) = e_J(\rho_J) + e_{\overline{J}}(\rho_{\overline{J}})$.

The number-of-particles-of-i-th-type of the overall system in state ρ and of each subsystem J are given by the mean-value state functionals $n_i(\rho) = \langle N_i \rangle = \mathrm{Tr}(\rho N_i)$ and $n_{iJ}(\rho_J) = \langle N_{iJ} \rangle = \mathrm{Tr}[\rho(N_{iJ} \otimes I_{\overline{J}})] = n_i^J(\rho) = \langle (N_{iJ} \otimes I_{\overline{J}})^J \rangle = \mathrm{Tr}(\rho_J N_{iJ}) = (\sqrt{\rho_J}|\sqrt{\rho_J} N_{iJ})$, respectively, and from (80) we clearly have $n_i(\rho) = \sum_{J=1}^M n_{iJ}(\rho_J)$. Indeed, it is noteworthy that, for any ρ,

$$(N_{iJ} \otimes I_{\overline{J}})^J = N_{iJ} . \tag{101}$$

For generality, we assume that the generators of the motion on $\mathcal{H}$ are I, H and G_i, with $[G_i, H] = 0$ but not necessarily with the separated structure of the number operators. Moreover, for each state ρ and each constituent J, we denote by $\{R_{iJ}, i = 0, 1, 2, \ldots, z_J(\rho)\}$ a set of hermitian operators in $\mathcal{L}(\mathcal{H})$ such that the operators in the set $\{\sqrt{\rho_J}(R_{iJ})^J\}$, where

$$(R_{iJ})^J = \mathrm{Tr}_{\overline{J}}[(I_J \otimes \rho_{\overline{J}}) R_{iJ}] \tag{102}$$

are linearly independent and span the linear manifold generated by the operators $\sqrt{\rho_J} I_J$, $\sqrt{\rho_J}(H)^J$, $\sqrt{\rho_J}(G_i)^J$. If the latter operators are linearly independent, for subsystem J, then the set $\{R_{iJ}\}$ may be chosen to coincide with the generators of the motion. If they are not independent, then it could be a smaller subset (in which it is convenient, though not necessary, to maintain $R_{0J} = I$ and, if possible, $R_{1J} = H$). In any case, we call the operators in the set $\{R_{iJ}\}$ the (instantaneous) "generators of the motion of subsystem J". The structure of the irreversible term $\hat{\mathbf{D}}_M$ is invariant under transformation from one set $\{R_{iJ}\}$ to any other $\{R'_{iJ}\}$ with the same defining properties.

For each instantaneous generator R_{iJ} of subsystem J, we define the "local perception" mean functional

$$r_{iJ}^J(\rho) = \mathrm{Tr}[\rho_J (R_{iJ})^J] = (\sqrt{\rho_J}|\sqrt{\rho_J}(R_{iJ})^J) . \tag{103}$$

We denote by

$$\mathcal{L}\{\sqrt{\rho_J}(R_{iJ})^J\} = \mathcal{L}\{\sqrt{\rho_J} I_J, \sqrt{\rho_J}(H)^J, \sqrt{\rho_J}(G_i)^J\} \tag{104}$$

the linear span of the operators $\{\sqrt{\rho_J}(R_{iJ})^J\}$ or, that is the same, the linear span of the operators $\sqrt{\rho_J} I_J$, $\sqrt{\rho_J}(H)^J$, $\sqrt{\rho_J}(G_i)^J$.

By definition of the set $\{R_{iJ}\}$, the Gram determinant

$$\begin{aligned}\Gamma(\{\sqrt{\rho_J}(R_{iJ})^J\}) &= \det[\{(\sqrt{\rho_J}(R_{iJ})^J|\sqrt{\rho_J} R_{jJ})^J)_J\}] \\ &= \det[\{\langle \Delta R_{iJ} \Delta R_{jJ} \rangle^J\}]\end{aligned} \tag{105}$$

is always strictly positive.

The set $\{R_{iJ}\}$ can be conveniently chosen so that $\{\sqrt{\rho_J}(R_{iJ})^J\}$ is (instantaneously) an orthonormal set, in which case we denote it by $\{\sqrt{\rho_J}(A_{iJ})^J\}$ and

$$(\sqrt{\rho_J}(A_{iJ})^J|\sqrt{\rho_J}A_{jJ})^J) = \delta_{ij} , \tag{106}$$

$$\Gamma_J(\{\sqrt{\rho_J}(A_{iJ})^J\}) = 1 . \tag{107}$$

11. The steepest-entropy-ascent ansatz for a composite system

Maintaining the validity of (3) and (18), we now assume

$$\sqrt{\rho}\, E_D = \sum_{J=1}^{M} \sqrt{\rho_J}\, E_{DJ} \otimes \rho_{\overline{J}} , \tag{108}$$

as a first step to guarantee the strong separability condition (77) [see Appendix G for a discussion related to the form of (108)].

The second step is to make sure that E_{DJ}, which in general may be a function of the operators ρ, H and G_i on $\mathcal{H}$, reduces to a function of the operators ρ_J, H_J and G_{iJ} on $\mathcal{H}^J$ only, whenever the constituent is, at the same time, uncorrelated ($\rho = \rho_J \otimes \rho_{\overline{J}}$), non-interacting ($H = H_J \otimes I_{\overline{J}} + I_J \otimes H_{\overline{J}}$) and not coupled through the non-Hamiltonian generators ($G_i = G_{iJ} \otimes I_{\overline{J}} + I_J \otimes G_{i\overline{J}}$).

To preserve the formal analogy with the notation for the indivisible system, we define the local "partial gradient" operators

$$\left[\frac{\partial r^J_{iJ}(\rho)}{\partial\sqrt{\rho_J}}\right]_{(R_{iJ})^J} = \sqrt{\rho_J}(R_{iJ})^J + (R_{iJ})^J\sqrt{\rho_J} , \tag{109}$$

$$\left[\frac{\partial s^J(\rho)}{\partial\sqrt{\rho_J}}\right]_{(S)^J} = 2\sqrt{\rho_J}(S)^J . \tag{110}$$

Now, it is easy to show that the structure of $\mathrm{d}\rho/\mathrm{d}t$ assumed with (3), (18) and (108) yields [37]

$$\frac{\mathrm{d}r_{iJ}(\rho)}{\mathrm{d}t} = \mathrm{Tr}(\frac{\mathrm{d}\rho}{\mathrm{d}t}R_{iJ}) = 2\sum_{J=1}^{M}\left(E_{DJ}\middle|\sqrt{\rho_J}(R_{iJ})^J\right) , \tag{111}$$

$$\begin{aligned}\frac{\mathrm{d}s(\rho)}{\mathrm{d}t} &= -k_{\mathrm{B}}\mathrm{Tr}(\frac{\mathrm{d}\rho}{\mathrm{d}t}) - k_{\mathrm{B}}\mathrm{Tr}(\frac{\mathrm{d}\rho}{\mathrm{d}t}B\ln\rho)\\ &= \sum_{J=1}^{M}\left[-2k_{\mathrm{B}}(E_{DJ}|\sqrt{\rho_J}I_J)_J + 2(E_{DJ}|\sqrt{\rho_J}(S)^J)\right]\\ &= \sum_{J=1}^{M}\left[-2k_{\mathrm{B}}(\sqrt{\rho_J}I_J|E_{DJ}) + \left(\left[\frac{\partial s^J(\rho)}{\partial\sqrt{\rho_J}}\right]_{(S)^J}\middle|E_{DJ}\right)\right].\end{aligned} \tag{112}$$

Finally, we assume that each E_{DJ} is: (A) orthogonal to $\mathcal{L}\{\sqrt{\rho_J}(R_{iJ})^J\}$, so that all the rates of change in (111) and the first term in the rate of entropy change in (112) are zero; and (B) in the direction of the local partial gradient of the "locally perceived" overall-system entropy functional, $s^J(\rho)$, i.e., we take

$$E_{DJ} = \frac{1}{4k_{\mathrm{B}}\tau_J(\rho)}\left[\left[\frac{\partial s^J(\rho)}{\partial\sqrt{\rho_J}}\right]_{(S)^J}\right]_{\perp\mathcal{L}\{\sqrt{\rho_J}(R_{iJ})^J\}} = \frac{1}{2\tau_J(\rho)}D_J , \tag{113}$$

where, similarly to what is done for the indivisible system,

$$D_J = \left[\sqrt{\rho_J}(B\ln\rho)^J\right]_{\perp\mathcal{L}\{\sqrt{\rho_J}(R_{iJ})^J\}} \tag{114a}$$

$$= \sqrt{\rho_J}(B\ln\rho)^J - \left[\sqrt{\rho_J}(B\ln\rho)^J\right]_{\mathcal{L}\{\sqrt{\rho_J}(R_{iJ})^J\}} \tag{114b}$$

$$= -\frac{\begin{vmatrix} \sqrt{\rho_J}(\Delta S)^J & \sqrt{\rho_J}(\Delta R_{1J})^J & \cdots & \sqrt{\rho_J}(\Delta R_{iJ})^J & \cdots \\ (\Delta S \Delta R_{1J})^J & (\Delta R_{1J}\Delta R_{1J})^J & \cdots & (\Delta R_{iJ}\Delta R_{1J})^J & \cdots \\ \vdots & \vdots & \ddots & \vdots & \ddots \\ (\Delta S \Delta R_{iJ})^J & (\Delta R_{1J}\Delta R_{iJ})^J & \cdots & (\Delta R_{iJ}\Delta R_{iJ})^J & \cdots \\ \vdots & \vdots & \ddots & \vdots & \ddots \end{vmatrix}}{k_B\,\Gamma(\{\sqrt{\rho_J}(R_{iJ})^J\})}\,. \tag{114c}$$

Most of the results found for the single constituent extend to the composite system in a straightforward way. In particular, using the properties of determinants it is easy to verify that

$$(E_{DJ}|\sqrt{\rho_J}) = \mathrm{Tr}(\sqrt{\rho_J}E_{DJ}^\dagger + E_{DJ}^\dagger\sqrt{\rho_J}) = 0 \tag{115}$$

and that the mean value of each R_{iJ} is independently conserved by each subsystem, i.e., $(E_{DJ}|\sqrt{\rho_J}(R_{iJ}^J) = 0$, and therefore $\mathrm{Tr}(\rho I)$, $\mathrm{Tr}(\rho H)$, and the $\mathrm{Tr}(\rho G_i)$'s are time invariant.

In general, by virtue of (19), (75), (108), and (115), the rate of change of the reduced state operator of subsystem J is given by the equation

$$\frac{\mathrm{d}\rho_J}{\mathrm{d}t} = -\frac{i}{\hbar}[H_J,\rho_J] - \frac{i}{\hbar}\mathrm{Tr}_{\overline{J}}([V,\rho]) + \left(\frac{\mathrm{D}\rho}{\mathrm{D}t}\right)^J \tag{116}$$

where we defined

$$\left(\frac{\mathrm{D}\rho}{\mathrm{D}t}\right)^J = \mathrm{Tr}_{\overline{J}}\left(\frac{\mathrm{D}\rho}{\mathrm{D}t}\right) = \sqrt{\rho_J}E_{DJ}^\dagger + E_{DJ}^\dagger\sqrt{\rho_J}\,. \tag{117}$$

Indeed, in general, we have

$$\frac{\mathrm{D}\rho}{\mathrm{D}t} = \frac{\mathrm{d}\rho}{\mathrm{d}t} + \frac{i}{\hbar}[H,\rho] = \sum_{J=1}^{M}\left(\frac{\mathrm{D}\rho}{\mathrm{D}t}\right)^J \otimes \rho_{\overline{J}}\,. \tag{118}$$

The rate of entropy change is given by the relations

$$\frac{\mathrm{d}s(\rho)}{\mathrm{d}t} = \sum_{J=1}^{M}\frac{k_B}{\tau_J(\rho)}(D_J|D_J) \tag{119a}$$

$$= \sum_{J=1}^{M} 4k_B\tau_J(\rho)\,(E_{DJ}|E_{DJ}) \tag{119b}$$

$$= \sum_{J=1}^{M}\frac{1}{k_B\tau_J(\rho)}\,\frac{\Gamma(\sqrt{\rho_J}(S)^J,\{\sqrt{\rho_J}(R_{iJ})^J\})}{\Gamma(\{\sqrt{\rho_J}(R_{iJ})^J\})}\,. \tag{119c}$$

The dynamics reduces to the Schrödinger-von Neumann unitary Hamiltonian dynamics when

$$\sqrt{\rho_J}(B\ln\rho)^J = \sqrt{\rho_J}(C)^J \qquad \text{with } C = \sum_i \lambda_{iJ}R_{iJ}\,, \tag{120}$$

for each J and, therefore, the state is nondissipative {equilibrium if $[B,H]=0$, limit cycle if $[B,H]\neq 0$, including the case of pure-state standard QM when $\mathrm{Tr}(B)=1$}.

The maximum entropy thermodynamic equilibrium states are given by (32).

Onsager's reciprocity relations follow again from (38) for the operator $B\ln\rho$. Eqs. (53) and (56) are still valid [of course, with $\hat{\mathbf{D}}_M$ in (52) instead of $\hat{\mathbf{D}}_1$], with the dissipative conductivities given by

$$L_{ij}(\rho) = L_{ji}(\rho) = \sum_{J=1}^{M} L_{ij}^J(\rho)\ , \tag{121a}$$

$$L_{ij}^J(\rho) = L_{ji}^J(\rho) = \frac{\begin{vmatrix} \langle\Delta X_i\Delta X_j\rangle^J & \langle\Delta R_1\Delta X_j\rangle^J & \cdots & \langle\Delta R_k\Delta X_j\rangle^J & \cdots \\ \langle\Delta X_i\Delta R_1\rangle^J & \langle\Delta R_1\Delta R_1\rangle^J & \cdots & \langle\Delta R_k\Delta R_1\rangle^J & \cdots \\ \vdots & \vdots & \ddots & \vdots & \ddots \\ \langle\Delta X_i\Delta R_k\rangle^J & \langle\Delta R_1\Delta R_k\rangle^J & \cdots & \langle\Delta R_k\Delta R_k\rangle^J & \cdots \\ \vdots & \vdots & \ddots & \vdots & \ddots \end{vmatrix}}{k_{\rm B}\tau_J(\rho)\,\Gamma(\{\sqrt{\rho_J}(R_{iJ})^J\})}\ . \tag{121b}$$

Callen's fluctuation-dissipation relations, implied by the explicit structure [(121)] of the Onsager dissipative conductivities, are greatly simplified if we choose the orthonormal set $\{\sqrt{\rho_J}(A_{iJ})^J\}$ instead of $\{\sqrt{\rho_J}(R_{iJ})^J\}$ and the set $\{\sqrt{\rho_J}(X_{jJ})^J\}$ to be an orthogonal extension of $\{\sqrt{\rho_J}(A_{iJ})^J\}$; then,

$$L_{ij}^J(\rho) = \frac{1}{k_{\rm B}\tau_J(\rho)}\langle\Delta X_i\Delta X_j\rangle^J\ . \tag{122}$$

The assumption about the time–energy Heisenberg uncertainty relation in section 7 can be extended as well. We assume that each E_{DJ} gives rise to a characteristic time $\tau_{DJ}(\rho)$ defined by (see Appendix C)

$$\frac{1}{\tau_{DJ}(\rho)^2} = 4\,(E_{DJ}|\,E_{DJ})\ , \tag{123}$$

which should independently satisfy the uncertainty relation

$$\tau_{DJ}(\rho)^2\langle\Delta H\Delta H\rangle^J \geq \hbar^2/4\ . \tag{124}$$

Therefore, using (113) and (114), we obtain the lower bounds to the internal relaxation times,

$$\tau_J(\rho)^2 \geq \frac{\hbar^2}{4k_{\rm B}^2}\frac{\Gamma(\sqrt{\rho_J}(S)^J,\{\sqrt{\rho_J}(R_{iJ})^J\})}{\langle\Delta H\Delta H\rangle^J\,\Gamma(\{\sqrt{\rho_J}(R_{iJ})^J\})} = \frac{\hbar^2(D_J|D_J)}{4\langle\Delta H\Delta H\rangle^J} \tag{125}$$

or, using, (119) and (56),

$$\tau_J(\rho) \geq \frac{\hbar^2}{4k_{\rm B}\langle\Delta H\Delta H\rangle^J}\sum_i\sum_j f_i f_j L_{ij}^J(\rho)\ , \tag{126}$$

and the upper bound to the entropy generation rate

$$\frac{{\rm d}s(\rho)}{{\rm d}t} \leq \sum_{J=1}^{M}\frac{2}{\hbar}\sqrt{\frac{\langle\Delta H\Delta H\rangle^J\,\Gamma(\sqrt{\rho_J}(S)^J,\{\sqrt{\rho_J}(R_{iJ})^J\})}{\Gamma(\{\sqrt{\rho_J}(R_{iJ})^J\})}}\ . \tag{127}$$

Taking strict equalities in each of (125) yields the maximum entropy generation rate compatible with the time–energy uncertainty relations, and corresponds to a possible "extreme" closure ansatz.

The variational formulation can be found, after assuming the structure of (108), by solving the constrained maximization problem

$$\max \frac{\mathrm{d}s}{\mathrm{d}t}(\{E_{DJ}\}) \tag{128a}$$

$$\text{subject to } \frac{\mathrm{d}r_{iJ}}{\mathrm{d}t}(\{E_{DJ}\}) = \dot{r}_{iJ} \text{ and } (E_{DJ}|E_{DJ}) = \frac{1}{4\tau_{DJ}^2}\ , \tag{128b}$$

where $\dot{r}_{iJ}$ and τ_{DJ}^2 are given real functionals of ρ. By virtue of (111) and (112), the necessary solving conditions in terms of Lagrange multipliers for each E_{DJ} become

$$\left[\frac{\partial s^J(\rho)}{\partial\sqrt{\rho_J}}\right]_{(S)^J} - 2\sum_i \lambda_i^J \sqrt{\rho_J}(R_{iJ})^J - \lambda_0^J E_{DJ} = 0\ , \tag{129}$$

clearly verified by the E_{DJ}'s given in (113) and (114).

If two subsystems A and B are non-interacting but in correlated states, the reduced state operators obey the equations

$$\frac{\mathrm{d}\rho_A}{\mathrm{d}t} = -\frac{i}{\hbar}[H_A,\rho_A] - \sum_{\substack{J=1\\J\in A}}^{M} \frac{1}{2\tau_J(\rho)}\left[\sqrt{\rho_J}D_J + D_J^\dagger\sqrt{\rho_J}\right]\otimes(\rho_A)_{\overline{J}}\ , \tag{130a}$$

$$\frac{\mathrm{d}\rho_B}{\mathrm{d}t} = -\frac{i}{\hbar}[H_B,\rho_B] - \sum_{\substack{J=1\\J\in B}}^{M} \frac{1}{2\tau_J(\rho)}\left[\sqrt{\rho_J}D_J + D_J^\dagger\sqrt{\rho_J}\right]\otimes(\rho_B)_{\overline{J}}\ , \tag{130b}$$

where $(\rho_A)_{\overline{J}} = \mathrm{Tr}_J(\rho_A)$ and $(\rho_B)_{\overline{J}} = \mathrm{Tr}_J(\rho_B)$.

The strong separability conditions (77) are satisfied provided the internal-relaxation-time functionals are either all constants or satisfy the following set of nontrivial conditions,

$$\tau_J(\rho_A\otimes\rho_B, H_A\otimes I_B + I_A\otimes H_B, G_{iA}\otimes I_B + I_A\otimes G_{iB}) = \begin{cases} \tau_J(\rho_A,H_A,G_{iA}) & \text{for } J\in A \\ \tau_J(\rho_B,H_B,G_{iB}) & \text{for } J\in B \end{cases} \quad \text{for any } \rho_A \text{ and } \rho_B, \tag{131a}$$

$$\tau_J(\rho, H_A\otimes I_B + I_A\otimes H_B, G_{iA}\otimes I_B + I_A\otimes G_{iB}) = \begin{cases} \tau_J{}'(\rho,H_A,G_{iA}) & \text{for } J\in A \\ \tau_J{}'(\rho,H_B,G_{iB}) & \text{for } J\in B \end{cases} \quad \text{for any } \rho\ . \tag{131b}$$

It is noteworthy that Conditions (131) are satisfied by the maximal-entropy-generation-rate ansatz, i.e., if we assume that each $\tau_J(\rho)$ is given by (125) with strict equality. Indeed, if A and B are noninteracting, the structure [114c)] of each operator D_J with $J \in A$ is such that any dependence on H_B cancels out, moreover $\langle\Delta H\Delta H\rangle^J = \langle\Delta H_A\Delta H_A\rangle^J$. Thus, any dependence on H_B cancels out in (11a) and, similarly, any dependence on H_A cancels out in (11b).

It is interesting, however, that $\mathrm{d}\rho_A/\mathrm{d}t$ in general may depend not only on the "local" (reduced) state operator ρ_A but also on the overall state ρ through the operators $(B\ln\rho)^J$ (with $J \in A$), thus determining a collective-behavior effect on the local dynamics originating from existing correlations. In fact, operators D_J are in terms of $(\Delta S)^J$ which may differ, when the subsystems are correlated, from operators $(\Delta S_A)^J$. In other words, the lack of interactions between two subsystems does guarantee that the energy of each subsystem is defined and conserved [(78)] and that the reduced dynamics of each subsystem is local, in the sense of independent of the Hamiltonian operator (and the other generators) of the other subsystem, but does not necessarily imply that each subsystem evolves independently of existing correlations with the other subsystem.

Regarding the time evolution of correlations, in [11] we defined the correlation functional between two subsystems A and B,

$$\sigma_{AB}(\rho) = \mathrm{Tr}(\rho\ln\rho) - \mathrm{Tr}_A(\rho_A\ln\rho_A) - \mathrm{Tr}_B(\rho_B\ln\rho_B)\ , \tag{132}$$

which is nonnegative definite in general, and zero only if $\rho = \rho_A \otimes \rho_B$. The rate of change of the correlation can be written as

$$\frac{\mathrm{d}\sigma_{AB}(\rho)}{\mathrm{d}t} = \dot{\sigma}_{AB}|_H - \dot{\sigma}_{AB}|_D \ . \tag{133}$$

Based on our understanding of the equation of motion, we conjectured that $\dot{\sigma}_{AB}|_D$ should always be nonnegative, if it is true that the irreversible term can only destroy correlations between subsystems, but cannot create them. However, this conjecture remains to be proved. The corresponding entropy inequality to our knowledge has not yet been studied.

12. Conclusions

Based on unpublished geometrical reasoning [10], on the ansatz that a broader quantum kinematics may unite mechanics and thermodynamics [7, 19], and on the ansatz of a maximal entropy generation quantum dynamics may entail and extend the traditional Schrödinger–von Neumann unitary dynamics, we rederived most previous results in [10, 11, 13, 14, 15, 16] on the well-behaved nonlinear equation of motion proposed by the present author in his 1981 doctoral thesis for quantum thermodynamics, which entails the second law and nonequilibrium steepest-entropy-ascent dynamics with Onsager's reciprocity and Callen's fluctuation-dissipation relations.

Together with the variational principle formulation derived in [3] and the observation in [7, 9] that only the functional $-k_B \mathrm{Tr}(\rho \ln \rho)$ can represent the physical entropy, we may conclude that all the results so far confirm that our equation of motion has all the necessary features to provide a self-consistent and conceptually-sound resolution of the century-old dilemma on the nature of entropy and irreversibility, alternative to Boltzmann's statistical approach and valid also for systems with few degrees of freedom.

The nonlinear dynamics encompasses within a unified framework all the successful results of quantum mechanics, equilibrium and nonequilibrium thermodynamics. It also holds the promise to provide a fundamental framework within which to address the currently unexplained evidence on loss of quantum coherence, to design new fundamental experiments, to examine new applications on the lines of those developed in [3, 15, 20], and to further investigate the dependence of the internal-relaxation-time functional on the state operator and physical constants, as well as possibly verify the ansatz proposed in this paper by which each indivisible subsystem follows the direction of steepest perceived entropy ascent at the highest rate compatible with the time–energy uncertainty principle.

The equation of motion satisfies the set of conditions discussed in Appendix A and, therefore, preserves most of the traditional conceptual keystones of physical thought, including a strongest form of the non-relativistic principle of causality, by which future states of a strictly isolated system should unfold deterministically from initial states along smooth unique trajectories in state domain defined for all times (future as well as past [38]). Interestingly, while the maximum entropy states are attractors in forward time, the unitary-solutions boundary limit cycles of standard quantum mechanics are attractors in backward time.

As pointed out by Onsager and Machlup [23], the fluctuation-dissipation relations cannot be derived in any rigorous way from the traditional Hamiltonian dynamical principles, unless these are complemented by some additional postulate closely related to the additional principles, assumptions, or approximations needed to derive the Onsager reciprocity relations. This is sometimes referred to as the irreversibility paradox. In other words, in order to infer any feature of irreversibility (including its very existence) from the irreducibly reversible dynamical principle of standard Hamiltonian mechanics, we must complement it with some additional postulate that seems to contradict it.

Within our nonlinear quantum (thermo)dynamics based on (1) and (2) the paradox is resolved. The augmented state domain ansatz broadens the set of conceivable states but includes the standard pure states, and the nonlinear equation of motion describes irreversibile time evolutions and entails reciprocity and fluctuation-dissipation relations, but maintains the standard unitary dynamics of pure states. In fact, the proof emerges not from the Hamiltonian term of the equation of motion but from the irreversible, maximum entropy production term and is, thus, accomplished without the aid of the usual empirical and heuristic assumptions (microscopic reversibility and small deviations from stable equilibrium).

Finally, from the Heisenberg time–energy uncertainty principle we derived a plausible lower bound for the internal-relaxation-time functionals from which follows an upper bound for the rate of entropy generation. Consequently, we proposed a rather extreme but physically intriguing maximal-entropy-generation-rate ansatz, by which each indivisible subsystem follows the direction of steepest perceived entropy ascent at the highest rate compatible with the time–energy uncertainty principle. If this ansatz is experimentally verified, the nonlinear dynamics is complete and self-consistent, with no need of new physical constants.

Subject, of course, to experimental and further theoretical validation, this extreme ansatz represents a possible closure of the original nonlinear dynamics with no need of new physical constants. Physically, it would imply for example that energy is spontaneously redistributed among the initially occupied energy levels of the system, at the fastest rate compatible with (a) the time–energy uncertainty limitations, (b) the system structure, and (c) the energy exchange rates among subsystems (as driven by the usual interaction terms in the Hamiltonian operator through the usual unitary-dynamics term in the equation of motion).

Acknowledgements

The author is grateful to Prof. Michael von Spakovsky and Prof. Dieter Schuch for suggestions and encouragement. The work was initially supported in part by a grant from the Italian PRIN04 and was completed under the Cariplo–UniBS–MIT-MechE faculty exchange program co-sponsored by the Brescia University, the Department of Mechanical Engineering of the Massachusetts Institute of Technology, and the CARIPLO Foundation, Italy under grant 2008-2290.

Appendix A. Criteria for a general (nonlinear) quantum dynamics compatible with thermodynamics

Within a quantum theory that accepts the augmented set of true quantum states described by state operators ρ without the restriction $\rho^2 = \rho$, and a nonlinear dynamical law for a strictly isolated system, the following demanding set of conditions should be satisfied in order for the theory to be compatible or, better, imply the second law of thermodynamics without contradicting the fundamental results of standard quantum mechanics (QM). Obviuosly these are the criteria we followed in designing (1) and (2), and discussed at length in [13, 18].

Conditions 6, 7 and 8 are closely related to the condition recently referred to as strong separability [2]. We also added a condition on correlations and locality to reflect the need to avoid, and possibly resolve, physical inconsistencies related to nonlocality issues, as well as a strong causality condition that is nontrivial and quite demanding both from the conceptual and the technical mathematical points of view.

Certainly, when viewed from different perspectives — e.g., different physical interpretations of the augmented state domain $\rho^2 \neq \rho$ ansatz, of the role of the nonlinear extension of the Schrödinger equation of motion, of the Shannon-von Neumann entropy functional $-k_{\rm B}{\rm Tr}(\rho\ln\rho)$ versus other nonextensive functionals, of the role of the system's environment and the measuring apparati, and so on — some authors might view this set of conditions as too strong in many respects. Nevertheless, our equation of motion demonstrates that at least a satisfactory dynamics exists which satisfies all such conditions and, in our view, features a number of intriguing, unifying and far-reaching implications.

1. Causality. Forward and backward in time

We consider the set $\mathcal{P}$ of all linear, hermitian, nonnegative-definite, unit-trace operators ρ on the standard QM Hilbert space $\mathcal{H}$ associated with the strictly isolated system[36]. Every solution of the equation of motion, i.e., every trajectory $u(t, \rho)$ passing at time $t = 0$ through state ρ in $\mathcal{P}$, should lie entirely in $\mathcal{P}$ for all times t, $-\infty < t < +\infty$.

2. Conservation of energy and number of particles

Since the system is isolated, the value of the energy functional $e(\rho) = \mathrm{Tr}(\rho H)$, where H is the standard QM Hamiltonian operator, must remain invariant along every trajectory. If $\mathcal{H}$ is the Fock space of an isolated system consisting of M types of elementary constituents (e.g., atoms and molecules if chemical and nuclear reactions are inhibited; or atomic nuclei and electrons for modelling chemical reactions) each with a number operator N_i ($[H, N_i] = 0$ and $[N_i, N_j] = 0$), then also the value of each number-of-constituents functional $n_i(\rho) = \mathrm{Tr}(\rho N_i)$ must remain invariant along every trajectory. Depending on the type of system, there may be other time-invariant functionals, e.g., the total momentum components $p_j(\rho) = \mathrm{Tr}(\rho P_j)$, with $j = x, y, z$, for a free particle (in which case Galileian invariance must also be verified, for $[H, P_j] = 0$ and $[P_i, P_j] = 0$). In what follows, we denote by $g_i(\rho) = \mathrm{Tr}(\rho G_i)$ the set of non-Hamiltonian time-invariant functionals, if any, with $[H, G_i] = 0$ and $[G_i, G_j] = 0$ (clearly, H and the G_i's have a common eigenbasis that we denote by $\{|\psi_\ell\rangle\}$).

3. Standard QM unitary evolution of pure ($\rho^2 = \rho$) states

The unitary time evolution of the states of QM according to the Schrödinger equation of motion must be compatible with the more general dynamical law. These trajectories, passing through any state ρ such that $\rho^2 = \rho$ and entirely contained in the state domain of Quantum Mechanics, must be solutions also of the more general dynamical law. In view of the fact that the states of QM are the extreme points of the augmented state domain $\mathcal{P}$, the trajectories of QM must be boundary solutions (limit cycles) of the dynamical law.

If the complete dynamics preserves the feature of uniqueness of solutions throughout the state domain $\mathcal{P}$, then pure states can only evolve according to the Schrödinger equation of motion and, therefore, $\hat{\mathbf{D}}_M(\rho, H, G_i, \ldots) = 0$ when $\rho^2 = \rho$. This feature may be responsible for hiding the presence of deviations from QM in experiments where the isolated system is prepared in a pure state. It also implies that no trajectory can enter or leave the state domain of QM. Thus, by continuity, there must be trajectories that approach indefinitely these boundary solutions (of course, this can only happen backward in time, as $t \to -\infty$, for otherwise the entropy of the isolated system would decrease in forward time).

4. Entropy nondecrease. Irreversibility

The principle of nondecrease of entropy for an isolated system must be satisfied, i.e., the rate of change of the entropy functional $-k_B\mathrm{Tr}(\rho \ln \rho)$ must be nonnegative along every trajectory, $-k_B\mathrm{Tr}[u(t, \rho) \ln u(t, \rho)] \geq -k_B\mathrm{Tr}(\rho \ln \rho)$.

5. Stability of the thermodynamic equilibrium states. Second law

A state operator ρ of the isolated system represents an equilibrium state if $\mathrm{d}\rho/\mathrm{d}t = 0$. For each given set of feasible values of the energy functional $e(\rho)$ and the number-of-particle functionals $n_i(\rho)$ (i.e., the functionals that must remain invariant according to Condition 2 above), among all the equilibrium states that the dynamical law may admit there must be one and only one which is globally stable (definition and discussion in Appendix B). This stable equilibrium state must represent the corresponding state of equilibrium thermodynamics and, therefore, must be of the form given by (32). All the other equilibrium states that the dynamical law may admit must not be globally stable.

6. Non-interacting subsystems. Separate energy conservation

For an isolated system composed of two distinguishable subsystems A and B with associated Hilbert spaces $\mathcal{H}^A$ and $\mathcal{H}^B$, so that the Hilbert space of the system is $\mathcal{H} = \mathcal{H}^A \otimes \mathcal{H}^B$, if the two subsystems are non-interacting, i.e., the Hamiltonian operator $H = H_A \otimes I_B + I_A \otimes H_B$, then the functionals $\mathrm{Tr}[(H_A \otimes I_B)\rho] = \mathrm{Tr}_A(H_A\rho_A)$ and $\mathrm{Tr}[(I_A \otimes H_B)\rho] = \mathrm{Tr}_B(H_B\rho_B)$ represent the energies of the two subsystems and must remain invariant along every trajectory, even if the states of A and B are correlated, i.e., even if $\rho \neq \rho_A \otimes \rho_B$. Of course, $\rho_A = \mathrm{Tr}_B(\rho)$, $\rho_B = \mathrm{Tr}_A(\rho)$, Tr_B denotes the partial trace over $\mathcal{H}^B$ and Tr_A the partial trace over $\mathcal{H}^A$.

7. Independent states. Weak separability. Separate entropy nondecrease

Two distinguishable subsystems A and B are in independent states if the state operator $\rho = \rho_A \otimes \rho_B$, so that the entropy operator $S = -k_B B \ln \rho = S_A \otimes I_B + I_A \otimes S_B = -k_B[B_A \ln \rho_A \otimes I_B + I_A \otimes B_B \ln \rho_B]$. For permanently non-interacting subsystems, every trajectory passing through a state in which the subsystems are in independent states must maintain the subsystems in independent states along the entire trajectory. When two uncorrelated systems do not interact with each other, each must evolve in time independently of the other.

In addition, if at some instant of time two subsystems A and B, not necessarily non-interacting, are in independent states, then the instantaneous rates of change of the subsystem's entropy functionals $-k_B\mathrm{Tr}(\rho_A \ln \rho_A)$ and $-k_B\mathrm{Tr}(\rho_B \ln \rho_B)$ must both be nondecreasing in time.

8. Correlations and locality. Strong separability

Two non-interacting subsystems A and B initially in correlated states (possibly due to a previous interaction that has then been turned off) should each proceed in time towards less correlated states or, at least, maintain the same level of quantum entanglement. The generation of quantum entanglement between interacting subsystems should emerge only through the Schrödinger-von Neumann term $-i[H,\rho]/\hbar$ of the equation of motion, whereas the other terms, that might entail loss of correlations between subsystems, must not be able to create them. This condition is perhaps too strong and its validity for our equation is still only conjectural. In any case, the dynamics should not generate locality problems, i.e., faster-than-light communication between noninteracting subsystems, even if in entangled or correlated states. In other words, when subsystem A is not interacting with subsystem B, it should never be possible to influence the local observables of A by acting only on the interactions within B, such as switching on and off parameters or measurement devices within B. This does not mean that existing correlations between A and B established by past interactions should have no influence whatsoever on the time evolution of the local observables of either A or B. In particular, we see no physical reason to request that two different states ρ and ρ' such that $\rho'_A = \rho_A$ should evolve in such a way that $\mathrm{d}\rho'_A/\mathrm{d}t = \mathrm{d}\rho_A/\mathrm{d}t$ whenever A is isolated (but not uncorrelated) from the rest of the overall system. For example, state ρ' could be the maximum entropy stable equilibrium state (and, therefore, $\rho' = \rho'_A \otimes \rho'_B$, $\mathrm{d}\rho'/\mathrm{d}t = 0$) whereas in state ρ subsystems A and B could be correlated and evolving in time towards the stable equilibrium state or ρ could even be a pure entangled state evolving along a unitary trajectory according to Condition 3 above, and therefore it would never reach stable equilibrium.

Appendix B. Lyapunov stability and thermodynamic stability

The condition concerning stability of the thermodynamic equilibrium states is extremely restrictive and requires further discussion.

In order to implement Condition 5 in Appendix A, we need to establish the relation between the notion of stability implied by the second law of thermodynamics [32, 33] and the mathematical concept of stability. An equilibrium state is stable, in the sense required by the second law, if it can be altered to a different state only by interactions that leave net effects in the state of the environment. We call this notion of stability *global stability*. The notion of stability according to Lyapunov is called *local stability*.

We denote the trajectories generated by the dynamical law on our state domain by $u(t,\rho)$, i.e., $u(t,\rho)$ denotes the state at time t along the trajectory that at time $t=0$ passes through state ρ. A state ρ_e is an equilibrium state if and only if $u(t,\rho_e) = \rho_e$ for all times t. An equilibrium state ρ_e is *locally stable (according to Lyapunov)* if and only if for every $\epsilon > 0$ there is a $\delta(\epsilon) > 0$ such that $d(\rho,\rho_e) < \delta(\epsilon)$ implies $d(u(t,\rho),\rho_e) < \epsilon$ for all $t > 0$ and every ρ, i.e., such that every trajectory that passes within the distance $\delta(\epsilon)$ from state ρ_e proceeds in time without ever exceeding the distance ϵ from ρ_e. Conversely, an equilibrium state ρ_e is unstable if and only if it is not locally stable, i.e., there is an $\epsilon > 0$ such that for every $\delta > 0$ there is a trajectory passing within distance δ from ρ_e and reaching at some later time farther than the distance ϵ from ρ_e.

The Lyapunov concept of instability of equilibrium is clearly equivalent to that of instability stated in thermodynamics according to which an equilibrium state is unstable if, upon experiencing a minute and short lived influence by some system in the environment (i.e., just enough to take it from state ρ_e to a neighbouring state at infinitesimal distance δ), proceeds from then on spontaneously to a sequence of entirely different states (i.e., farther than some finite distance ϵ).

It follows that the concept of stability in thermodynamics implies that of Lyapunov local stability. However, it is stronger because it also excludes the concept of *metastability*. Namely, the states of equilibrium thermodynamics are *global* stable equilibrium states in the sense that not only they are locally stable but they cannot be altered to entirely different states even by means of interactions which leave temporary but finite effects in the environment. Mathematically, the concept of metastability can be defined as follows. An equilibrium state ρ_e is *metastable* if and only if it is locally stable but there is an $\eta > 0$ and an $\epsilon > 0$ such that for every $\delta > 0$ there is a trajectory $u(t,\rho)$ passing at $t = 0$ between distance η and $\eta+\delta$ from ρ_e, $\eta < d(u(0,\rho),\rho_e) < \eta+\delta$, and reaching at some later time $t > 0$ a distance farther than $\eta+\epsilon$, $d(u(t,\rho),\rho_e) \geq \eta+\epsilon$. Thus, the concept of global stability implied by the second law is as follows. An equilibrium state ρ_e is *globally stable* if for every $\eta > 0$ and every $\epsilon > 0$ there is a $\delta(\epsilon,\eta) > 0$ such that every trajectory $u(t,\rho)$ with $\eta < d(u(0,\rho),\rho_e) < \eta+\delta(\epsilon,\eta)$, i.e., passing at time $t = 0$ between distance η and $\eta+\delta$ from ρ_e, remains within $d(u(t,\rho),\rho_e) < \eta+\epsilon$ for every $t > 0$, i.e., proceeds in time without ever exceeding the distance $\eta+\epsilon$.

The second law requires that for each set of values of the invariants $\mathrm{Tr}(\rho H)$ and $\mathrm{Tr}(\rho G_i)$ (as many G_i's as required by the structure of the system), and of the parameters embedded in the Hilbert space $\mathcal{H}$ and the Hamiltonian H describing the external forces (such as the size of a container), there is one and only one globally stable equilibrium state. Thus, the dynamical law may admit many equilibrium states that all share the same values of the invariants and the parameters, but among all these only one is globally stable, i.e., all the other equilibrium states are either unstable or metastable.

Interestingly, we may use this condition to show that a unitary (Hamiltonian) dynamical law would be inconsistent with the second-law stability requirement. A unitary dynamical law in the augmented kinematics would be expressed by an equation of motion $i\hbar\dot{\rho} = [H,\rho]$ with trajectories $u(t,\rho) = U(t)\rho U^{-1}(t)$ with $U(t) = \exp(-itH/\hbar)$. Such a dynamical law would admit as equilibrium states all the states ρ_e such that $\rho_e H = H\rho_e$. Of these states there are more than just one for each set of values of the invariants. With respect to the metric $d(\rho_1,\rho_2) = \mathrm{Tr}|\rho_1-\rho_2|$, it is easy to show that every trajectory $u(t,\rho)$ would be equidistant from any given equilibrium state ρ_e, i.e., $d(u(t,\rho),\rho_e) = d(u(0,\rho),\rho_e)$ for all t and all ρ. Therefore, all the equilibrium states would be globally stable and there would be more than just one for each set of values of the invariants, thus violating the second-law requirement.

The entropy functional $-k_{\mathrm{B}}\mathrm{Tr}(\rho\ln\rho)$ plays a useful role in proving the stability of the states of equilibrium thermodynamics [(32)] provided that the dynamical law guarantees that $-k_{\mathrm{B}}\mathrm{Tr}[u(t,\rho)\ln u(t,\rho)] \geq -k_{\mathrm{B}}\mathrm{Tr}(\rho\ln\rho)$ for every trajectory, i.e., provided Condition 4 above is satisfied. The proof of this is nontrivial and is given in [32] where, however, we also show that the entropy functional is not a Lyapunov function, even if, in a strict sense that depends on the continuity and the conditional stability of the states of equilibrium thermodynamics, it does provide a criterion for the stability of these states. Anyway, even if the entropy were a Lyapunov function, this would suffice only to guarantee the local stability of the states of equilibrium thermodynamics but not to guarantee, as required by the second law, the instability or metastability of all the other equilibrium states.

Appendix C. Characteristic times

Using (3), the rate of change of the mean functional $f(\rho) = \mathrm{Tr}(\rho F)$ may be written as

$$\frac{\mathrm{d}f(\rho)}{\mathrm{d}t} = \mathrm{Tr}\left(\frac{\mathrm{d}\rho}{\mathrm{d}t}F\right) = 2\,(\sqrt{\rho}F\,|E\,)\ . \tag{C.1}$$

For the Schrödinger–von Neumann evolution, the characteristic time of change of $f(\rho)$ may be defined as [39]

$$\frac{1}{\tau_{FH}^2} = \frac{[\mathrm{d}f(\rho)/\mathrm{d}t]^2}{\langle\Delta F\Delta F\rangle} = \frac{4\left(\sqrt{\rho}\Delta F\,|E_H\,\right)^2}{\langle\Delta F\Delta F\rangle}\ . \tag{C.2}$$

Because operators $\sqrt{\rho}\Delta F/\sqrt{\langle\Delta F\Delta F\rangle}$ are unit norm, in the sense that $(\sqrt{\rho}\Delta F|\sqrt{\rho}\Delta F)/\langle\Delta F\Delta F\rangle = 1$, it follows that the characteristic times τ_{FH} are bounded by the value attained for an operator F such that $\sqrt{\rho}\Delta F$ is in the same direction as E_H, i.e., such that

$$\frac{\sqrt{\rho}\Delta F}{\sqrt{\langle\Delta F\Delta F\rangle}} = \frac{E_H}{\sqrt{(E_H\,|E_H\,)}}\ . \tag{C.3}$$

Therefore,

$$\frac{1}{\tau_{FH}^2} \le 4\,(E_H\,|E_H\,) = \frac{1}{\tau_H^2}\ . \tag{C.4}$$

For this reason, in (14) we take τ_H equal to the lower bound of the τ_{FH}'s.

By analogy, but considering separately the contribution of each subsystem to the overall-system dynamics embedded in our nonlinear dynamics, we define the characteristic time of the irreversible part of the rate of change of the mean functional $f(\rho) = \mathrm{Tr}(\rho F)$ due to the J-th constituent subsystem as

$$\frac{1}{\tau_{FDJ}^2} = \frac{[\mathrm{D}f(\rho)/\mathrm{D}t]_J^2}{\langle\Delta F\Delta F\rangle^J} = \frac{4\left(\sqrt{\rho_J}(\Delta F)^J\,|E_{DJ}\,\right)^2}{\langle\Delta F\Delta F\rangle^J}\ . \tag{C.5}$$

Again, because operators $\sqrt{\rho_J}(\Delta F)^J/\sqrt{\langle\Delta F\Delta F\rangle^J}$ are unit norm, in the sense that $(\sqrt{\rho_J}(\Delta F)^J|\sqrt{\rho_J}(\Delta F)^J)/\langle\Delta F\Delta F\rangle^J = 1$, it follows that the characteristic times τ_{FDJ} are bounded by the value attained for an operator F such that $\sqrt{\rho_J}(\Delta F)^J$ is in the same direction as E_{DJ}, i.e., such that

$$\frac{\sqrt{\rho_J}(\Delta F)^J}{\sqrt{\langle\Delta F\Delta F\rangle^J}} = \frac{E_{DJ}}{\sqrt{(E_{DJ}\,|E_{DJ}\,)}}\ . \tag{C.6}$$

Therefore,

$$\frac{1}{\tau_{FDJ}^2} \le 4\,(E_{DJ}\,|E_{DJ}\,) = \frac{1}{\tau_{DJ}^2}\ . \tag{C.7}$$

For this reason, in (62) and (123) we take τ_D and τ_{DJ} equal to the respective lower bounds of the τ_{FDJ}'s.

Appendix D. Special form of the equation of motion

If, for a given state operator ρ, we construct the set $\{\sqrt{\rho}X_j\}$ so as to be an orthogonal extension of the orthonormal subset $\{\sqrt{\rho}A_i\}$, i.e., with $X_i = A_i$ for $i \le a$ and $(\sqrt{\rho}X_i|\sqrt{\rho}X_j) = \delta_{ij}$ for all i and j, then $D = -\sum_{j>a} f_j \sqrt{\rho}X_j$ [(45)] and the one-constituent equation of motion reduces to the (only apparently linear) form

$$\frac{\mathrm{d}\rho}{\mathrm{d}t} = -\frac{i}{\hbar}[H,\rho] + \frac{1}{2\tau(\rho)}\sum_{j>a} f_j(\rho)\,\{X_j(\rho),\rho\}\ , \tag{D.1}$$

where the dependences of f_j, X_j and τ on ρ are evidenced in order to emphasize the nonlinearity.

For such special choice of the X_j's, assuming $X_1 = I$, we have $x_1(\rho) = \mathrm{Tr}(\rho) = 1$, $x_{i\neq 1}(\rho) = \mathrm{Tr}(\rho X_{i\neq 1}) = 0$, $\langle\Delta X_1 \Delta X_i\rangle = 0$, $\langle \Delta X_{i\neq 1}\Delta X_{j\neq 1}\rangle = \delta_{ij}$, $L_{ij}(\rho) = 0$ and $\mathrm{D}x_i(\rho)/\mathrm{D}t = 0$ for i or $j \le a$, $L_{ij}(\rho) = \delta_{ij}/\tau(\rho)$ and $\mathrm{D}x_i(\rho)/\mathrm{D}t = f_i(\rho)/\tau(\rho)$ for i and $j > a$, so that

$$\frac{\mathrm{d}s(\rho)}{\mathrm{d}t} = \frac{k_\mathrm{B}}{\tau(\rho)}\sum_{k>a} f_k(\rho)^2\ , \tag{D.2}$$

$$\tau(\rho)|_{\min} = \frac{\hbar}{2\langle\Delta H \Delta H\rangle}\sqrt{\sum_{k>a} f_k(\rho)^2}\ , \tag{D.3}$$

$$\left.\frac{\mathrm{d}s(\rho)}{\mathrm{d}t}\right|_{\max} = \frac{2k_\mathrm{B}}{\hbar}\sqrt{\langle\Delta H\Delta H\rangle}\sqrt{\sum_{k>a} f_k(\rho)^2}\ , \tag{D.4}$$

where in the last two equations we made use of (26), (65) and (66) with strict equality.

In view of the dependence of the X_j's on ρ and, therefore, on time, the apparently simple form of (D.1) may not be as useful as it seems.

Appendix E. Extension to time-varying rates of the generators of the motion

For the purposes of quantum thermodynamics, in our view the irreversile part of the equation of motion should not account for rates of change of the mean values of the generators of the motion other than through the Hamiltonian term, consistently with all the results of standard QM.

Nevertheless, the mathematical extension of (1) and (2) to (artificially) imposed rates $\dot{r}_i \neq 0$ (arbitrarily specified as functions of time) is straightforward and may be useful in modeling applications or other frameworks [6, 17]. It does emerge naturally from the maximization problem (47); it does so, however, implicitly, through substitution of (70) back into the constraint Eqs. (68b).

The explicit form in terms of projections and, therefore, the equivalent expressions by means of Gram determinants amount to assuming, for the operator D in the equation of motion, instead of (1c) or the equivalent (21),

$$D = [\sqrt{\rho}\ln\rho]_{\perp\mathcal{L}\{\sqrt{\rho}R_i\}} + \sum_j \alpha_j[\sqrt{\rho}R_j]_{\perp\mathcal{L}\{\sqrt{\rho}\ln\rho,\sqrt{\rho}R_{i\neq j}\}}\ , \tag{E.1}$$

where the explicit expression for the j-th term in the summation is obtained from the ratio of determinants in (21c) by interchanging everywhere $\sqrt{\rho}\ln\rho$ with $\sqrt{\rho}R_j$ so that

$$\alpha_j = \tau(\rho)\,\dot{r}_j\,\frac{\Gamma(\sqrt{\rho}\ln\rho,\{\sqrt{\rho}R_{i\neq j}\})}{\Gamma(\sqrt{\rho}\ln\rho,\{\sqrt{\rho}R_i\})}\ . \tag{E.2}$$

Geometrically, the additional terms are in the steepest-r_j-ascent direction compatible with maintaining constant the mean values of the other generators and the entropy.

Analogous obvious extension to the composite system case can be obtained by assuming for the operator D_J, instead of (2c),

$$D_J = [\sqrt{\rho_J}(B\ln\rho)^J]_{\perp\mathcal{L}\{\sqrt{\rho_J}(R_i)^J\}} + \sum_j \alpha_j^J [\sqrt{\rho_J}(R_j)^J]_{\perp\mathcal{L}\{\sqrt{\rho_J}(B\ln\rho)^J, \sqrt{\rho_J}(R_{i\neq j})^J\}} , \tag{E.3}$$

with the α_j^J's such that

$$\dot{r}_j = \sum_{J=1}^{M} \frac{\alpha_j^J}{\tau_J(\rho)} \frac{\Gamma(\sqrt{\rho_J}(B\ln\rho)^J, \{\sqrt{\rho_J}(R_i)^J\})}{\Gamma(\sqrt{\rho_J}(B\ln\rho)^J, \{\sqrt{\rho_J}(R_{i\neq j})^J\})} . \tag{E.4}$$

Appendix F. Extension to other entropy or mean value functionals

For the purposes of quantum thermodynamics, in our view the necessary entropy functional is $-k_B \mathrm{Tr}(\rho\ln\rho)$, for the reasons given in [7, 9, 11, 13, 16, 18, 24].

However, in view of the recent literature on nonextensive quantum theories, as suggested also in [3], it may be useful to note that the entire formalism of our equation of motion can be readily reformulated in the case of any other well-behaved entropy functional $s(\rho)$ and set of nonlinear generator functionals $r_i(\rho)$.

For the single constituent system it suffices to substitute throughout the operator $(-2k_B\sqrt{\rho}I) + 2\sqrt{\rho}S = (-2k_B\sqrt{\rho}) - 2k_B\sqrt{\rho}\ln\rho$ (notice that in the Gram determinants the addenda in parantheses cancel out) with the new entropy gradient operator, $\partial s(\rho)/\partial\sqrt{\rho}$, and the operators $2\sqrt{\rho}R_i$ with [operators which when symmetrized ($\{A, A^\dagger\}/2$) are equal to] the gradient operators of the new generator functionals, $\partial r_i(\rho)/\partial\sqrt{\rho}$.

However, for a composite system, consistently with the nonextensivity of these theories, it may be difficult or not at all possible to identify the operators corresponding to $(S)^J$ and $(R_i)^J$ representing the subsystems' local perceptions of the entropy and the generators.

Appendix G. A noteworthy equation for a composite system that fails to meet a separability condition

It is interesting to note that the role of the $\sqrt{\rho_J}$ operators in (2) [and that of $\sqrt{\rho}$ in (1)] is formally useful but only auxiliary, because wherever there is a $\sqrt{\rho_J}$, another $\sqrt{\rho_J}$ comes in front or behind it. In fact, in [10] the equation of motion was in terms of ρ_J only. We realized the usefulness of the $\sqrt{\rho_J}$ formalism only later, in connection with the proof of the steepest-entropy-ascent geometric property [13].

By allowing a more relevant role of the $\sqrt{\rho}$ operator [30], an alternative to the construction in (108) might appear formally better, and suggest the apparently alternative equation of motion based on the following definitions

$$E_D = -\sum_{J=1}^{M} \frac{1}{2\tau_J(\rho)} D'_J \otimes \sqrt{\rho_{\overline{J}}} , \tag{G.1}$$

$$(\sqrt{\rho}F)^J_{\surd} = \mathrm{Tr}_{\overline{J}}[(I_J \otimes \sqrt{\rho_{\overline{J}}})\sqrt{\rho}F] , \tag{G.2}$$

$$D'_J = [(\sqrt{\rho}\ln\rho)^J_{\surd}]_{\perp\mathcal{L}\{(\sqrt{\rho})^J_{\surd}, (\sqrt{\rho}H)^J_{\surd}, [(\sqrt{\rho}G_i)^J_{\surd}]\}} . \tag{G.3}$$

In fact, it can be readily verified that for a given F on $\mathcal{H}$ with $[F, H] = 0$ the rate of change of $\mathrm{Tr}(\rho F)$ would be zero if and only if $(\sqrt{\rho}F)^J_{\surd}$ is in $\mathcal{L}\{(\sqrt{\rho})^J_{\surd}, (\sqrt{\rho}H)^J_{\surd}, [(\sqrt{\rho}G_i)^J_{\surd}]\}$, and so $\mathrm{Tr}(\rho)$, $\mathrm{Tr}(\rho H)$, and $\mathrm{Tr}(\rho G_i)$ would be conserved. The expressions for the rate of entropy generation, the Onsager relations and the other results would be almost identical to those obtained from our equation, except for the substitution throughout of $\sqrt{\rho_J}(F)^J$ with $(\sqrt{\rho}F)^J_{\surd}$.

However, the resulting dynamics would fail to satisfy at least the important property expressed by (78), because it can be verified that if $H = H_J \otimes I_{\overline{J}} + I_J \otimes H_{\overline{J}}$ but $\rho \neq \rho_J \otimes \rho_{\overline{J}}$ then separate conservation of the non-interacting subsystem's energy would not be guaranteed.

References

[1] Domokos G and Kovesi-Domokos S 1999 *J. Phys.* A **32** 4105
[2] Czachor M 1998 *Phys. Rev.* A **57** 4122; Czachor M and Kuna M 1998 *Phys. Rev.* A **58** 128; Czachor M and Naudts J 1999 *Phys. Rev.* E **59** R2497
[3] Gheorghiu-Svirschevski S 2001 *Phys. Rev.* A **63** 022105 and the Addendum, Gheorghiu-Svirschevski S 2001 *Phys. Rev.* A **63** 054102
[4] For an account of the vaste literature on these applications see the following papers and references therein: Weinberg S 1989 *Phys. Rev. Lett.* **62** 485; Stern A, Aharonov Y and Imry Y 1990 *Phys. Rev.* A **41** 3436; Ekert A K 1991 *Phys. Rev. Lett.* **67** 661; Holyst J A and Turski L A 1992 *Phys. Rev.* A **45** 6180; Vidal G and Werner R F 1993 *Phys. Rev.* A **65** 032314; Unruh W G and Wald R M 1995 *Phys. Rev.* D **52** 2176; Bennett C H et al 1996 *Phys. Rev. Lett.* **76** 722; Grigorescu M 1998 *Physica* A **256** 149; Miranowicz A, Matsueda H and Wahiddin M R B 2000 *J. Phys.* A **33** 5159
[5] Englman R 2002 appendix in Lemanska M and Jaeger Z 2002 *Physica* D **170** 72
[6] Aerts D et al 2003 *Phys. Rev.* E **67** 051926
[7] Hatsopoulos G N and Gyftopoulos E P 1976 *Found. Phys.* **6** 15; Hatsopoulos G N and Gyftopoulos E P 1976 *Found. Phys.* **6** 127; Hatsopoulos G N and Gyftopoulos E P 1976 *Found. Phys.* **6** 439; Hatsopoulos G N and Gyftopoulos E P 1976 *Found. Phys.* **6** 561
[8] We use the term "revolutionary" in the sense introduced by Kuhn T S 1970 *The Structure of Scientific Revolutions* (Chicago: The University of Chicago Press)
[9] Gyftopoulos E P and Çubukçu E 1997 *Phys. Rev.* E **55** 3851
[10] Beretta G P 1981 *On the General Equation of Motion of Quantum Thermodynamics and the Distinction between Quantal and Nonquantal Uncertainties* (Cambridge, MA: Massachusetts Institute of Technology Sc.D. thesis, unpublished); Beretta G P 2005 *arXiv:quant-ph* 0509116
[11] Beretta G P, Gyftopoulos E P, Park J L and Hatsopoulos G N 1984 *Nuovo Cimento* B **82** 169; Beretta G P, Gyftopoulos E P and Park J L 1985 *Nuovo Cimento* B **87** 77
[12] Beretta G P 2009 *Rep. Math. Phys.* **64** 139
[13] Beretta G P 1986 *Frontiers of Nonequilibrium Statistical Physics (NATO Advanced Study Institute Series B: Physics vol 135) (Proc. June 1984 NATO ASI Conf. Santa Fe, NM, June 1984)* ed G T Moore and M O Scully (New York: Plenum Press) p 193 and p 205; Beretta G P 1986 *The Physics of Phase Space (Lecture Notes in Physics vol 278)* ed Y S Kim and W W Zachary (New York: Springer-Verlag) p 441
[14] Beretta G P 1985 *Int. J. Theor. Phys.* **24** 119
[15] Beretta G P 1985 *Int. J. Theor. Phys.* **24** 1233
[16] Beretta G P 1987 *Found. Phys.* **17** 365
[17] Beretta G P 1986 *Computer-Aided Engineering of Energy Systems (ASME Book H0341C)* ed R A Gaggioli (New York: ASME) p 129; Beretta G P 1987 *Second Law Analysis of Thermal Systems (ASME Book I00236)* ed M J Moran and E Sciubba (New York: ASME) p 17; Beretta G P 1987 *Second Law Analysis of Heat Transfer in Energy Systems (ASME Book G00390 HTD vol 80)* ed R F Boehm and N Lior (New York: ASME) p 31
[18] For interpretative discussions see also: Beretta G P 1987 *Proc. ANAE-ONERA-CERT Int. Symp. on Stability (Paris)* (Toulose: Centre d'Etudes et de Recherches), p 87; Beretta G P 1992 *Proc. 1991 Taormina Conf. on Thermodynamics,* edited by P V Giaquinta, M Vicentini Missoni and F Wanderlingh (Messina: Atti dell'Accademia Peloritana dei Pericolanti vol LXX-1) p 61
[19] Maddox J 1985 *Nature* **316** 11
[20] Korsch H J and Steffen H 1987 *J. Phys.* A **20** 3787; Hensel M and Korsch H J 1992 *J. Phys.* A **25** 2043
[21] Janussis A 1985 *Lett. Nuovo Cimento* **43** 309
[22] Onsager L 1931 *Phys. Rev.* **37** 405; Onsager L 1931 *Phys. Rev.* **38** 2265; Casimir H B G 1945 *Rev. Mod. Phys.* **17** 34; Kubo R 1957 *J. Phys. Soc. Jpn.* **12** 570
[23] Callen H B 1947 (Cambridge, MA: Massachusetts Institute of Technology Sc.D. thesis, unpublished); Callen H B, Barasch M L and Jackson J L 1952 *Phys. Rev.* **88** 1382; Callen H B and Welton T A 1951 *Phys. Rev.* **83** 34; Greene R F and Callen H B 1951 *Phys. Rev.* **83** 1231; Greene R F and Callen H B 1952 *Phys. Rev.* **88** 1387; Callen H B and Greene R F 1952 *Phys. Rev.* **86** 702; Onsager L and Machlup S 1953 *Phys. Rev.* **91** 1505.
[24] Gyftopoulos E P 2001 *J. Energy Resources Technology (ASME Transactions)* **123** 110
[25] Ellis J, Lopez J L, Mavromatos N E and Nanopoulos D V 1996 *Phys. Rev.* D **53** 3846 and references therein.
[26] See, e.g., Gallis M R 1991 *Phys. Rev.* A **45** 47; Reznik B 1996 *Phys. Rev. Lett.* **76** 1192; Grigorescu M 1998 *Physica* A **256** 149 and references therein.
[27] See, e.g., Kostin M D 1972 *J. Chem. Phys.* **57** 3589; Bialinicki-Birula I and Mycielski J 1976 *Ann. Phys. (N.Y.)* **100** 62; Hasse R W 1975 *J. Math. Phys.* **16** 2005; Gisin N 1981 *J. Phys. A: Math. Gen.* **14** 2259; Schuch D, Chung K-M and Hartmann H 1983 *J. Math. Phys.* **24** 1652; van Wonderen A J and Lendi K 2000 *J. Phys.* A **33** 5757; Tuszynski J A and Dixon J M 2001 *Physica* A **290** 69
[28] Öttinger H C 2001 *Phys. Rev.* E **62** 4720
[29] Equation (21d) is obtained from (21c) by subtracting from each raw (except the second) the second raw multiplied by second element of that raw. As a result, the second column entries become all zero except the second which is unity, and so it can be eliminated together with the second raw.
[30] The $\rho = \gamma\gamma^\dagger$ formalism proposed by Reznik [26] and adopted by Gheorghiu-Svirschevski [3], extends

our $\rho = (\sqrt{\rho})^2$ formalism but, as shown in the present paper, is not necessary to derive our equation. Conceptually, it is troublesome because it leaves unspecified the choice of γ for a given ρ, and risks to imply the existence of yet unidentified physical observables that would render the state operator insufficient. Formally, it may suggest the adoption of (G.1-G.3) for a composite system (properly extended to the complex square root formalism), which has many correct properties but fails to meet a necessary separability requirement (see Appendix G).

[31] See references [1] in [3].

[32] Beretta G P 1986 *J. Math. Phys.* **27** 305. The technical conjecture proposed therein was later found proved in Hiai F, Ohya M and Tsukada M 1981 *Pacific J. Math.* **96** 99

[33] This equivalence is well known and must be traced to Hatsopoulos G N and Keenan J H 1965 *Principles of General Thermodynamics* (New York: Wiley). It has been further clarified in Keenan J H, Hatsopoulos G N and Gyftopoulos E P 1972 *Principles of Thermodynamics* (Chicago: Encyclopaedia Britannica) and fully developed in Gyftopoulos E P and Beretta G P 2005 *Thermodynamics: Foundations and Applications* (Mineola: Dover) (first edition 1991 New York: Macmillan)

[34] See, e.g., Margenau H 1936 *Phys. Rev.* **49** 240; Margenau H 1963 *Philosophy of Science* **30** 6; Park J L 1968 *Philosophy of Science* **35** 3; Park J L 1968 *Am. J. Phys.* **36** 211; Gyftopoulos E P and von Spakovsky M R 2007 *Proc. ECOS07: Keynotes* (Padova: ASME) p 69

[35] This notation includes the special cases in which subsystem J consists of only, say, the 4-th type of particle (in which case N_{iJ} is the null operator for every $i \neq 4$) possibly with a fixed amount, say, 5 particles (in which case $N_{4J} = 5I_J$). Note that some N_{iJ} may be the null operator if subsystem J does not contain particles of type i. Even in this case the set of generators $\{R_{iJ}\}$ remains well-defined.

[36] By ***strictly isolated*** we mean that the system interacts with no other systems and at some time (and, hence, at all times) is in an independent state when viewed as a subsystem of any conceivable composite system containing it.

[37] As shown in [11], $\mathrm{Tr}(\rho \dot{B} \ln \rho) = 0$ by the definition of B.

[38] For rigorous proofs, the reader is referred to the original papers, where only some technical general aspects of the mathematics of the nonlinear equation for $\dim \mathcal{H} = \infty$ were left unresolved or based on technical conjectures, especially those related to the global stability of the thermodynamic equilibrium states, the instability or metastability of the other equilibrium states and all boundary solutions and limit cycles, and the existence and uniqueness in the general case of all solutions both forward and backwards in time, i.e., for $-\infty < t < \infty$. For a two-level system, instead, all these proofs are complete [14]. The work in [3] provides a further proof of preservation of positivity. That in [32] completes the proofs regarding stability.

[39] See, e.g., Messiah A 1976 *Quantum Mechanics* (New York: Wiley) p 320

Supersymmetry in nuclear physics

Roelof Bijker
Instituto de Ciencias Nucleares, Universidad Nacional Autónoma de México, A.P. 70-543, 04510 México, D.F., México

E-mail: bijker@nucleares.unam.mx

Abstract. I discuss several new developments in nuclear supersymmetry, in particular the identification of a new supersymmetric quartet of nuclei in the $A \sim 190$ region of the nuclear mass table consisting of the ^{192,193}Os and ^{193,194}Ir nuclei, and a study of correlations between different transfer reactions by means of generalized F-spin and $SU(3)$ isoscalar factors. The relevant $SU(3)$ isoscalar factors are derived in explicit form.

1. Introduction

Nuclear supersymmetry was proposed more than twenty five years ago [1] in the context of the interacting boson model (IBM) and the interacting boson-fermion model (IBFM) which have proved remarkably successful in providing a unified framework of even-even [2] and odd-even nuclei [3], respectively. One of its most attractive features is that it gives rise to a simple algebraic description, in which dynamical symmetries and supersymmetries play a central role, both as a way to improve our basic understanding of the importance of (super)symmetry in nuclear dynamics, and as a starting point for more precise calculations. Nuclear supersymmetry provides a theoretical framework in which different nuclei are treated as members of the same supermultiplet and whose spectroscopic properties are described by a single Hamiltonian and a single set of transition and transfer operators.

Aside from their esthetic appeal, (super)symmetries provide energy formula, selection rules and closed expressions for electromagnetic transition rates and transfer strengths which can be used as benchmarks to study and interpret the experimental data, even if these symmetries may be valid only approximately. Historically, symmetries have played an important role in nuclear physics. Examples are isospin symmetry, the Wigner supermultiplet theory, special solutions to the Bohr Hamiltonian, the Elliott model, pseudo-spin symmetries and the dynamical symmetries and supersymmetries of the IBM and its extensions.

Nuclear supersymmetry is a composite-particle phenomenon that should not be confused with fundamental supersymmetry, as used in particle physics and quantum field theory where it is postulated as a generalization of the Lorentz-Poincaré invariance as a fundamental symmetry of Nature and predicts the existence of supersymmetric particles, such as the photino and the selectron, for which experimental evidence is yet to be found. If experiments about to start at the LHC at CERN find evidence of supersymmetric particles, supersymmetry would be badly broken, as their masses must be much higher than those of their normal partners. In contrast to particle physics, nuclear supersymmetry has been verified experimentally.

Originally, nuclear supersymmetry was formulated as a symmetry among pairs of nuclei consisting of an even-even and an odd-even nucleus [1, 4, 5]. Subsequently, by including the neutron-proton degree of freedom it was extended to quartets of nuclei, in which an even-even, an odd-proton, an odd-neutron and an odd-odd nucleus form a supermultiplet [6]. A detailed study using state-of-the-art experimental techniques [7, 8, 9] showed that the odd-odd nucleus ^{196}Au, together with the odd-neutron nucleus ^{195}Pt, the odd-proton nucleus ^{195}Au and the even-even nucleus ^{194}Pt constitute an example of a supersymmetric quartet of nuclei [6, 10]. Actually, the interpretation of these four nuclei as members of a supersymmetric quartet made it possible to predict [6] the properties of ^{196}Au almost 15 years before they were measured experimentally [7, 8].

The purpose of this contribution is twofold. First, the concept of nuclear supersymmetry is reviewed briefly before discussing several new developments, in particular the identification of a new supersymmetric quartet in the $A \sim 190$ mass region, consisting of the ^{192,193}Os and ^{193,194}Ir nuclei, and the study of correlations between different one- and two-nucleon transfer reactions.

2. Nuclear supersymmetry

Supersymmetry was introduced [1] in nuclear physics in 1980 by Iachello in the context of the Interacting Boson Model (IBM) [2] and its extensions. The spectroscopy of atomic nuclei is characterized by the interplay between collective (bosonic) and single-particle (fermionic) degrees of freedom. The IBM describes collective excitations in even-even nuclei in terms of a system of interacting monopole and quadrupole bosons with angular momentum $l = 0, 2$. The bosons are associated with the number of correlated valence proton and neutron pairs, and hence the number of bosons N is half the number of valence nucleons. Since it is convenient to express the Hamiltonian and other operators of interest in second quantized form, one introduces creation, $s^\dagger$ and $d_m^\dagger$, and annihilation, s and d_m, operators, which altogether can be denoted by $b_i^\dagger$ and b_i with $i = l, m$ ($l = 0, 2$ and $-l \leq m \leq l$). The operators $b_i^\dagger$ and b_i satisfy the commutation relations

$$[b_i, b_j^\dagger] = \delta_{ij} , \qquad [b_i^\dagger, b_j^\dagger] = [b_i, b_j] = 0 . \tag{1}$$

The bilinear products

$$B_{ij} = b_i^\dagger b_j , \tag{2}$$

generate the algebra of $U(6)$ the unitary group in 6 dimensions

$$[B_{ij}, B_{kl}] = B_{il}\, \delta_{jk} - B_{kj}\, \delta_{il} . \tag{3}$$

The IBM Hamiltonian and other operators of interest are expressed in terms of the generators of $U(6)$. In general, the Hamiltonian has to be diagonalized numerically to obtain the energy eigenvalues and wave functions. There exist, however, special situations in which the eigenvalues can be obtained in closed, analytic form. These special solutions provide a framework in which energy spectra and other nuclear properties (such as quadrupole transitions and moments) can be interpreted in a qualitative way. These situations correspond to dynamical symmetries of the Hamiltonian [2]. The concept of dynamical symmetry has been shown to be a very useful tool in different branches of physics. A well-known example in nuclear physics is the Elliott $SU(3)$ model [11] to describe the properties of light nuclei in the *sd* shell. Another case is the $SU(3)$ flavor symmetry of Gell-Mann and Ne'eman [12] to classify the baryons and mesons into flavor octets, decuplets and singlets and to describe their masses with the Gell-Mann-Okubo mass formula.

A dynamical symmetry arises, when the Hamiltonian is expressed in terms of Casimir invariants of a chain of subgroups of $G = U(6)$, $G \supset G_1 \supset G_2 \supset \ldots$ only. The eigenstates can then be classified uniquely according to the irreducible representations of G and its subgroups G_1, G_2, ... The different representations of G, G_1, G_2 ... are split but not admixed by the Hamiltonian. The energy eigenvalues are given by the expectation values of the Casimir operators. In addition, by using standard group theoretical techniques it is possible to obtain analytic expressions for electromagnetic transition rates and quadrupole moments, etc.

For odd-mass nuclei the IBM has been extended to include single-particle degrees of freedom of the odd nucleon [3]. The Interacting Boson-Fermion Model (IBFM) has as its building blocks a set of N bosons with $l = 0, 2$ and $M = 1$ fermion (either a proton or a neutron) occupying single-particle orbits with angular momenta $j = j_1, j_2, \ldots$ The components of the fermion angular momenta span the Ω-dimensional space of the group $U(\Omega)$ with $\Omega = \sum_j (2j+1)$. We introduce, in addition to the boson operators for the collective degrees of freedom, fermion creation $a_i^\dagger$ and annihilation a_i operators for the extra nucleon. The fermion operators satisfy anti-commutation relations

$$\{a_i, a_j^\dagger\} = \delta_{ij} , \qquad \{a_i^\dagger, a_j^\dagger\} = \{a_i, a_j\} = 0 . \tag{4}$$

The bilinear products

$$A_{ij} = a_i^\dagger a_j , \tag{5}$$

generate the algebra of $U(m)$, the unitary group in m dimensions

$$[A_{ij}, A_{kl}] = A_{il}\,\delta_{jk} - A_{kj}\,\delta_{il} . \tag{6}$$

By construction, the fermion operators commute with the boson operators.

$$[B_{ij}, A_{kl}] = 0 . \tag{7}$$

The operators B_{ij} and A_{ij} generate the Lie algebra of the symmetry group $G = U^B(6) \otimes U^F(\Omega)$ of the IBFM. The dynamical symmetries that can arise in the IBFM are known under the name of dynamical boson-fermion symmetries for odd-mass nuclei.

Boson-fermion symmetries can further be extended by introducing the concept of supersymmetries [4], in which states in both even-even and odd-even nuclei are treated in a single framework. So far, I have discussed the symmetry properties of a mixed system of boson and fermion degrees of freedom for a fixed number of bosons N and one fermion $M = 1$. The operators B_{ij} and A_{ij} can only change bosons into bosons and fermions into fermions. In addition to B_{ij} and A_{ij}, one can introduce operators that change a boson into a fermion and *vice versa*, but conserve the total number of bosons and fermions

$$F_{ij} = b_i^\dagger a_j , \qquad G_{ij} = a_i^\dagger b_j . \tag{8}$$

The enlarged set of operators B_{ij}, A_{ij}, F_{ij} and G_{ij} forms a closed algebra which consists of both commutation and anticommutation relations

$$\begin{aligned}
[B_{ij}, B_{kl}] &= B_{il}\,\delta_{jk} - B_{kj}\,\delta_{il} , \\
[B_{ij}, A_{kl}] &= 0 , \\
[B_{ij}, F_{kl}] &= F_{il}\,\delta_{jk} , \\
[B_{ij}, G_{kl}] &= -G_{kj}\,\delta_{il} , \\
[A_{ij}, A_{kl}] &= A_{il}\,\delta_{jk} - A_{kj}\,\delta_{il} , \\
[A_{ij}, F_{kl}] &= -F_{kj}\,\delta_{il} , \\
[A_{ij}, G_{kl}] &= G_{il}\,\delta_{jk} , \\
\{F_{ij}, F_{kl}\} &= 0 , \\
\{F_{ij}, G_{kl}\} &= B_{il}\,\delta_{jk} + A_{kj}\,\delta_{il} , \\
\{G_{ij}, G_{kl}\} &= 0 .
\end{aligned} \tag{9}$$

Table 1. Overview of algebraic models

Model	Generators	Invariant	Symmetry
IBM	$b_i^\dagger b_j$	N	$U(6)$
IBFM	$b_i^\dagger b_j$, $a_\mu^\dagger a_\nu$	N, M	$U(6) \otimes U(\Omega)$
SUSY	$b_i^\dagger b_j$, $a_\mu^\dagger a_\nu$, $b_i^\dagger a_\mu$, $a_\mu^\dagger b_i$	$\mathcal{N}$	$U(6/\Omega)$

This algebra can be identified with that of the graded Lie group $G = U(6/\Omega)$. It provides an elegant scheme in which the IBM and IBFM can be unified into a single framework [4]

$$G = U(6/\Omega) \supset U^B(6) \otimes U^F(\Omega) \ . \tag{10}$$

In this supersymmetric framework, even-even and odd-mass nuclei form the members of a supermultiplet which is characterized by $[\mathcal{N}\}$ where $\mathcal{N} = N + M$, *i.e.* the total number of bosons and fermions. Thus, supersymmetry distinguishes itself from other symmetries in that it includes, in addition to transformations among fermions and among bosons, also transformations that change a boson into a fermion and *vice versa* (see Table 1).

The Hamiltonian of nuclear supersymmetry is written in terms of the generators of the graded Lie algebra of $U(6/\Omega)$ of Eq. (9). A dynamical supersymmetry arises when the Hamiltonian can be expressed in terms of Casimir operators of a chain of subgroups of $U(6/\Omega)$. Dynamical nuclear supersymmetries correspond to very special cases of the Hamiltonian which may not be applicable to all regions of the nuclear chart, but nevertheless several nuclei in the Os-Ir-Pt-Au region have been found to provide experimental evidence for the approximate occurrence of supersymmetries in nuclei.

2.1. $U(6/4)$ supersymmetry

The Os-Ir-Pt-Au mass region provides ample experimental evidence for the occurrence of dynamical (super)symmetries in nuclei. The even-even nuclei ^{194,196}Pt are the standard examples of the $SO(6)$ limit of the IBM [13] and the odd proton, in first approximation, occupies the single-particle level $2d_{3/2}$. In this special case, the boson and fermion groups, $SO^B(6)$ and $SU^F(4)$, are locally isomorphic and can be combined into the spinor group $Spin(6)$. The odd-proton nuclei ^{191,193}Ir and ^{193,195}Au were suggested as examples of the $Spin(6)$ limit of the IBFM [1, 14]. The appropriate extension to a supersymmetry is by means of the graded Lie group $U(6/4)$

$$U(6/4) \supset U^B(6) \otimes U^F(4) \supset SO^B(6) \otimes SU^F(4) \supset Spin(6) \supset Spin(5) \supset Spin(3) \ . \tag{11}$$

A dynamical supersymmetry arises when the Hamiltonian is expressed in terms of the Casimir invariants of the subgroups of $U(6/4)$

$$H = a\,\mathcal{C}_{2Spin(6)} + b\,\mathcal{C}_{2Spin(5)} + c\,\mathcal{C}_{2Spin(3)} \ . \tag{12}$$

The energy spectrum is given by the expectation value of the Casimir invariants of the spinor groups

$$E = a\left[\sigma_1(\sigma_1+4) + \sigma_2(\sigma_2+2) + \sigma_3^2\right] + b\left[\tau_1(\tau_1+3) + \tau_2(\tau_2+1)\right] + c\,J(J+1) \ , \tag{13}$$

which simultaneously describes the spectra of both the even-even and the odd-proton nucleus with a single set of parameters a, b and c. The pairs of nuclei ^{190}Os - ^{191}Ir, ^{192}Os - ^{193}Ir, ^{192}Pt - ^{193}Au and ^{194}Pt - ^{195}Au were analyzed as examples of a $U(6/4)$ supersymmetry with $\mathcal{N} = 9$, 8, 8 and 7, respectively [4].

2.2. $U(6/12)$ supersymmetry
Another example of a dynamical supersymmetry in this mass region is that of the Pt isotopes. The even-even nuclei are well described by the $SO(6)$ limit of the IBM and the odd neutron mainly occupies the negative parity orbits $3p_{1/2}$, $3p_{3/2}$ and $3f_{5/2}$ which can be decomposed into an pseudo-orbital part $k = 0, 2$ and a pseudo-spin part $s = 1/2$. Since the pseudo-orbital part of the odd neutron has the same values of angular momentum as the bosons, the boson and fermion chains can be combined into a single one. The corresponding graded Lie group is $U(6/12)$ which can be reduced into

$$\begin{aligned} U(6/12) &\supset U^B(6) \otimes U^F(12) \supset U^B(6) \otimes U^F(6) \otimes U^F(2) \\ &\supset U^{BF}(6) \otimes U^F(2) \supset SO^{BF}(6) \otimes U^F(2) \\ &\supset SO^{BF}(5) \otimes U^F(2) \supset SO^{BF}(3) \otimes SU^F(2) \\ &\supset Spin(3) \; . \end{aligned} \quad (14)$$

In this case, the Hamiltonian

$$H = \alpha\, \mathcal{C}_{2U^{BF}(6)} + \beta\, \mathcal{C}_{2SO^{BF}(6)} + \gamma\, \mathcal{C}_{2SO^{BF}(5)} + \delta\, \mathcal{C}_{2SO^{BF}(3)} + \epsilon\, \mathcal{C}_{2Spin(3)} \; , \quad (15)$$

simultaneously describes the excitation spectra of both the even-even and the odd-neutron nucleus with a single set of parameters α, β, γ, δ and ϵ. The energy spectrum is given by the eigenvalues of the Casimir operators

$$\begin{aligned} E &= \alpha\, [N_1(N_1+5) + N_2(N_2+3)] + \beta\, \left[\sigma_1(\sigma_1+4) + \sigma_2(\sigma_2+2) + \sigma_3^2\right] \\ &\quad +\gamma\, [\tau_1(\tau_1+3) + \tau_2(\tau_2+1)] + \delta\, L(L+1) + \epsilon\, J(J+1) \; . \end{aligned} \quad (16)$$

The pair of nuclei ^{194}Pt - ^{195}Pt was studied as an example of a $U(6/12)$ supersymmetry with $\mathcal{N} = 7$ [5, 15, 16, 17].

3. Supersymmetric quartets of nuclei
The concept of nuclear SUSY was extended in 1985 to include the neutron-proton degree of freedom [6]. In this case, a supermultiplet consists of an even-even, an odd-proton, an odd-neutron and an odd-odd nucleus. The first experimental evidence of a supersymmetric quartet was found in the $A \sim 190$ mass region in the ^{194,195}Pt and ^{195,196}Au nuclei as an example of the $U(6/12)_\nu \otimes U(6/4)_\pi$ supersymmetry [7, 8, 9, 18, 19], in which the odd neutron is allowed to occupy the $3p_{1/2}$, $3p_{3/2}$ and $2f_{5/2}$ orbits of the 82-126 shell, and the odd proton the $2d_{3/2}$ orbit of the 50-82 shell. The interpretation of these four nuclei as members of a supersymmetric quartet made it possible to predict [6] the properties of ^{196}Au almost 15 years before they were measured experimentally [7, 8].

The relevant subgroup chain of the $U(6/12)_\nu \otimes U(6/4)_\pi$ supersymmetry is a combination of the group reductions in Eqs. (11) and (14) (see Table 2). If the Hamiltonian is expressed in terms of Casimir invariants of the subgroups appearing in the reduction shown in Table 2, a dynamical supersymmetry arises. For example, the Hamiltonian

$$H = A\,\mathcal{C}_{2U^{BF_\nu}(6)} + B\,\mathcal{C}_{2SO^{BF_\nu}(6)} + B'\,\mathcal{C}_{2Spin(6)} + C\,\mathcal{C}_{2Spin(5)} + E\,\mathcal{C}_{2Spin(3)} + F\,\mathcal{C}_{2SU(2)} \; , \quad (17)$$

describes the excitation spectra of a quartet of nuclei characterized by $\mathcal{N}_\nu$ and $\mathcal{N}_\pi$ [6]. The quartet consists of an even-even nucleus with $N_\nu = \mathcal{N}_\nu$, $M_\nu = 0$ and $N_\pi = \mathcal{N}_\pi$, $M_\pi = 0$, an odd-proton nucleus with $N_\nu = \mathcal{N}_\nu$, $M_\nu = 0$ and $N_\pi = \mathcal{N}_\pi - 1$, $M_\pi = 1$, an odd-neutron nucleus with $N_\nu = \mathcal{N}_\nu - 1$, $M_\nu = 1$ and $N_\pi = \mathcal{N}_\pi$, $M_\pi = 0$, and an odd-odd nucleus with $N_\nu = \mathcal{N}_\nu - 1$, $M_\nu = 1$ and $N_\pi = \mathcal{N}_\pi - 1$, $M_\pi = 1$. Here, all terms that only contribute to binding energies were

Table 2. Group lattice for the $U(6/12)_\nu \otimes U(6/4)_\pi$ supersymmetry

$$\begin{array}{ccccccccc}
 & & & U(6/12)_\nu & & \otimes & & U(6/4)_\pi & \\
 & & & [\mathcal{N}_\nu\} & & & & [\mathcal{N}_\pi\} & \\
 & & & \downarrow & & & & \downarrow & \\
 & U^{F_\nu}(12) & & \otimes & U^{B_\nu}(6) & \otimes & U^{B_\pi}(6) & \otimes & U^{F_\pi}(4) \\
 & [1^{M_\nu}] & & & [N_\nu] & & [N_\pi] & & [1^{M_\pi}] \\
 & \downarrow & & & & \searrow \quad \swarrow & & & \\
U^{F_\nu}(2) & \otimes & U^{F_\nu}(6) & \otimes & & U^B(6) & & & \\
[n_1,n_2] & & [2^{n_2},1^{n_1-n_2}] & & & [N-i,i] & & & \\
 & & \searrow & & \swarrow & & & & \downarrow \\
 & & & U^{BF_\nu}(6) & & & & & \\
 & & & [N_1,\ldots,N_6] & & & & & \\
 & & & \downarrow & & & & & \\
 & & & SO^{BF_\nu}(6) & & \otimes & & & SU^{F_\pi}(4) \\
 & & & (\Sigma_1,\Sigma_2,\Sigma_3) & & & & & \\
\downarrow & & & & \searrow & & \swarrow & & \\
 & & & & & Spin(6) & & & \\
 & & & & & (\sigma_1,\sigma_2,\sigma_3) & & & \\
 & & & & & \downarrow & & & \\
 & & & & & Spin(5) & & & \\
 & & & & & (\tau_1,\tau_2) & & & \\
 & & & & & \downarrow & & & \\
SU^{F_\nu}(2) & & & \otimes & & Spin(3) & & & \\
S & & & & & J & & & \\
 & & \searrow & & \swarrow & & & & \\
 & & & SU(2) & & & & & \\
 & & & L & & & & &
\end{array}$$

neglected. The energy spectra of the four nuclei belonging to the supersymmetric quartet are described simultaneously by a single energy formula in terms of the eigenvalues of the Casimir operators

$$\begin{aligned}
E \;=\; & A\,[N_1(N_1+5)+N_2(N_2+3)+N_1(N_1+1)] \\
& +B\left[\Sigma_1(\Sigma_1+4)+\Sigma_2(\Sigma_2+2)+\Sigma_3^2\right] \\
& +B'\left[\sigma_1(\sigma_1+4)+\sigma_2(\sigma_2+2)+\sigma_3^2\right] \\
& +C\,[\tau_1(\tau_1+3)+\tau_2(\tau_2+1)] \\
& +D\,L(L+1)+E\,J(J+1)\;. \qquad (18)
\end{aligned}$$

The coefficients A, B, B', C, D, and E can be determined in a fit of the experimental excitation energies.

Table 3. The number of bosons and fermions in the supersymmetric quartets of Pt-Au and Os-Ir nuclei which are characterized by $\mathcal{N}_\pi = 2$, $\mathcal{N}_\nu = 5$ and $\mathcal{N}_\pi = 3$, $\mathcal{N}_\nu = 5$, respectively.

Nucleus	N_π	M_π	N_ν	M_ν	Nucleus	N_π	M_π	N_ν	M_ν
$^{194}_{78}\mathrm{Pt}_{116}$	2	0	5	0	$^{192}_{76}\mathrm{Os}_{116}$	3	0	5	0
$^{195}_{78}\mathrm{Pt}_{117}$	2	0	4	1	$^{193}_{76}\mathrm{Os}_{117}$	3	0	4	1
$^{195}_{79}\mathrm{Au}_{116}$	1	1	5	0	$^{193}_{77}\mathrm{Ir}_{116}$	2	1	5	0
$^{196}_{79}\mathrm{Au}_{117}$	1	1	4	1	$^{194}_{77}\mathrm{Ir}_{117}$	2	1	4	1

The number of bosons and fermions are related to the number of valence nucleons, *i.e.* the number of protons and neutrons outside the closed shells. For nuclei in the $A \sim 190$ mass region, the relevant closed shells are $Z = 82$ for protons and $N = 126$ for neutrons. The number of bosons and fermions for the ^{194,195}Pt and ^{195,196}Au quartet of nuclei are obtained as follows. For the even-even nucleus $^{194}_{78}\mathrm{Pt}_{116}$ the number of bosons are $N_\pi = (82 - 78)/2 = 2$ and $N_\nu = (126 - 116)/2 = 5$. There are no unpaired nucleons $M_\pi = M_\nu = 0$. For the odd-neutron nucleus $^{195}_{78}\mathrm{Pt}_{117}$ there are 9 valence neutrons which leads to $N_\nu = 4$ neutron bosons and $M_\nu = 1$ unpaired neutron. The $_{79}$Au isotopes have 3 valence protons which are divided over $N_\pi = 1$ proton boson and $M_\pi = 1$ unpaired proton. This supersymmetric quartet of nuclei is characterized by $\mathcal{N}_\pi = N_\pi + M_\pi = 2$ and $\mathcal{N}_\nu = N_\nu + M_\nu = 5$. The number of bosons and fermions are summarized in Table 3.

The odd-odd nucleus ^{196}Au, together with the odd-neutron nucleus ^{195}Pt, the odd-proton nucleus ^{195}Au and the even-even nucleus ^{194}Pt, have been verified experimentally using state-of-the-art techniques [7, 8, 9] to closely fulfill the rules that define a supersymmetric quartet [6, 10]. Figure 1 shows a comparison of the experimental spectrum of the odd-odd nucleus ^{196}Au with the theoretical spectrum for the $U(6/12)_\nu \otimes U(6/4)_\pi$ supersymmetry calculated with the energy formula of Eq. (18). The values of the parameters are given in Table 4.

The $A \sim 190$ mass region is a particularly complex one, displaying transitional behavior such as prolate-oblate deformed shapes, γ-unstability, triaxial deformation and/or coexistence

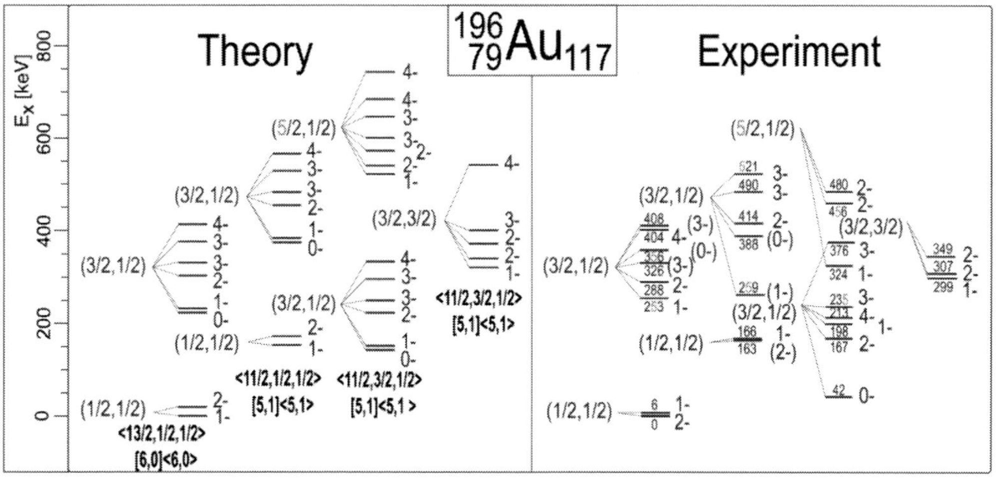

Figure 1. Comparison between the theoretical and experimental spectrum of ^{196}Au.

Table 4. Values of the parameters in keV

	A	B	B'	C	D	E	Ref.
Pt-Au	52.5	8.7	−53.9	48.8	8.8	4.5	[8]
Os-Ir	41.0	−6.0	−29.0	38.0	6.3	4.5	[20]

of different configurations which present a daunting challenge to nuclear structure models. Nevertheless, despite its complexity, this mass region has been a rich source of empirical evidence for the existence of dynamical symmetries in nuclei both for even-even, odd-proton, odd-neutron and odd-odd nuclei, as well as supersymmetric pairs [1, 5] and quartets of nuclei [6, 7].

Recently, the structure of the odd-odd nucleus ^{194}Ir was investigated by a series of transfer and neutron capture reactions [21]. The odd-odd nucleus ^{194}Ir differs from ^{196}Au by two protons, the number of neutrons being the same. The latter is crucial, since the dominant interaction between the odd neutron and the core nucleus is of quadrupole type, which arises from a more general interaction in the IBFM for very special values of the occupation probabilities of the $3p_{1/2}$, $3p_{3/2}$ and $2f_{5/2}$ orbits, *i.e.* to the location of the Fermi surface for the neutron orbits [22]. This situation is satisfied to a good approximation by the ^{195}Pt and ^{196}Au nuclei which both have the 117 neutrons. The same is expected to hold for the isotones ^{193}Os and ^{194}Ir. For this reason, it is reasonable to expect the odd-odd nucleus ^{194}Ir to provide another example of

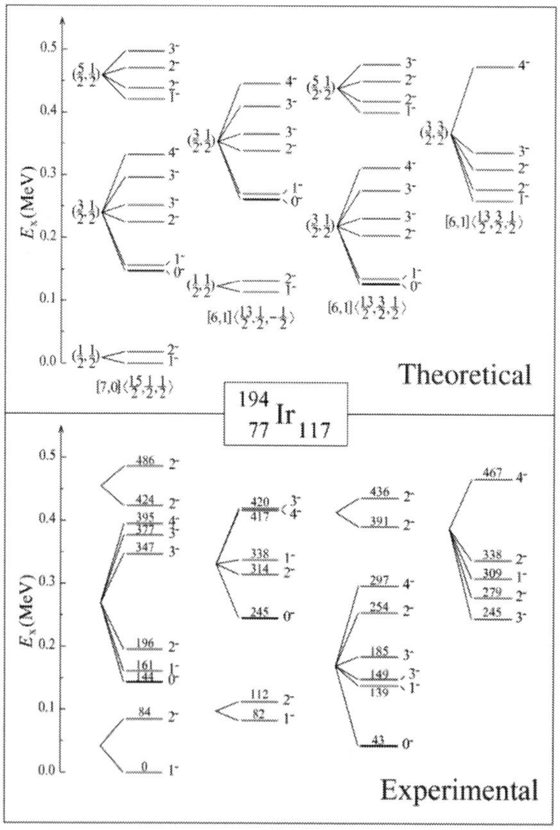

Figure 2. Comparison between the theoretical and experimental spectrum of ^{194}Ir.

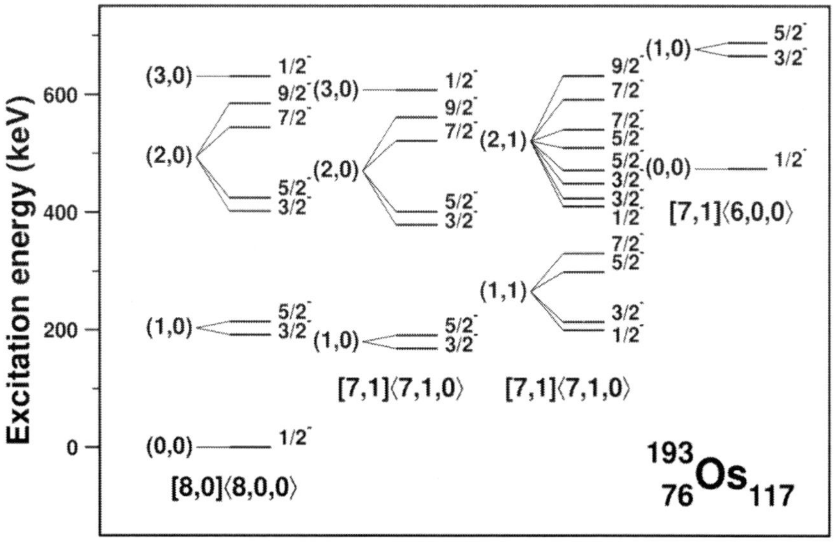

Figure 3. Prediction of the spectrum of ^{193}Os for the $U_\nu(6/12) \otimes U_\pi(6/4)$ supersymmetry.

a dynamical symmetry in odd-odd nuclei.

The new data from the polarized $(\vec{d},\alpha)$ transfer reaction provided crucial new information about and insight into the structure of the spectrum of ^{194}Ir which led to significant changes in the assignment of levels [21, 20] as compared to previous work [23]. Figure 2 shows the negative parity levels of ^{194}Ir in comparison with the theoretical spectrum in which it is assumed that these levels originate from the $\nu 3p_{1/2}$, $\nu 3p_{3/2}$, $\nu 2f_{5/2} \otimes \pi 2d_{3/2}$ configuration. The theoretical energy spectrum is calculated using the energy formula of Eq. (18) with the parameter values of Table 4. This parameter set is a lot closer to the parameter values used for ^{196}Au [8] than the ones in [23], indicating systematics in this zone of the nuclear chart. Given the complex nature of the spectrum of heavy odd-odd nuclei, the agreement is remarkable. There is an almost one-to-one correlation between the experimental and theoretical level schemes [21].

The successful description of the odd-odd nucleus ^{194}Ir opens the possibility of identifying a second quartet of nuclei in the $A \sim 190$ mass region with $U(6/12)_\nu \otimes U(6/4)_\pi$ supersymmetry. The new quartet consists of the nuclei ^{192,193}Os and ^{193,194}Ir and is characterized by $\mathcal{N}_\pi = 3$ and $\mathcal{N}_\nu = 5$ (see Table 3). Whereas the ^{192}Os and ^{193,194}Ir nuclei are well-known experimentally, the available data for ^{193}Os is rather scarce. Figure 3 shows the predicted spectrum for ^{193}Os obtained from Eq. (18) using the same parameter set as for ^{194}Ir [21]. The ground state of ^{193}Os has spin and parity $J^P = \frac{3}{2}^-$, which seems to imply that the second band with labels $[7,1]$, $(7,1,0)$ is the ground state band, rather than $[8,0]$, $(8,0,0)$. This ordering of bands is supported by preliminary results from the one-neutron transfer reaction ^{192}Os$(\vec{d},p)^{193}$Os [24].

An analysis of the energy spectra of the four nuclei that make up the quartet shows that the parameter set obtained in 1981 for the pair ^{192}Os-^{193}Ir [4] is very close to that of ^{194}Ir [21], which indicates that the nuclei ^{192,193}Os and ^{193,194}Ir may be interpreted in terms of a quartet of nuclei with $U(6/12)_\nu \otimes U(6/4)_\pi$ supersymmetry.

4. Correlations

The nuclei belonging to a supersymmetric quartet are described by a single Hamiltonian, and hence the wave functions, transition and transfer rates are strongly correlated. As an example of these correlations, I consider transfer reactions between the ^{194,195}Pt and ^{192,193}Os nuclei. The Pt and Os nuclei are connected by one-neutron transfer reactions within the same supersymmetric quartet ^{194}Pt $\leftrightarrow$ ^{195}Pt and ^{192}Os $\leftrightarrow$ ^{193}Os, whereas the transitions between the Pt and Os nuclei involve the transfer of a proton pair between different quartets ^{194}Pt $\leftrightarrow$ ^{192}Os and ^{195}Pt $\leftrightarrow$ ^{193}Os.

4.1. Generalized F-spin

The correlations between different transfer reactions can be derived in an elegant and explicit way by a generalization of the concept of F-spin which was introduced in the neutron-proton IBM [25] in order to distinguish between proton and neutron bosons.

The eigenstates of the $U(6/12)_\nu \otimes U(6/4)_\pi$ supersymmetry are characterized by the irreducible representations $[N_1, N_2, N_3]$ of $U^{BF_\nu}(6)$ which arise from the coupling of three different $U(6)$ representations, $[N_\nu]$ for the neutron bosons, $[N_\pi]$ for the proton bosons and $[N_\rho]$ for the pseudo-orbital angular momentum of the odd neutron ($N_\rho = 0$ for the even-even and odd-proton nucleus of the quartet, and $N_\rho = 1$ for the odd-neutron and the odd-odd nucleus). In analogy with the three quark flavors in the quark model (u, d and s), also in this case there are three different types of identical objects (π, ν and ρ), which can be distinguished by F-spin and hypercharge Y. The two kinds of bosons form an F-spin doublet, $F = \frac{1}{2}$, with charge states $F_z = \frac{1}{2}$ for protons (π) and $F_z = -\frac{1}{2}$ for neutrons (ν) [25]. In addition, the bosons carry hypercharge $Y = \frac{1}{3}$. The pseudo-orbital part (ρ) of the angular momentum of the odd neutron has $F = F_z = 0$ and $Y = -\frac{2}{3}$ [26].

Group theoretically, the generalized F-spin is defined by the reduction

$$\begin{array}{ccccc} U(18) & \supset & U(6) & \otimes & U(3) \\ \downarrow & & \downarrow & & \downarrow \\ [N] & & [N_1, N_2, N_3] & & [N_1, N_2, N_3] \end{array} \tag{19}$$

Here $U(6)$ is to be identified with the $U^{BF_\nu}(6)$ of the group reduction of Table 2 which is the result of first coupling the bosons at the level of $U(6)$, followed by coupling the orbital part

$$|[N_\nu], [N_\pi]; [N_\nu + N_\pi - i, i], [N_\rho]; [N_1, N_2, N_3]\rangle \ . \tag{20}$$

This sequence of $U(6)$ couplings can be described in a completely equivalent way by the three-dimensional index group $U(3)$ of Eq. (19) which can be reduced to

$$\begin{array}{ccccccccc} U(3) & \supset & SU(3) & \supset & [SU(2) & \supset & SO(2)] & \otimes & U(1) \\ \downarrow & & \downarrow & & \downarrow & & \downarrow & & \downarrow \\ [N_1, N_2, N_3] & & (\lambda, \mu) & & F & & F_z & & Y \end{array} \tag{21}$$

The relation between the two sets of quantum numbers is given by

$$\begin{aligned} (\lambda, \mu) &= (N_1 - N_2, N_2 - N_3) \ , \\ F &= \frac{1}{2}(N_\pi + N_\nu - 2i) \ , \\ F_z &= \frac{1}{2}(N_\pi - N_\nu) \ , \\ Y &= \frac{1}{3}(N_\pi + N_\nu - 2N_\rho) \ . \end{aligned} \tag{22}$$

As a result, matrix elements between states with the same quantum numbers but different $U(6)$ couplings are then related by $SU(3)$ isoscalar factors, and hence correlations between different transfer reactions can be derived in terms of these isoscalar factors by means of the concept of generalized F-spin. Some $SU(3)$ isoscalar factors relevant for nuclear supersymmetry are derived in Appendix A.

4.2. One-neutron transfer

In a study of the ^{194}Pt $\rightarrow$ ^{195}Pt stripping reaction it was found [15] that one-neutron $j = 3/2$, $5/2$ transfer reactions can be described by the operator

$$P_{\nu}^{(j)\dagger} \;=\; \frac{\alpha_j}{\sqrt{2}} \left[\left(\tilde{s}_{\nu} \times a^{\dagger}_{\nu,j} \right)^{(j)} - \left(\tilde{d}_{\nu} \times a^{\dagger}_{\nu,\frac{1}{2}} \right)^{(j)} \right] . \tag{23}$$

It is convenient to take ratios of intensities, since they do not depend on the value of the coefficient α_j and hence provide a direct test of the wave functions. For the stripping reaction ^{194}Pt $\rightarrow$ ^{195}Pt (ee $\rightarrow$ on) the ratio of intensities for the excitation of the $(\tau_1, \tau_2) = (1, 0)$, $L = 2$ doublet with $J = 3/2$, $5/2$ belonging to the first excited band with $[\mathcal{N} - 1, 1]$, $(\mathcal{N} - 1, 1, 0)$ relative to that of the ground state band $[\mathcal{N}]$, $(\mathcal{N}, 0, 0)$ is given by [15, 20]

$$R(\text{ee} \rightarrow \text{on}) = \frac{(\mathcal{N}-1)(\mathcal{N}+1)(\mathcal{N}+4)}{2(\mathcal{N}+2)} , \tag{24}$$

which gives $R = 29.3$ for ^{194}Pt $\rightarrow$ ^{195}Pt ($\mathcal{N} = \mathcal{N}_{\nu} + \mathcal{N}_{\pi} = 7$), to be compared to the experimental value of 19.0 for $j = 5/2$, and $R = 37.8$ for ^{192}Os $\rightarrow$ ^{193}Os ($\mathcal{N} = \mathcal{N}_{\nu} + \mathcal{N}_{\pi} = 8$). The equivalent ratio for the inverse pick-up reaction is given by

$$R(\text{on} \rightarrow \text{ee}) = R(\text{ee} \rightarrow \text{on}) \frac{\mathcal{N}_{\pi}}{(\mathcal{N}_{\nu} + \mathcal{N}_{\pi} - 1)\mathcal{N}_{\nu}} . \tag{25}$$

which gives $R = 1.96$ for ^{195}Pt $\rightarrow$ ^{194}Pt ($\mathcal{N}_{\pi} = 2$ and $\mathcal{N}_{\nu} = 5$) and $R = 3.24$ for ^{193}Os $\rightarrow$ ^{192}Os ($\mathcal{N}_{\pi} = 3$ and $\mathcal{N}_{\nu} = 5$). This means that the mixed symmetry $L = 2$ state is predicted to be excited more strongly than the first excited $L = 2$ state.

This correlation between pick-up and stripping reactions has been derived in a general way only using the symmetry relations that exist between the wave functions of the even-even and odd-neutron nuclei of the supersymmetric quartet. The factor in the right-hand side of Eq. (25) can be expressed in terms of $SU(3)$ isoscalar factors. It is important to emphasize, that Eqs. (24) and (25) are parameter-independent predictions which are a direct consequence of nuclear SUSY and which can be tested experimentally.

4.3. Two-proton transfer

The two supersymmetric quartets in the mass $A \sim 190$ region differ by two protons. In principle, the connection between the two quartets can be studied by two-proton transfer reactions. In the IBM, two-proton transfer operator is, in first order, given by

$$P_{\pi}^{\dagger} = \alpha\, s_{\pi}^{\dagger} , \qquad P_{\pi} = \alpha\, s_{\pi} . \tag{26}$$

Whereas the operator s_{π} only excites the ground state of the final nucleus, $s_{\pi}^{\dagger}$ can also populate excited states.

Table 5 shows the results for ratios of spectroscopic strengths between even-even nuclei. The selection rules of the operator $s_{\pi}^{\dagger}$ allow the excitation of states with with $(\tau_1, \tau_2) = (0, 0)$ and $L = 0$ belonging to the ground band $(\Sigma_1, \Sigma_2, \Sigma_3) = (\mathcal{N} + 1, 0, 0)$ and excited bands with $(\mathcal{N} - 1, 0, 0)$. The corresponding ratios for the odd-neutron nuclei are strongly correlated to those of the even-even nuclei (see Tables 5 and 6)

$$\begin{aligned} S_1 &= R_1 \\ S_2 &= R_2 , \\ S_{3a} &= R_3 \frac{\mathcal{N}_{\pi}+1}{\mathcal{N}_{\nu}\mathcal{N}} , \\ S_{3b} &= R_3 \frac{(\mathcal{N}_{\nu}-1)(\mathcal{N}+1)}{\mathcal{N}_{\nu}\mathcal{N}} . \end{aligned} \tag{27}$$

Table 5. Ratios of spectroscopic strengths for two-proton transfer reactions between even-even nuclei $R_{\mathrm{ee}}(\mathcal{N}_\pi, \mathcal{N}_\nu \to \mathcal{N}_\pi+1, \mathcal{N}_\nu)$ to final states with $(\tau_1, \tau_2) = (0,0)$. $\mathcal{N} = \mathcal{N}_\nu + \mathcal{N}_\pi$ refers to the initial nucleus ($\mathcal{N} = 7$ for ^{194}Pt $\to$ ^{192}Os).

n	$[N_1, N_2]$	$(\Sigma_1, \Sigma_2, \Sigma_3)$	R_n
1	$[\mathcal{N}+1, 0]$	$(\mathcal{N}+1, 0, 0)$	1
2	$[\mathcal{N}+1, 0]$	$(\mathcal{N}-1, 0, 0)$	$\frac{\mathcal{N}(\mathcal{N}+3)}{(\mathcal{N}+1)^2(\mathcal{N}+4)}$
3	$[\mathcal{N}, 1]$	$(\mathcal{N}-1, 0, 0)$	$\frac{\mathcal{N}_\nu(\mathcal{N}-1)(\mathcal{N}+2)}{(\mathcal{N}_\pi+1)(\mathcal{N}+1)^2(\mathcal{N}+4)}$

Table 6. Ratios of spectroscopic strengths for two-proton transfer reactions between odd-neutron nuclei $S_{\mathrm{on}}(\mathcal{N}_\pi, \mathcal{N}_\nu \to \mathcal{N}_\pi+1, \mathcal{N}_\nu)$ to final states with $(\tau_1, \tau_2) = (0,0)$. $\mathcal{N} = \mathcal{N}_\nu + \mathcal{N}_\pi$ refers to the initial nucleus ($\mathcal{N} = 7$ for ^{195}Pt $\to$ ^{193}Os).

n	$[\mathcal{N}-i, i]$	$[N_1, N_2]$	$(\Sigma_1, \Sigma_2, \Sigma_3)$	S_n
1	$[\mathcal{N}, 0]$	$[\mathcal{N}+1, 0]$	$(\mathcal{N}+1, 0, 0)$	1
2	$[\mathcal{N}, 0]$	$[\mathcal{N}+1, 0]$	$(\mathcal{N}-1, 0, 0)$	R_2
3a	$[\mathcal{N}, 0]$	$[\mathcal{N}, 1]$	$(\mathcal{N}-1, 0, 0)$	$R_3\frac{\mathcal{N}_\pi+1}{\mathcal{N}_\nu\mathcal{N}}$
3b	$[\mathcal{N}-1, 1]$	$[\mathcal{N}, 1]$	$(\mathcal{N}-1, 0, 0)$	$R_3\frac{(\mathcal{N}_\nu-1)(\mathcal{N}+1)}{\mathcal{N}_\nu\mathcal{N}}$

The coefficients in the right-hand side correspond to the ratio of two $SU(3)$ isoscalar factors. Moreover, the summed strength for even-even nuclei is the same as that for odd-neutron nuclei $\sum_n R_n = \sum_n S_n$.

5. Summary and conclusions

In this contribution, I discussed the concept of supersymmetry as used in nuclear physics. In this application, a supermultiplet consists of a pair or a quartet of nuclei whose wave functions are related by supersymmetry. This has important consequences for energies, transition and transfer strengths.

It was suggested that there exists a second quartet of nuclei in the $A \sim 190$ mass region with $U(6/12)_\nu \otimes U(6/4)_\pi$ supersymmetry, consisting of the ^{192,193}Os and ^{193,194}Ir nuclei. The analysis is based largely on new experimental information on ^{194}Ir. Given the complexity of the $A \sim 190$ mass region, the simple yet detailed description of ^{194}Ir in a supersymmetry scheme is remarkable.

Nuclear supersymmetry establishes precise links among the spectroscopic properties of different nuclei. This relation has been used to predict the energies of ^{193}Os. In order to establish the existence of a second supersymmetric quartet of nuclei in the $A \sim 190$ mass region, it is crucial that the nucleus ^{193}Os be studied in more detail experimentally. The predictions for correlations between one-neutron transfer reactions in Pt and Os can be tested experimentally by combining for example $(\vec{d}, p)$ stripping and (p, d) pick-up reactions.

Since the wave functions of the members of a supermultiplet are connected by symmetry, there exists a high degree of correlations between different one- and two-nucleon transfer reactions not only between nuclei belonging to the same quartet, but also for nuclei from different multiplets. In order to discuss these correlations induced by nuclear supersymmetry, the concept of generalized F-spin was developed. As a consequence, the correlations can be expressed in an elegant and explicit way in terms of $SU(3)$ isoscalar factors. It was shown that a special class of these isoscalar factors can be expressed in terms of the usual $SU(2)$ Clebsch-Gordan coefficients. As an example, the correlations between one-neutron transfer reactions and two-proton transfer reactions were studied. In the former case, nuclear supersymmetry predicts that the $L = 2$ mixed symmetry states in the even-even nuclei ^{194}Pt and ^{192}Os are excited much stronger (two to three times as strong) than the first excited $L = 2$ state.

It is important to note, that the technique of the generalized F-spin is valid for any system whose wave functions are described in terms of couplings of three different symmetric representations of $U(6)$, as for example in nuclear supersymmetry (this contribution), in an algebraic description of the neutron skin in very neutron-rich nuclei [27] and in an isospin-invariant extension of the IBM for light nuclei called IBM-3 [28].

Appendix A. SU(3) isoscalar factors

The isoscalar factors can be derived in the standard way from the recursion relations for the $SU(3)$ step operators combined with the phase convention of De Swart [29] and Racah's factorization lemma

$$\left\langle \begin{array}{cc|c} (\lambda_1,\mu_1) & (\lambda_2,\mu_2) & (\lambda,\mu) \\ F_1,F_{1z},Y_1 & F_2,F_{2z},Y_2 & F,F_z,Y \end{array} \right\rangle = \left\langle \begin{array}{cc|c} (\lambda_1,\mu_2) & (\lambda_2,\mu_2) & (\lambda,\mu) \\ F_1,Y_1 & F_2,Y_2 & F,Y \end{array} \right\rangle \langle F_1,F_{1z},F_2,F_{2z}|F,F_z\rangle \ . \tag{A.1}$$

Inspection of the wave functions in nuclear supersymmetry shows that the ground state wave functions of the nuclei that belong to a supersymmetric quartet of nuclei are characterized by the symmetric representation $[N_1, N_2, N_3] = [N, 0, 0]$ of $U(6)$ or, equivalently, by the $SU(3)$ labels $(\lambda, \mu) = (N, 0)$. For the examples discussed in this contribution, the relevant $SU(3)$ couplings are

$$\begin{aligned} (N,0)\otimes(1,0) &= (N+1,0)\oplus(N-1,1) \ , \\ (N,0)\otimes(0,1) &= (N,1)\oplus(N-1,0) \ . \end{aligned} \tag{A.2}$$

In the special case of the first coupling, the $SU(3)$ isoscalar factors can be derived in terms of $SU(2)$ Clebsch-Gordan coefficients. This can be seen by employing two different realizations of the reduction $SU(3) \supset SU(2) \otimes U(1)$. The one used in Eq. (21) gives rise to the basis $|(\lambda,\mu),F,F_z,Y_F\rangle$ (here I use $Y_F = Y$ to indicate that this is the hypercharge operator associated

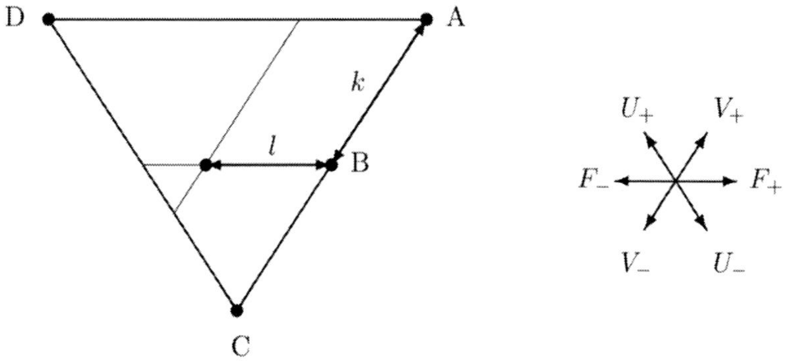

Figure A1. Weight diagram of the $SU(3)$ multiplet $(\lambda, 0)$

with F-spin). Another basis is provided by V-spin and the corresponding hypercharge Y_V which leads to $|(\lambda,\mu),V,V_z,Y_V\rangle$. The generators of the two realizations of $SU(2)\otimes U(1)$ are given by

$$\begin{array}{rclrcl} F_+ &=& \pi^\dagger\nu & V_+ &=& \pi^\dagger\rho \\ F_- &=& \nu^\dagger\pi & V_- &=& \rho^\dagger\pi \\ F_z &=& (\pi^\dagger\pi-\nu^\dagger\nu)/2 & V_z &=& (\pi^\dagger\pi-\rho^\dagger\rho)/2 \\ Y_F &=& (\pi^\dagger\pi+\nu^\dagger\nu-2\rho^\dagger\rho)/3 & Y_V &=& (\pi^\dagger\pi+\rho^\dagger\rho-2\nu^\dagger\nu)/3 \end{array} \tag{A.3}$$

The weights in the two realizations are related to one another by

$$\begin{aligned} V_z &= \frac{3}{4}Y_F+\frac{1}{2}F_z \;, \\ Y_V &= -\frac{1}{2}Y_F+F_z \;. \end{aligned} \tag{A.4}$$

For the symmetric representation $(\lambda,\mu)=(N,0)$, the relation between the two sets of basis states is one-to-one

$$\left|(N,0),\frac{N-k}{2},\frac{N-k}{2}-l,\frac{N}{3}-k\right\rangle_F = \left|(N,0),\frac{N-l}{2},\frac{N-l}{2}-k,\frac{N}{3}-l\right\rangle_V \;, \tag{A.5}$$

with $k=0,1,\ldots,N$ and $l=0,1,\ldots,N-k$.

Consider now the coupling of $(N,0)$ and $(1,0)$ to $(N+1,0)$. The state $|A\rangle$ in Figure A1 is characterized by

$$\begin{aligned} |\psi_A\rangle &= \left|(N+1,0),\frac{N+1}{2},\frac{N+1}{2},\frac{N+1}{3}\right\rangle_F = \left|(N,0),\frac{N}{2},\frac{N}{2},\frac{N}{3}\right\rangle_F \left|(1,0),\frac{1}{2},\frac{1}{2},\frac{1}{3}\right\rangle_F \\ &= \left|(N+1,0),\frac{N+1}{2},\frac{N+1}{2},\frac{N+1}{3}\right\rangle_V = \left|(N,0),\frac{N}{2},\frac{N}{2},\frac{N}{3}\right\rangle_V \left|(1,0),\frac{1}{2},\frac{1}{2},\frac{1}{3}\right\rangle_V \;, \end{aligned} \tag{A.6}$$

i.e. it has the same labels in either of the bases. Next, the state $|\psi_B\rangle$ in Figure A1 can be obtained by applying the operator V_- k times. In the V-basis, the expansion coefficients are the usual $SU(2)$ Clebsch-Gordan coefficients

$$\begin{aligned} |\psi_B\rangle &= \left|(N+1,0),\frac{N+1}{2},\frac{N+1}{2}-k,\frac{N+1}{3}\right\rangle_V \\ &= \left\langle\frac{N}{2},\frac{N}{2}-k,\frac{1}{2},\frac{1}{2}\right|\left.\frac{N+1}{2},\frac{N+1}{2}-k\right\rangle \\ &\qquad \left|(N,0),\frac{N}{2},\frac{N}{2}-k,\frac{N}{3}\right\rangle_V \left|(1,0),\frac{1}{2},\frac{1}{2},\frac{1}{3}\right\rangle_V \\ &+ \left\langle\frac{N}{2},\frac{N}{2}+1-k,\frac{1}{2},-\frac{1}{2}\right|\left.\frac{N+1}{2},\frac{N+1}{2}-k\right\rangle \\ &\qquad \left|(N,0),\frac{N}{2},\frac{N}{2}+1-k,\frac{N}{3}\right\rangle_V \left|(1,0),\frac{1}{2},-\frac{1}{2},\frac{1}{3}\right\rangle_V \;. \end{aligned} \tag{A.7}$$

Finally, if we rewrite this equation in the F-basis, it is easy to see that its expansion coefficients (or $SU(3)$ isoscalar factors) in the F-basis can be expressed in terms of $SU(2)$ Clebsch-Gordan coefficients as

$$\left\langle \begin{array}{cc} (N,0) & (1,0) \\ \frac{N-k}{2},\frac{N}{3}-k & \frac{1}{2},\frac{1}{3} \end{array} \right| \left. \begin{array}{c} (N+1,0) \\ \frac{N+1-k}{2},\frac{N+1}{3}-k \end{array} \right\rangle$$

$$= \left\langle \frac{N}{2}, \frac{N}{2} - k, \frac{1}{2}, \frac{1}{2} \middle| \frac{N+1}{2}, \frac{N+1}{2} - k \right\rangle = \sqrt{\frac{N+1-k}{N+1}},$$

$$\left\langle \begin{matrix} (N,0) \\ \frac{N+1-k}{2}, \frac{N}{3}+1-k \end{matrix} \begin{matrix} (1,0) \\ 0, -\frac{2}{3} \end{matrix} \middle| \begin{matrix} (N+1,0) \\ \frac{N+1-k}{2}, \frac{N+1}{3} - k \end{matrix} \right\rangle$$

$$= \left\langle \frac{N}{2}, \frac{N}{2} + 1 - k, \frac{1}{2}, -\frac{1}{2} \middle| \frac{N+1}{2}, \frac{N+1}{2} - k \right\rangle = \sqrt{\frac{k}{N+1}}. \quad \text{(A.8)}$$

The isoscalar factors for the states with $(\lambda, \mu) = (N-1, 1)$ are obtained by the orthogonality condition

$$\left\langle \begin{matrix} (N,0) \\ \frac{N-k}{2}, \frac{N}{3} - k \end{matrix} \begin{matrix} (1,0) \\ \frac{1}{2}, \frac{1}{3} \end{matrix} \middle| \begin{matrix} (N-1,1) \\ \frac{N+1-k}{2}, \frac{N+1}{3} - k \end{matrix} \right\rangle$$

$$= \left\langle \frac{N}{2}, \frac{N}{2} - k, \frac{1}{2}, \frac{1}{2} \middle| \frac{N-1}{2}, \frac{N+1}{2} - k \right\rangle = -\sqrt{\frac{k}{N+1}},$$

$$\left\langle \begin{matrix} (N,0) \\ \frac{N+1-k}{2}, \frac{N}{3}+1-k \end{matrix} \begin{matrix} (1,0) \\ 0, -\frac{2}{3} \end{matrix} \middle| \begin{matrix} (N-1,1) \\ \frac{N+1-k}{2}, \frac{N+1}{3} - k \end{matrix} \right\rangle$$

$$= \left\langle \frac{N}{2}, \frac{N}{2} + 1 - k, \frac{1}{2}, -\frac{1}{2} \middle| \frac{N-1}{2}, \frac{N+1}{2} - k \right\rangle = \sqrt{\frac{N+1-k}{N+1}}. \quad \text{(A.9)}$$

The $SU(3)$ isoscalar factors for the couplings of Eq. (A.2) are presented in Table A1.

Table A1. $SU(3)$ isoscalar factors

F_1	Y_1	F_2	Y_2	(λ,μ)	F	Y	$\left\langle \begin{matrix} (N,0) \\ F_1, Y_1 \end{matrix} \begin{matrix} (1,0) \\ F_2, Y_2 \end{matrix} \middle\| \begin{matrix} (\lambda,\mu) \\ F, Y \end{matrix} \right\rangle$
$\frac{N-k}{2}$	$\frac{N}{3}-k$	$\frac{1}{2}$	$\frac{1}{3}$	$(N+1,0)$	$\frac{N+1-k}{2}$	$\frac{N+1}{3}-k$	$\sqrt{\frac{N+1-k}{N+1}}$
$\frac{N+1-k}{2}$	$\frac{N}{3}+1-k$	0	$-\frac{2}{3}$	$(N+1,0)$	$\frac{N+1-k}{2}$	$\frac{N+1}{3}-k$	$\sqrt{\frac{k}{N+1}}$
$\frac{N-k}{2}$	$\frac{N}{3}-k$	$\frac{1}{2}$	$\frac{1}{3}$	$(N-1,1)$	$\frac{N+1-k}{2}$	$\frac{N+1}{3}-k$	$-\sqrt{\frac{k}{N+1}}$
$\frac{N+1-k}{2}$	$\frac{N}{3}+1-k$	0	$-\frac{2}{3}$	$(N-1,1)$	$\frac{N+1-k}{2}$	$\frac{N+1}{3}-k$	$\sqrt{\frac{N+1-k}{N+1}}$
$\frac{N-k}{2}$	$\frac{N}{3}-k$	$\frac{1}{2}$	$\frac{1}{3}$	$(N-1,1)$	$\frac{N-1-k}{2}$	$\frac{N+1}{3}-k$	-1
F_1	Y_2	F_2	Y_2	(λ,μ)	F	Y	$\left\langle \begin{matrix} (N,0) \\ F_1, Y_1 \end{matrix} \begin{matrix} (0,1) \\ F_2, Y_2 \end{matrix} \middle\| \begin{matrix} (\lambda,\mu) \\ F, Y \end{matrix} \right\rangle$
$\frac{N+1-k}{2}$	$\frac{N}{3}+1-k$	$\frac{1}{2}$	$-\frac{1}{3}$	$(N,1)$	$\frac{N+2-k}{2}$	$\frac{N+2}{3}-k$	1
$\frac{N-k}{2}$	$\frac{N}{3}-k$	0	$\frac{2}{3}$	$(N,1)$	$\frac{N-k}{2}$	$\frac{N+2}{3}-k$	$\sqrt{\frac{N+2-k}{N+2}}$
$\frac{N+1-k}{2}$	$\frac{N}{3}+1-k$	$\frac{1}{2}$	$-\frac{1}{3}$	$(N,1)$	$\frac{N-k}{2}$	$\frac{N+2}{3}-k$	$\sqrt{\frac{k}{N+2}}$
$\frac{N-k}{2}$	$\frac{N}{3}-k$	0	$\frac{2}{3}$	$(N-1,0)$	$\frac{N-k}{2}$	$\frac{N+2}{3}-k$	$-\sqrt{\frac{k}{N+2}}$
$\frac{N+1-k}{2}$	$\frac{N}{3}+1-k$	$\frac{1}{2}$	$-\frac{1}{3}$	$(N-1,0)$	$\frac{N-k}{2}$	$\frac{N+2}{3}-k$	$\sqrt{\frac{N+2-k}{N+2}}$

Journal of Physics: Conference Series **237** (2010) 012005 doi:10.1088/1742-6596/237/1/012005

Acknowledgments
It is a pleasure to thank José Barea, Alejandro Frank and Franco Iachello for interesting discussions on supersymmetric nuclei. This work was supported in part by PAPIIT-UNAM (grant IN113808).

References
[1] Iachello F 1980 *Phys. Rev. Lett.* **44** 772
[2] Iachello F and Arima A 1987 *The interacting boson model* (Cambridge: Cambridge University Press)
[3] Iachello F and Van Isacker P 1991 *The interacting boson-fermion model* (Cambridge: Cambridge University Press)
[4] Balantekin A B, Bars I and Iachello F 1981 *Phys. Rev. Lett.* **47** 19
Balantekin A B, Bars I and Iachello F 1981 *Nucl. Phys.* A **370** 284
[5] Balantekin A B, Bars I, Bijker R and Iachello F 1983 *Phys. Rev.* C **27** 1761
[6] Van Isacker P, Jolie J, Heyde K and Frank A 1985 *Phys. Rev. Lett.* **54** 653
[7] Metz A, Jolie J, Graw G, Hertenberger R, Gröger J, Günther C, Warr N and Eisermann Y 1999 *Phys. Rev. Lett.* **83** 1542
[8] Gröger J, Jolie J, Krücken R, Beausang C W, Caprio M, Casten R F, Cederkall J, Cooper J R, Corminboeuf F, Genilloud L, Graw G, Günther C, De Huu M, Levon A I, Metz A, Novak J R, Warr N and Wendel T 2000 *Phys. Rev.* C **62** 064304
[9] Wirth H F, Graw G, Christen S, Eisermann Y, Gollwitzer A, Hertenberger R, Jolie J, Metz A, Möller O, Tonev D and Valnion B D 2004 *Phys. Rev.* C **70** 014610
[10] Jolie J 2002 *Scientific American* July 70
[11] Elliott J P 1958 *Proc. Roy. Soc.* A **245** 128
Elliott J P 1958 *Proc. Roy. Soc.* A **245** 562
[12] Gell-Mann M and Ne'eman Y 1964 *The Eightfold Way* (New York: Benjamin)
[13] Cizewski J A, Casten R F, Smith G J, Stelts M L, Kane W R, Börner H G and Davidson W F 1978 *Phys. Rev. Lett.* **40** 167
Arima A and Iachello F 1978 *Phys. Rev. Lett.* **40** 385
[14] Iachello F and Kuyucak S 1981 *Ann. Phys.* **136** 19
[15] Bijker R and Iachello F 1985 *Ann. Phys.* **161** 360
[16] Sun H Z, Frank A and Van Isacker P 1983 *Phys. Rev.* C **27** 2430
Sun H Z, Frank A and Van Isacker P 1984 *Ann. Phys.* **157** 183
[17] Metz A, Eisermann Y, Gollwitzer A, Hertenberger R, Valnion B D, Graw G and Jolie J 2000 *Phys. Rev.* C **61** 064313
[18] Barea J, Bijker R and Frank A 2004 *J. Phys. A: Math. Gen.* **37** 10251
[19] Barea J, Bijker R and Frank A 2005 *Phys. Rev. Lett.* **94** 152501
[20] Barea J, Bijker R, Frank A, Graw G, Hertenberger R, Wirth H F, Christen S, Jolie J, Tonev D, Balodis M, Bērziņš J, Krāmere N and Von Egidy T 2009 *Phys. Rev.* C **79** 031304(R)
[21] Balodis M, Wirth H F, Graw G, Hertenberger R, Bērziņš J, Krāmere N, Jolie J, Christen S, Möller O, Tonev D, Barea J, Bijker R, Frank A and Von Egidy T 2008 *Phys. Rev.* C **77** 064602
[22] Bijker R and Scholten O 1985 *Phys. Rev.* C **32** 591
[23] Jolie J and Garrett P 1996 *Nucl. Phys.* A **596** 234
[24] Eisermann Y, Graw G, Metz A, Hertenberger R and Wirth H F 2002 *Czech. J. Phys.* **52** C627
[25] Arima A, Otsuka T, Iachello F and Talmi I 1977 *Phys. Lett.* B **66** 205
[26] Bijker R, Barea J and Frank A 2009 *Rev. Mex. Fís.* S **55** (2) 30 [arXiv:0902.4863]
[27] Warner D D and Van Isacker P 1997 *Phys. Lett.* B **395** 145
[28] García-Ramos J E and Van Isacker P 1999 *Ann. Phys.* **274** 45
[29] De Swart J J 1963 *Rev. Mod. Phys.* **35** 916

Journal of Physics: Conference Series **237** (2010) 012006 doi:10.1088/1742-6596/237/1/012006

Symplectic geometry of Maxwell theory and the photon concept

Ernst Binz and Alfred Rieckers

Inst. f. Mathematik und Informatik, A5/6, D–68131 Mannheim, Germany,
Inst. f. Theor. Physik, Auf der Morgenstelle 14, D–72076 Tübingen, Germany

E-mail: binz@rumms.uni-mannheim.de, alfred.rieckers@uni-tuebingen.de

Abstract. For an arbitrary multiply connected cavity a solution theory for the Maxwell equations is developed under ideal conductor boundary conditions. The trajectories of the electromagnetic fields are decomposed into their Helmholtz–Hodge components, including so–called cohomological fields which correspond to the harmonic differential forms. A canonical formalism is formulated in the Coulomb gauge, which comprises only the transversal and cohomological components. The canonical vector potentials and their conjugate fields are smeared in terms of smooth test functions. The Poisson bracket for the smeared fields is based on a symplectic form in the test function space and the corresponding infinite dimensional symplectic Lie algebra is introduced. The conservation quantities of Maxwell theory arise in this manner as co–momentum images of the Lie algebra. By means of a symplectic transformation the transversal dynamical generator is diagonalized and defines a special complexification of the test function space. A diagonalization of the cohomological dynamical generator is proved impossible.

Since the Weyl quantization provides a symplectic equivariant map of the classical fields onto the quantum fields (affiliated with our C*–Weyl algebra), we have in quantum electrodynamics the same freedom of symplectic transformations and complexifications in the test function space (and not in the representing Hilbert space). Using the diagonalizing complexification we introduce the (smeared) transversal creation and annihilation operators and arrive at a unique particle structure for the transversal fields in a Fock representation. The symplectic Lie algebra elements, multiplied by $i\hbar$, constitute the one–photon observables, where especially the diagonal dynamical Maxwell generator generalizes Einstein's photon energy to our general set up. The quantum co–momentum map acquires the form of a second quantization of the one–photon observables and leads to particle conserving multi–photon Hamiltonians for the complex linear Lie algebra elements, whereas the anti–linear Lie algebra elements generate particle non–conserving squeezing Hamiltonians. Since there is no particle structure for the quantized cohomological fields their collective nature is confirmed, illustrating the difference between quantization and particle discretization.

1. Introduction

As is well known, the photon concept has been introduced by Einstein 1905 in the framework of a nonrelativistic, statistical theory of light.

In the very first paper on canonical commutation relations (the latter not appearing in Heisenberg's paper), namely in *Zur Quantenmechanik* by Born and Jordan (1925), it is stated that the infinitely many degrees of freedom of the electromagnetic field should be quantized analogously to the material ones. This point of view is sharpened by Dirac's principle, formulated shortly afterwards, by which one has to replace the Poisson bracket by the scaled commutator (see also our Section on *deformation quantization* below), since then Poisson commuting classical variables lead to commuting quantum observables.

After the introduction of creation and annihilation operators for photons by Dirac (1927) (inappropriately in terms of number and phase variables), relativistic quantum electrodynamics (QED) started (1929/30) with two papers by Heisenberg and Pauli. In his book *Anschauliche Quantentheorie* (1936) Jordan writes that it be time to finally clarify the connection between electromagnetic fields and photons.
Since then there have still been published many articles and books on this subject, where it is surprising that in several texts on Quantum Optics the authors explicitly renounce to define one–photon observables. Also in perturbation theoretic QED the photon states appear only in the degenerated form of plane waves.
In the present work we take up again the quantization of nonrelativistic ED making use of the mathematical methods which are at hand in our time, especially of infinite dimensional Poisson manifolds, group theory, and C*–algebraic Weyl quantization.
We smear the classical electromagnetic fields by test functions to increase the solution manifold of the Maxwell equations and to gain a greater similarity to the smeared quantized fields.
In view of the recent developments on trapped photons and on sophisticated opto–electronic devices it is necessary to formulate the theory in arbitrary, open, possibly multiply connected regions $\Lambda \subset \mathbb{R}^3$ (cavities).
In this general setup we are going to demonstrate that the photon observables have their formal roots in a diagonalized form of deterministic Maxwell theory.

The (open, connected) Λ is called an *interior domain* if it is bounded, and an *exterior domain* if its complement $\Lambda^c \equiv \mathbb{R}^3 \backslash \Lambda$ is bounded. By $\bar{\Lambda}$ we denote the closure of Λ, and by $\partial\Lambda$ its boundary. Often we require some smoothness conditions on the boundary of Λ. Note that also an exterior domain has a bounded boundary $\partial\Lambda$.
As is well known, the physical meaning of the *longitudinal* and *transversal* parts of the fields is very different, and for multiply connected cavities Λ there arise also *additional types* of electric and magnetic fields, which combine properties of longitudinal and transversal fields. Since they depend on the cohomological features of Λ, especially on the first and second Betti numbers, we term them *cohomological fields*. (They correspond to the so–called harmonic differential forms of differential geometry.)
They got into focus rather recently, in spite of the following remark of Maxwell [1]:
We are led here to considerations belonging to the Geometry of Position, a subject which, though its importance was pointed out by Leibniz and illustrated by Gauss, has been little studied. (Maxwell called the second Betti number $b_2(\Lambda)$ *periphractic number*.)
Nowadays the "Geometry of Position" is widely studied in calculational electrotechnics and is the basis for the *finite elements method*. There are computer programs to determine the cuts for making Λ simply connected. The mathematical aspects are pursued in *topological electromagnetism* (cf. [2]), which mainly is concerned with quasi–stationary situations.

To formulate a photon theory we demonstrate that the evolution of the classical (and then of the quantized) fields is decomposable into *separated dynamical systems* associated with the different field types. By exploiting the reduction techniques for Hilbert space operators we reduce in the main text the topological considerations, connected with differentiable manifolds, to a minimum and gain the mathematical control of non–smooth boundaries at the same time. (Proofs which are missing in the present Article are elaborated in collaboration with R. Honegger in still unpublished material.)
To apply qualitative methods the connection with differential forms is however useful, what we describe in a forthcoming paper.

2. Fields and potentials

We assume at first *finite electromagnetic energy* in Λ and formulate the dynamics in L^2–Hilbert space, where we define the *unbounded vector–differential operators* with their domains of definition as follows:

$$\begin{aligned}
&\operatorname{grad}_0 \varphi := \nabla\varphi\,, \quad \forall\varphi \in \operatorname{dom}(\operatorname{grad}_0) := W_0^1(\Lambda,\mathbb{R})\,,\\
&\operatorname{grad} \varphi := \nabla\varphi\,, \quad \forall\varphi \in \operatorname{dom}(\operatorname{grad}) := W^1(\Lambda,\mathbb{R})\,,\\
&\operatorname{div}_0 \phi := \nabla\cdot\phi\,, \quad \forall\phi \in \operatorname{dom}(\operatorname{div}_0) := W_0(\operatorname{div};\Lambda,\mathbb{R}^3)\,,\\
&\operatorname{div} \phi := \nabla\cdot\phi\,, \quad \forall\phi \in \operatorname{dom}(\operatorname{div}) := W(\operatorname{div};\Lambda,\mathbb{R}^3)\,,\\
&\operatorname{curl}_0 \phi := \nabla\times\phi\,, \quad \forall\phi \in \operatorname{dom}(\operatorname{curl}_0) := W_0(\operatorname{curl};\Lambda,\mathbb{R}^3)\,,\\
&\operatorname{curl} \phi := \nabla\times\phi\,, \quad \forall\phi \in \operatorname{dom}(\operatorname{curl}) := W(\operatorname{curl};\Lambda,\mathbb{R}^3)\,.
\end{aligned} \tag{2.1}$$

We make use of certain Sobolev spaces, which we characterize only for the special case that there exists a normal vector almost everywhere at the boundary. The derivatives are meant in the weak sense.

$$\begin{aligned}
\operatorname{dom}(\operatorname{grad}_0) &= W_0^1(\Lambda,\mathbb{R}) = \{\varphi\in L^2(\Lambda,\mathbb{R}) \,|\, \frac{\partial\varphi}{\partial x_i} \in L^2(\Lambda,\mathbb{R}),\ \varphi|_{\partial\Lambda} = 0\}\,,\\
\operatorname{dom}(\operatorname{grad}) &= W^1(\Lambda,\mathbb{R}) = \{\varphi\in L^2(\Lambda,\mathbb{R}) \,|\, \frac{\partial\varphi}{\partial x_i} \in L^2(\Lambda,\mathbb{R}),\ 1\le i\le 3\}\,.
\end{aligned} \tag{2.2}$$

$$\begin{aligned}
\operatorname{dom}(\operatorname{div}_0) &= W_0(\operatorname{div};\Lambda,\mathbb{R}^3) = \{\phi\in L^2 \,|\, \nabla\cdot\phi \in L^2(\Lambda,\mathbb{R}),\ \phi\cdot n|_{\partial\Lambda} = 0\}\\
\operatorname{dom}(\operatorname{div}) &= W(\operatorname{div};\Lambda,\mathbb{R}^3) = \{\phi\in L^2(\Lambda,\mathbb{R}^3) \,|\, \nabla\cdot\phi \in L^2(\Lambda,\mathbb{R})\}
\end{aligned} \tag{2.3}$$

$$\begin{aligned}
\operatorname{dom}(\operatorname{curl}_0) &= W_0(\operatorname{curl};\Lambda,\mathbb{R}^3) = \{\phi\in L^2 \,|\, \nabla\times\phi \in L^2(\Lambda,\mathbb{R}),\ \phi\times n|_{\partial\Lambda} = 0\}\,,\\
\operatorname{dom}(\operatorname{curl}) &= W_0(\operatorname{curl};\Lambda,\mathbb{R}^3) = \{\phi\in L^2(\Lambda,\mathbb{R}^3) \,|\, \nabla\times\phi \in L^2(\Lambda,\mathbb{R})\}\,.
\end{aligned} \tag{2.4}$$

As a first gratification for specifying the operator domains we have nice adjointing relations for the closed operators in the various L^2–spaces:

$$\begin{aligned}
&\operatorname{grad}_0^* = -\operatorname{div} && \operatorname{div}^* = -\operatorname{grad}_0\,,\\
&\operatorname{grad}^* = -\operatorname{div}_0 && \operatorname{div}_0^* = -\operatorname{grad}\,,\\
&\operatorname{curl}_0^* = \operatorname{curl} && \operatorname{curl}^* = \operatorname{curl}_0\ .
\end{aligned}$$

We work with the direct real Hilbert sum

$$\mathcal{R} := L^2(\Lambda,\mathbb{R}^3) \boxplus L^2(\Lambda,\mathbb{R}^3)\,, \tag{2.5}$$

in which we combine the electric and magnetic fields to the 6–tuples

$$\psi := (\mathbf{E},\mathbf{B}) \equiv \mathbf{E}\boxplus\mathbf{B} \in \mathcal{R}\,.$$

(We use $\boxplus$ for the orthogonal decomposition into the electric and magnetic parts, whereas other orthogonal decompositions are indicated by $\oplus$.)

2.1. Boundary conditions and Maxwell operator

We arrange the Maxwell equations as follows. The first two equations

$$\nabla \cdot \mathbf{E}_t(x) = \epsilon_0^{-1} \rho_t(x) , \tag{2.6}$$
$$\nabla \cdot \mathbf{B}_t(x) = 0 , \tag{2.7}$$

do not describe the dynamical evolution, which rather is expressed by the remaining two:

$$\begin{aligned} \frac{\partial \mathbf{E}_t(x)}{\partial t} &= c^2 \nabla \times \mathbf{B}_t(x) - \frac{1}{\epsilon_0} \mathbf{j}_t(x) , \\ \frac{\partial \mathbf{B}_t(x)}{\partial t} &= -\nabla \times \mathbf{E}_t(x) . \end{aligned} \tag{2.8}$$

If we concentrate on the tuple of dynamical equations we supplement it by the continuity equation (derivable from (2.8) combined with (2.6))

$$\frac{\partial \rho_t(x)}{\partial t} + \nabla \cdot \mathbf{j}_t(x) = 0 , \quad \forall x \in \Lambda , \quad \forall t \in \mathbb{R} . \tag{2.9}$$

Throughout our exposition the boundary surface $\partial\Lambda$ of Λ is supposed to consist physically of a perfect conductor. In a perfect conductor the skin region is zero and the surface charge and current densities accommodate without delay to the arbitrarily fast varying fields.

On the surface of a perfect conductor wall, enclosing vacuum, the continuous field components have to vanish. This yields the boundary conditions

$$\mathbf{E} \times n|_{\partial\Lambda} = 0 , \quad \mathbf{B} \cdot n|_{\partial\Lambda} = 0 , \quad \forall t \in \mathbb{R} . \tag{2.10}$$

The extension of $L^2(\Lambda, \mathbb{R}^3)$–classes of fields to the boundary is achieved by means of a boundary operator. If the system of the Maxwell differential Eqs. (2.6) to (2.8) is solved under the boundary condition (2.10), leading to the solution trajectory $\mathbb{R} \ni t \mapsto (\mathbf{E}_t, \mathbf{B}_t)$, then the surface current density $\mathbf{j}_t^{\partial\Lambda} : \partial\Lambda \to \mathbb{R}^3$ and the surface charge density $\rho_t^{\partial\Lambda} : \partial\Lambda \to \mathbb{R}$ may be calculated from the relations

$$\rho_t^{\partial\Lambda} = -\epsilon_0 \mathbf{E}_t \cdot n|_{\partial\Lambda} , \quad \mathbf{j}_t^{\partial\Lambda} = -\mu_0^{-1} \mathbf{B}_t \times n|_{\partial\Lambda} , \quad \forall t \in \mathbb{R} . \tag{2.11}$$

We assume now that $\mathbf{j}_t \in L^2(\Lambda, \mathbb{R}^3)$, for all $t \in \mathbb{R}$. So we rewrite the dynamical Maxwell Eqs. (2.8) in the mathematically concise form

$$\underbrace{\frac{d}{dt} \begin{pmatrix} \mathbf{E}_t \\ \mathbf{B}_t \end{pmatrix}}_{=: \psi_t} = \underbrace{\begin{pmatrix} 0 & \mathrm{curl} \\ -\mathrm{curl}_0 & 0 \end{pmatrix}}_{=: \mathbb{A}} \underbrace{\begin{pmatrix} \mathbf{E}_t \\ \mathbf{B}_t \end{pmatrix}}_{=: \psi_t} + \underbrace{\begin{pmatrix} -\mathbf{j}_t \\ 0 \end{pmatrix}}_{=: \gamma_t} . \tag{2.12}$$

(For convenience we have momentarily set $\epsilon_0 = \mu_0 = c = 1$.)

Lemma 2-1 *$\mathbb{A}$ is a closed operator in the real Hilbert space $\mathcal{R}$ and satisfies $\mathbb{A}^* = -\mathbb{A}$. (This is equivalent to $i\mathbb{A}$ being selfadjoint in the complexified Hilbert space $\mathcal{R} + i\mathcal{R}$.)*
Consequently, the free Maxwell dynamics $\exp\{t\mathbb{A}\}$, $t \in \mathbb{R}$, constitutes a strongly continuous orthogonal group in $\mathcal{R}$.

2.2. Solution of the Maxwell Cauchy problem

A Cauchy problem is an initial–value problem, where in general – as in the present case – also spatial boundary conditions have to be taken into account, what then is called an *initial boundary value problem* (e.g. [3]).

Theorem 2-2 *Let Λ be an arbitrary domain in $\mathbb{R}^3$ and suppose a current density $t \mapsto \mathbf{j}_t$, which is weakly locally integrable. Then the Maxwell weak Cauchy problem, is uniquely solvable for every initial value $\psi_{t_0} \in \mathcal{R}$, which satisfies the boundary conditions. Its solution trajectory $\mathbb{R} \ni t \mapsto \psi_t \in \mathcal{R}$ is $\|.\|_{\mathcal{R}}$–continuous and is uniquely given by*

$$\psi_t = \exp\{(t-t_0)\mathbb{A}\}\psi_{t_0} + \int_{t_0}^{t} \exp\{(t-s)\mathbb{A}\}\gamma_s \, ds\,, \quad \forall t \in \mathbb{R}\,. \tag{2.13}$$

This is the first form of our *magic formula*, which discloses important features of the solution trajectory.

2.3. The Helmholtz–Hodge decomposition

The subsequent result is a modification of material in [4].

Theorem 2-3 *We have for an arbitrary (also unbounded) domain the following unique decomposition of quadrat integrable vector fields into mutually orthogonal components*

$\mathcal{R}$	$=$	$L^2(\Lambda,\mathbb{R}^3)$	$\boxplus$	$L^2(\Lambda,\mathbb{R}^3)$
		$\overbrace{\qquad\qquad}^{=\ker(\mathrm{div})}$ over $\mathbb{H}_2 \oplus \overline{\mathrm{ran}(\mathrm{curl})}$		$\overbrace{\qquad\qquad}^{=\ker(\mathrm{div}_0)}$ over $\mathbb{H}_1 \oplus \overline{\mathrm{ran}(\mathrm{curl}_0)}$
	$=$	$\overline{\mathrm{ran}(\mathrm{grad}_0)} \oplus \mathbb{H}_2 \oplus \overline{\mathrm{ran}(\mathrm{curl})}$	$\boxplus$	$\overline{\mathrm{ran}(\mathrm{grad})} \oplus \mathbb{H}_1 \oplus \overline{\mathrm{ran}(\mathrm{curl}_0)}$
		$\underbrace{\qquad\qquad}_{=\ker(\mathrm{curl}_0)}$ under $\overline{\mathrm{ran}(\mathrm{grad}_0)} \oplus \mathbb{H}_2$		$\underbrace{\qquad\qquad}_{=\ker(\mathrm{curl})}$ under $\overline{\mathrm{ran}(\mathrm{grad})} \oplus \mathbb{H}_1$
ψ	$=$	$\mathbf{E}$	$\boxplus$	$\mathbf{B}$
	$=$	$\mathbf{E}^{\parallel} \oplus \mathbf{E}^{co} \oplus \mathbf{E}^{\top}$	$\boxplus$	$\mathbf{B}^{\parallel} \oplus \mathbf{B}^{co} \oplus \mathbf{B}^{\top}$
γ_t	$=$	$-\mathbf{j}_t$	$\boxplus$	0
	$=$	$-\mathbf{j}_t^{\parallel} \oplus -\mathbf{j}_t^{co} \oplus -\mathbf{j}_t^{\top}$	$\boxplus$	$0^{\parallel} \oplus 0^{co} \oplus 0^{\top}$

We now assume as our "Standard Assumption" that Λ is an interior domain, satisfying the following conditions: The boundary $\partial\Lambda$ consists of a finite number of connected C^2–smooth surfaces, Λ is locally situated on one side of $\partial\Lambda$, and Λ is made simply connected by a finite numer of regular, non–intersecting cuts, which are non–tangential to $\partial\Lambda$ (cf. [4], Vol. III). We may omit then the closure operations for defining the Helmholtz–Hodge subspaces, and obtain the following spaces of cohomological fields

$$\mathbb{H}_1 = \{\phi \in L^2(\Lambda,\mathbb{C}^3) \mid \nabla\cdot\phi = 0\,,\ \nabla\times\phi = 0\,,\ \phi\cdot n|_{\partial\Lambda} = 0\} \subset W^1(\Lambda,\mathbb{C}^3)$$
$$\mathbb{H}_2 = \{\psi \in L^2(\Lambda,\mathbb{C}^3) \mid \nabla\cdot\psi = 0\,,\ \nabla\times\psi = 0\,,\ \psi\times n|_{\partial\Lambda} = 0\} \subset W^1(\Lambda,\mathbb{C}^3)\,.$$

For the connection with differential geometry we assume an interior domain Λ such as described in the second part of the preceding Theorem. Its closure is a *manifold with boundary* as described in [5], where also vector analysis is translated into the language of differential forms. According to this, we associate the vector fields of $\mathbb{H}_1$ by means of Hilbert space duality with one–forms, and those of $\mathbb{H}_2$ at first with one–forms and then by Hodge duality with two–forms. In this manner we obtain the connection of $\mathbb{H}_1$ and $\mathbb{H}_2$ with the first resp. second cohomology group of de Rham (the quotients of the closed forms over the exact forms, of the first resp. second degree).

The geometric meaning is elucidated by using the duality for the transition of the n–differential forms (cochains) to n–dimensional objects (chains) (caused by integrating the n–cochains over the n–chains). The 1–chains (modulo closed paths contractible to a point) determine the multiplicity of connectedness of Λ. Thus $\dim \mathbb{H}_1$ equals the first Betti number $b_1(\Lambda)$. The 2–chains (modulo closed 2–dimensional surfaces contractible to a point) give the number of the disconnected components of $\partial\Lambda$ without counting the surface of the infinite, connected component of Λ^c. Thus $\dim \mathbb{H}_2$ equals the second Betti number $b_2(\Lambda)$.
In this manner we get the rule

(1) $b_1(\Lambda) = 0 \iff \mathbb{H}_1 = \{0\} \iff \Lambda$ is simply connected.
(2) $b_2(\Lambda) = 0 \iff \mathbb{H}_2 = \{0\} \iff \partial\Lambda$ is connected.

In the drawing of a cubic cavity Λ, the two columns prevent two classes of closed paths to be

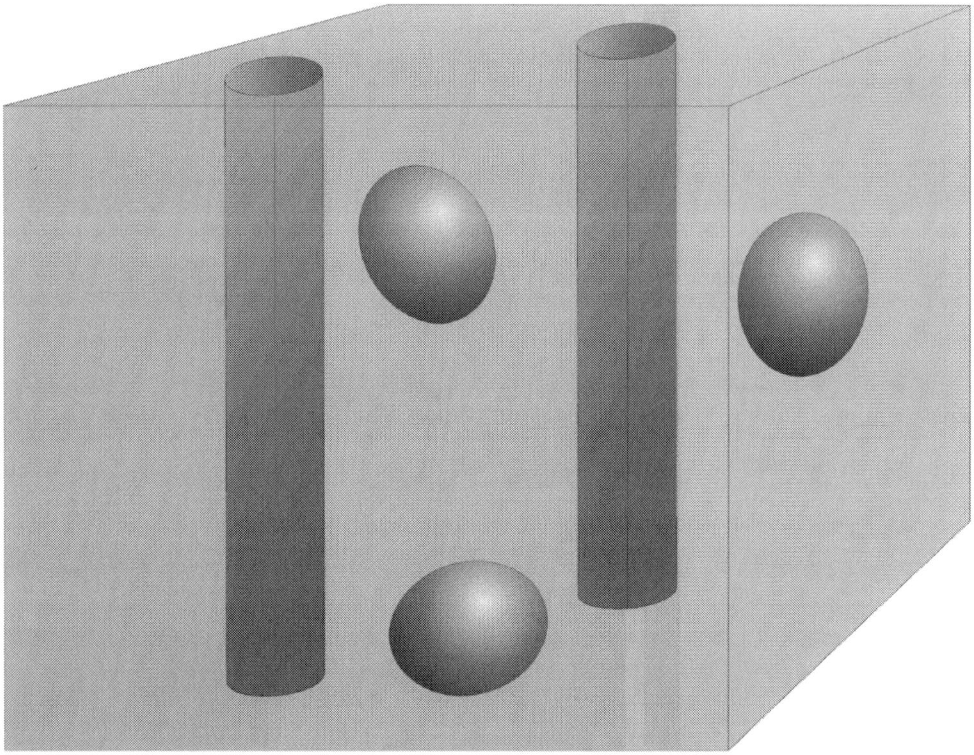

Cubic cavity Λ_a with inserted conducting obstacles.

contracted to a point and we have $b_1(\Lambda) = 2$. (And one needs in fact two cuts to make Λ simply connected.) The three inserted egg–shaped conductors prevent three classes of closed surfaces to be shrinked to a point. The surfaces of the conductors constitute the surfaces of the bounded components of Λ^c (whereas the interior regions of the columns is part of the infinite component of Λ^c). We see that here $b_2(\Lambda) = 3$.

From the Helmholtz–Hodge decomposition of the magnetic field it follows

$$\mathbf{B} \in \ker(\mathrm{div}_0) \iff \left[\mathbf{B}^{\|} = 0 \quad \& \quad \mathbf{B}^{\mathrm{co}}, \mathbf{B}^{\top} \in \ker(\mathrm{div}_0)\right],$$

Thus the magnetic field (where magnetic monopoles are excluded by vanishing divergence) has in general not only a transversal but also a *cohomological component* $\mathbf{B}^{co}$.

2.4. Decomposition of dynamical operators

A reducing subspace for $\mathbb{A}$ is a *closed* subspace $\mathcal{K}$ of $\mathcal{R}$ satisfying

$$\exp\{t\mathbb{A}\}(\mathcal{K}) = \mathcal{K}, \quad \forall t \in \mathbb{R}.$$

The reduction $\mathbb{A}^\top$ of $\mathbb{A}$ from $\mathcal{R}$ to the transversal sub–Hilbert space

$$\mathcal{R}_f^\top = \overline{\mathrm{ran}(\mathrm{curl})} \boxplus \overline{\mathrm{ran}(\mathrm{curl}_0)} = \ker(\mathbb{A})^\perp = \overline{\mathrm{ran}(\mathbb{A})}$$

is the matrix operator

$$\mathbb{A}^\top = \begin{pmatrix} 0 & \mathrm{curl}| \\ -\mathrm{curl}_0| & 0 \end{pmatrix}. \tag{2.14}$$

$\mathrm{curl}|$ and $\mathrm{curl}_0|$ are the restrictions to the orthogonal complements of their kernels (and thus possess an inverse). We have

$$\begin{array}{|ll|}\hline \mathcal{R} & = \overbrace{\left(\overline{\mathrm{ran}(\mathrm{grad}_0)} \boxplus \overline{\mathrm{ran}(\mathrm{grad})}\right)}^{=\mathcal{R}_f^{\|}} \oplus \overbrace{\left(\mathbb{H}_2 \boxplus \mathbb{H}_1\right)}^{=\mathcal{R}_f^{co}} \oplus \overbrace{\left(\overline{\mathrm{ran}(\mathrm{curl})} \boxplus \overline{\mathrm{ran}(\mathrm{curl}_0)}\right)}^{=\mathcal{R}_f^\top} \\ \hline \exp\{t\mathbb{A}\} = & \qquad\qquad \mathbb{1} \qquad\qquad \oplus \qquad \mathbb{1} \qquad \oplus \qquad \exp\{t\mathbb{A}^\top\} \\ \hline \end{array}.$$

The material current being omitted, just the circulations of the transversal fields give rise to their temporal variations (and vice versa).

The injectivity of $\mathbb{A}^\top$ ensures that the diagonal matrix–operator

$$(\mathbb{A}^\top)^*\mathbb{A}^\top = -(\mathbb{A}^\top)^2 = \begin{pmatrix} \mathrm{curl}^2_{t0}| & 0 \\ 0 & \mathrm{curl}^2_{n0}| \end{pmatrix} \tag{2.15}$$

is *strictly positive*, where

$$\mathrm{curl}^2_{t0}| = \mathrm{curl}|\,\mathrm{curl}_0|, \qquad \mathrm{curl}^2_{n0}| = \mathrm{curl}_0|\,\mathrm{curl}|.$$

The two curlcurl operators consist each of a product of the restricted, unbounded curl operators (what improves the formal notation $\nabla\times\nabla\times$).

2.5. Helmholtz–Hodge decomposition of the field trajectories

Theorem 2-4 *A solution trajectory of Theorem 2-2 (with the magic formula for the total fields), is equivalent to the Helmholtz–Hodge decomposed system of trajectories*

$$\mathbf{E}_t^{\|} = \mathbf{E}_{t_0}^{\|} - \int_{t_0}^t \mathbf{j}_s^{\|}\, ds \quad \in \overline{\mathrm{ran}(\mathrm{grad}_0)},$$

$$\mathbf{E}_t^{co} = \mathbf{E}_{t_0}^{co} - \int_{t_0}^t \mathbf{j}_s^{co} ds \quad \in \mathbb{H}_2,$$

$$\mathbf{B}_t^{\|} = 0 \quad (\textit{equivalently} \quad \mathbf{B}_t \in \ker(\mathrm{div}_0)),$$

$$\mathbf{B}_t^{co} = \mathbf{B}_{t_0}^{co} \quad \in \mathbb{H}_1 \quad \forall t, \ \textit{having no driving current},$$

$$\begin{pmatrix} \mathbf{E}_t^\top \\ \mathbf{B}_t^\top \end{pmatrix} = \exp\{(t-t_0)\mathbb{A}^\top\}\begin{pmatrix} \mathbf{E}_{t_0}^\top \\ \mathbf{B}_{t_0}^\top \end{pmatrix} + \int_{t_0}^t \exp\{(t-s)\mathbb{A}^\top\}\begin{pmatrix} -\mathbf{j}_s^\top \\ 0^\top \end{pmatrix} ds \quad \in \mathcal{R}_f^\top.$$

This is our transversally reduced magic formula.

3. Introduction of the potentials

As usual we look for potentials which satisfy

$$\mathbf{E}_t = -\nabla u_t - \frac{d\mathbf{A}_t}{dt}\,, \qquad \mathbf{E}_t^{\|} = -\operatorname{grad}_0 u_t - \dot{\mathbf{A}}_t^{\|}\,, \tag{3.1}$$

$$\mathbf{B}_t = \nabla \times \mathbf{A}_t\,, \text{ for all } t \in \mathbb{R}\,. \tag{3.2}$$

More precisely, we intend the following Helmholtz–Hodge decomposition of the vector potential, fitting to that for $\mathbf{E}$:

$$\begin{array}{|ccc|}\hline \mathrm{L}^2(\Lambda,\mathbb{R}^3) & = & \underbrace{\mathrm{ran}(\mathrm{grad}_0) \oplus \overbrace{\mathbb{H}_2 \oplus \mathrm{ran}(\mathrm{curl})}^{=\ \ker(\mathrm{div})}}_{=\ \ker(\mathrm{curl}_0)} \\ \hline \mathbf{E} & = & \mathbf{E}^{\|} \oplus \mathbf{E}^{\mathrm{co}} \oplus \mathbf{E}^{\top} \\ \hline \mathbf{A} & = & \mathbf{A}^{\|} \oplus \mathbf{A}^{\mathrm{co}} \oplus \mathbf{A}^{\top} \\ \hline \end{array}\,. \tag{3.3}$$

Since Poincaré's Lemma is globally valid in simply connected regions only, the existence of the potentials in more general cavities is not trivial. The more so, as we have to care about the boundary conditions and the Helmholtz–Hodge decomposition. Only part of the Maxwell equations are needed for the existence of the potentials.

Assumption 3-1 *We consider electromagnetic field trajectories* $[t \mapsto (\mathbf{E}_t, \mathbf{B}_t)]$ *with values in the real Hilbert space* $\mathcal{R} = \mathrm{L}^2(\Lambda,\mathbb{R}^3) \boxplus \mathrm{L}^2(\Lambda,\mathbb{R}^3)$, *which split according to the Helmholtz–Hodge decomposition of Theorem 2-4, and which satisfy the following requirements:*

(a) $\mathbf{E}_t^{\|} \in \mathrm{ran}(\mathrm{grad}_0)$ *for all* $t \in \mathbb{R}$, *and* $[t \mapsto \mathrm{grad}_0^{-1}\mathbf{E}_t^{\|}] \in \mathrm{C}(\mathbb{R}, \mathrm{W}_0^1(\Lambda,\mathbb{R}))$.

(b) $[t \mapsto \mathbf{E}_t^{co}] \in \mathrm{C}(\mathbb{R}, \mathbb{H}_2)$.

(c) $\mathbf{B}_t^{\top} \in \mathrm{ran}(\mathrm{curl}_0)$ *for all* $t \in \mathbb{R}$, *meaning the existence condition for the transversal vector potential* $\mathbf{A}_t^{\top} := \mathrm{curl}_0|^{-1}\mathbf{B}_t^{\top}$.

(d) Those Maxwell equations which do not involve sources

$$\mathbf{B}_t \in \ker(\mathrm{div}_0)\,, \qquad \frac{d\mathbf{B}_t^{co}}{dt} = 0\,, \qquad \frac{d\mathbf{B}_t^{\top}}{dt} = -\,\mathrm{curl}_0\,\mathbf{E}_t^{\top}\,; \tag{3.4}$$

are fulfilled (in the weak formulation).

Theorem 3-2 *For an arbitrary (open connected) domain* Λ *let be given a force field trajectory* $[t \mapsto (\mathbf{E}_t, \mathbf{B}_t)]$ *which satisfies Assumption 3-1.*
Assume additionally an arbitrary scalar potential trajectory $[t \mapsto u_t] \in \mathrm{C}(\mathbb{R}, \mathrm{W}_0^1(\Lambda,\mathbb{R}))$ *(including the special case* $u_t \equiv 0$*).*
Then there exists a vector potential trajectory $[t \mapsto \mathbf{A}_t \in \mathrm{W}_0(\mathrm{curl};\Lambda,\mathbb{R})]$ *such that we have in terms of weak time derivatives*

$$\begin{aligned}
\dot{\mathbf{A}}_t^{\|} &= -(\mathbf{E}_t^{\|} + \mathrm{grad}_0\, u_t)\,, && \textit{thus}\quad \mathbf{A}_t^{\|} = \mathbf{A}_0^{\|} - \int_0^t (\mathbf{E}_s^{\|} + \mathrm{grad}_0\, u_s)\,ds\,, \\
\dot{\mathbf{A}}_t^{co} &= -\mathbf{E}_t^{co}\,, && \textit{thus}\quad \mathbf{A}_t^{co} = \mathbf{A}_0^{co} - \int_0^t \mathbf{E}_s^{co}\,ds\,, \\
\dot{\mathbf{A}}_t^{\top} &= -\mathbf{E}_t^{\top} = \frac{d\,\mathrm{curl}_0|^{-1}\mathbf{B}_t^{\top}}{dt}\,, && \textit{thus}\quad \mathbf{A}_t^{\top} = \mathbf{A}_0^{\top} - \int_0^t \mathbf{E}_s^{\top}\,ds\,.
\end{aligned} \tag{3.5}$$

Here the Helmholtz–Hodge components of $\mathbf{A}_t$ are meant in the sense of an $\mathbf{E}$*–field, as indicated in (3.3). For the transversal magnetic field we assume* $\mathbf{B}_t^\top = \mathrm{curl}_0\, \mathbf{A}_t^\top \in \mathrm{ran}(\mathrm{curl}_0)$*, what we can prove under the Standard Assumption of Theorem 2-3.*
The time independent cohomological magnetic field $\mathbf{B}_t^{co}$ *is not covered by the above Helmholtz–Hodge compatible vector potential* $\mathbf{A}_t$*. At least under the Standard Assumption there exists an additional cohomological vector potential* A_{co}*, such that*

$$\mathbf{B}_t^{co} = \mathrm{curl}\, \mathsf{A}_{co} = constant\,. \tag{3.6}$$

Whereas $\mathbf{A}_t^{\|}$ can be completely gauged away adding the gradient of a gauge function $[t \mapsto \lambda_t] \in C^1(\mathbb{R}, W_0^1(\Lambda, \mathbb{R}))$, $\mathbf{A}_t^{co}$ is the gradient of a scalar function with different boundary conditions. A_{co} is the gradient only in the region Λ minus the cuts which make Λ simply connected. Thus it can maximally be gauged away only in the cutted region, leaving a singular expression for the gradient (not for A_{co}) on the cuts. If $\mathbf{B}$ vanishes, physical conditions may require nevertheless a non–vanishing A_{co} which is then for itself a cohomological vector field. Its couplings to the electromagnetic current (cf. the Lagrange and Hamilton functions) may be gauged away, but it influences the phases in the quantized theory in terms of non–vanishing line integrals around the columns of Figure 2.3 (generalized Aharanov–Bohm effects).

Note that the Dirac monopole requires a region Λ with one (interior) point omitted, which does not satisfy the Standard Assumption.

4. Lagrange and Hamilton formalism

4.1. Lagrange theory of charged fluids

Consider M charged point particles in Λ with positions $\mathbf{q}_t^k$, $1 \leq k \leq M$. The transition to continuous matter, called *charged fluid*, is performed by $k \longrightarrow x$ at $t = 0$ such that $\mathbf{q}_t^k \longrightarrow \mathbf{q}_t(x)$ with $\mathbf{q}_0(x) = x$, $x \in \Lambda$.
The *position map (at time t)* $\mathbf{q}_t : \Lambda \to \Lambda$ is bijective and volume preserving, if the fluid is incompressible. In dependence on t it is in general no group, but one always has $\mathbf{q}_{-t} \circ \mathbf{q}_t = \mathbf{q}_0 = $ id.
The material properties of the fluid elements are given by the matter density $m(x)$ and charge density $\varrho(x)$ at $t = 0$. As the time proceeds (forwards or backwards) these data of the fluid elements are at the positions $\mathbf{q}_t(x)$. Thus $\varrho_t(x) = \varrho(\mathbf{q}_{-t}(x))$.

For a fluid element with index x at time t (at position $\mathbf{q}_t(x)$) we have the velocity $\dot{\mathbf{q}}_t(x) = d\mathbf{q}_t(x)/dt$ and current $\mathbf{j}_t(x) = \varrho(x)\dot{\mathbf{q}}_t(x)$. (The current at x is $\mathbf{j}_t(\mathbf{q}_{-t}(x))$.)
The return to the discrete case is achieved by setting $m_t(x) = \sum_{k=1}^M m_k\, \delta(\mathbf{q}_t^k - x)$, $\rho_t(x) = \sum_{k=1}^M e_k\, \delta(\mathbf{q}_t^k - x)$. $\mathbf{j}_t(x) = \sum_{k=1}^M e_k\, \dot{\mathbf{q}}_t^k\, \delta(\mathbf{q}_t^k - x)$.

For the total system "charged fluid plus field" we have the position and velocity variables: $\phi = (\mathbf{q}, \mathbf{A}, u) \in \mathcal{Q}$, $\dot{\phi} = (\dot{\mathbf{q}}, \dot{\mathbf{A}}, \dot{u}) \in \mathcal{V}$, the values of which constitute the *velocity phase space* $\mathcal{M} = \mathcal{Q} \times \mathcal{V}$.
$\mathbf{A} = (\mathbf{A}^{\|}, \mathbf{A}^{co}, \mathbf{A}^\top)$ is the Helmholtz–Hodge compatible vector potential and u the scalar potential.
We assume here *square integrability* for all fields over Λ.
The functions $m(x)$, $\varrho(x)$, and $\mathsf{A}_{co}(x)$, which are prepared at $t = 0$, characterize as a kind of external data the system and are termed *system functions*. The velocity phase space for M point charges is found by reading $\mathbf{q}_t = (\mathbf{q}_t^1, \ldots, \mathbf{q}_t^M)$.
We generalize the *standard Lagrangian*, as it is used for discrete charges especially in Quantum

Optics, to the form

$$\begin{aligned}
L^{\mathrm{st}}[\phi_t,\dot\phi_t] =& \frac{1}{2}\int_\Lambda m(x)\dot{\mathbf{q}}_t(x)^2 d^3x + \frac{\epsilon_0}{2}\int_\Lambda \left[(-\nabla u_t(x) - \dot{\mathbf{A}}_t(x))^2 - c^2(\nabla\times\mathbf{A}_t(x))^2\right] d^3x \\
&- \frac{\epsilon_0 c^2}{2}\int_\Lambda (\nabla\times \mathsf{A}_{\mathrm{co}}(x))^2\, d^3x \\
&- \int_\Lambda \varrho(x) u_t(\mathbf{q}_t(x))\, d^3x + \int_\Lambda \varrho(x)\dot{\mathbf{q}}_t(x)\cdot \mathsf{A}_t(\mathbf{q}_t(x))\, d^3x \\
=& \frac{1}{2}(\dot{\mathbf{q}}_t|m\dot{\mathbf{q}}_t) + \frac{\epsilon_0}{2}\left[\|\operatorname{grad}_0 u_t + \dot{\mathbf{A}}_t^{\|}\|^2 + \|\dot{\mathbf{A}}_t^{\mathrm{co}}\|^2 + \|\dot{\mathbf{A}}_t^{\top}\|^2\right] \\
&- \frac{1}{2\mu_0}\left[\|\operatorname{curl}_0 \mathbf{A}_t^{\top}\|^2 + \|\operatorname{curl}\mathsf{A}_{\mathrm{co}}\|^2\right] - (\rho_t|u_t) + (\mathbf{j}_t|\mathbf{A}_t) + (\mathbf{j}_t|\mathsf{A}_{\mathrm{co}})\,.
\end{aligned} \tag{4.1}$$

For M charged point particles we evaluate formally the inner products and arrive at

$$(\rho_t|u_t) = \int_\Lambda \rho_t(x)u_t(x)\, d^3x = \sum_{k=1}^{M} e_k u_t(\mathbf{q}_t^k)\,.$$

$$(\mathbf{j}_t|\mathbf{A}_t) = \int_\Lambda \mathbf{j}_t(x)\cdot \mathsf{A}_t(x)\, d^3x = \sum_{k=1}^{M} e_k\, \dot{\mathbf{q}}_t^k \cdot \mathsf{A}_t(\mathbf{q}_t^k)\,.$$

We obtain for the *canonical momenta*

$$\begin{aligned}
\mathbf{p} &:= \frac{\delta L^{\mathrm{st}}}{\delta\dot{\mathbf{q}}} = m\dot{\mathbf{q}} + \varrho\left[\mathbf{A}\circ\mathbf{q} + \mathsf{A}_{\mathrm{co}}\circ\mathbf{q}\right], \\
\mathbf{p}^k &:= \frac{\delta L^{\mathrm{st}}}{\delta\dot{\mathbf{q}}^k} = m_k\dot{\mathbf{q}}^k + e_k\left[\mathbf{A}(\mathbf{q}^k) + \mathsf{A}_{\mathrm{co}}(\mathbf{q}^k)\right] \quad \text{(discrete case)}, \\
\mathbf{Y}^{\|} &:= \frac{\delta L^{\mathrm{st}}}{\delta\dot{\mathbf{A}}^{\|}} = \epsilon_0(\operatorname{grad}_0 u + \dot{\mathbf{A}}^{\|}) = -\epsilon_0\mathbf{E}^{\|}, \\
\mathbf{Y}^{\mathrm{co}} &:= \frac{\delta L^{\mathrm{st}}}{\delta\dot{\mathbf{A}}^{\mathrm{co}}} = \epsilon_0\dot{\mathbf{A}}^{\mathrm{co}} = -\epsilon_0\mathbf{E}^{\mathrm{co}}, \\
\mathbf{Y}^{\top} &:= \frac{\delta L^{\mathrm{st}}}{\delta\dot{\mathbf{A}}^{\top}} = \epsilon_0\dot{\mathbf{A}}^{\top} = -\epsilon_0\mathbf{E}^{\top}.
\end{aligned} \tag{4.2}$$

Because of $\frac{\delta L^{\mathrm{st}}}{\delta\dot u} = 0$ the standard Lagrangian is non–regular [6].
Let us therefore omit u and work in the *momentum phase space* with the thus reduced position variables and associated canonical momenta

$$\phi^{\mathrm{r}} = (\mathbf{q}, \mathbf{A}^{\|}, \mathbf{A}^{\mathrm{co}}, \mathbf{A}^{\top}) \in \mathcal{Q}^{\mathrm{r}}, \qquad \pi = (\mathbf{p}, \mathbf{Y}^{\|}, \mathbf{Y}^{\mathrm{co}}, \mathbf{Y}^{\top}) \in \mathcal{P}.$$

The momentum phase space is denoted $\mathcal{N} := \mathcal{Q}^{\mathrm{r}} \times \mathcal{P}$.

4.2. The Coulomb gauge Hamilton formalism

While we work under the *temporal gauge condition* $u \equiv 0$ with the full phase space $\mathcal{N}$, we drop in the *Coulomb gauge*, defined here by

$$\dot{\mathbf{A}}^{\|} \equiv 0\,,$$

the longitudinal fields $\mathbf{A}^{\|}$, $\mathbf{Y}^{\|}$. Thus we use the reduced phase space $\mathcal{N}^{\mathrm{r}} := \mathcal{Q}^{\mathrm{rr}} \times \mathcal{P}^{\mathrm{r}}$ with the canonical variables

$$\phi^{\mathrm{rr}} = (\mathbf{q}, \mathbf{A}^{\mathrm{co}}, \mathbf{A}^{\top}) \in \mathcal{Q}^{\mathrm{rr}}, \qquad \pi^{\mathrm{r}} = (\mathbf{p}, \mathbf{Y}^{\mathrm{co}}, \mathbf{Y}^{\top}) \in \mathcal{P}^{\mathrm{r}}\,.$$

Because of the Coulomb gauge condition we have $\mathbf{A}_t^{\|} = \mathbf{A}_{t_0}^{\|}$. Since $\mathbf{Y}^{\|}$ is proportional to $\mathbf{E}^{\|}$ we may get hold on it by (2.6) and (3.1) and use the invertibility of the divergence operator div, when restricted to the longitudinal fields, to obtain

$$\mathbf{Y}^{\|} := -\,\mathrm{div}\,|^{-1}\rho =: \epsilon_0\,\mathrm{grad}_0\, u_{\mathrm{Coul}} * \varrho\,, \tag{4.3}$$

where we have introduced the Green's function of the Laplacian, namely the Coulomb potential u_{Coul}, arising from the Poisson equation

$$\rho = -\epsilon_0\,\mathrm{div}\,\mathrm{grad}_0\, u_{\mathrm{Coul}} = -\epsilon_0 \Delta_\infty u_{\mathrm{Coul}} \tag{4.4}$$

with the Dirichlet Laplacian $-\Delta_\infty = -\,\mathrm{div}\,\mathrm{grad}_0$ in Λ.
This gives $\mathbf{Y}_t^{\|}$ as the convolution with the charge density ρ_t, without propagational delay.
We obtain the *Coulomb Hamilton function* $H_{\mathrm{Coul}} = H_{\mathrm{mat}} + H_{\mathrm{rad}} + H_{\mathrm{int}} :\ \mathcal{N}^{\mathrm{r}} \to \mathbb{R}$, including now the four system functions $m, \varrho, \mathsf{A}_{\mathrm{co}}$, and u_{Coul}:

$$H_{\mathrm{mat}}[\phi, \pi 1] := (\mathbf{p}|\frac{1}{2m}\mathbf{p}) + \frac{1}{2\epsilon_0}(\varrho|u_{\mathrm{Coul}} * \varrho) \tag{4.5}$$

$$H_{\mathrm{rad}}[\phi, \pi] := \frac{1}{2\epsilon_0}\Big[\|\mathbf{Y}^{\mathrm{co}}\|^2 + \|\mathbf{Y}^{\top}\|^2\Big] + \frac{1}{2\mu_0}\Big[\|\,\mathrm{curl}_0\,\mathbf{A}^{\top}\|^2 + \|\,\mathrm{curl}\,\mathsf{A}_{\mathrm{co}}\|^2\Big], \tag{4.6}$$

$$H_{\mathrm{int}}[\phi, \pi] := \underbrace{-(\frac{\varrho}{m}\mathbf{p}|\left[\mathbf{A}\circ\mathbf{q} + \mathsf{A}_{\mathrm{co}}\circ\mathbf{q}\right]) +}_{=:\, H_\ell,\ \text{linear interaction}} \underbrace{\frac{\varrho^2}{2m}\|\mathbf{A}\circ\mathbf{q} + \mathsf{A}_{\mathrm{co}}\circ\mathbf{q}\|^2}_{=:\, H_{\mathrm{q}},\ \text{quadratic interaction}}, \tag{4.7}$$

where we omit from now on the superscript r in $(\phi, \pi) \in \mathcal{N}^{\mathrm{r}} \equiv \mathcal{N}$.
The canonical equations from the Coulomb Hamiltonian yield indeed the Newton–Lorentz and Maxwell equations, expressed in terms of continuous force and current densities. They are highly non–linear in the position function $\mathbf{q}$. They split into the Helmholtz–Hodge components. Especially, by calculating the functional derivatives of the Coulomb Hamiltonian to the canonical variables $\mathbf{A}_t^{\mathrm{co}}$ and to $\mathbf{Y}_t^{\mathrm{co}}$ one derives the *dynamics for the cohomological canonical fields*

$$\underbrace{\frac{d}{dt}\begin{pmatrix}\mathbf{A}_t^{\mathrm{co}}\\ \mathbf{Y}_t^{\mathrm{co}}\end{pmatrix}}_{=\,F_t^{\mathrm{co}}} = \underbrace{\begin{pmatrix}0 & \mathbb{1}\\ 0 & 0\end{pmatrix}}_{=\,\mathbb{I}'}\underbrace{\begin{pmatrix}\mathbf{A}_t^{\mathrm{co}}\\ \mathbf{Y}_t^{\mathrm{co}}\end{pmatrix}}_{=\,F_t^{\mathrm{co}}} + \underbrace{\begin{pmatrix}0^{\mathrm{co}}\\ \mathbf{j}_t^{\mathrm{co}}\end{pmatrix}}_{=:\,\mathbf{J}_t^{\mathrm{co}}} \tag{4.8}$$

(Observe the introduction of the cohomological phase space points F_t^{co} and modified current $\mathbf{J}_t^{\mathrm{co}}$.)
By calculating the functional derivatives of the Coulomb Hamiltonian to $\mathbf{A}_t^{\top}$ and to $\mathbf{Y}_t^{\top}$ one derives the *dynamics for the transversal canonical fields*

$$\underbrace{\frac{d}{dt}\begin{pmatrix}\mathbf{A}_t^{\top}\\ \mathbf{Y}_t^{\top}\end{pmatrix}}_{=\,F_t^{\top}} = \underbrace{\begin{pmatrix}0 & \mathbb{1}\\ -\,\mathrm{curl}_{t0}^2| & 0\end{pmatrix}}_{=:\,\mathbb{B}'=(S')^{-1}\mathbb{A}^{\top}S'}\underbrace{\begin{pmatrix}\mathbf{A}_t^{\top}\\ \mathbf{Y}_t^{\top}\end{pmatrix}}_{=\,F_t^{\top}} + \underbrace{\begin{pmatrix}0^{\top}\\ \mathbf{j}_t^{\top}\end{pmatrix}}_{=:\,\mathbf{J}_t^{\top}}. \tag{4.9}$$

(Observe the introduction of the transversal phases space points $F_t^{\top}$ and modified current $\mathbf{J}_t^{\top}$.)
In the following we shall often write $F_t^{\top} \equiv F_t$ for shortness. The transformation S' connects the transversal canonical fields with the transversal force fields

$$\underbrace{\begin{pmatrix}\mathbf{E}^{\top}\\ \mathbf{B}^{\top}\end{pmatrix}}_{=\,\psi^{\top}} = \underbrace{\begin{pmatrix}0 & -\mathbb{1}\\ \mathrm{curl}_0| & 0\end{pmatrix}}_{=:\,S'}\underbrace{\begin{pmatrix}\mathbf{A}^{\top}\\ \mathbf{Y}^{\top}\end{pmatrix}}_{=\,F} \tag{4.10}$$

where the inverse relation is given by

$$\underbrace{\begin{pmatrix}\mathbf{A}^{\top}\\ \mathbf{Y}^{\top}\end{pmatrix}}_{=F} = \underbrace{\begin{pmatrix}0 & \mathrm{curl}_0|^{-1}\\ -\mathbb{1} & 0\end{pmatrix}}_{=(S')^{-1}} \underbrace{\begin{pmatrix}\mathbf{E}^{\top}\\ \mathbf{B}^{\top}\end{pmatrix}}_{=\psi^{\top}} . \tag{4.11}$$

By this we gain directly the indicated relation (4.9) between the transversal force–field Maxwell operator $\mathbb{A}^{\top}$ and the canonical Maxwell operator $\mathbb{B}'$. The dash indicates duality with respect to the corresponding quantity in the test function space, which we introduce below.
This part of the canonical dynamical equations may suffice to illustrate, that we have for charged fluids, with square integrable sources, a well defined infinite dimensional Hamilton formalism. In this Hamilton formulation the original Maxwell theory is supplemented by the kinetic energy and the Coulomb interaction of the fluid elements.
In order to arrive at specific matter models one goes over to the singular point charge distributions. Dropping constant selfenergies we find formally the usual Coulomb Hamiltonian for *M point particles*

$$H_{\mathrm{mat}}[\phi,\pi] := \sum_{k=1}^{M} \frac{1}{2m_k}(\mathbf{p}^k)^2 + \frac{1}{2\epsilon_0}\sum_{k,l=1;k\neq l}^{M} e_k e_l\, V(\mathbf{q}^k,\mathbf{q}^l) \tag{4.12}$$

$$H_{\mathrm{rad}}[\phi,\pi] := \frac{1}{2\epsilon_0}\Big[\|\mathbf{Y}^{\mathrm{co}}\|^2 + \|\mathbf{Y}^{\top}\|^2\Big] + \frac{1}{2\mu_0}\Big[\|\,\mathrm{curl}_0\,\mathbf{A}^{\top}\|^2 + \|\,\mathrm{curl}\,\mathsf{A}_{\mathrm{co}}\|^2\Big], \tag{4.13}$$

$$H_{\mathrm{int}}[\phi,\pi] := \underbrace{-\sum_{k=1}^{M}\frac{e_k}{m_k}\mathbf{p}^k\cdot\big[\mathbf{A}(\mathbf{q}^k)+\mathsf{A}_{\mathrm{co}}(\mathbf{q}^k)\big]}_{=:\,H_\ell,\ \text{linear interaction}} + \underbrace{\sum_{k=1}^{M}\frac{e_k^2}{2m_k}\big[\mathbf{A}(\mathbf{q}^k)+\mathsf{A}_{\mathrm{co}}(\mathbf{q}^k)\big]^2}_{=:\,H_{\mathrm{q}},\ \text{quadratic interaction}} . \tag{4.14}$$

From Eq. (4.9) we conclude the *magic formula for canonical transversal fields*:

$$\underbrace{\begin{pmatrix}\mathbf{A}_t^{\top}\\ \mathbf{Y}_t^{\top}\end{pmatrix}}_{=F_t} = \exp\{(t-t_0)\mathbb{B}'\}\underbrace{\begin{pmatrix}\mathbf{A}_{t_0}^{\top}\\ \mathbf{Y}_{t_0}^{\top}\end{pmatrix}}_{=F_{t_0}} + \int_{t_0}^{t}\exp\{(t-s)\mathbb{B}'\}\underbrace{\begin{pmatrix}0^{\top}\\ \mathbf{j}_s^{\top}\end{pmatrix}}_{=\mathbf{J}_s^{\top}}\, ds\ \in E_{\mathrm{cf}}^{\top\,\prime} . \tag{4.15}$$

We see immediately that the fields are not square integrable, if we insert the singular point current, since the free (canonical) Maxwell dynamics $\exp\{(t-s)\mathbb{B}'\}$ does not spread the sharp current distribution. We have no finite field energy and the norms in (4.13) are not well defined!

On the other hand, fundamental models work with point charges. So we are forced to smear the current and the fields by test functions. For the canonical fields the *test function space* is denoted E_{cf}. It should be introduced as a locally convex topological vector space (LC–space), in a way that makes the operators of interest continuous.

5. Poisson formalism and phase space dynamics

We describe first the phase space in the Coulomb gauge, for which we have to introduce the test function space. We do this separately for each Helmholtz–Hodge component. We begin with the test function spaces E_e^{κ} for the **E**–fields and make the same choice for the test spaces E_a^{κ} and E_y^{κ} for the **A**– and **Y**–fields, where κ assumes the values $\{\|, \mathrm{co}, \top\}$.
We introduce the total electric test function space, equipped with an LC–topology, according to the following scheme of a smooth Helmholtz–Hodge decomposition (without closures) into

gradients, cohomological fields, and curls

$$
\boxed{\begin{array}{rcl} E_e & := & \underbrace{E_e^{\|} \oplus \overbrace{E_e^{co}}^{} }_{\subseteq \ker(\mathrm{curl}_0)} \overbrace{\oplus E_e^{\top}}^{\subseteq \ker(\mathrm{div})} \\ & = & \mathrm{grad}_0(E_u) \oplus \mathbb{H}_2 \oplus \mathrm{curl}(E_b^{\top}) \end{array}} \tag{5.1}
$$

and observe that it is $\|.\|$–dense in $\mathrm{L}^2(\Lambda, \mathbb{R}^3)$ and constitutes a Hilbert space core for the unrestricted operators div and curl_0.
The test function space for the canonical fields is then $E_{\mathrm{cf}} = E_a \boxplus E_y \ni f = (f_a, f_y)$. The dual spaces contain the respective L^2–spaces

$$
E_{\mathrm{cf}} \subseteq \mathcal{R} \subseteq E'_{\mathrm{cf}} = \left[E_a \subseteq \mathrm{L}^2(\Lambda, \mathbb{R}^3) \subseteq E'_a\right] \boxplus \left[E_y \subseteq \mathrm{L}^2(\Lambda, \mathbb{R}^3) \subseteq E'_y\right].
$$

In the Coulomb gauge the longitudinal fields are not dynamical variables, rather may considered as indices for a foliation of the total phase space as Poisson space into nondegenerated leaves. Since we consider M point particles the material part of the phase space is finite dimensional, namely $\mathbb{R}^{6M} \ni (\mathbf{q}, \mathbf{p})$. Thus each nondegenerated leave is isomorphic to the reduced phase space $\mathcal{N} \ni (\phi, \pi) = (\mathbf{q}, \mathbf{A}^{\mathrm{co}}, \mathbf{A}^{\top}, \mathbf{p}, \mathbf{Y}^{\mathrm{co}}, \mathbf{Y}^{\top})$.
The elements of the predual space

$$
E = (\mathbb{R}^{6M}, E_{\mathrm{cf}}) = (\mathbb{R}^{6M}, E_{\mathrm{cf}}^{\mathrm{co}}, E_{\mathrm{cf}}^{\top}), \quad \text{with} \quad E' = \mathcal{N}, \tag{5.2}
$$

are written $(\phi^{\cdot}, \pi^{\cdot}) = (\mathbf{q}, f_a, \mathbf{p}, f_y) = (\mathbf{q}, f_a^{\mathrm{co}}, f_y^{\top}, \mathbf{p}, f_y^{\mathrm{co}}, f_y^{\top})$. The symplectic form on E is

$$
\begin{aligned}
&\sigma((\phi^{\cdot}_1, \pi^{\cdot}_1), (\phi^{\cdot}_2, \pi^{\cdot}_2)) := \\
&= \sum_{k=1}^{M} [\mathbf{q}_1^k \cdot \mathbf{p}_2^k - \mathbf{p}_1^k \cdot \mathbf{q}_2^k] \\
&\quad + \left[(f_{a,1}^{\mathrm{co}} | f_{y,2}^{\mathrm{co}}) - (f_{y,1}^{\mathrm{co}} | f_{a,2}^{\mathrm{co}})\right] + \left[(f_{a,1}^{\top} | f_{y,2}^{\top}) - (f_{y,1}^{\top} | f_{a,2}^{\top})\right].
\end{aligned} \tag{5.3}
$$

σ is nondegenerated on E, but would be degenerated on the predual of the non–reduced phase space. Note that $E_{\mathrm{cf}}^{\mathrm{co}}$ and $E_{\mathrm{cf}}^{\top}$ are symplectically orthogonal.

Quite generally, let (E, σ) denote an arbitrary pre–symplectic space, that is a real vector space equipped with a pre–symplectic form and with an LC–topology for which σ is separately continuous (and for which the relevant operators are continuous).

The topological dual E' of E is taken as the flat *phase space manifold*, equipped with the weak*–topology, or $\sigma(E', E)$–topology, in which we have $E'' = E$.

5.1. Poisson bracket
Because our phase space is flat, we have for the tangent and cotangent spaces

$$
T_F E' = E', \qquad T_F^* E' = E'' = E \tag{5.4}
$$

at each $F \in E'$. The total differential of $A : E' \to \mathbb{R}$ is given by

$$
d_F A[G] := \left. \frac{dA[F + tG]}{dt} \right|_{t=0}, \quad \forall G \in T_F E', \tag{5.5}
$$

if (required even for Banach Poisson manifolds)

$$d_F A \in T_F^* E' = E, \quad \forall F \in E'. \tag{5.6}$$

The complexified cotangent space is

$$T_F^* E' \oplus i T_F^* E' = E \oplus iE.$$

The most relevant observables are the **smeared fields**

$$\Phi^0(f) : E' \to \mathbb{R}, \quad \Phi^0(f)[F] := F(f), \ \forall F \in E'. \tag{5.7}$$

(The superscript 0 indicates $\hbar = 0$.) Since the smeared fields are unbounded functions on the phase space E' one uses often the **classical Weyl elements**

$$W^0(f) := \exp\{i\Phi^0(f)\}, \quad W^0(f)[F] = \exp\{iF(f)\}, \ \forall F \in E'. \tag{5.8}$$

Their total differentials are calculated to

$$\begin{aligned} d_F\Phi^0(f) &= f \in T_F^* E' = E, \\ d_F W^0(f) &= i\exp\{iF(f)\}f = iW^0(f)[F]f \quad \in E \oplus iE. \end{aligned} \tag{5.9}$$

We define the *Poisson bracket*

$$\{A, B\}[F] := \Sigma_F(d_F A, d_F B) \tag{5.10}$$
$$\begin{aligned} &= -\sigma(d_F A_1, d_F B_1) \\ &- i\sigma(d_F A_1, d_F B_2) - i\sigma(d_F A_2, d_F B_1) + \sigma(d_F A_2, d_F B_2), \end{aligned} \tag{5.11}$$

where $A = A_1 + iA_2$ and $B = B_1 + iB_2$ are smooth $\mathbb{C}$–valued functions, still called classical observables. One finds the fundamental Poisson brackets

$$\{\Phi^0(f), \Phi^0(g)\} = \sigma(f, g)\mathbb{1}, \tag{5.12}$$
$$\{W^0(f), W^0(g)\} = \sigma(f, g)W^0(f + g). \tag{5.13}$$

(Equation (5.12) generalizes the usual Poisson bracket for canonical point fields involving δ–functions.)
As a *Poisson algebra* $(\mathcal{P}, \{.,.\})$ we take the following small commutative *–algebra (and not $C^\infty(E')$, common for finite dimensional phase space):

$$\Delta(E, 0) := \mathrm{LH}\{W^0(f) \mid f \in E\}, \tag{5.14}$$

in which one has beside the Poisson product the commutative point–wise multiplication and the complex conjugation. The supremum norm is a C*–norm in $\Delta(E, 0)$, but we cannot form the norm–closure, since then the Poisson product would fail.

5.2. Hamiltonian phase space flow
For each $f \in E$ the $\mathbb{R}$–linear mapping

$$\sigma_\sharp f : E \to \mathbb{R}, \quad \text{where} \quad (\sigma_\sharp f)(g) := \sigma(f, g) \quad \forall g \in E, \tag{5.15}$$

is continuous, implying $\sigma_\sharp(f) \in E'$. Thus we have an $\mathbb{R}$–linear mapping $\sigma_\sharp : E \to E'$, $f \mapsto \sigma_\sharp f$, which is LC–weak* continuous.

If the form σ is non–degenerate then $\sigma_\sharp : E \to E'$ is injective and the image $\sigma_\sharp(E)$ is $\sigma(E', E)$–dense in E' (a proper subspace for infinite dimensions).
The Hamiltonian vector field $F \mapsto X_H[F] \in T_F E' = E'$, associated with $H : E' \to \mathbb{R}$ is defined by

$$X_H[F] := -\sigma_\sharp(d_F H), \quad \forall F \in E'. \tag{5.16}$$

(Recall that the vectors are in E' and the allowed co–vectors in E.) Equivalently we may express this relation in terms of a Poisson bracket with the smeared field

$$X_H[F](f) = -\sigma(d_F H, f) = \{H, \Phi^0(f)\}[F], \quad f \in E. \tag{5.17}$$

The Hamiltonian flow satisfies the differential equation

$$\frac{d}{dt}(\varphi_t^H F)(f) = -\sigma(d_{\varphi_t^H F} H, f), \quad \forall f \in E, \quad \forall F \in E'. \tag{5.18}$$

Proposition 5-1 *Provided a smooth Hamiltonian function H, then*

$$\frac{d}{dt} A \circ \varphi_t^H = \{H, A\} \circ \varphi_t^H \tag{5.19}$$

for every smooth function $A : E' \to \mathbb{C}$. Furthermore, we have the energy conservation

$$H[\varphi_t^H F] = H[F], \quad \forall F \in E', \quad \forall t \in \mathbb{R}. \tag{5.20}$$

Definition 5-2 *Let σ be jointly continuous symplectic form. Denote by* $\mathrm{symp}(E, \sigma)$ *the set of* symplectic transformations $T : E \overset{bijective}{\longrightarrow} E$ *satisfying*

$$\sigma(Tf, Tg) = \sigma(f, g), \quad \forall f, g \in E. \tag{5.21}$$

Clearly $\mathrm{symp}(E, \sigma)$ *is a group.*
A strongly LC–continuous symplectic one–parameter (semi–) group $\{T_t \mid t \in \mathbb{R}_+\} \subset$ $\mathrm{symp}(E, \sigma)$, *for which by definition $\mathbb{R} \ni t \mapsto T_t = \exp\{tG\} f$ LC–continuous for each $f \in E$, is termed* strongly LC–differentiable *if it has a generator G which is LC–continuous and defined on all of E.*
We denote by $\mathcal{G}_{\mathrm{symp}}(E, \sigma)$ the set of all these strong symplectic generators G, endowed with the $\mathbb{R}$–linear operations and the commutator product, and term it the **symplectic Lie algebra** .

It seems natural to equip $\mathcal{G}_{\mathrm{symp}}(E, \sigma)$ with the weak operator topology, where $G_i \to G$, if and only if $F(G_i f) \to F(Gf)$, $\forall F \in E'$, $\forall f \in E$. The Lie product in $\mathcal{G}_{\mathrm{symp}}(E, \sigma)$ is then separately continuous but in general not bi–continuous. $\mathcal{G}_{\mathrm{symp}}(E, \sigma)$ has an isomorphic realization by the dual transformations acting in the phase space E', that is by the set $\{G' \mid G \in \mathcal{G}_{\mathrm{symp}}(E, \sigma)\}$.

Theorem 5-3 *For each $G \in \mathcal{G}_{\mathrm{symp}}(E, \sigma)$ we introduce the* quadratic Hamilton function *H_G by*

$$H_G[\sigma_\sharp f] := \frac{1}{2}\sigma(f, Gf), \quad \forall f \in E, \tag{5.22}$$

defined on $\sigma_\sharp E \subset E'$. The map $J : \mathcal{G}_{\mathrm{symp}}(E, \sigma) \to C^\infty(\sigma_\sharp E)$, $J_G = H_G$, is called the co–momentum map.
We have for each $G \in \mathcal{G}_{\mathrm{symp}}(E, \sigma)$ that the Hamiltonian vector field associated with H_G is effectuated by the linear phase space map

$$X_{H_G} = G'. \tag{5.23}$$

Consequently, the global phase space flow $\varphi_t^{H_G}$ associated with H_G exists, and is given by the weak–continuous one–parameter group*

$$\varphi_t^{H_G} = \exp\{tG'\}, \quad t \in \mathbb{R}, \quad \textit{defined everywhere on } E' \,. \tag{5.24}$$

This is the dual group to the original $\{\exp\{tG\} \mid t \in \mathbb{R}\} \subset \mathrm{symp}(E,\sigma)$ (the strongly LC–differentiable symplectic group).
Moreover, the dynamics for the classical field observable $\Phi^0(f)$ as well as for the classical Weyl function $W^0(f)$ (defined in the Eqs. (5.7) and (5.8)) may be written

$$\mathbb{R} \ni t \mapsto \alpha_t^{H_G}(\Phi^0(f)) = \Phi^0(f) \circ \varphi_t^{H_G} = \Phi^0(\exp\{tG\}f)\,, \tag{5.25}$$

$$\mathbb{R} \ni t \mapsto \alpha_t^{H_G}(W^0(f)) = W^0(f) \circ \varphi_t^{H_G} = W^0(\exp\{tG\}f) \tag{5.26}$$

for each test function $f \in E$.
In terms of the co–momentum map we obtain the Lie homomorphism

$$\{J_{G_1}, J_{G_2}\} = J_{[G_1,G_2]}, \quad G_1, G_2 \in \mathcal{G}_{\mathrm{symp}}(E,\sigma)\,. \tag{5.27}$$

Only as a side remark let us introduce also in our infinite dimensional case the *momentum map associated with the action of the Lie algebra $\mathcal{G}_{\mathrm{symp}}(E,\sigma)$ on E'* by means of

$$\mathbf{J} : \sigma_\sharp(E) \mapsto \mathcal{G}_{\mathrm{symp}}(E,\sigma)^*, \; \mathbf{J}[F](G) := J_G[F], \;\; F \in \sigma_\sharp(E), \; G \in \mathcal{G}_{\mathrm{symp}}(E,\sigma)\,, \tag{5.28}$$

where $\mathcal{G}_{\mathrm{symp}}(E,\sigma)^*$ denotes the continuous linear forms on $\mathcal{G}_{\mathrm{symp}}(E,\sigma)$.
In other words, the momentum map is the evaluation of the Hamilton function H_G at each $\sigma_\sharp f$, viewed as a dual element to the Lie algebra in virtue of its linear dependence on $G \in \mathcal{G}_{\mathrm{symp}}(E,\sigma)$. The interpretation of this evaluation as an element in $\mathcal{G}_{\mathrm{symp}}(E,\sigma)^*$ gives the discussion of group orbits a more abstract setting in form of the co–adjoint orbits.

A Lie–Poisson structure on $\mathcal{G}_{\mathrm{symp}}(E,\sigma)^*$ is introduced by the Poisson bracket for the special functions $\tilde{G}(\eta) := \eta(G)$, $\forall \eta \in \mathcal{G}_{\mathrm{symp}}(E,\sigma)^*, G \in \mathcal{G}_{\mathrm{symp}}(E,\sigma)$, as follows

$$\{\tilde{G_1}, \tilde{G_2}\} := \widetilde{[G_1, G_2]}\,. \tag{5.29}$$

Thus, the momentum map transforms (part of) the phase space E' into the more abstract Poisson manifold $\mathcal{G}_{\mathrm{symp}}(E,\sigma)^*$, preserving the Poisson structure. That means that the pullback $\mathbf{J}^*\tilde{G}$ leads to the phase space function $J_G = H_G$ for all $G \in \mathcal{G}_{\mathrm{symp}}(E,\sigma)$ and satisfies

$$\mathbf{J}^*\{\tilde{G_1}, \tilde{G_2}\} = \{\mathbf{J}^*\tilde{G_1}, \mathbf{J}^*\tilde{G_2}\}, \quad G_1, G_2 \in \mathcal{G}_{\mathrm{symp}}(E,\sigma)\,. \tag{5.30}$$

6. Maxwell Generators and their Hamilton Functions

We return to the phase space in the Coulomb gauge. The dynamical operators in the test function space are the Hermitian adjoints of the original operators. In order to possess a dual operator they should be defined on the whole test function space and be continuous in the LC–topology of the test function space.
By means of the Helmholtz–Hodge decomposition we have a separated *cohomological free Maxwell dynamics* which we deduce from Eq. (4.8) to be given by $\{\exp\{t\mathbb{I}\} \mid t \in \mathbb{R}\} \subseteq \mathrm{symp}(E^{\mathrm{co}}_{\mathrm{cf}}, \sigma)$, with the strong symplectic generator (inserting for the moment the physical constants)

$$\mathbb{I} = \begin{pmatrix} 0 & 0 \\ \frac{1}{\epsilon_0}\mathbb{1} & 0 \end{pmatrix} \quad \in \mathcal{G}_{\mathrm{symp}}(E^{\mathrm{co}}_{\mathrm{cf}}, \sigma)\,. \tag{6.1}$$

It leads with Eq. (5.22) to the Hamilton function

$$H_{\mathbb{I}}[\mathbf{A}^{\text{co}},\mathbf{Y}^{\text{co}}] = \frac{1}{2\epsilon_0}\|\mathbf{Y}^{\text{co}}\|^2 = \frac{\epsilon_0}{2}\|\mathbf{E}^{\text{co}}\|^2 = \frac{\epsilon_0}{2}\int_\Lambda \mathbf{E}^{\text{co}}(x)^2\,d^3x\,, \tag{6.2}$$

where we have denoted the test functions f embedded via $\sigma_\sharp f$ into $E_{\text{cf}}^{\text{co}\prime}$ by the original field symbols.
The *transversal dynamical pre–dual operator* is

$$\mathbb{B} = \begin{pmatrix} 0 & -\frac{1}{\mu_0}\,\text{curl}^2_{t0}| \\ \frac{1}{\epsilon_0}\mathbb{1} & 0 \end{pmatrix}$$

and this gives rise with Eq. (5.22) to the strictly positive quadratic Hamiltonian function corresponding to the *transversal field energy* in the domain Λ,

$$\begin{aligned} H_{\mathbb{B}}[\mathbf{A}^\top,\mathbf{Y}^\top] &= \frac{1}{2\epsilon_0}\|\mathbf{Y}^\top\|^2 + \frac{1}{2\mu_0}\|\,\text{curl}_0\,\mathbf{A}^\top\|^2 \\ &= \frac{\epsilon_0}{2}\|\mathbf{E}^\top\|^2 + \frac{1}{2\mu_0}\|\mathbf{B}^\top\|^2 = \frac{\epsilon_0}{2}\int_\Lambda \left(\mathbf{E}^\top(x)^2 + c^2\mathbf{B}^\top(x)^2\right)d^3x\,. \end{aligned} \tag{6.3}$$

If Λ is invariant under rotations about a fixed axis, then these rotations constitute a symplectic transformation in the field phase space and their generator leads to a quadratic Hamiltonian, the angular momentum of the field about the axis. Similarly the invariance of the (unbounded) Λ under translations in a fixed direction provides the field momentum as quadratic Hamiltonian. These considerations disclose the symplectic invariant structure of Maxwell theory.
We make sure, that $\mathbb{B}'$ indeed gives rise to a mathematically reasonable weak*–continuous one–parameter group $t \mapsto \exp\{t\mathbb{B}'\}$, if we prove that the predual transformation group

$$t \mapsto \exp\{t\mathbb{B}\} = S\exp\{-t\mathbb{A}^\top\}S^{-1}, \quad \text{acting on} \quad E_{\text{cf}}^\top, \tag{6.4}$$

is strongly continuous in the following sense: Each $\exp\{t\mathbb{B}\}$ is an LC–homeomorphism on $E_{\text{cf}}^\top$, $t \mapsto \exp\{t\mathbb{B}\}f$ is LC–continuous for every $f \in E_{\text{cf}}^\top$, and for each continuous semi–norm q on $E_{\text{cf}}^\top$ there exists a continuous semi–norm p on $E_{\text{cf}}^\top$ together with a continuous map $t \mapsto a_t \geq 0$ such that

$$q(\exp\{t\mathbb{B}\}f) \leq a_t p(f)\,, \quad \forall f \in E_{\text{cf}}^\top\,, \quad \forall t \in \mathbb{R}\,.$$

The mere continuity of $\mathbb{B}$ is seen directly from the fact that each of the following predual matrix operators is an LC–continuous homeomorphism

$$\underbrace{\begin{pmatrix} 0 & -\text{curl}^2_{t0}| \\ \mathbb{1} & 0 \end{pmatrix}}_{=\mathbb{B}} = -\underbrace{\begin{pmatrix} 0 & \text{curl}| \\ -\mathbb{1} & 0 \end{pmatrix}}_{=S}\underbrace{\begin{pmatrix} 0 & \text{curl}| \\ -\text{curl}_0| & 0 \end{pmatrix}}_{=\mathbb{A}^\top}\underbrace{\begin{pmatrix} 0 & -\mathbb{1} \\ \text{curl}|^{-1} & 0 \end{pmatrix}}_{=S^{-1}}. \tag{6.5}$$

By transposing the transversal magic formula Eq. (4.15) into the predual and applying it under use of Eq. (5.25) to the classical Weyl elements we gain the following form

$$\alpha^{\text{curr}}_{t_0\to t}(W^0(f)) = \exp\{i\textstyle\int_{t_0}^t \mathbf{J}_s^\top(\exp\{(t-s)\mathbb{B}\}f)\,ds\}\,W^0(\exp\{(t-t_0)\mathbb{B}\}f)\,. \tag{6.6}$$

Note that the influence of the current is expressed by means of a phase factor, mathematically a character on the additive group E_{cf}. Since the time dependence of the current is rather arbitrary the dynamical operators in the classical Heisenberg picture do not form a group, but only a groupoid, in which the partial composition law

$$\alpha^{\text{curr}}_{t_0\to t}\circ\alpha^{\text{curr}}_{t\to t'} = \alpha^{\text{curr}}_{t_0\to t'}, \quad \forall\, t_0, t, t' \in \mathbb{R}\,, \tag{6.7}$$

requires the equality of intermediate time.

6.1. Diagonalization of the Transversal Free Maxwell Dynamics

For the following it is essential that $\mathrm{curl}^2_{t0}| = \mathrm{curl}|\,\mathrm{curl}_0|$ is strictly positive. By spectral calculus the $\sqrt{\mathrm{curl}^2_{t0}|}^{\pm 1/2}$ are well defined.

Definition 6-1 *Let (E, σ) be a symplectic space. A real–linear operator $j : E \to E$ is called a complex structure in (E, σ), if it satisfies*

$$\begin{aligned} & j^2 f = -f\,, \\ & \sigma(jf, jg) = \sigma(f, g)\,, \\ & \sigma(f, jf) \geq 0\,, \end{aligned} \tag{6.8}$$

for all $f, g \in E$.

It follows that $j \in \mathrm{symp}(E, \sigma)$ with inverse $j^{-1} = -j$. If j exists, it allows for the multiplication by complex numbers $z = \mathrm{Re}(z) + i\,\mathrm{Im}(z)$, where $if := jf$, so that

$$zf := \mathrm{Re}(z)f + \mathrm{Im}(z)jf\,, \quad \forall z \in \mathbb{C}\,, \quad f \in E\,. \tag{6.9}$$

In this manner E becomes a complex vector space. A complex inner product is introduced by

$$(f|g)_j := \sigma(f, jg) + i\sigma(f, g)\,, \quad \forall f, g \in E\,. \tag{6.10}$$

The associated norm is denoted $\|.\|_j$.
If j is any complex structure in (E, σ), then

$$j_T := T^{-1} j T\,, \quad T \in \mathrm{symp}(E, \sigma)\,, \tag{6.11}$$

is also a complex structure in (E, σ), for which we have

$$\|f\|_{j_T} = \|Tf\|_j\,, \quad \mathrm{Re}(f|g)_{j_T} = \mathrm{Re}(Tf|Tg)_j\,, \quad \forall f, g \in E\,. \tag{6.12}$$

Our transversal test function space $(E^{\top}_{\mathrm{cf}}, \sigma)$ of smooth canonical fields (f_a, f_y) has as symplectic form σ the transversal part of (5.3): $\sigma((f_a, f_y), (f'_a, f'_y)) = (f_a|f'_y) - (f_y|f'_a)$, which involves the $L^2(\Lambda, \mathbb{R}^3)$–scalar products. The most simple complex structure in $(E^{\top}_{\mathrm{cf}}, \sigma)$, satisfying the requirements (6.8), is given by

$$j\,(f_a, f_y) := (-f_y, f_a), \text{ expressed in matrix notation by } \quad j = \begin{pmatrix} 0 & -\mathbb{1} \\ \mathbb{1} & 0 \end{pmatrix}. \tag{6.13}$$

In terms of the function realization of $(E^{\top}_{\mathrm{cf}}, \sigma)$ we get the isomorphic complexification if we write (f_a, f_y) as $f_a + if_y$ and define the complex unit as the multiplication of the function values by $i \in \mathbb{C}$. The symplectic form is then realized by $\mathrm{Im}(.|.)$, making use of the scalar product in $L^2(\Lambda, \mathbb{C}^3)$. This complexified symplectic space is denoted by $(E^{\top}, \mathrm{Im}(.|.))$, where $E^{\top}$ is a dense subspace of $\mathcal{H}^{\top}$, the transversal part of $L^2(\Lambda, \mathbb{C}^3)$. So the most natural isomorphism v between the symplectic spaces is

$$(E^{\top}_{\mathrm{cf}}, \sigma) \ni (f_a, f_y) \to v(f_a, f_y) := f_a + if_y \in (E^{\top}, \mathrm{Im}(.|.))\,, \tag{6.14}$$

which gives

$$j = v^{-1} i\, v, \quad \text{with } i \text{ the complex unit in } \mathcal{H}^{\top}\,. \tag{6.15}$$

Remark 6-2 *The qualification "natural" of the foregoing complexification of* $(E^{\top}_{cf}, \sigma)$ *is substantiated by the fact that the physically induced splitting* $E^{\top}_{cf} = E_a \boxplus E_y$ *(in a fixed inertial system) is expressible as* E_a *being the Lagrangian subspace of* $(E^{\top}_{cf}, \sigma)$ *corresponding to* j*. That means that* E_a *is the* σ*–orthogonal complement of* $jE_a = E_y$*. Changing the* j *would lead to other Lagrangian subspaces.*

Let us now introduce the diagonalization transformation V as the $\mathbb{R}$–linear, injective mapping from the real space $E^{\top}_{\rm cf} \ni (f_a, f_y)$ into the complex Hilbert space $\mathcal{H}^{\top} \ni f_a + if_y$, given by

$$V(f_a, f_y) := w^{-1} f_a + iwf_y = Vf \quad \text{with} \quad w := (\epsilon_0 c\sqrt{\mathrm{curl}^2_{t0}|})^{+1/2}, \tag{6.16}$$

so that

$$V = v \circ W \quad \text{with} \quad W = \begin{pmatrix} w^{-1} & 0 \\ 0 & w \end{pmatrix}. \tag{6.17}$$

Theorem 6-3 *Let again* $E^{\top} \subset \mathcal{H}^{\top}$ *be the image of* $E^{\top}_{cf}$ *under the above diagonalization transformation* V*. Then it holds:*

(a) The diagonalization V *from Eq. (6.16) is a symplectic LC–homeomorphism from* $(E^{\top}_{cf}, \sigma)$ *onto* $(E^{\top}, \mathrm{Im}(.|.))$*, satisfying*

$$\mathrm{Im}(Vf|Vg) = \sigma(f, g), \quad \forall f, g \in E^{\top}_{cf}. \tag{6.18}$$

(b) $(\mathcal{H}^{\top}, E^{\top}, c\sqrt{\mathrm{curl}^2_{t0}|}, V)$ *is the unique (up to unitary equivalence)* diagonalization of the transversal canonical Maxwell dynamics, *given by the one–parameter unitary group* $\exp\{t\mathbb{B}\}$*,* $t \in \mathbb{R}$*, on* $(E^{\top}_{cf}, \sigma)$*:*

$$V \exp\{t\mathbb{B}\}V^{-1} = \exp\{itc\sqrt{\mathrm{curl}^2_{t0}|}\}|_{E^{\top}}, \quad \forall t \in \mathbb{R}. \tag{6.19}$$

That means that any other unitary group in another Hilbert space realizing $\exp\{t\mathbb{B}\}$ *is reached by a unitary transformation from the given one.*

(c) $V\mathbb{B}V^{-1} = ic\sqrt{\mathrm{curl}^2_{t0}|} =: i\mathbb{D}$ *is an LC–homeomorphism in* $E^{\top} \subset \mathcal{H}^{\top}$*.*
(By (b) $\mathbb{D}$ *is a symmetric operator which has a unique extension to a selfadjoint operator in* $\mathcal{H}^{\top}$*. and is thus a complex linear operator.)*

(d) There is no diagonalization *for the longitudinal and cohomological fields.*

The form of $\mathbb{D}$ is obtained by direct calculation. The domain of definition of $\mathbb{D}$ is a subspace of the domain of $\mathrm{curl}_0|_{E^{\top}}$, that are "smooth" transversal complex fields with vanishing tangential component at the boundary $\partial\Lambda$.
If for the nilpotent generator $\mathbb{I}$ of the longitudinal and cohomological dynamics there would be any diagonalization map onto an $i\mathbb{J}$, with $\mathbb{J}$ a bounded selfadjoint complex–linear operator in a complex Hilbert space, then $\mathbb{J}$ would be also nilpotent, and thus equal to 0. (See however our Observation 8-7.)
It is remarkable that the quadratic Hamilton function Eq. (6.3) (the transversal Maxwell energy, originally of the type (5.22)) may by (6.18) be written in terms of a scalar product in the complex diagonalization Hilbert space, as reveals the following calculation

$$H_{\mathbb{B}}[\sigma_{\sharp}f] = \frac{1}{2}\sigma(f, \mathbb{B}f) = \frac{1}{2}\mathrm{Im}(Vf|i\mathbb{D}Vf) = \frac{1}{2}(Vf|\mathbb{D}Vf), \quad \forall f \in E^{\top}_{\rm cf}. \tag{6.20}$$

In the last step the essential selfadjointness of $\mathbb{D}$ has been observed, which makes the last scalar product real.

If we choose a real orthonormal basis $\{e_n \,|\, n \in \mathbb{N}\}$ in the complex $E^\top \subset \mathcal{H}^\top$, then the real linear combination of $\{e_n, ie_n \,|\, n \in \mathbb{N}\}$ are dense in $\mathcal{H}^\top$. Observe that $\Phi^0(V^{-1}e)[\sigma_\sharp f]$ $= (\sigma_\sharp f)(V^{-1}e) = \sigma(f, V^{-1}e) = \mathrm{Im}(Vf|e)$, for all $f \in E^\top_{\mathrm{cf}}$ and all $e \in E^\top$. We then get the *diagonal representation* of the Hamiltonian in terms of quadratic field expressions

$$\begin{aligned} H_{\mathbb{B}}[\sigma_\sharp f] &= \frac{1}{2}\sum_{n\in\mathbb{N}}(Vf|e_n)\,(e_n|\mathbb{D}Vf) \\ &= \frac{1}{2}\sum_{n\in\mathbb{N}}\left(\mathrm{Im}(Vf|e_n)\;\mathrm{Im}(Vf|\mathbb{D}e_n) + \mathrm{Im}(Vf|ie_n)\;\mathrm{Im}(Vf|i\mathbb{D}e_n)\right) \\ &= \frac{1}{2}\sum_{n\in\mathbb{N}}\left(\Phi^0(V^{-1}e_n)\Phi^0(V^{-1}\mathbb{D}e_n) + \Phi^0(V^{-1}ie_n)\Phi^0(V^{-1}i\mathbb{D}e_n)\right)[\sigma_\sharp f]\,. \end{aligned} \tag{6.21}$$

Provided the orthonormal basis $\{e_n \mid n \in \mathbb{N}\}$ consists of eigenvectors of $\mathbb{D}$ with eigenvalues $\varepsilon_n \in \mathbb{R}$ (supposing a pure point spectrum for $\mathbb{D}$, valid in bounded cavities), then $H_{\mathbb{B}}$ admits the simpler diagonal representation on $\sigma_\sharp(E^\top_{\mathrm{cf}})$ which has a similarity to a sum of harmonic oscillators ($\Phi^0(V^{-1}e_n)$ corresponds to the n–th position and $\Phi^0(V^{-1}ie_n)$ to the n–th momentum coordinate):

$$H_{\mathbb{B}} = \frac{1}{2}\sum_{n\in\mathbb{N}}\varepsilon_n\left(\Phi^0(V^{-1}e_n)^2 + \Phi^0(V^{-1}ie_n)^2\right). \tag{6.22}$$

The $V^{-1}e_n$ and $V^{-1}ie_n$ are the back transforms of the complex eigen modes to the eigen value ε_α into the real field formalism (where we assume again a finite cavity). Since $\mathrm{curl}^2_{t0}| = \mathrm{curl}|\,\mathrm{curl}_0|$ is strictly positive we know $\varepsilon_n > 0$. One also knows that the ε_n converge to $+\infty$ (and are thus at most finitely degenerate). We see that $H_{\mathbb{B}}[\sigma_\sharp(V^{-1}e_n)] + H_{\mathbb{B}}[\sigma_\sharp(V^{-1}ie_n)] = \varepsilon_\alpha$, so that the evaluation of the Hamilton function at these two special phase space points sums up to an eigen energy of $\mathbb{D}$. $\mathbb{D}$ describes the filtering of Vf into eigen modes, and $\mathbb{B}$ does the same for $f \in E^\top_{\mathrm{cf}}$. But neither $\mathbb{D}$ nor $\mathbb{B}$ are observables in the proper sense for the present classical theory, since one must strictly adhere to defining a classical observable as a phase space function. Note for further illustration that neither the frequency nor the polarization of a plane wave in a cubic box, which we write out below, are classical observables.

7. QED as Weyl quantized electrodynamics

7.1. The C–Weyl algebra and field operators*

Our developments employ the very flexible and universal construction of the Weyl algebra $\mathcal{W}(E, \hbar\sigma)$ for an arbitrary pre–symplectic test function space (E, σ) [7]. The C*–Weyl algebra $\mathcal{W}(E, \hbar\sigma)$ is generated by the Weyl elements $W^\hbar(f)$, $f \in E$, satisfying the *Weyl relations*

$$\begin{aligned} W^\hbar(f)W^\hbar(g) &= \exp\{-\tfrac{i}{2}\hbar\sigma(f,g)\}W^\hbar(f+g)\,, \\ W^\hbar(f)^* &= W^\hbar(-f)\,, \end{aligned} \qquad \forall f, g \in E\,. \tag{7.1}$$

We use the same test function space for all values of the Planck parameter $\hbar$, where the latter varies in $\mathbb{R}_+$ (including $\hbar = 0$).
The set of all finite linear combinations

$$\Delta(E, \hbar\sigma) := \mathrm{LH}\{W^\hbar(f) \mid f \in E\} \tag{7.2}$$

is a *–algebra.

There exists a *unique* C^**–norm* $\|.\|$ ([8], [9]) in $\Delta(E,\hbar\sigma)$, such that every representation and every state on $\Delta(E,\hbar\sigma)$ is $\|.\|$–continuous:

$$\|A\| = \sup\{\|\Pi(A)\| \mid \Pi \text{ representation of } \Delta(E,\hbar\sigma)\}\,, \quad A \in \Delta(E,\hbar\sigma)\,,$$

the so–called minimal regular norm.

Definition 7-1 *For all $\hbar \in \mathbb{R}_+$ the $\|.\|$–completion of $\Delta(E,\hbar\sigma)$ by means of the minimal regular norm*

$$\mathcal{W}(E,\hbar\sigma) := \overline{\Delta(E,\hbar\sigma)}^{\|\|} \tag{7.3}$$

is called the Weyl algebra *over the pre–symplectic space $(E,\hbar\sigma)$ [7].*

Definition 7-2 *Suppose $(\Pi,\mathcal{H}_\Pi)$ to be a regular representation of the C^*–Weyl algebra $\mathcal{W}(E,\hbar\sigma)$, in which by definition $t \to W_\Pi(tf)$ is a strongly continuous unitary group. Then we associate with it the* quantum field system $(\Phi_\Pi,\mathcal{H}_\Pi)$

$$\Phi_\Pi(f) = -i\,\frac{dW_\Pi(tf)}{dt}\bigg|_{t=0}\,, \quad f \in E\,. \tag{7.4}$$

according to Stone's Theorem.

The field operators satisfy the (smeared) *canonical commutation relations*

$$[\Phi_\Pi(f),\Phi_\Pi(g)] \subseteq i\hbar\sigma(f,g)\mathbb{1}_\Pi\,. \tag{7.5}$$

The most suggestive choice for a *quantization map* $Q_\hbar$ is now given by

$$Q_\hbar^{\text{Weyl}}(W^0(f)) := W^\hbar(f)\,, \quad \forall f \in E\,, \tag{7.6}$$

and its $\mathbb{C}$–linear resp. continuous extension to $\Delta(E,0\cdot\sigma) \subset \mathcal{W}(E,0)$.
But we may take also other prescriptions $Q_\hbar = Q_\hbar^w$, namely

$$Q_\hbar^w(W^0(f)) := w(\hbar,f)W^\hbar(f)\,, \quad \forall f \in E\,, \tag{7.7}$$

for certain quantization factors $w(\hbar,f) \in \mathbb{C}$. The choice of w determines the *operator ordering* of the field operators. For example $w(\hbar,f) = \exp\{-\frac{\hbar(1\mp 2)}{4}\|f\|_j^2\}$ leads to normal resp. anti–normal ordering of the creation and annihilation operators defined via the complexification j (see (8.1) below). Beside other things one has to postulate $w(\hbar,f) \to 1$ for $\hbar \to 0$ ([10]).

Theorem 7-3 *The modified Weyl quantization maps of Eq. (7.7), decorated with an appropriate quantization factor $w(\hbar,f)$, lead to strict deformation quantizations of the Poisson algebra $\Delta(E,0)$, as characterized by the subsequent Definition below [11].*
These strict deformation quantizations all are equivalent to each other in the sense of [12]) (by approaching smoothly the same classical limit in the norm topology).

Definition 7-4 *Let $I \subseteq \mathbb{R}$ be the range of values for $\hbar$, including 0 as accumulation point. A strict quantization $(\mathcal{A}^\hbar,Q_\hbar)_{\hbar\in I}$ of a Poisson algebra $(\mathcal{P},\{.,.\})$ (which owes also a commutative product $\cdot$) consists for each value $\hbar \in I$ of a C^*–algebra $\mathcal{A}^\hbar$ with norm $\|.\|_\hbar$ and of a linear, $*$–preserving map*

$$Q_\hbar : \mathcal{P} \longrightarrow \mathcal{A}^\hbar\,, \qquad \text{the quantization map}\,,$$

such that Q_0 is the identical embedding of $\mathcal{P}$ into the commutative $\mathcal{A}^0$, and such that the following conditions are satisfied:

(a) *[Dirac's Condition] The $\hbar$–scaled commutator approaches the Poisson bracket in the norm topology as $I_0 \ni \hbar \to 0$, that means,*

$$\lim_{\hbar \to 0} \|[Q_\hbar(A), Q_\hbar(B)]/i\hbar - Q_\hbar(\{A, B\})\|_\hbar = 0\,, \quad \forall A, B \in \mathcal{P}\,.$$

(b) *[von Neumann's Condition] In the limit $\hbar \to 0$ one has the asymptotic product homomorphism*

$$\lim_{\hbar \to 0} \|Q_\hbar(A)Q_\hbar(B) - Q_\hbar(AB)\|_\hbar = 0\,, \quad \forall A, B \in \mathcal{P}\,.$$

(c) *[Rieffel's Condition] $I \ni \hbar \mapsto \|Q_\hbar(A)\|_\hbar$ is continuous for each $A \in \mathcal{P}$.*

The strict quantization $(\mathcal{A}^\hbar, Q_\hbar)_{\hbar \in I}$ is called a strict deformation quantization, *if*

(d) *[Deformation Condition] The map $Q_\hbar : \mathcal{P} \to \mathcal{A}^\hbar$ is injective. And moreover, its image $Q_\hbar(\mathcal{P})$ is closed with respect to the product of $\mathcal{A}^\hbar$ (i.e. $Q_\hbar(A)Q_\hbar(B) \in Q_\hbar(\mathcal{P})$ for all $A, B \in \mathcal{P}$), or equivalently, $Q_\hbar(\mathcal{P})$ is a sub–*–algebra of $\mathcal{A}^\hbar$.*

One gets the connection with the usual deformation quantization [13], [14] by equipping the *–algebra of classical observables $\mathcal{P}$ with the *deformed, non–commutative product* $\cdot_\hbar$ according to

$$A \cdot_\hbar B := Q_\hbar^{-1}(Q_\hbar(A)Q_\hbar(B))\,, \quad \forall A, B \in \mathcal{P}\,.$$

Theorem 7-3 tells us that the decorated Weyl quantizations satisfy all of the foregoing conditions and may especially be formulated in terms of a deformed product for phase space functions. From Eq. (7.1) we obtain the commutator

$$[W^\hbar(f), W^\hbar(g)] = [\exp\{-i\hbar\sigma(f,g)/2\} - \exp\{i\hbar\sigma(f,g)/2\}]W^\hbar(f+g)\,, \quad \forall f, g \in E\,, \qquad (7.8)$$

which approaches 0 (commutativity!) for small $\hbar$, whereas the scaled commutator is seen to approach the classical Poisson bracket Eq. (5.13). For the unbounded field operators we obtain the corresponding classical asymptotic behavior only in terms of certain expectation values. For a foundational discussion it is noteworthy that a quantized (field) theory, obtained by strict deformation quantization, may be formulated in terms of functions over the classical phase space.

We understand henceforth under *Weyl quantization* the quantization map Eq. (7.6), in which the trivial quantization factor leads to symmetric products for the field operators.

7.2. Symplectic–affine equivariance of Weyl quantization

The linear test function space determines not only the symplectic group $\mathrm{symp}(E, \sigma)$ but also the additive group of (affine) translations. The Bohr compactification [15] of the latter, and also that of E'_τ, is isomorphic to the group $\widehat{E}$ of all characters on E, independently of the LC topology τ.

The semi–direct product of $\mathrm{symp}(E, \sigma)$ with $\widehat{E}$ forms the symplectic–affine group.

Since the test function space is the same for all $\hbar$ one observes by inspection that we have the following commutative diagram, where $\beta_T^\hbar$ denotes the *–automorphism in the Weyl algebra $\mathcal{W}(E, \hbar\sigma)$, which is induced by the symplectic transformation $T \in \mathrm{symp}(E, \sigma)$ (called *Bogoliubov automorphism* in [8]; note that the symplectic form in the Weyl relations is invariant under T):

$$\begin{array}{ccc} \text{Classical Theory} & & \text{Quantum Theory} \\ \mathcal{W}(E, 0) & \overset{\text{quantization}}{\Longrightarrow} & \mathcal{W}(E, \hbar\sigma) \\ \downarrow \beta_T^0 & & \downarrow \beta_T^\hbar \\ \mathcal{W}(E^\top, 0) & \overset{\text{quantization}}{\Longrightarrow} & \mathcal{W}(E^\top, \hbar\sigma)\,. \end{array} \qquad (7.9)$$

As for the affine translations, they are given by the multiplication of the Weyl elements by the associated character and commute thus also with the linear quantization map.
The application of the Weyl quantization to ED proceeds by adapting the test function space. In Coulomb gauge the degenerate pre–symplectic form causes the longitudinal fields to stay classical. On the other side, not only the transversal fields but also the cohomological fields are transformed into canonical quantum fields.
Especially for the Weyl quantization of the transversal canonical fields we have the following commutative diagram, where $\beta_V^\hbar$ denotes now the *–isomorphism between $\mathcal{W}(E_{\rm cf}^\top, \hbar\sigma)$ and $\mathcal{W}(E^\top, \hbar\,\mathrm{Im}(.|.))$, which is induced by the special symplectic diagonalization transformation V (6.16) (leading from the real $(E_{\rm cf}^\top, \hbar\sigma)$ to the complex $(E^\top, \hbar\,\mathrm{Im}(.|.))$):

$$\begin{array}{ccc} \text{Classical ED} & & \text{QED} \\ \mathcal{W}(E_{\rm cf}^\top, 0) & \overset{\text{quantization}}{\Longrightarrow} & \mathcal{W}(E_{\rm cf}^\top, \hbar\sigma) \\ \downarrow \beta_V^0 & & \downarrow \beta_V^\hbar \\ \mathcal{W}(E^\top, 0) & \overset{\text{quantization}}{\Longrightarrow} & \mathcal{W}(E^\top, \hbar\,\mathrm{Im}(.|.))\,. \end{array} \tag{7.10}$$

We learn from the commutativity of the diagram that the complexification of the test function space, and thus of the dual phase space, is not a necessary pre–requisite for quantization. But it has, in fact, a special meaning in the *Fock representation*, as we discuss later on. As is common usage in physics [16], we employ the quantized ED only after diagonalization (and thus after complexification).
If $R \in E_{\rm cf}^{\top\prime}$ we write for the "diagonalized" linear form $\hat{R}(Vf) := R(f)$ resp. $\hat{R}(h) = R(V^{-1}h)$ for $f = (f_a, f_y) \in E_{\rm cf}^\top$ and $h = h_1 + ih_2 \in E^\top$. The diagonalized smeared current is therefore

$$\hat{\mathbf{J}}_t^\top(h) = (0, \mathbf{j}_t^\top)(V^{-1}h) = \mathbf{j}_t^\top(\epsilon_0 c\sqrt{\mathrm{curl}_{t0}^2|})^{-1/2}h_2)\,. \tag{7.11}$$

We may apply the symplectic–affine equivariance of quantization to the magic formula Eq. (6.6) under the condition that the current remains classical and obtain the quantized version.

Theorem 7-5 *The direct quantization of the transversal Maxwell dynamics (in Weyl form for the canonical fields) leads, under the condition that the current stays classical, to the* algebraic, quantized magic formula *in diagonal form*

$$\alpha_{t_0\to t}^{\hbar,\mathrm{curr}}(\hat{W}^\hbar(h)) = \exp\{i\textstyle\int_{t_0}^t \hat{\mathbf{J}}_s^\top(\exp\{(t-s)i\mathbb{D}\}h)\,ds\}\,\hat{W}^\hbar(\exp\{(t-t_0)\mathbb{D}\}h)\,, \tag{7.12}$$

where $\hat{\mathbf{J}}_t^\top$ is the diagonalized current and $h \in E^\top \subset \mathcal{H}^\top$.
The expression (7.12) gives, in dependence on t, a trajectory in the abstract Weyl algebra $\mathcal{W}(E^\top, \hbar\,\mathrm{Im}(.|.))$, provided $\hat{\mathbf{J}}_s^\top \in E^{\top\prime}$, $\forall s \in \mathbb{R}$.
*By linear and norm continuous extension the $\alpha_{t_0\to t}^{\hbar,\mathrm{curr}}$ constitute a groupoid of *–automorphisms in $\mathcal{W}(E^\top, \hbar\,\mathrm{Im}(.|.))$.*
Even for the free case $\hat{\mathbf{J}}_t^\top \equiv 0$ the $\alpha_{t_0\to t}^{\hbar,\mathrm{curr}}$ do not depend (point–wise norm) continuously on the time parameters (since the abstract Weyl operators do not depend continuously on the test functions).

In the following we work with the quantities only after having applied the diagonalization map and drop the hat over the symbols!
For a (mesoscopic) system of lasing two–level atoms with level splitting $\Delta\eta$ the current *is* classical, namely the polarization current of the phase–locked, oscillating atoms, and has the form $\mathbf{J}_t^\top(h) = \exp\{it\Delta\eta/\hbar\}\mathbf{J}_0^\top(h)$. (Note that $\Delta\eta/\hbar = \omega_{\rm atom}$ is independent of $\hbar$.) In this case $t \to \alpha_{0\to t}^{\hbar,\mathrm{curr}}$ is a group of *–automorphisms in $\mathcal{W}(E^\top, \hbar\,\mathrm{Im}(.|.))$.

Quite generally, the current is determined in mesoscopic quantum mechanical radiation models by the coupling between matter and quantized field and becomes classical in the weak coupling limit ([17]). In this manner one gets the *diagonalized* current and one needs Eq. (7.11) to find the current for Maxwell theory. In this context the Heisenberg dynamics is always a group.
For explicit discussions one looks for a regular representation $(\Pi, \mathcal{H}_\Pi)$ of $\mathcal{W}(E^\top, \hbar\,\mathrm{Im}(.|.))$, which contains – beside $W^\hbar_\Pi(h) := \Pi(W^\hbar(h))$ – also $\exp\{i\int_{t_0}^t \mathbf{J}_s^\top(h)\,ds\}\,W^\hbar_\Pi(h)$ for all $h \in E^\top$ and for all $t, t_0 \in \mathbb{R}$, and in which the Heisenberg dynamics may be implemented by a strongly continuous group of unitaries. In the representation is then a selfadjoint operator H^Π such that

$$\Pi\left(\alpha^{\hbar,\mathrm{curr}}_{t_0\to t}(W^\hbar(h))\right) = \exp\{i(t-t_0)H^\Pi/\hbar\}\; W^\hbar_\Pi(f)\; \exp\{-i(t-t_0)H^\Pi/\hbar\}\,. \tag{7.13}$$

H^Π is not uniquely determined by this requirement and should be renormalized with respect to a reference state (ground or equilibrium state). By differentiation one formulates the dynamics for the quantized, representation dependent fields $\Phi^\hbar_\Pi(h)$.
Even for the free case $\mathbf{J}_t^\top \equiv 0$ it is important to perform the discussion in the outlined manner, in order to obtain the correct form for H^Π and $\Phi^\hbar_\Pi(h)$.
Note that up to now, we did not introduce a particle structure for the quantized fields.

8. Particle structure

Let us go back to the Weyl algebra over the real transversal test function space $\mathcal{W}(E^\top_{\mathrm{cf}}, \hbar\sigma)$, where $\hbar$ may range in $\mathbb{R}_+$. The first step for a particle structure is to define the annihilation and creation operators

$$a^\hbar_\Pi(g) := \frac{1}{\sqrt{2}}\left(\Phi^\hbar_\Pi(g) + i\Phi^\hbar_\Pi(jg)\right), \qquad a^{\hbar\,*}_\Pi(g) := \frac{1}{\sqrt{2}}\left(\Phi^\hbar_\Pi(g) - i\Phi^\hbar_\Pi(jg)\right), \tag{8.1}$$

for any $g \in E^\top_{\mathrm{cf}}$ and for any regular representation $(\Pi, \mathcal{H}_\Pi)$ of $\mathcal{W}(E^\top_{\mathrm{cf}}, \hbar\sigma)$. These are closed operators which are Hermitian adjoint to each other. The decisive point is that we need for their definition a *complexification* of the test function space with imaginary unit j in the sense of Def. 6-1. (whereas i is the imaginary unit of the representation Hilbert space $\mathcal{H}_\Pi$).
We read from (8.1) $a^\hbar_\Pi(jg) = -ia^\hbar_\Pi(g)$ (anti–linearity in the test function) and $a^{\hbar\,*}_\Pi(jg) = ia^{\hbar\,*}_\Pi(g)$ (linearity in the test function). These relations elucidate, by the way, the usual splitting of the Hermitian quantum field into "positive and negative frequency parts".
From the field commutation relations we obtain, observing (6.10),

$$\begin{aligned}[a^\hbar_\Pi(g), a^{\hbar\,*}_\Pi(g')] &\subset \frac{1}{2}[i\hbar\sigma(g,g') + i\hbar\sigma(jg,jg') - \hbar\sigma(jg,g') + \hbar\sigma(g,jg')]\mathbb{1}_\Pi \\ &= \hbar(g|g')_j\mathbb{1}_\Pi, \quad g, g' \in E^\top\,.\end{aligned} \tag{8.2}$$

But there are infinitely many complexifications (cf. (6.11)) , fitting to the given symplectic form. So it is important that the uniqueness of the diagonalization of the free Maxwell dynamics provides us with a distinguished $j = i$, already used in $E^\top \subset \mathcal{H}^\top$ after the diagonalization. (The two just mentioned i's are discriminated by the vectors on which they act from the left.)
Given a complexification j of the test function space there is a distinguished representation $(\Pi_F, \mathcal{H}_F)$ of the Weyl algebra, which is called the *Fock representation*. For notational economy we describe it only for our special complexification $j = i$ and use our Weyl algebra in the form $\mathcal{W}(E^\top, \hbar\,\mathrm{Im}(.|.))$.
The Fock representation $(\Pi_F, \mathcal{H}_F)$ is *uniquely* characterized by the existence of a cyclic vector Ω_{vac} with $a_F(h)\Omega_{\mathrm{vac}} := 0, \forall h \in E^\top$ and by the family of products of creation operators $\{a^*_F(h_1)\ldots a^*_F(h_n)\Omega_{\mathrm{vac}}\,|\, h_i \in E^\top, n \in \mathbb{N}_0\}$ applied to the vacuum, spanning the total space $\mathcal{H}_F$.

We denote by $a_F(h), a^*_F(h_i)$ the creation and annihilation operators in the Fock representation divided by $\sqrt{\hbar}$, satisfying

$$[a_F(h), a_F{}^*(h')] \subset (h|h')\mathbb{1}_F, \quad h, h' \in E^\top . \tag{8.3}$$

Let us denote by $\mathcal{H}_F^{(n)}, n \in \mathbb{N}_0$, the sub–Hilbert space, generated by just n creation operators. $\mathcal{H}_F^{(0)}$ is then the one–dimensional $\mathbb{C}\Omega_{\text{vac}}$. $\mathcal{H}_F^{(1)}$ is spanned by the vectors $a^*_F(h)\Omega_{\text{vac}}, h \in E^\top$, which satisfy by (8.3)

$$(a_F{}^*(h)\Omega_{\text{vac}}|a_F{}^*(h')\Omega_{\text{vac}}) = (h|h'), \quad h, h' \in E^\top . \tag{8.4}$$

Thus we obtain the same scalar product as in $\mathcal{H}^\top$ and may identify the $a_F{}^*(h)\Omega_{\text{vac}} \in \mathcal{H}_F^{(1)}$ with $h \in \mathcal{H}^\top$ (and may identify the two i's). This is a hint for the fact that we can smear the creation and annihilation operators actually by test functions from all of $\mathcal{H}^\top$ and that the Weyl operators $W_F^\hbar(h)$ in the Fock representation depend continuously (in the strong operator topology) on the test functions (in the norm topology).

Corollary 8-1 *The quantized Maxwell dynamics* (7.12) *stays (for finite times) in Fock space, if and only if the diagonalized classical current* $\hat{\mathbf{J}}_t^\top$ *is square integrable.*

If U is a unitary in $\mathcal{H}^\top$ we define its second quantization $\Gamma(U)$ by application on the total set of vectors as follows

$$\Gamma(U)a^*_F(h_1)\dots a^*_F(h_n)\Omega_{\text{vac}} := a^*_F(Uh_1)\dots a^*_F(Uh_n)\Omega_{\text{vac}}, \; h_i \in E^\top, n \in \mathbb{N}_0 . \tag{8.5}$$

For $n = 0$ Eq. (8.5) expresses the invariance of Ω_{vac} under $\Gamma(U)$.

Observation 8-2 *A unitary U in $\mathcal{H}^\top$, which leaves $E^\top$ invariant, is a special symplectic transformation in $(E^\top, \hbar\,\text{Im}(.|.))$ and leads to the Bogoliubov isomorphism $\beta_U^\hbar$ in $\mathcal{W}(E^\top, \hbar\,\text{Im}(.|.))$.*
$\Gamma(U)$ is the unique implementation of β_U in $\mathcal{H}_F$ which leaves the vacuum vector invariant.
$\Gamma(U)$ leaves each subspace $\mathcal{H}_F^{(n)}$ invariant.

For a strongly continuous unitary group $U_t = \exp\{itA\}$ we define the second quantization of its selfadjoint generator by

$$d\Gamma(A) := \frac{d\,\Gamma(\exp\{itA\})}{i\,dt}|_{t=0}, \quad \text{on a suitable domain in } \mathcal{H}_F . \tag{8.6}$$

Observe that $d\Gamma(A)\Omega_{\text{vac}} = 0$, for all selfadjoint A in $\mathcal{H}^\top$. Especially the diagonalized free Maxwell dynamics has the form $\exp\{it\mathbb{D}\}$, so that the corresponding Bogoliubov automorphisms have in Fock space the implementing unitaries $\Gamma(\exp\{it\mathbb{D}\})$ with generator $d\Gamma(\mathbb{D})$. Since in quantum theory we write the dynamics as $\exp\{it\hbar\mathbb{D}/\hbar\}$, the physical relevant generator is $d\Gamma(\hbar\mathbb{D}) =: H^F$. This is already the systematic introduction of the free electrodynamic Hamiltonian in Fock space (independently of any basis system). Since $\mathbb{D}$ is strictly positive (and may have 0 as spectral value, but not as eigen value) the spectrum $\sigma(H^F)$ is positive and Ω_{vac} is the only eigen vector of H^F to the eigenvalue 0.

Starting from the unit operator $\mathbb{1}$ in $\mathcal{H}^\top$, which generates the phase rotations $\exp\{it\mathbb{1}\}$, we introduce by second quantization the unitaries $\Gamma(\exp\{it\mathbb{1}\})$, which implement the gauge transformations of the first kind in Fock space. They commute with all other $\Gamma(\exp\{itA\})$.
The second quantized generator is

$$d\Gamma(\mathbb{1}) =: N_F, \quad \text{the particle number operator in } \mathcal{H}_F . \tag{8.7}$$

The selfadjoint N_F has the eigen spaces $\mathcal{H}_F^{(n)}$ with the eigen values $n \in \mathbb{N}_0$. Thus the spectrum $\sigma(N_F)$ equals $\mathbb{N}_0$ and Ω_{vac} is also a ground state for the gauge automorphisms.
There are representations of $\mathcal{W}(E^\top, \hbar\,\text{Im}(.|.))$ in which the generators of the free Maxwell dynamics and of the first kind gauge transformations have different spectra from those of the Fock generators, or do not commute with each other, or do not exist at all.

Theorem 8-3 *The Fock representation with the special complexification $j = i$, defined by the diagonalization of the free Maxwell dynamics, is the only representation with a particle structure. By this we mean that it is the only representation in which the Bogoliubov automorphisms of the free Maxwell dynamics and of the first kind gauge transformations are unitarily implemented so that the following conditions are satisfied:*
The unitaries for the time and gauge transformations commute with each other, and the corresponding selfadjoint generators have a positive spectrum, where the particle number spectrum equals $\mathbb{N}_0$.
They have a cyclic vector, namely Ω_{vac}, as common lowest eigenvector with eigenvalue 0.

The requirements on a representation with particle structure, formulated in the foregoing Theorem, demand the existence of a vacuum, void of particles and energy, a so-called *bare vacuum*.
The fact that the cohomological quantized fields satisfy the canonical commutation relations, but do not possess a bare vacuum, illustrates the difference between quantum behavior and particle structure for fields. Thus it is a special feature of the Maxwell dynamics, that the transversal quantized fields exhibit a particle structure in a Fock representation, which is associated with a uniquely determined complexification of the theory.

8.1. Photon observables

In the distinguished Fock representation with particle structure, the corresponding quanta, the **photons**, display all features of quantum particles.

Definition 8-4 *The vectors $\psi^{(n)} \in \mathcal{H}_F^{(n)}$ are called n–photon wave functions, $n \in \mathbb{N}$, and the selfadjoint operators $A^{(n)}$ in $\mathcal{H}_F^{(n)}$ are called n–photon observables.*
Especially $d\Gamma(\hbar\mathbb{D})|_{\mathcal{H}_F^{(1)}} \equiv \hbar\mathbb{D} = \hbar c\sqrt{\text{curl}_{t0}^2|}$ is the one–photon energy operator, and correspondingly $d\Gamma(\hbar\mathbb{D})|_{\mathcal{H}_F^{(n)}}$ (without interaction terms!) the n–photon energy operator.

Since $\mathcal{H}_F^{(1)}$ is unitarily isomorphic to $\mathcal{H}^\top \supset E^\top = VE_{\text{cf}}^\top$ the "smooth" one–photon wave functions have the form $\psi^{(1)} \equiv \psi = w^{-1}f_a + iwf_y = Vf$ with $w := (\epsilon_0 c\sqrt{\text{curl}_{t0}^2|})^{+1/2}$.

Definition 8-5 *If i is the complex unit in $\mathcal{H}^\top$, then let ι be the "un–diagonalized" complex unit $V^{-1}iV$ (which equals according to (6.17) $W^{-1}jW$, with $j = v^{-1}iv$ the natural complex unit in the real $(E_{cf}^\top, \sigma)$).*
Then we denote by $\mathcal{G}_{\text{symp}}^l(E^\top, \sigma)$ the sub–Lie algebra of $\mathcal{G}_{\text{symp}}(E^\top, \sigma)$ of all elements which are linear for ι (i.e. commute with ι).
We denote by $\mathcal{G}_{\text{symp}}^a(E^\top, \sigma)$ the subset of $\mathcal{G}_{\text{symp}}(E^\top, \sigma)$ of all elements G which are anti–linear for ι (i.e. which satisfy $G\iota = -\iota G$).
The operators in $\mathcal{G}_{\text{symp}}^{l,a}(E^\top, \hbar\,\text{Im}(.|.))$ are linear resp. antilinear with respect to i.

Corollary 8-6 *Let $A^{(1)}$ be a selfadjoint one–photon observable, which leaves the dense domain $E^\top \subset \mathcal{H}_F^{(1)}$ invariant. Then the "un–diagonalized" symplectic generator $G := V^{-1}iA^{(1)}V$ is an element in the sub–Lie algebra $\mathcal{G}_{\text{symp}}^l(E_{cf}^\top, \sigma)$ of complex linear elements (with respect to ι).*

Reversely each $G \in \mathcal{G}^l_{\rm symp}(E^\top_{cf}, \sigma)$ leads via the diagonalization map V to a anti–symmetric operator $i\hat{G}$ with domain $E^\top \subset \mathcal{H}^\top$, so that $\hat{G}$ determines a selfadjoint one–photon observable $A^{(1)}$.
Like in (6.20) we calculate the Hamilton function for $G \in \mathcal{G}^l_{\rm symp}(E^\top_{cf}, \sigma)$

$$H_G[\sigma_\sharp f] = \frac{1}{2}\sigma(f, Gf) = \frac{1}{2}\,{\rm Im}(Vf|i\hat{G}Vf) = \frac{1}{2}\|Vf\|^2(\psi_f|\hat{G}\psi_f)\,, \quad \forall f \in E^\top_{cf}\,, \tag{8.8}$$

where in the last equality we used the realness of the scalar product with the essentially selfadjoint $\hat{G}$ and defined $\psi_f := Vf/\|Vf\|$. In the quantum interpretation the ψ_f are (smooth, complex, normalized) one–photon wave functions, that are probability amplitudes.

The preceding formula expresses the values of certain classical quadratic Hamilton functions in terms of the statistical expectation values of the corresponding one–photon observables. This will only say, that also in the presence of many photons, described by an arbitrary vector in Fock space, their energy may be expressed by a one–photon wave function with scaled amplitude.
Nevertheless we take this as an indication, that for detecting the quantum features of photons the measurement of n–photon observables, with $n \geq 2$, resp. of correlation functions, are required. Recall in this context that in EPR–experiments always correlations between one–photon observables of different particles, like *spin* (classically: angular momentum of a polarized beam) are measured.
Interesting is the counting of photons, which measures the eigenvalues of the particle number operator N_F, and not of the corresponding trivial one–photon observable $\mathbb{1}$. Taken as $\hat{G}$, the latter leads to the undiagonalized Lie algebra element $G = V^{-1}i\mathbb{1}V = \iota$, with classical Hamilton function $H_\iota[\sigma_\sharp f] = \frac{1}{2}\sigma(f, \iota f) = \frac{1}{2}[(w^{-1}f_a|w^{-1}f_a) + (wf_y|wf_y)]$. Thus we have also classically a Hamilton function for the Poisson generator of the first kind gauge transformations.

For illustration let us discuss the one–photon energy $\hbar\mathbb{D}$ in the cubic box Λ_a of side a, depicted in our figure (2.3). Λ_a is assumed to sit in the first quadrant, with lower left vertex at 0. If there are no conductors within Λ_a a complete system of transversal eigen functions e_n, with vanishing tangential components at the boundary, is given by the formula

$$e_{n,i}(x) = \varepsilon^\alpha(k)\sin(k_1\cdot x)\cos(k_i\cdot x)\sin(k_3\cdot x)\,, \quad i = 1,2,3,\ k = k(n),\ \alpha = 1,2\,, \tag{8.9}$$

which indicates that the i–th factor is the cos–function. The vector $k(n)$ has the components $k_j(n) = \pi n_j/a$, where $n_j = 0,1,2,\ldots,\ j = 1,2,3$. (Actually $n_j = 0$ is only allowed for $j = i$.)
The two polarization vectors $\varepsilon^\alpha(k)$, which contain also the normalization constants, are orthogonal to k, so that $\nabla\cdot e_n(x) = -k\cdot\varepsilon^\alpha(k)\sin(k_1\cdot x)\sin(k_2\cdot x)\sin(k_3\cdot x) = 0$. Therefore, on the linear hull of this basis, the operator ${\rm curl}^2_{t0}|$ coincides with $-\Delta$, acting component–wise, and we recover **Einstein's photon energy**

$$\hbar\mathbb{D}\,e_n(x) = \hbar c\sqrt{-\Delta}\,e_n(x) = \hbar c|k(n)|\,e_n(x)\,. \tag{8.10}$$

Since the spacing between the eigen values tends to 0 for increasing box volumina, the spectrum is called in physics a *quasi–continuum*, and summation over the spectrum is often replaced by integration (as in the Planck formula).
Any kind of conducting obstacles within Λ_a makes the cavity a more complicated manifold and deforms the spectrum of $\hbar\mathbb{D}$. It is a challenging question, which topological features of the cavity are encoded in the deformed spectrum of the one–photon Hamiltonian, the latter displaying some similarity to an elliptic Dirac operator (?).

8.2. General symplectic generators in Fock space

We have seen that each $G \in \mathcal{G}_{\rm symp}(E_{\rm cf}^{\top}, \sigma)$, which generates the strongly differentiable one–parameter symplectic group $\{\exp\{tG\} \mid t \in \mathbb{R}\} \subset {\rm symp}(E_{\rm cf}^{\top}, \sigma)$, is by the transformation V mapped onto $i\hat{G} \in \mathcal{G}_{\rm symp}(E^{\top}, \hbar\,{\rm Im}(.|.))$. In the latter form it generates the complex symplectic transformations $\{\exp\{it\hat{G}\} \mid t \in \mathbb{R}\} \subset {\rm symp}(E^{\top}, \hbar\,{\rm Im}(.|.))$.

So we have to analyze the general $\mathbb{R}$–linear $T \in {\rm symp}(E^{\top}, \hbar\,{\rm Im}(.|.))$, which decompose uniquely into the ($\mathbb{C}$–) linear part T_l and the antilinear part T_a according to

$$T = T_l + T_a\,, \quad \text{where} \quad T_l := \frac{1}{2}(T - iTi)\,, \quad T_a := \frac{1}{2}(T + iTi)\,. \tag{8.11}$$

By the embedding $E^{\top} \subset \mathcal{H}^{\top}$ one knows [18] that there are a positive selfadjoint (linear) operator S, a (linear) unitary U, and an antilinear involution J (that is, $J = J^* = J^{-1}$), so that $T_l = U\cosh(S)|_E$ and $T_a = UJ\sinh(S)|_E$. We see that T is linear, iff $S = 0$ and iff it is unitary. Thus $\exp\{it\hat{G}\}$ is linear iff $\hat{G}$ extends to a selfadjoint A in $\mathcal{H}^{\top}$.

The association $\mathcal{G}_{\rm symp}(E^{\top}, \hbar\,{\rm Im}(.|.)) \ni \hat{G} \to d\Gamma(A)$ is a Lie homomorphism and may be considered a *quantum mechanical co–momentum map*. In the end, each $G \in \mathcal{G}^{l}_{\rm symp}(E^{\top}, \sigma)$ yields a second quantized $d\Gamma(A)$ in Fock space, for which one knows the series representation

$$d\Gamma(A) = \sum_{k=1}^{\infty} a_F^{\hbar\,*}(A\mathbf{u}_k)\, a_F^{\hbar}(\mathbf{u}_k)\,, \tag{8.12}$$

where $\{\mathbf{u}_k \mid k \in \mathbb{N}\}$ is a complete orthonormal system in the domain of A, and the series converges in the strong resolvent sense. Note that $\Omega_{\rm vac}$ is annihilated in virtue of the normally ordered form of $d\Gamma(A)$.

The connection with the classical diagonal representation (6.21) is achieved by noting that $\Phi_F^{\hbar}(h)^2 + \Phi_F^{\hbar}(ih)^2 - \|h\|^2 \mathbb{1}_F = 2a_F^{\hbar\,*}(h)\, a_F^{\hbar}(h)$. This allows for (8.12) the form

$$d\Gamma(A) = \sum_{k=1}^{\infty} \lambda_k\, a_F^{\hbar\,*}(\mathbf{u}_k)\, a_F^{\hbar}(\mathbf{u}_k) = \frac{1}{2}\sum_{k=1}^{\infty} \lambda_k \left[\Phi_F^{\hbar}(\mathbf{u}_k)^2 + \Phi_F^{\hbar}(i\mathbf{u}_k)^2 - \mathbb{1}_F\right], \tag{8.13}$$

if $A\mathbf{u}_k = \lambda_k\, \mathbf{u}_k$ for all $k \in \mathbb{N}$. A direct formal quantization of (6.22), the popular "harmonic oscillator quantization", would not produce the infinite renormalization constant $\frac{1}{2}\sum_{k=1}^{\infty} \lambda_k$ and would not at all lead to a meaningful operator in Fock space.

More complicated are the anti–linear symplectic generators $G \in \mathcal{G}^{a}_{\rm symp}(E^{\top}, \sigma)$. We sketch only an example. Assume

$$T_t = \exp\{-itD\} = \exp\{-itJ|D|\} = \underbrace{\cosh(t|D|)}_{=(T_t)_l}\underbrace{-iJ\sinh(t|D|)}_{=(T_t)_a}\,, \quad t \in \mathbb{R}\,, \tag{8.14}$$

with the selfadjoint, $\mathbb{C}$–antilinear operator D, the $\mathbb{C}$–linear positive selfadjoint operator $|D| = \sqrt{D^*D}$ in $\mathcal{H}^{\top}$ and with an $\mathbb{C}$–antilinear involution J in $\mathcal{H}^{\top}$ (i.e., $J = J^* = J^{-1}$), which respectively are given in terms of an orthonormal system by

$$\begin{aligned} D &= \sum_{k=1}^{N} \zeta_k(.\,|e_k)e_k\,, \qquad |D| = \sum_{k=1}^{N} |\zeta_k|\, \underbrace{|e_k)(e_k|}_{=(e_k|\,.)e_k}\,, \\ Je_k &= \frac{\zeta_k}{|\zeta_k|}e_k\,, \quad \forall k \in \{1,\ldots,N\}\,, \end{aligned} \tag{8.15}$$

(where we define $\frac{\zeta_k}{|\zeta_k|} := 1$ for $\zeta_k = 0$). $D = J|D|$ is a kind of polar decomposition of the antilinear selfadjoint D.
Since the T_t are symplectic they lead to Bogoliubov transformations in $\mathcal{W}(E^\top, \hbar\,\mathrm{Im}(.|.))$. For finite N the latter are implemented by unitaries of the form $\exp\{itH_D\}$ with

$$H_D = \frac{1}{2}\sum_{k=1}^{N}\left(\zeta_k a^*(e_k)^2 + \overline{\zeta_k}\, a(e_k)^2\right), \tag{8.16}$$

involving complex coefficients with

$$\zeta_{2m-1} = -\zeta_{2m}, \quad m \in \{1,\ldots,M\}, \quad N := 2M. \tag{8.17}$$

The Fock–implemented T_t transform the field operator as

$$\exp\{itH_D\}\Phi_F(f)\exp\{-itH_D\} = \Phi(T_t f), \quad \forall f \in E^\top, \tag{8.18}$$

that is like the complex linear symplectic generators. But, since the T_t have now also an anti–linear part (the odd powers in the exponential series), we derive for the transformed annihilation operators

$$a_{T_t}(f) := \exp\{itH_D\}a_F(f)\exp\{-itH_D\} = a_F(\cosh(t|D|)f) - ia_F^*(J\sinh(t|D|)f), \tag{8.19}$$

for all $f \in E^\top$. The condition for a finite Hilbert–Schmidt norm

$$\|H_D\Omega_{\mathrm{vac}}\|^2 = \frac{1}{2}\sum_{k=1}^{\infty}|\zeta_k|^2 = \frac{1}{2}\|D\|_{\mathrm{HS}}^2 = \frac{1}{2}\|\,|D|\,\|_{\mathrm{HS}}^2\,. \tag{8.20}$$

is necessary and sufficient that the H_D generate the symplectic transformations in Fock space also for infinite N. Note that D is then an antilinear Hilbert–Schmidt operator, a special bounded operator in $\mathcal{H}^\top$.
The physical interest for symplectic transformations of this kind, i.e. with generators $G \in \mathcal{G}^a_{\mathrm{symp}}(E^\top, \sigma)$, stems from the fact that the $\exp\{it\hat{G}\}$ leave only the imaginary part of the scalar product invariant, but may decrease (or increase) the real part (cf. (6.12)). Since the real part of the scalar product shows up in the variance of the field operators in certain states (squeezed states) one may have reduced quantum noise in optical communication.

If condition (8.20) is not satisfied, then there are however other representations (GNS–representations over the squeezed vacuum), in which the infinite–mode squeezing Hamiltonians exist. Thus the quantum co–momentum map requires many Fock representations for representing general strong symplectic generators.

The physical realization of a squeezing Hamiltonian is based on a form, which arises by changing the basis system as follows

$$e_{2m-1} =: \frac{1}{\sqrt{2}}\left(u_{s,m} + u_{i,m}\right), \quad e_{2m} =: \frac{1}{\sqrt{2}}\left(u_{s,m} - u_{i,m}\right), \quad m \in \{1,\ldots,M\}.$$

The so–called signal and idler modes $\{u_{s,1},\ldots,u_{s,M},u_{i,1},\ldots,u_{i,M}\}$ generate a new orthonormal system in the test function space $E^\top$.
From the $\mathbb{C}$–linearity of $f \mapsto a^*(f)$ and $\mathbb{C}$–antilinearity of $f \mapsto a(f)$ we obtain the relations $a^\#(u_{s,m})a^\#(u_{i,m}) = \frac{1}{2}\left(a^\#(e_{2m-1})^2 - a^\#(e_{2m})^2\right)$, which are valid for both the smeared creation

operators ($a^{\#}(h) \equiv a^*(h)$) and annihilation operators ($a^{\#}(h) \equiv a(h)$). This leads to the following "non–degenerate" form of the squeezing Hamiltonian (e.g. [19], [20], [21],)

$$H_{\rm Dnd} = \sum_{m=1}^{M} \left(\eta_m a^*(u_{s,m}) a^*(u_{i,m}) + \overline{\eta_m}\, a(u_{s,m}) a(u_{i,m}) \right) . \tag{8.21}$$

For each term m one has so–called four–wave mixing: A non–linear crystal is pumped by the idler mode $u_{i,m}$ of a laser. The incoming laser mode interacts with the incoming signal mode in the first term, whereas the second term describes this matter mediated interaction for the two outgoing modes. The microscopically detailed description would require the raising and lowering operators of the material. The complex $\eta_m \in \mathbb{C}$ result from averaging over the non–Hermitian matter operators and contain also the interaction strength, which is, for itself, dependent from the laser intensity. Therefore, the η_m are called *pumping parameters.* Only by this averaging procedure it appears as if one could have a direct interaction between different photon rays.

Altogether we see that the anti–linear symplectic generators $G \in \mathcal{G}^a_{\rm symp}(E^\top, \sigma)$ lead to quantum co–momenta, which do not commute with the particle number and they describe creation and annihilation processes.

Observation 8-7 *A simple but physically relevant generator with anti–linear part is given by* $\mathbb{I}$ *of Eq. (6.1). We may perform the natural complexification* v *of the cohomological test function space* E^{co}_{cf} *from (5.2) analogously as in* $E^\top_{cf}$ *and obtain* $v\mathbb{I}v^{-1} = i\mathbb{J} = i(\mathbb{J}_l + \mathbb{J}_a) = \frac{i}{2\epsilon_0}(\mathbb{1} + \mathcal{C})$, *where* $\mathbb{1}$ *and* $\mathcal{C}$ *are the unit and complex conjugation in the finite dimensional* $E^{co}_a \boxplus iE^{co}_y$, *the complexification of* $\mathbb{H}_2$. *Related to this complexification we may introduce creation and annihilation operators for the quantized cohomological canonical fields, as prescribed in (8.1), and then a Fock space and a particle number operator. Since* $\mathbb{J}_l$ *and* $\mathbb{J}_a$ *commute with each other, a Hamiltonian can easily be implemented in the Fock space. It is of the form* $d\Gamma(\mathbb{J}_l) + H_D$, $D = \mathbb{J}_a$ *(cf. (8.16)). Because of* H_D *the Hamiltonian is not lower bounded and and does not commute with the particle number operator. This illustrates the fact that* **the cohomological fields have no particle structure** *in the sense of Theorem 8-3.*
The quantized cohomological canonical fields, which commute with all transversal fields, have nevertheless an interesting dynamics, which seems to describe collective excitations.

We end our discussion with a theorem on strongly continuous symplectic groups, adapted to the complexified and completed version $(\mathcal{H}^\top, {\rm Im}(.|.))$ of our test function space $(E^\top_{\rm cf}, \sigma)$ for the transversal canonical electromagnetic fields.

Theorem 8-8 *The following assertions are valid:*

(a) *Let* $C = C^*$ *be a linear operator in* $\mathcal{H}^\top$, *and let* $D = D^*$ *be an antilinear bounded operator in* $\mathcal{H}^\top$. *Then*

$$\{\exp\{it(C + D)\} \mid t \in \mathbb{R}\} \subset {\rm symp}(\mathcal{H}^\top, {\rm Im}(.|.))$$

constitutes a strongly continuous symplectic one–parameter group with the $(c = 1)$*–growth*

$$\|\exp\{it(C + D)\}\| \leq \exp\{\|D\|\,|t|\}, \quad \forall t \in \mathbb{R},$$

(meaning that the coefficient in front of the right hand exponential is unity).

(b) *Let* $\{T_t \mid t \in \mathbb{R}\} \subset {\rm symp}(\mathcal{H}^\top, {\rm Im}(.|.))$ *be a strongly continuous one–parameter group with* $(c = 1)$*–growth, i.e. with* $\|T_t\| \leq \exp\{\delta|t|\}$ *for all* $t \in \mathbb{R}$ *for some* $\delta \geq 0$. *Then there exists a unique linear* $C = C^*$ *and a unique antilinear bounded* $D = D^*$ *in* $\mathcal{H}^\top$ *such that*

$$T_t = \exp\{it(C + D)\}, \quad \forall t \in \mathbb{R}.$$

References

[1] Maxwell J C 1873 *A Treatise on Electricity and Magnetism* vol 1,2 (Oxford: Clarendon Press)
[2] Gross P W and Kotiuga P R 2004 *Electromagnetic Theory and Computation: A Topological Approach* (Cambridge: University Press)
[3] Leis R 1986 *Initial Boundary Value Problems in Mathematical Physics* (Stuttgart: Teubner)
[4] Dautray R and Lions J L 1990 *Mathematical Analysis and Numerical Methods for Science and Technology* vol 3 (New York: Springer)
[5] Schwarz G 1995 *Hodge Decompositions – A Method for Solving Boundary Value Problems* (Lecture Notes in Mathematics vol no 1607) (New York: Springer)
[6] Abraham R and Marsden J E 1978 *Foundations of Mechanics* 2nd ed (London: Benjamin)
[7] Binz E, Honegger R and Rieckers A 2004 *J. Math. Phys.* **45** 2885
[8] Bratteli O and Robinson D W 1987 *Operator Algebras and Quantum Statistical Mechanics* vol 1 (New York: Springer)
[9] Pedersen G K 1979 *C*-Algebras and their Automorphism Groups* (London: Academic Press)
[10] Honegger R and Rieckers A 2005 *Publ. RIMS Kyoto Univ.* **41** 113
[11] Binz E, Honegger R and Rieckers A 2004 *Ann. Henri Poincaré* **5** 327
[12] Landsman N P 1998 *Mathematical Topics Between Classical and Quantum Mechanics* (New York: Springer)
[13] Bayen F, Flato M, Fronsdal C, Lichnerovicz A and Sternheimer D 1978 *Ann. Phys.* **111** 61
[14] Sternheimer D 2003 *Poisson Geometry, Deformation Quantization and Group Representations* **4** 1
[15] Hewitt E and Ross K A 1963 *Abstract Harmonic Analysis* vol 1 (New York: Springer)
[16] Cohen–Tannoudji C, Dupont–Roc J and Grynberg G 1989 *Photons & Atoms, Introduction to QED* (New York: John Wiley & Sons)
[17] Honegger R and Rieckers A 1994 *Publ. RIMS Kyoto Univ.* **30** 111
[18] Honegger R and Rieckers A 1998 *J. Math. Phys.* **39** 777
[19] Huang J and Kumar P 1989 *Phys. Rev.* A **40** 1670
[20] Ma X and Rhodes W 1990 *Phys. Rev.* A **41** 4625
[21] Ekert A K and Knight P L 1991 *Phys. Rev.* A **43** 3934

Journal of Physics: Conference Series **237** (2010) 012007 doi:10.1088/1742-6596/237/1/012007

On entanglement in neutrino mixing and oscillations

Massimo Blasone[†1,2], **Fabio Dell'Anno**[1,2,3], **Silvio De Siena**[1,2,3] **and Fabrizio Illuminati**[1,2,3,4]

[1] Dipartimento di Matematica e Informatica, Università degli Studi di Salerno, Via Ponte don Melillo, I-84084 Fisciano (SA), Italy
[2] INFN Sezione di Napoli, Gruppo collegato di Salerno, Baronissi (SA), Italy
[3] CNR-INFM Coherentia, Napoli, and CNISM, Unità di Salerno, Italy
[4] ISI Foundation for Scientific Interchange, Viale Settimio Severo 65, I-10133 Torino, Italy

E-mail: [†]`blasone@sa.infn.it`

Abstract. We report on recent results about entanglement in the context of particle mixing and oscillations. We study in detail single-particle entanglement arising in two-flavor neutrino mixing. The analysis is performed first in the context of Quantum Mechanics, and then for the case of Quantum Field Theory.

1. Introduction

Entanglement has been widely investigated in a number of physical systems, ranging from condensed matter to atomic physics, and quantum optics [1]. Also in the context of particle physics, the role of entanglement has been considered, see for instance [2].

In this paper, we consider the entanglement associated to neutrino mixing and oscillations. A detailed study of such a topic has been performed recently in the context of quantum mechanics [3, 4]. Here we review the main results of these studies in the simplest case of two flavors. We show that these results suggest a simple extension of the analysis to the relativistic domain, thus providing the physical basis and the mathematical tools for the quantification of entanglement in quantum field theory.

The phenomenon of particle mixing, associated with a mismatch between flavor and mass of the particle, appears in several instances: quarks, neutrinos, and the neutral K-meson system [5, 6]. Particle mixing is at the basis of important effects as neutrino oscillations and CP violation [7]. Flavor mixing for the case of three generations is described by the Pontecorvo–Maki–Nakagawa–Sakata (PMNS) matrix in the lepton instance [8, 9].

In the following we consider only the simplest case of two flavors. In such a case, the PMNS matrix reduces to the 2×2 rotation Pontecorvo matrix $\mathbf{U}(\theta)$,

$$\mathbf{U}(\theta) = \begin{pmatrix} \cos\theta & \sin\theta \\ -\sin\theta & \cos\theta \end{pmatrix}, \tag{1}$$

which connects the neutrino states with definite flavor with those with definite masses:

$$|\underline{\nu}^{(f)}\rangle = \mathbf{U}(\theta)\,|\underline{\nu}^{(m)}\rangle \tag{2}$$

where $|\underline{\nu}^{(f)}\rangle = (|\nu_e\rangle, |\nu_\mu\rangle)^T$ and $|\underline{\nu}^{(m)}\rangle = (|\nu_1\rangle, |\nu_2\rangle)^T$.

From Eq. (2), we see that each flavor state is given by a superposition of mass eigenstates, i.e. $|\nu_\alpha\rangle = U_{\alpha 1}|\nu_1\rangle + U_{\alpha 2}|\nu_2\rangle$. Let us recall that both $\{|\nu_\alpha\rangle\}$ and $\{|\nu_i\rangle\}$ are orthonormal, i.e. $\langle\nu_\alpha|\nu_\beta\rangle = \delta_{\alpha,\beta}$ and $\langle\nu_i|\nu_j\rangle = \delta_{i,j}$, where where $\alpha, \beta = e, \mu$ and $i, j = 1, 2$.

We now establish the following correspondence with two-qubit states:

$$|\nu_1\rangle \equiv |1\rangle_1|0\rangle_2 \equiv |10\rangle, \qquad |\nu_2\rangle \equiv |0\rangle_1|1\rangle_2 \equiv |01\rangle, \tag{3}$$

where $|\rangle_i$ denotes states in the Hilbert space for neutrinos with mass m_i. Thus, the occupation number allows to interpret the flavor states as constituted by entangled superpositions of the mass eigenstates. Quantum entanglement emerges as a direct consequence of the superposition principle. It is important to remark that the Fock space associated with the neutrino mass eigenstates is physically well defined. In fact, at least in principle, the mass eigenstates can be produced or detected in experiments performing extremely precise kinematical measurements [10]. In this framework, as discussed in [3], the quantum mechanical state (2) is entangled in the field modes, although being a single-particle state.

Mode entanglement defined for single-photon states of the radiation field or associated with systems of identical particles has been discussed in [11]. The concept of mode entanglement in single-particle states has been widely discussed and is by now well established [11, 12]. Successful experimental realizations using single-photon states have been reported as well [13]. Moreover, remarkably, the nonlocality of single-photon states has been experimentally demonstrated [14], verifying a theoretical prediction [15]. Furthermore, the existing schemes to probe nonlocality in single-particle states have been generalized to include massive particles of arbitrary type [16].

In the dynamical regime, flavor mixing (and neutrino mass differences) generates the phenomenon of neutrino oscillations. The mass eigenstates $|\nu_j\rangle$ have definite masses m_j and definite energies ω_j. Their propagation can be described by plane wave solutions of the form $|\nu_j(t)\rangle = e^{-i\omega_j t}|\nu_j\rangle$. The time evolution of the flavor neutrino states Eq.(2) is given by:

$$|\underline{\nu}^{(f)}(t)\rangle = \widetilde{\mathbf{U}}(t)|\underline{\nu}^{(f)}\rangle\,, \qquad \widetilde{\mathbf{U}}(t) \equiv \mathbf{U}(\theta)\,\mathbf{U}_0(t)\,\mathbf{U}(\theta)^{-1}\,, \tag{4}$$

where $|\underline{\nu}^{(f)}\rangle$ are the flavor states at $t = 0$, $\mathbf{U}_0(t) = diag(e^{-i\omega_1 t}, e^{-i\omega_2 t})$, and $\widetilde{\mathbf{U}}(t = 0) = 1\!\!1$.

At time t, the probability associated with the transition $\nu_\alpha \to \nu_\beta$ is

$$P_{\nu_\alpha\to\nu_\beta}(t) \;=\; |\langle\nu_\beta|\nu_\alpha(t)\rangle|^2 \;=\; |\widetilde{\mathbf{U}}_{\alpha\beta}(t)|^2\,, \tag{5}$$

where $\alpha, \beta = e, \mu$. The explicit form for the transition probabilities in the two flavor case is:

$$P_{\nu_e\to\nu_e}(t) \;=\; 1 - \sin^2(2\theta)\sin^2\left(\frac{\omega_2 - \omega_1}{2}t\right), \tag{6}$$

$$P_{\nu_e\to\nu_\mu}(t) \;=\; \sin^2(2\theta)\sin^2\left(\frac{\omega_2 - \omega_1}{2}t\right). \tag{7}$$

Flavor neutrino states are well defined in the context of Quantum Field Theory (QFT), where they are obtained as eigenstates of the flavor neutrino charges [17, 18]. In the relativistic limit, the exact QFT flavor states reduce to the usual Pontecorvo flavor states Eq.(2): flavor modes are thus legitimate and physically well-defined individual entities and mode entanglement can be defined and studied in analogy with the static case of [3]. We can thus establish the following correspondence with two-qubit states:

$$|\nu_e\rangle \equiv |1\rangle_e|0\rangle_\mu, \quad |\nu_\mu\rangle \equiv |0\rangle_e|1\rangle_\mu. \tag{8}$$

States $|0\rangle_\alpha$ and $|1\rangle_\alpha$ correspond, respectively, to the absence and the presence of a neutrino in mode α. Entanglement is thus established among flavor modes, in a single-particle setting. Eq. (4) can then be recast as

$$|\nu_\alpha(t)\rangle = \widetilde{\mathbf{U}}_{\alpha e}(t)|1\rangle_e|0\rangle_\mu + \widetilde{\mathbf{U}}_{\alpha\mu}(t)|0\rangle_e|1\rangle_\mu\,, \tag{9}$$

with the normalization condition $\sum_\beta |\widetilde{\mathbf{U}}_{\alpha\beta}(t)|^2 = 1$ $(\alpha, \beta = e, \mu)$. The time-evolved states $|\underline{\nu}^{(f)}(t)\rangle$ are entangled superpositions of the two flavor eigenstates with time-dependent coefficients. Thus, flavor oscillations can be related to bipartite (flavor) entanglement of single-particle states [4].

2. Entanglement in neutrino mixing – Quantum Mechanics

As discussed in the Introduction, the flavor neutrino state at a given time, say $|\nu_e(t)\rangle$ for definiteness, can be regarded as an entangled state either in terms of the mass eigenstates or in terms of the flavor eigenstates (at a fixed time). In the first instance, which was studied in detail for the multipartite case in [3], we have a static entanglement, in the sense that the result of the entanglement measures on the state $|\nu_e(t)\rangle$ do not depend on time. In the second case, considered for the general three flavor case in [4], the entanglement varies with time as it is related to the oscillations of flavor(s).

In this Section, we discuss these two forms of entanglement in the simple case of two flavors, by means of entropy measures first, and then using the characterization of entanglement in terms of quantum uncertainties. The second approach turns out to be suitable for the generalization of our discussion to the case of mixing among relativistic quantum fields – see next Section.

2.1. Neutrino entanglement via linear entropy

In terms of the mass eigenstates, the electron neutrino state at time t reads:

$$|\nu_e(t)\rangle \;=\; e^{-i\omega_1 t}\cos\theta\,|\nu_1\rangle + e^{-i\omega_2 t}\sin\theta\,|\nu_2\rangle\,, \tag{10}$$

where $|\nu_i\rangle$ are interpreted as the qubits, see Eq.(3).

Following the usual procedure, we construct the density operator $\rho^{(\alpha)} = |\nu_\alpha(t)\rangle\langle\nu_\alpha(t)|$ corresponding to the pure state $|\nu_\alpha(t)\rangle$. Then we consider the density matrix $\rho_i^{(\alpha)} = Tr_j[\rho^{(\alpha)}]$ reduced with respect to the index j. For the specific case of the state Eq.(10), we have $\rho^{(e)} = |\nu_e(t)\rangle\langle\nu_e(t)|$ and

$$\rho_1^{(e)} = Tr_2[\rho^{(e)}] = \cos^2\theta\,|1\rangle_1\,{}_1\langle 1| + \sin^2\theta\,|0\rangle_1\,{}_1\langle 0|\,, \tag{11}$$

$$\rho_2^{(e)} = Tr_1[\rho^{(e)}] = \cos^2\theta\,|0\rangle_2\,{}_2\langle 0| + \sin^2\theta\,|1\rangle_2\,{}_2\langle 1|\,. \tag{12}$$

It is then easy to calculate the corresponding linear entropies, which turn out to be equal:

$$S_L^{(1;2)}(\rho_e) = 2\left(1 - Tr_1[(\rho_1^{(e)})^2]\right) = \sin^2(2\theta), \tag{13}$$

$$S_L^{(2;1)}(\rho_e) = 2\left(1 - Tr_2[(\rho_2^{(e)})^2]\right) = \sin^2(2\theta), \tag{14}$$

Similar results are found for the muon neutrino state. Note that the above results are particular cases of the more general ones obtained for the three flavor neutrino states in [3], where it was found that such states can be classified as generalized W states. In the present (two-flavor) case, the form of the entangled state is simply that of a Bell state.

Eqs.(13)-(14) express the fact that flavor neutrino states at any time can be regarded as entangled superpositions of the mass qubits $|\nu_i\rangle$, where the entanglement is a function of the mixing angle only.

Let us now turn to the dynamic entanglement arising in connection with flavor oscillations. To this aim, we rewrite the electron neutrino state $|\nu_e(t)\rangle$ as

$$|\nu_e(t)\rangle \;=\; \widetilde{\mathbf{U}}_{ee}(t)\,|\nu_e\rangle + \widetilde{\mathbf{U}}_{e\mu}(t)|\nu_\mu\rangle\,, \tag{15}$$

Journal of Physics: Conference Series **237** (2010) 012007 doi:10.1088/1742-6596/237/1/012007

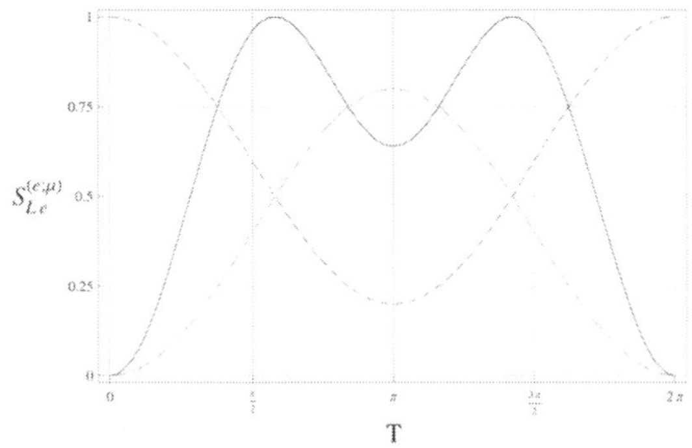

Figure 1. (Color online) Linear entropy $S_L^{(e;\mu)}(\rho_e)$ (full) as a function of the scaled time $T = \frac{2Et}{\Delta m_{12}^2}$. The mixing angle θ is fixed at the experimental value $\sin^2\theta = 0.314$. The transition probabilities $P_{\nu_e\to\nu_e}$ (dashed) and $P_{\nu_e\to\nu_\mu}$ (dot-dashed) are reported as well for comparison.

where $|\nu_e\rangle$, $|\nu_\mu\rangle$ are the flavor neutrino states at time $t = 0$ and are now taken as the relevant qubits (see Eq.(8)). By proceeding in a similar way as done for the static case, we arrive at the following expression for the linear entropies associated to the above state:

$$S_L^{(\mu;e)}(\rho_e) = S_L^{(e;\mu)}(\rho_e) = 4|\tilde{\mathbf{U}}_{ee}(t)|^2\,|\tilde{\mathbf{U}}_{e\mu}(t)|^2 = 4|\tilde{\mathbf{U}}_{ee}(t)|^2\,(1-|\tilde{\mathbf{U}}_{ee}(t)|^2)\,. \tag{16}$$

Eq.(16) establishes that the linear entropy of the reduced state is equal to the product of the two-flavor transition probabilities given in Eqs.(5)-(7). It is remarkable that simple expressions similar to those of Eq. (16) hold also for the three flavor case [4].

Note also that, for any reduced state ρ of a two-level system one has that $S_L = 2[1-Tr(\rho^2)] = 4Det\rho = 4\lambda_1(1-\lambda_1)$, where λ_1 is one of the two non-negative eigenvalues of ρ, and the relation $\lambda_1+\lambda_2 = 1$ has been exploited. Comparing with Eq. (16), one sees that the transition probabilities coincide with the eigenvalues of the reduced state density matrix.

In figure 1 we show the behavior of$S_L^{(e;\mu)}(\rho_e)$ as a function of the scaled, dimensionless time $T = \frac{2Et}{\Delta m_{12}^2}$. In the same figure, we also report the behavior of the transition probabilities $P_{\nu_e\to\nu_e}$ and $P_{\nu_e\to\nu_\mu}$. The plots have a clear physical interpretation. At time $T = 0$, the entanglement is zero, the global state of the system is factorized, and the two flavors are not mixed. For $T > 0$, flavors start to oscillate and the entanglement is maximal at largest mixing: $P_{\nu_e\to\nu_e} = P_{\nu_e\to\nu_\mu} = 0.5$, and minimum at $T = \pi$.

2.2. Neutrino entanglement via uncertainties

An alternative characterization of entanglement can be given in terms of quantum uncertainties [21, 22]. In order to apply this formalism to the case of neutrino mixing and oscillations we introduce the (fermionic) annihilation operator α_i for a neutrino with mass m_i, with anti-commutator $\{\alpha_i, \alpha_j\} = \delta_{ij}$. We then define neutrino states with definite masses as:

$$|\nu_i\rangle \equiv \alpha_i^\dagger|0\rangle_m\,, \quad i = 1,2 \tag{17}$$

where $|0\rangle_m \equiv |0\rangle_1 \otimes |0\rangle_2$ is the vacuum for the mass eigenstates.

Next we define the flavor annihilation operators by means of the following relations:

$$\alpha_e(t) = \cos\theta\,\alpha_1(t) + \sin\theta\,\alpha_2(t)\,, \tag{18}$$

$$\alpha_\mu(t) = -\sin\theta\,\alpha_2(t) + \cos\theta\,\alpha_1(t)\,, \tag{19}$$

where $\alpha_i(t) = e^{i\omega_i t}\alpha_i$, with $i = 1, 2$.

The flavor states are given by:

$$|\nu_\sigma(t)\rangle \equiv \alpha_\sigma^\dagger(t)|0\rangle_m, \quad \sigma = e, \mu. \tag{20}$$

We use in the following the notation $|\nu_\sigma\rangle \equiv |\nu_\sigma(t=0)\rangle$.

The hamiltonian for the system is

$$\begin{aligned} H &= \omega_{ee}\alpha_e^\dagger(t)\alpha_e(t) + \omega_{\mu\mu}\alpha_\mu^\dagger(t)\alpha_\mu(t) + \omega_{e\mu}\left(\alpha_e^\dagger(t)\alpha_\mu(t) + \alpha_\mu^\dagger(t)\alpha_e(t)\right) \\ &= \omega_1\alpha_1^\dagger\alpha_1 + \omega_2\alpha_2^\dagger\alpha_2 \end{aligned} \tag{21}$$

where we used the relations $\omega_{ee} = \omega_1\cos^2\theta + \omega_2\sin^2\theta$, $\omega_{\mu\mu} = \omega_1\sin^2\theta + \omega_2\cos^2\theta$, $\omega_{e\mu} = (\omega_2 - \omega_1)\sin\theta\cos\theta$.

In the mass basis, we can introduce the $SU(2)$ operators (the superscript m stands for mass):

$$J_+^m = \alpha_1^\dagger\alpha_2 \ , \qquad J_-^m = \alpha_2^\dagger\alpha_1, \tag{22}$$

$$J_3^m = \frac{1}{2}\left(\alpha_1^\dagger\alpha_1 - \alpha_2^\dagger\alpha_2\right), \tag{23}$$

$$\mathcal{C} = \frac{1}{2}\left(\alpha_1^\dagger\alpha_1 + \alpha_2^\dagger\alpha_2\right). \tag{24}$$

We also have

$$N_1 = \mathcal{C} + J_3^m, \tag{25}$$

$$N_2 = \mathcal{C} - J_3^m, \tag{26}$$

$$H = \omega_1 N_1 + \omega_2 N_2. \tag{27}$$

The static entanglement of the electron neutrino state $|\nu_e(t)\rangle$ defined in Eq.(20), is characterized, in the present formalism, by the variances associated with the numbers N_i, relative to the mass qubits. Thus we have:

$$\begin{aligned} \Delta N_i(\nu_e) &\equiv \langle\nu_e(t)|N_i^2|\nu_e(t)\rangle - \langle\nu_e(t)|N_i|\nu_e(t)\rangle^2 \\ &= \frac{1}{4}\sin^2(2\theta), \qquad i = 1, 2. \end{aligned} \tag{28}$$

This result differs by a factor 4 from that obtained by means of the linear entropy, Eqs.(13)-(14).

In order to discuss the dynamical entanglement of the state $|\nu_e(t)\rangle$, we need to introduce flavor oscillations, which can be seen both in terms of overlaps of states at different times:

$$P_{\nu_e\to\nu_e}(t) = |\langle\nu_e|\nu_e(t)\rangle|^2, \tag{29}$$

$$P_{\nu_e\to\nu_\mu}(t) = |\langle\nu_\mu|\nu_e(t)\rangle|^2, \tag{30}$$

with $P_{\nu_e\to\nu_e}(t) + P_{\nu_e\to\nu_\mu}(t) = 1$, or equivalently in terms of expectation values of number operators at time t:

$$P_{\nu_e\to\nu_e}(t) = \langle\nu_e|N_e(t)|\nu_e\rangle, \tag{31}$$

$$P_{\nu_e\to\nu_\mu}(t) = \langle\nu_e|N_\mu(t)|\nu_e\rangle, \tag{32}$$

$$N_\sigma(t) = \alpha_\sigma^\dagger(t)\alpha_\sigma(t), \qquad \sigma = e, \mu. \tag{33}$$

The explicit expressions of the transition probabilities are given in Eqs.(6),(7).

In the flavor basis, we can again introduce $SU(2)$ operators (at time t):

$$J^f_+(t) = \alpha^\dagger_e(t)\alpha_\mu(t) \quad , \quad J^f_-(t) = \alpha^\dagger_\mu(t)\alpha_e(t) , \tag{34}$$

$$J^f_3(t) = \frac{1}{2}\left(\alpha^\dagger_e(t)\alpha_e(t) - \alpha^\dagger_\mu(t)\alpha_\mu(t)\right) , \tag{35}$$

$$\mathcal{C} = \frac{1}{2}\left(\alpha^\dagger_e(t)\alpha_e(t) + \alpha^\dagger_\mu(t)\alpha_\mu(t)\right) = \frac{1}{2}\left(\alpha^\dagger_1\alpha_1 + \alpha^\dagger_2\alpha_2\right) , \tag{36}$$

where the superscript f stands for flavor. The flavor number operators are:

$$N_e(t) = \mathcal{C} + J^f_3(t) , \tag{37}$$

$$N_\mu(t) = \mathcal{C} - J^f_3(t) . \tag{38}$$

Note that the Hamiltonian can be written as:

$$H = \omega_{ee}N_e(t) + \omega_{\mu\mu}N_\mu(t) + \omega_{e\mu}(J^f_+(t) + J^f_-(t)) . \tag{39}$$

Flavor entanglement is given by the variances of the above flavor numbers. Consider first the variances of the $SU(2)$ operators in the flavor basis, which give:

$$\Delta J_1(\nu_e)(t) = \frac{1}{4}\left[1 - \sin^2(4\theta)\sin^4\left(\frac{\omega_2-\omega_1}{2}t\right)\right] , \tag{40}$$

$$\Delta J_2(\nu_e)(t) = \frac{1}{4} - \sin^2(2\theta)\sin^2[(\omega_2-\omega_1)t] , \tag{41}$$

$$\Delta J_3(\nu_e)(t) = \sin^2(2\theta)\sin^2\left(\frac{\omega_2-\omega_1}{2}t\right)\left[1 - \sin^2(2\theta)\sin^2\left(\frac{\omega_2-\omega_1}{2}t\right)\right] , \tag{42}$$

$$\Delta\mathcal{C}(\nu_e) = 0 . \tag{43}$$

When we restrict to a given flavor (e.g. electron neutrinos), we find that the entanglement is given by

$$\Delta N_e(\nu_e)(t) \equiv \langle\nu_e(t)|N^2_e(t)|\nu_e(t)\rangle - \langle\nu_e(t)|N_e(t)|\nu_e(t)\rangle^2 \tag{44}$$

$$= P_{\nu_e\to\nu_e}(t)\, P_{\nu_e\to\nu_\mu}(t) , \tag{45}$$

with the same result for $\Delta N_\mu(\nu_e)(t)$. The above result coincides (again up to a factor 4) with the one obtained in Eq.(16) by means of the linear entropy.

Analogous results are easily obtained for the state $|\nu_\mu(t)\rangle$.

3. Neutrino mixing in Quantum Field Theory

We now look for an extension of the above discussion to a relativistic context. To this aim, we need to consider neutrino mixing at level of fields [17]. For two flavors, the mixing transformations are

$$\nu_e(x) = \cos\theta\,\nu_1(x) + \sin\theta\,\nu_2(x) , \tag{46}$$

$$\nu_\mu(x) = -\sin\theta\,\nu_1(x) + \cos\theta\,\nu_2(x) , \tag{47}$$

where $\nu_e(x)$ and $\nu_\mu(x)$ are the Dirac neutrino fields with definite flavors. Here, $\nu_1(x)$ and $\nu_2(x)$ are the free neutrino fields with definite masses m_1 and m_2, respectively. For the purpose of discussing the quantization of flavor fields and the phenomenon of neutrino oscillations, it is sufficient to consider only the free Lagrangian term for these two free fields:

$$\mathcal{L}_\nu(x) \,=\, \bar{\nu}_m(x)\left(i\,\not\partial - M_\nu^d\right)\nu_m(x)\,, \tag{48}$$

where $\nu_m^T = (\nu_1, \nu_2)$ and $M_\nu^d = diag(m_1, m_2)$. $\mathcal{L}_\nu(x)$ is invariant under global $U(1)$ phase transformations of the type $\nu'_m(x) = e^{i\alpha}\nu_m(x)$. This implies the conservation of the Noether charge $Q = \int I^0(x)d^3\mathbf{x}$ (with $I^\mu(x) = \bar{\nu}_m(x)\gamma^\mu\nu_m(x)$) which is indeed the total charge of the system, i.e. the total lepton number of neutrinos.

Consider now the global $SU(2)$ transformation [19]:

$$\nu'_m(x) = e^{i\alpha_j\cdot\tau_j}\,\nu_m(x)\,, \qquad\qquad j = 1,2,3\,, \tag{49}$$

with α_j real constants, $\tau_j = \sigma_j/2$ with σ_j being the Pauli matrices.

$\mathcal{L}_\nu$ is not invariant under the transformations (49) since $m_1 \neq m_2$. By use of the equations of motion, we obtain

$$\delta\mathcal{L}_\nu = i\alpha_j\bar{\nu}_m(x)\left[\tau_j,\, M_\nu^d\right]\nu_m(x) = -\alpha_j\partial_\mu J^\mu_{m,j}(x)\,, \tag{50}$$

where the currents are:

$$J^\mu_{m,j}(x) = \bar{\nu}_m(x)\,\gamma^\mu\,\tau_j\,\nu_m(x)\,, \qquad\qquad j = 1,2,3. \tag{51}$$

The related charges

$$Q_{m,j}(t) = \int d^3\mathbf{x}\,J^0_{m,j}(x)\,, \tag{52}$$

satisfy the $su(2)$ algebra:

$$[Q_{m,i}(t), Q_{m,j}(t)] = i\varepsilon_{ijk}Q_{m,k}(t). \tag{53}$$

The Casimir operator is proportional to the total (conserved) charge: $Q_{m,0} = \frac{1}{2}Q_\nu$ and also $Q_{m,3}$ is conserved, due to the fact that M_ν^d is diagonal. This implies that the charges for ν_1 and ν_2 are separately conserved. The $U(1)$ Noether charges associated with ν_1 and ν_2 can be then expressed as

$$Q_1 \equiv \frac{1}{2}Q + Q_{m,3}\,, \qquad\qquad Q_2 \equiv \frac{1}{2}Q - Q_{m,3}\,, \tag{54}$$

$$Q_i = \int d^3\mathbf{x}\,\nu_i^\dagger(x)\,\nu_i(x)\,, \tag{55}$$

with Q total (conserved) charge and $i = 1,2$.

Let us now consider the Lagrangian $\mathcal{L}_\nu(x)$ written in the flavor basis

$$\mathcal{L}_\nu(x) \,=\, \bar{\nu}_f(x)\,(i\,\not\partial - M_\nu)\,\nu_f(x)\,, \tag{56}$$

where $\nu_f^T = (\nu_e,\, \nu_\mu)$ and $M_\nu = \begin{pmatrix} m_{\nu_e} & m_{\nu_{e\mu}} \\ m_{\nu_{e\mu}} & m_{\nu_\mu} \end{pmatrix}$.

The variation of the Lagrangian (56) under the $SU(2)$ transformation:

$$\nu'_f(x) = e^{i\alpha_j\cdot\tau_j}\,\nu_f(x) \qquad\qquad j = 1,2,3\,, \tag{57}$$

is given by

$$\delta\mathcal{L}_\nu(x) = i\alpha_j\bar{\nu}_f(x)\,[\tau_j, M_\nu]\,\nu_f(x) = -\alpha_j\partial_\mu J^\mu_{f,j}(x)\,, \tag{58}$$

where

$$J^\mu_{f,j}(x) = \bar{\nu}_f(x)\,\gamma^\mu\,\tau_j\,\nu_f(x)\,, \qquad j = 1,2,3\,. \tag{59}$$

Again, the charges

$$Q_{f,j}(t) = \int d^3\mathbf{x}\, J^0_{f,j}(x)\,, \tag{60}$$

close the $su(2)$ algebra, however, because of the off-diagonal (mixing) terms in M_ν, $Q_{f,3}(t)$ is time dependent. This implies an exchange of charge between ν_e and ν_μ, resulting in the phenomenon of neutrino oscillations. The (time dependent) flavor charges for mixed fields are then defined as [19]:

$$Q_e(t) = \frac{1}{2}Q + Q_{f,3}(t)\,, \qquad Q_\mu(t) = \frac{1}{2}Q - Q_{f,3}(t)\,, \tag{61}$$

$$Q_\sigma(t) = \int d^3\mathbf{x}\,\nu_\sigma^\dagger(x)\,\nu_\sigma(x)\,, \tag{62}$$

where $\sigma = e, \mu$ and $Q_e(t) + Q_\mu(t) = Q$.

3.1. Flavor states for mixed neutrinos

Till now our considerations have been essentially classical. We now quantize the fields with definite masses as usual (see Appendix), and consider the eigenstates of the above defined charges.

The normal ordered charge operators for free neutrinos ν_1, ν_2 are:

$$:Q_i: \equiv \int d^3\mathbf{x}\,:\nu_i^\dagger(x)\,\nu_i(x): = \sum_r \int d^3\mathbf{k}\left(\alpha^{r\dagger}_{\mathbf{k},i}\alpha^r_{\mathbf{k},i} - \beta^{r\dagger}_{-\mathbf{k},i}\beta^r_{-\mathbf{k},i}\right), \tag{63}$$

where $i = 1,2$ and $:..:$ denotes normal ordering with respect to the vacuum $|0\rangle_{1,2}$. The neutrino states with definite masses defined as

$$|\nu^r_{\mathbf{k},i}\rangle = \alpha^{r\dagger}_{\mathbf{k},i}|0\rangle_{1,2}, \qquad i = 1,2, \tag{64}$$

are clearly eigenstates of Q_1 and Q_2, which can be identified with the lepton charges of neutrinos in the absence of mixing.

The situation is more delicate when mixing is present. In such a case, the flavor neutrino states have to be defined as the eigenstates of the flavor charges $Q_\sigma(t)$ (at a given time). The relation between the flavor charges in the presence of mixing and those in the absence of mixing is:

$$Q_e(t) = \cos^2\theta\,Q_1 + \sin^2\theta\,Q_2 + \sin\theta\cos\theta\int d^3\mathbf{x}\left[\nu_1^\dagger(x)\nu_2(x) + \nu_2^\dagger(x)\nu_1(x)\right], \tag{65}$$

$$Q_\mu(t) = \sin^2\theta\,Q_1 + \cos^2\theta\,Q_2 - \sin\theta\cos\theta\int d^3\mathbf{x}\left[\nu_1^\dagger(x)\nu_2(x) + \nu_2^\dagger(x)\nu_1(x)\right]. \tag{66}$$

Notice that the last term in these expressions is proportional to the charge $Q_{m,1}$ defined above (cf. Eq.(52)). The presence of such a term forbids the construction of eigenstates of the $Q_\sigma(t)$ in the Hilbert space $\mathcal{H}_{1,2}$. One then is led [17] to define another Hilbert space, $\mathcal{H}_{e,\mu}$ for the flavor neutrino fields. The flavor vacuum state $|0\rangle_{e,\mu}$ and the mass vacuum state $|0\rangle_{1,2}$ are orthogonal to each other (see Appendix).

The normal ordered flavor charge operators for mixed neutrinos are then written as

$$
\begin{aligned}
:: Q_\sigma(t) :: \; &\equiv \int d^3\mathbf{x} :: \nu_\sigma^\dagger(x)\, \nu_\sigma(x) :: \\
&= \sum_r \int d^3\mathbf{k} \left(\alpha^{r\dagger}_{\mathbf{k},\sigma}(t)\alpha^r_{\mathbf{k},\sigma}(t) \; - \; \beta^{r\dagger}_{-\mathbf{k},\sigma}(t)\beta^r_{-\mathbf{k},\sigma}(t) \right) ,
\end{aligned} \tag{67}
$$

where $\sigma = e, \mu$, and $::\ ...\ ::$ denotes normal ordering with respect to $|0\rangle_{e,\mu}$. Thus, the flavor charges are diagonal in the flavor annihilation/creation operators constructed by means of the mixing generator G_θ defined in the Appendix. The definition of the normal ordering $::\ ...\ ::$ for any operator A, is the usual one:

$$
:: A :: \equiv A \; - \; {}_{e,\mu}\langle 0|A|0\rangle_{e,\mu} \, . \tag{68}
$$

Note that $:: Q_\sigma(t) :: \; = \; G_\theta^{-1}(t) : Q_j : G_\theta(t)$, with $(\sigma, j) = (e, 1), (\mu, 2)$, and

$$
:: Q :: \; = \; :: Q_e(t) :: \; + \; :: Q_\mu(t) :: \; = \; : Q_1 : \; + \; : Q_2 : \; = \; : Q : \, . \tag{69}
$$

We define the flavor states as eigenstates of the flavor charges Q_σ at a reference time $t = 0$:

$$
|\nu^r_{\mathbf{k},\sigma}\rangle \equiv \alpha^{r\dagger}_{\mathbf{k},\sigma}(0)|0(0)\rangle_{e,\mu}, \qquad \sigma = e, \mu \tag{70}
$$

and similar ones for antiparticles. We have

$$
:: Q_e(0) :: |\nu^r_{\mathbf{k},e}\rangle \; = \; |\nu^r_{\mathbf{k},e}\rangle \, , \qquad\qquad :: Q_\mu(0) :: |\nu^r_{\mathbf{k},\mu}\rangle \; = \; |\nu^r_{\mathbf{k},\mu}\rangle \, , \tag{71}
$$

$$
:: Q_e(0) :: \; |\nu^r_{\mathbf{k},\mu}\rangle \; = \; 0 \; = \; :: Q_\mu(0) :: \; |\nu^r_{\mathbf{k},e}\rangle \, , \qquad :: Q_\sigma(0) :: \; |0\rangle_{e,\mu} \; = \; 0. \tag{72}
$$

The explicit form of the flavor states $|\nu^r_{\mathbf{k},e}\rangle$ and $|\nu^r_{\mathbf{k},\mu}\rangle$ at time $t = 0$ is given in the Appendix.

3.2. Oscillation formulas in QFT

Flavor oscillation formulas are derived by computing, in the Heisenberg representation, the expectation value of the flavor charge operators on the flavor state. We have

$$
\mathcal{Q}^{\mathbf{k}}_{\nu_e\to\nu_\sigma}(t) \;\; \equiv \;\; \langle \nu^r_{\mathbf{k},e}|Q_\sigma(t)|\nu^r_{\mathbf{k},e}\rangle \; = \; \left|\left\{\alpha^r_{\mathbf{k},\sigma}(t), \alpha^{r\dagger}_{\mathbf{k},e}(0)\right\}\right|^2 \; + \; \left|\left\{\beta^{r\dagger}_{-\mathbf{k},\sigma}(t), \alpha^{r\dagger}_{\mathbf{k},e}(0)\right\}\right|^2 , \tag{73}
$$

and

$$
{}_{e,\mu}\langle 0| :: Q_e(t) :: |0\rangle_{e,\mu} = \; {}_{e,\mu}\langle 0|Q_\mu(t)|0\rangle_{e,\mu} = 0 \, . \tag{74}
$$

The oscillation formulas are [18]:

$$
\mathcal{Q}^{\mathbf{k}}_{\nu_e\to\nu_e}(t) \;\; = \;\; 1 - \sin^2(2\theta)\Big[\, |U_\mathbf{k}|^2 \sin^2\left(\frac{\omega_{k,2} - \omega_{k,1}}{2} t\right) + |V_\mathbf{k}|^2 \sin^2\left(\frac{\omega_{k,2} + \omega_{k,1}}{2} t\right) \Big], \tag{75}
$$

$$
\mathcal{Q}^{\mathbf{k}}_{\nu_e\to\nu_\mu}(t) \;\; = \;\; \sin^2(2\theta)\Big[\, |U_\mathbf{k}|^2 \sin^2\left(\frac{\omega_{k,2} - \omega_{k,1}}{2} t\right) + |V_\mathbf{k}|^2 \sin^2\left(\frac{\omega_{k,2} + \omega_{k,1}}{2} t\right) \Big]. \tag{76}
$$

The charge conservation is ensured at any time:

$$
\mathcal{Q}^{\mathbf{k}}_{\nu_e\to\nu_e}(t) + \mathcal{Q}^{\mathbf{k}}_{\nu_e\to\nu_\mu}(t) = 1. \tag{77}
$$

The differences with respect to the Pontecorvo formulas are: the energy dependence of the amplitudes, and the additional oscillating term. In the relativistic limit: $|\mathbf{k}| \gg \sqrt{m_1 m_2}$, we have $|U_\mathbf{k}|^2 \longrightarrow 1$ and $|V_\mathbf{k}|^2 \longrightarrow 0$ and the traditional formulas are recovered.

4. Entanglement in neutrino mixing: Quantum Field Theory

Following what done in the QM case, we now calculate the entanglement associated to an electron neutrino state at time t, by means of the variances of the above discussed charge operators.

Let us start with the Noether charges Q_{ν_i}, which are expected to characterize the amount of static entanglement present in the states Eq.(70). We obtain:

$$\begin{aligned} \Delta Q_i(\nu_e)(t) &= \langle \nu^r_{\mathbf{k},e}|Q_i^2|\nu^r_{\mathbf{k},e}\rangle - \langle \nu^r_{\mathbf{k},e}|Q_i|\nu^r_{\mathbf{k},e}\rangle^2 \\ &= \frac{1}{4}\sin^2(2\theta)\,, \qquad i=1,2\,, \end{aligned} \tag{78}$$

in perfect agreement with the quantum mechanical result Eq.(28).

Next we consider dynamic entanglement, which is described by the variances of the flavor charges. We have:

$$\begin{aligned} \Delta Q_e(\nu_e)(t) &= \langle \nu^r_{\mathbf{k},e}|Q_e^2(t)|\nu^r_{\mathbf{k},e}\rangle - \langle \nu^r_{\mathbf{k},e}|Q_e(t)|\nu^r_{\mathbf{k},e}\rangle^2 \\ &= \langle \nu^r_{\mathbf{k},e}|\left[\sum_s \int d^3\mathbf{p}\left(\alpha^{s\dagger}_{\mathbf{p},e}(t)\alpha^s_{\mathbf{p},e}(t) - \beta^{s\dagger}_{-\mathbf{p},e}(t)\beta^s_{-\mathbf{p},e}(t)\right)\right]^2|\nu^r_{\mathbf{k},e}\rangle - \left[\mathcal{Q}^{\mathbf{k}}_{\nu_e\to\nu_e}(t)\right]^2 \\ &= \langle \nu^r_{\mathbf{k},e}|\alpha^{r\dagger}_{\mathbf{k},e}(t)\alpha^r_{\mathbf{k},e}(t)|\nu^r_{\mathbf{k},e}\rangle + \langle \nu^r_{\mathbf{k},e}|\beta^{r\dagger}_{-\mathbf{k},e}(t)\beta^r_{-\mathbf{k},e}(t)|\nu^r_{\mathbf{k},e}\rangle \\ &- 2\,\langle \nu^r_{\mathbf{k},e}|\alpha^{r\dagger}_{\mathbf{k},e}(t)\alpha^r_{\mathbf{k},e}(t)\beta^{r\dagger}_{-\mathbf{k},e}(t)\beta^r_{-\mathbf{k},e}(t)|\nu^r_{\mathbf{k},e}\rangle - \left[\mathcal{Q}^{\mathbf{k}}_{\nu_e\to\nu_e}(t)\right]^2 . \end{aligned} \tag{79}$$

We now consider the third term in Eq.(79). We have:

$$\begin{aligned} \langle \nu^r_{\mathbf{k},e}|\alpha^{r\dagger}_{\mathbf{k},e}(t)\alpha^r_{\mathbf{k},e}(t)\beta^{r\dagger}_{-\mathbf{k},e}(t)\beta^r_{-\mathbf{k},e}(t)|\nu^r_{\mathbf{k},e}\rangle &= {}_{e,\mu}\langle 0|\alpha^{r\dagger}_{\mathbf{k},e}(t)\alpha^r_{\mathbf{k},e}(t)\beta^{r\dagger}_{-\mathbf{k},e}(t)\beta^r_{-\mathbf{k},e}(t)|0\rangle_{e,\mu} \\ &+\left|\left\{\alpha^r_{\mathbf{k},e}(t),\alpha^{r\dagger}_{\mathbf{k},e}(0)\right\}\right|^2 {}_{e,\mu}\langle 0|\beta^{r\dagger}_{-\mathbf{k},e}(t)\beta^r_{-\mathbf{k},e}(t)|0\rangle_{e,\mu} \\ &-\left|\left\{\alpha^r_{\mathbf{k},e}(0),\beta^r_{-\mathbf{k},e}(t)\right\}\right|^2 {}_{e,\mu}\langle 0|\alpha^{r\dagger}_{\mathbf{k},e}(t)\alpha^r_{\mathbf{k},e}(t)|0\rangle_{e,\mu} \\ -\left\{\alpha^{r\dagger}_{\mathbf{k},e}(0),\beta^{r\dagger}_{-\mathbf{k},e}(t)\right\}&\left\{\alpha^r_{\mathbf{k},e}(0),\alpha^{r\dagger}_{\mathbf{k},e}(t)\right\} {}_{e,\mu}\langle 0|\alpha^r_{\mathbf{k},e}(t)\beta^r_{-\mathbf{k},e}(t)|0\rangle_{e,\mu} \\ +\left\{\alpha^r_{\mathbf{k},e}(0),\beta^r_{-\mathbf{k},e}(t)\right\}&\left\{\alpha^r_{\mathbf{k},e}(t),\alpha^{r\dagger}_{\mathbf{k},e}(0)\right\} {}_{e,\mu}\langle 0|\alpha^{r\dagger}_{\mathbf{k},e}(t)\beta^{r\dagger}_{-\mathbf{k},e}(t)|0\rangle_{e,\mu} . \end{aligned} \tag{80}$$

Explicit calculation of the above quantity shows that:

$$\langle \nu^r_{\mathbf{k},e}|\alpha^{r\dagger}_{\mathbf{k},e}(t)\alpha^r_{\mathbf{k},e}(t)\beta^{r\dagger}_{-\mathbf{k},e}(t)\beta^r_{-\mathbf{k},e}(t)|\nu^r_{\mathbf{k},e}\rangle = \langle \nu^r_{\mathbf{k},e}|\beta^{r\dagger}_{-\mathbf{k},e}(t)\beta^r_{-\mathbf{k},e}(t)|\nu^r_{\mathbf{k},e}\rangle\,, \tag{81}$$

so that we have

$$\Delta Q_e(\nu_e)(t) = \mathcal{Q}^{\mathbf{k}}_{\nu_e\to\nu_e}(t)\,\mathcal{Q}^{\mathbf{k}}_{\nu_e\to\nu_\mu}(t)\,, \tag{82}$$

which formally resembles the quantum mechanical result Eq.(44). The differences are now due to the presence of the flavor condensate, which affect the oscillation formulas (see Eqs.(75),(76)). In figure 2, the flavor entanglement formula is plotted in the QFT case against the corresponding QM case.

5. Conclusions

On the basis of recent results, we have discussed some aspect of entanglement in the phenomenon of neutrino mixing and oscillations. In the simple case of two flavor mixing, we have shown how to generalize previous results obtained in the context of quantum mechanics, to the case of

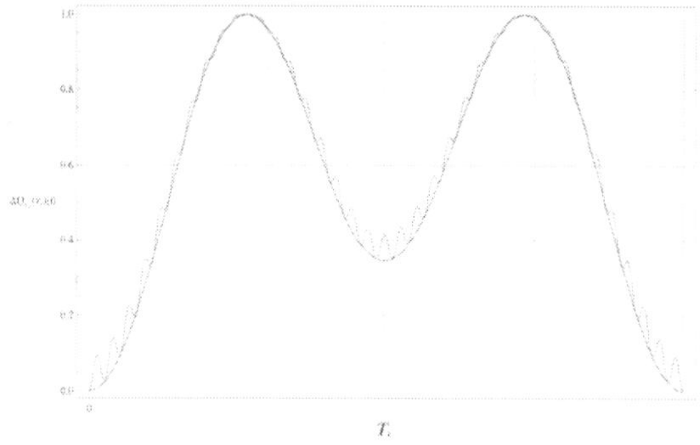

Figure 2. QM vs. QFT flavor entanglement for $|\nu_e(t)\rangle$ as a function of the scaled time $T = \frac{2Et}{\Delta m_{12}^2}$ with θ fixed at the value $\sin^2\theta = 0.314$.

quantum field theory. The difference between the QFT and the QM cases are related to the condensate vacuum structure associated to neutrino mixing in QFT.

Our study provides a simple, exactly solvable example for a possible extension of entanglement to quantum field theory. Apart from the differences with QM due to the flavor vacuum contributions, the QFT result is interesting from a more conceptual point of view. Indeed, both the static and the dynamical entanglement arise in connection with unitarily inequivalent representations: in the case of the static entanglement, the flavor Hilbert space at time t to which the entangled state $|\nu_\sigma(t)\rangle$ belongs, is unitarily inequivalent to the Hilbert space for the qubit states $|\nu_i\rangle$ [17]; on the other hand, in the case of dynamical entanglement, where the qubits are taken to be the flavor states at time $t = 0$, the inequivalence is among the flavor Hilbert space at different times [23]. In the first case, the relevant orthogonality relation is $\lim_{V\to\infty}\, {}_m\langle 0|0(t)\rangle_f = 0$, in the second $\lim_{V\to\infty}\, {}_f\langle 0(t')|0(t)\rangle_f = 0$, with $t \neq t'$.

Since the inequivalent representations are associated with a non-trivial condensate vacuum structure, the above conjecture suggests that, in the context of QFT, many interpretational issues connected with entanglement could be revisited in this new light.

Appendix A. QFT formalism for mixed fields

The fields $\nu_e(x)$ and $\nu_\mu(x)$ are defined through the mixing relations (47), in terms of the free fields $\nu_1(x)$ and $\nu_2(x)$ which are expanded as

$$\nu_i(x) = \frac{1}{\sqrt{V}}\sum_{\mathbf{k},r}\left[u^r_{\mathbf{k},i}\,\alpha^r_{\mathbf{k},i}(t) + v^r_{-\mathbf{k},i}\,\beta^{r\dagger}_{-\mathbf{k},i}(t)\right]e^{i\mathbf{k}\cdot\mathbf{x}}, \qquad i = 1,2 \tag{A.1}$$

with $\alpha^r_{\mathbf{k},i}(t) = \alpha^r_{\mathbf{k},i}\,e^{-i\omega_{k,i}t}$, $\beta^{r\dagger}_{\mathbf{k},i}(t) = \beta^{r\dagger}_{\mathbf{k},i}\,e^{i\omega_{k,i}t}$, and $\omega_{k,i} = \sqrt{\mathbf{k}^2 + m_i^2}$. The operator $\alpha^r_{\mathbf{k},i}$ and $\beta^r_{\mathbf{k},i}$, $i = 1,2$, $r = 1,2$ are the annihilator operators for the vacuum state $|0\rangle_m \equiv |0\rangle_1 \otimes |0\rangle_2$: $\alpha^r_{\mathbf{k},i}|0\rangle_m = \beta^r_{\mathbf{k},i}|0\rangle_m = 0$. The anticommutation relations are: $\left\{\nu_i^\alpha(x), \nu_j^{\beta\dagger}(y)\right\}_{t=t'} = \delta^3(\mathbf{x}-\mathbf{y})\delta_{\alpha\beta}\delta_{ij}$, with $\alpha,\beta = 1,...4$, and $\left\{\alpha^r_{\mathbf{k},i}, \alpha^{s\dagger}_{\mathbf{q},j}\right\} = \delta_{\mathbf{kq}}\delta_{rs}\delta_{ij}$; $\left\{\beta^r_{\mathbf{k},i}, \beta^{s\dagger}_{\mathbf{q},j}\right\} = \delta_{\mathbf{kq}}\delta_{rs}\delta_{ij}$, with $i,j = 1,2$. All other anticommutators vanish. The orthonormality and completeness relations are given by $u^{r\dagger}_{\mathbf{k},i}u^s_{\mathbf{k},i} = v^{r\dagger}_{\mathbf{k},i}v^s_{\mathbf{k},i} = \delta_{rs}$, $u^{r\dagger}_{\mathbf{k},i}v^s_{-\mathbf{k},i} = v^{r\dagger}_{-\mathbf{k},i}u^s_{\mathbf{k},i} = 0$, and $\sum_r(u^r_{\mathbf{k},i}u^{r\dagger}_{\mathbf{k},i} + v^r_{-\mathbf{k},i}v^{r\dagger}_{-\mathbf{k},i}) = 1\!\!1$.

The generator of the mixing transformations is given by [17]:

$$G_\theta(t) = \exp\left[\theta\int d^3\mathbf{x}\left(\nu_1^\dagger(x)\nu_2(x) - \nu_2^\dagger(x)\nu_1(x)\right)\right], \tag{A.2}$$

so that

$$\nu_\sigma^\alpha(x) = G_\theta^{-1}(t)\, \nu_i^\alpha(x)\, G_\theta(t)\,, \qquad (\sigma, i) = (e,1), (\mu,2)\,. \tag{A.3}$$

At finite volume, this is a unitary operator, $G_\theta^{-1}(t) = G_{-\theta}(t) = G_\theta^\dagger(t)$, preserving the canonical anticommutation relations. The generator $G_\theta^{-1}(t)$ maps the Hilbert space for free fields $\mathcal{H}_{1,2}$ to the Hilbert space for mixed fields $\mathcal{H}_{e,\mu}$: $G_\theta^{-1}(t) : \mathcal{H}_{1,2} \mapsto \mathcal{H}_{e,\mu}$. In particular, the flavor vacuum is given by $|0(t)\rangle_{e,\mu} = G_\theta^{-1}(t)\, |0\rangle_{1,2}$ at finite volume V. We denote by $|0\rangle_{e,\mu}$ the flavor vacuum at $t = 0$. In the infinite volume limit, the flavor and the mass vacua are unitarily inequivalent [17]. The explicit expression for $|0\rangle_{e,\mu}$ at time $t = 0$ in the reference frame for which $\mathbf{k} = (0, 0, |\mathbf{k}|)$ is

$$\begin{aligned} |0\rangle^{\mathbf{k}}_{e,\mu} &= \prod_r \Big[(1 - \sin^2\theta\, |V_{\mathbf{k}}|^2) - \epsilon^r \sin\theta\, \cos\theta\, |V_{\mathbf{k}}|(\alpha^{r\dagger}_{\mathbf{k},1}\beta^{r\dagger}_{-\mathbf{k},2} + \alpha^{r\dagger}_{\mathbf{k},2}\beta^{r\dagger}_{-\mathbf{k},1}) + \\ &+ \epsilon^r \sin^2\theta\, |V_{\mathbf{k}}||U_{\mathbf{k}}|(\alpha^{r\dagger}_{\mathbf{k},1}\beta^{r\dagger}_{-\mathbf{k},1} - \alpha^{r\dagger}_{\mathbf{k},2}\beta^{r\dagger}_{-\mathbf{k},2}) + \sin^2\theta\, |V_{\mathbf{k}}|^2 \alpha^{r\dagger}_{\mathbf{k},1}\beta^{r\dagger}_{-\mathbf{k},2}\alpha^{r\dagger}_{\mathbf{k},2}\beta^{r\dagger}_{-\mathbf{k},1}\Big]|0\rangle_{1,2}\,. \end{aligned} \tag{A.4}$$

The condensation density is given by

$$_{e,\mu}\langle 0|\alpha^{r\dagger}_{\mathbf{k},i}\alpha^r_{\mathbf{k},i}|0\rangle_{e,\mu} = {}_{e,\mu}\langle 0|\beta^{r\dagger}_{\mathbf{k},i}\beta^r_{\mathbf{k},i}|0\rangle_{e,\mu} = \sin^2\theta\, |V_{\mathbf{k}}|^2\,, \qquad i = 1,2\,. \tag{A.5}$$

The flavor fields are written as:

$$\nu_\sigma(\mathbf{x}, t) = \frac{1}{\sqrt{V}}\sum_{\mathbf{k},r} e^{i\mathbf{k}.\mathbf{x}}\left[u^r_{\mathbf{k},i}\, \alpha^r_{\mathbf{k},\sigma}(t) + v^r_{-\mathbf{k},i}\, \beta^{r\dagger}_{-\mathbf{k},\sigma}(t)\right], \quad (\sigma, i) = (e,1), (\mu,2). \tag{A.6}$$

The flavor annihilation operators are [17]:

$$\begin{aligned} \alpha^r_{\mathbf{k},e}(t) &= \cos\theta\, \alpha^r_{\mathbf{k},1}(t) + \sin\theta \sum_s \left[u^{r\dagger}_{\mathbf{k},1}u^s_{\mathbf{k},2}\, \alpha^s_{\mathbf{k},2}(t) + u^{r\dagger}_{\mathbf{k},1}v^s_{-\mathbf{k},2}\, \beta^{s\dagger}_{-\mathbf{k},2}(t)\right], \\ \alpha^r_{\mathbf{k},\mu}(t) &= \cos\theta\, \alpha^r_{\mathbf{k},2}(t) - \sin\theta \sum_s \left[u^{r\dagger}_{\mathbf{k},2}u^s_{\mathbf{k},1}\, \alpha^s_{\mathbf{k},1}(t) + u^{r\dagger}_{\mathbf{k},2}v^s_{-\mathbf{k},1}\, \beta^{s\dagger}_{-\mathbf{k},1}(t)\right], \\ \beta^r_{-\mathbf{k},e}(t) &= \cos\theta\, \beta^r_{-\mathbf{k},1}(t) + \sin\theta \sum_s \left[v^{s\dagger}_{-\mathbf{k},2}v^r_{-\mathbf{k},1}\, \beta^s_{-\mathbf{k},2}(t) + u^{s\dagger}_{\mathbf{k},2}v^r_{-\mathbf{k},1}\, \alpha^{s\dagger}_{\mathbf{k},2}(t)\right], \\ \beta^r_{-\mathbf{k},\mu}(t) &= \cos\theta\, \beta^r_{-\mathbf{k},2}(t) - \sin\theta \sum_s \left[v^{s\dagger}_{-\mathbf{k},1}v^r_{-\mathbf{k},2}\, \beta^s_{-\mathbf{k},1}(t) + u^{s\dagger}_{\mathbf{k},1}v^r_{-\mathbf{k},2}\, \alpha^{s\dagger}_{\mathbf{k},1}(t)\right]. \end{aligned} \tag{A.7}$$

In the reference frame where $\mathbf{k} = (0, 0, |\mathbf{k}|)$, we have

$$\begin{aligned} \alpha^r_{\mathbf{k},e}(t) &= \cos\theta\, \alpha^r_{\mathbf{k},1}(t) + \sin\theta\, \Big(|U_{\mathbf{k}}|\, \alpha^r_{\mathbf{k},2}(t) + \epsilon^r\, |V_{\mathbf{k}}|\, \beta^{r\dagger}_{-\mathbf{k},2}(t)\Big), \\ \alpha^r_{\mathbf{k},\mu}(t) &= \cos\theta\, \alpha^r_{\mathbf{k},2}(t) - \sin\theta\, \Big(|U_{\mathbf{k}}|\, \alpha^r_{\mathbf{k},1}(t) - \epsilon^r\, |V_{\mathbf{k}}|\, \beta^{r\dagger}_{-\mathbf{k},1}(t)\Big), \\ \beta^r_{-\mathbf{k},e}(t) &= \cos\theta\, \beta^r_{-\mathbf{k},1}(t) + \sin\theta\, \Big(|U_{\mathbf{k}}|\, \beta^r_{-\mathbf{k},2}(t) - \epsilon^r\, |V_{\mathbf{k}}|\, \alpha^{r\dagger}_{\mathbf{k},2}(t)\Big), \\ \beta^r_{-\mathbf{k},\mu}(t) &= \cos\theta\, \beta^r_{-\mathbf{k},2}(t) - \sin\theta\, \Big(|U_{\mathbf{k}}|\, \beta^r_{-\mathbf{k},1}(t) + \epsilon^r\, |V_{\mathbf{k}}|\, \alpha^{r\dagger}_{\mathbf{k},1}(t)\Big). \end{aligned} \tag{A.8}$$

Here $\epsilon^r = (-1)^r$ and

$$\begin{aligned} |U_{\mathbf{k}}| &\equiv u^{r\dagger}_{\mathbf{k},i}u^r_{\mathbf{k},j} = v^{r\dagger}_{-\mathbf{k},i}v^r_{-\mathbf{k},j} = \frac{|\mathbf{k}|^2 + (\omega_{k,1} + m_1)(\omega_{k,2} + m_2)}{2\sqrt{\omega_{k,1}\omega_{k,2}(\omega_{k,1} + m_1)(\omega_{k,2} + m_2)}}, \qquad i,j = 1,2, \quad i \neq j, \\ |V_{\mathbf{k}}| &\equiv \epsilon^r\, u^{r\dagger}_{\mathbf{k},1}v^r_{-\mathbf{k},2} = -\epsilon^r\, u^{r\dagger}_{\mathbf{k},2}v^r_{-\mathbf{k},1} = \frac{(\omega_{k,1} + m_1) - (\omega_{k,2} + m_2)}{2\sqrt{\omega_{k,1}\omega_{k,2}(\omega_{k,1} + m_1)(\omega_{k,2} + m_2)}}\, |\mathbf{k}|\,, \end{aligned} \tag{A.9}$$

with $|U_{\mathbf{k}}|^2 + |V_{\mathbf{k}}|^2 = 1$.

The explicit expressions for the flavor states $|\nu^r_{\mathbf{k},e}\rangle$ and $|\nu^r_{\mathbf{k},\mu}\rangle$ at time $t = 0$, in the reference frame for which $\mathbf{k} = (0, 0, |\mathbf{k}|)$ are

$$\begin{aligned}
|\nu^r_{\mathbf{k},e}\rangle &\equiv \alpha^{r\dagger}_{\mathbf{k},e}(0)|0\rangle_{e,\mu} && \text{(A.10)}\\
&= \left[\cos\theta\,\alpha^{r\dagger}_{\mathbf{k},1} + |U_{\mathbf{k}}|\,\sin\theta\,\alpha^{r\dagger}_{\mathbf{k},2} - \epsilon^r\,|V_{\mathbf{k}}|\,\sin\theta\,\alpha^{r\dagger}_{\mathbf{k},1}\alpha^{r\dagger}_{\mathbf{k},2}\beta^{r\dagger}_{-\mathbf{k},1}\right] G^{-1}_{\mathbf{k},s\neq r}(\theta) \prod_{\mathbf{p}\neq\mathbf{k}} G^{-1}_{\mathbf{p}}(\theta)|0\rangle_{1,2},\\
|\nu^r_{\mathbf{k},\mu}\rangle &\equiv \alpha^{r\dagger}_{\mathbf{k},\mu}(0)|0\rangle_{e,\mu} && \text{(A.11)}\\
&= \left[\cos\theta\,\alpha^{r\dagger}_{\mathbf{k},2} - |U_{\mathbf{k}}|\,\sin\theta\,\alpha^{r\dagger}_{\mathbf{k},1} + \epsilon^r\,|V_{\mathbf{k}}|\,\sin\theta\,\alpha^{r\dagger}_{\mathbf{k},1}\alpha^{r\dagger}_{\mathbf{k},2}\beta^{r\dagger}_{-\mathbf{k},2}\right] G^{-1}_{\mathbf{k},s\neq r}(\theta) \prod_{\mathbf{p}\neq\mathbf{k}} G^{-1}_{\mathbf{p}}(\theta)|0\rangle_{1,2},
\end{aligned}$$

where $G(\theta, t) = \prod_{\mathbf{p}} \prod_{s=1}^{2} G_{\mathbf{p},s}(\theta, t)$. In these states a multiparticle component is present, disappearing in the relativistic limit $|\mathbf{k}| \gg \sqrt{m_1 m_2}$: in this limit, since $|U_{\mathbf{k}}|^2 \longrightarrow 1$ and $|V_{\mathbf{k}}|^2 \longrightarrow 0$, the (quantum-mechanical) Pontecorvo states are recovered.

Acknowledgments

We acknowledge partial financial support from MIUR, INFN, INFM and CNISM. M.B. thanks the organizers of Symmetries in Science XIV, for the very nice and creative atmosphere in which the workshop was held.

References

[1] Nielsen M A and Chuang I L 2000, *Quantum Information and Quantum Computation* (Cambridge U. Press)
[2] Lee T D and Yang C N 1960 *Internal report, Argonne National Laboratory (unpublished)*; Inglis D R 1961 *Rev. Mod. Phys.* **33** 1; Day T B 1961 *Phys. Rev.* **121** 1204; Lipkin H J 1968 *Phys. Rev.* **176** 1715; Zralek M 1998 *Acta Phys. Polon.* B **29** 3925; Bertlmann R A 2006 *Lect. Notes Phys.* **689** 1; Li J L and Qiao C F *arXiv:0708.0291*
[3] Blasone M, Dell'Anno F, De Siena S, Di Mauro M, and Illuminati F 2008 *Phys. Rev.* D **77** 096002
[4] Blasone M, Dell'Anno F, De Siena S, and Illuminati F 2009 *EPL* **85** 50002
[5] Cheng T and Li L 1989 *Gauge Theory of Elementary Particle Physics* (Clarendon Press)
[6] C. Amsler et al. (Particle Data Group) 2008 *Phys. Lett.* B **667** 1
[7] Pontecorvo B 1967 *Zh. Eksp. Teor. Fiz.* **53** 1717 [1968 *Sov. Phys. JETP* **26** 984]
[8] Maki Z, Nakagawa M, and Sakata S 1962 *Prog. Theor. Phys.* **28** 870
[9] Okun L and Pontecorvo B 1957 *Zh. Eksp. Teor. Fiz.* **32** 1587; Pontecorvo B 1958 *Sov. Phys. JETP* **7** 172
[10] Kayser B 1981 *Phys. Rev.* D **24** 110
[11] Zanardi P 2002 *Phys. Rev.* A **65** 042101
[12] Björk G, Jonsson P, and Sánchez-Soto L L 2001 *Phys. Rev.* A **64** 042106; van Enk S J 2005 *Phys. Rev.* A **72** 064306; van Enk S J 2006 *Phys. Rev.* A **74** 026302; Terra Cunha M O, Dunningham J A, and Vedral V 2007 *Proc. of the Royal Soc.* A **463** 2277
[13] Lombardi E, Sciarrino F, Popescu S, and De Martini F 2002 *Phys. Rev. Lett.* **88** 070402; Lvovsky A I, Hansen H, Aichele T, Benson O, Mlynek J, and Schiller S 2001 *Phys. Rev. Lett.* **87** 050402; Babichev S A, Appel J, and Lvovsky A I 2004 *Phys. Rev. Lett.* **92** 193601
[14] Hessmo B, Usachev P, Heydari H, and Björk G 2004 *Phys. Rev. Lett.* **92** 180401
[15] Tan S M, Walls D F, and Collett M J 1991 *Phys. Rev. Lett.* **66** 252; Hardy L 1994 *Phys. Rev. Lett.* **73** 2279
[16] Dunningham J and Vedral V 2007 *Phys. Rev. Lett.* **99** 180404
[17] Blasone M and Vitiello G 1995 *Ann. Phys. (N.Y.)* **244** 283
[18] Blasone M, Henning P A, and Vitiello G 1999 *Phys. Lett.* B **451** 140
[19] Blasone M, Jizba P, and Vitiello G 2001 *Phys. Lett.* B **517** 471
[20] Blasone M, Dell'Anno F, De Siena S, and Illuminati F 2008 *Phys. Rev.* A **77** 062304
[21] Klyachko A, Öztop B, and Shumovsky A S 2006 *Appl. Phys. Lett.* **88** 124102
[22] Klyachko A, Öztop B, and Shumovsky A S 2007 *Phys. Rev.* A **75** 032315; Klyachko A and Shumovsky A S 2004 *J. Opt.* B **6** S29
[23] Blasone M, Capolupo A, Terranova F, and Vitiello G 2005 *Phys. Rev.* D **72** 013003

Journal of Physics: Conference Series **237** (2010) 012008 doi:10.1088/1742-6596/237/1/012008

Large extra dimensions and small black holes at the LHC

Marcus Bleicher and Piero Nicolini

Frankfurt Institute for Advanced Studies (FIAS), Johann Wolfgang Goethe-Universität, Ruth-Moufang-Str. 1, D-60438 Frankfurt am Main, Germany

Institut für Theoretische Physik, Johann Wolfgang Goethe-Universität, Max-von-Laue-Str. 1, D-60438 Frankfurt am Main, Germany

E-mail: bleicher@th.physik.uni-frankfurt.de, nicolini@th.physik.uni-frankfurt.de

Abstract. We present an overview of the conjectured production of microscopic black holes as a consequence of high energy hadronic collisions at the Large Hadron Collider (LHC), CERN (Geneva, Switzerland) from this year on. Provided the presence of large extra-dimensions, we analyze some possible scenarios that could turn out to be outstanding experimental discoveries. We also discuss some new models which have been recently proposed on the ground of quantum gravity arguments. The final comments are devoted to supposed potential risks connected with the formation of black holes in particle detectors.

1. Introduction

If we could look back at the scientific community in the 60's or in the 70's, we would observe something very unusual for the current time perspective: a clear separation between physicists working on the standard model of particle physics and those working on the theory of gravitation. At that time the main stream of particle physics experiments were carried out at energies far below the supposed energy at which the fundamental interactions, gravity included, would have been unified. In other words gravity was simply assumed to be too weak with respect to the other fundamental interactions and its long standing problem of lack of renormalization was somehow underestimated or at least considered relevant only at an unreachable energy scale. Therefore gravity was kept classical and the major efforts were directed towards particle physics, calculating Feynman diagrams, cross sections and decay rates. On the other hand, General Relativity became the domain of mathematical physicists, who were mainly interested in the formal aspects of the theory, "spoiling" it from any physical content. Against this background, in 1975 a paper by Steven Hawking ignited the interest for the reconciliation of particle physics and gravity [1]. Indeed combining General Relativity and Quantum Field Theory, Hawking showed that black holes can emit thermal radiation like a black body.

After more than 30 years of intensive research, the scenario is now drastically different and we have at least two plausible candidate theories of Quantum Gravity: Superstrings [2] and Loop Quantum Gravity[3]. Both of them have great merits and still-open problems, but at present both suffer from a basic limitation: the absence of any support from experiments. For this reason they cannot be considered as being more than theoretical speculations, even if they are both physically very promising, aesthetically very attractive and mathematically elegant and well defined. The situation is even more puzzling if we think that the LHC, the world's largest and highest-energy particle accelerator, can reach just 14 TeV, an energy 16 orders of magnitude lower than the Planck scale, the energy at which, without further hypotheses, we think that gravity will be quantized and unified with the other fundamental interactions. On the side of cosmic ray observations, there are also nontrivial problems. On the one hand cosmic rays can reach energies over 10^{20} eV, on the other, their detection requires large detectors with areas of at least many square kilometers to have a reasonable number of events per year. Given this point, there are only two ways out: on the experimental side to further improve the acceptance of cosmic ray detectors (as currently done e.g. with the AUGER observatory or ICECUBE) or to increase the energy of particle accelerators, while on the theoretical one can reformulate the scenario with further assumptions in order to predict the presence of quantum gravity effects at the current observable energy scales. Since for now the "experimental way out" is slow, expensive and difficult we will analyze the "theoretical" one in the next section.

2. The large extra-dimension scenario

The new idea is that our universe might have more than the well known three spatial dimensions, height, width and length, but up to seven additional space dimensions that are usually unobservable. This hypothesis is currently considered to be the unique viable solution to the long-standing hierarchy problem, namely the presence of two fundamental scales in nature, the electroweak scale and the Planck scale, separated by 16 orders of magnitude [4, 5, 6, 7, 8, 9, 10, 11, 12, 13]. Conversely, in the absence of any extra dimensions, Quantum Gravity maybe lost forever: it is very likely that at the current rate of technological progress, mankind will probably become extinct before any experimental evidence of Quantum Gravity would become accessible. According to the Brane World scenario, the main point in this potential resolution of the problem is that extra dimensions can be assumed to be as large as around a fraction of a millimeter. To this purpose, we have to suppose that Standard Model fields are spatially constrained on a 3-brane, actually living in a four dimensional sub-manifold of the higher dimensional spacetime, while only gravity would be allowed to probe the transverse "bulk" dimensions. These additional dimensions are curled up to tiny "doughnuts" to make them hard to observe with a probe that is bigger than the extension of the "doughnut" - i.e. the extra dimensions are compactified on a torus of radius R. What makes these models especially interesting is the fact that the gravitational interaction increases strongly (by factors on the order of 10^{32}) on distances below the extension of the extra dimensions. The strict way to show this is to start from the Einstein-Hilbert action for $(k+4)$-dimensional gravity and to integrate out the additional space like extra dimensions. However, there is an easier way to understand this. Consider a particle of mass M located in the bulk, a space time with $k+3$ dimensions. The general solution of the Poisson's equation yields its potential as a function of the radial distance r from the source

$$\phi(r) = \frac{1}{M_\star^{2+k}} \frac{M}{r^{1+k}}. \tag{1}$$

Here we have introduced a new fundamental mass-scale $M_\star$ with a suitable exponent. Remembering that the additional k space-time dimensions are compactified on radii R, then, at distances $r \gg R$, the extra dimensions should be hidden and the potential Eq. (1) should turn into the well known $1/r$ potential, however with a pre-factor including the volume of the extra dimensions

$$\phi(r) \to \frac{1}{M_\star^{2+k}} \frac{1}{R^k} \frac{M}{r} \equiv \frac{1}{M_P^2} \frac{M}{r} \quad \text{as } r \gg R. \tag{2}$$

In the limit of large distances, this should be identical to Newton's gravitational law, yielding the relation

$$M_\star^{(2+k)} = M_P^2/R^k \tag{3}$$

Assuming that $M_\star$ has the right order of magnitude to be compatible with observed physics, it can be seen that the volume of the extra dimensions suppresses the fundamental scale and thus, explains the huge value of the Planck mass

$$M_P \sim 1.22 \times 10^{28}\text{eV}. \tag{4}$$

Today one expects this new fundamental scale to be of the order of $M_\star \sim 1$ TeV, allowing sizes of extra dimensions between 1/10 mm to 1/1000 fm for k from 2 to 7. The case of one or two extra dimensions in the ADD scenario [4] is actually ruled out by Cavendish like experiments, which confirm the validity of Newton's law in four dimensions down to 1/10 mm, while one would need an extension of at least 10 km for $k = 1$. More in detail, according to the latest results [14, 15], the limit for the thickness of extra-dimension is $R \leq 44$ μm which implies that the fundamental mass is

$$M_\star \geq 3.2 \text{ TeV}. \tag{5}$$

This would mean that, according to the LHC schedule, we will be reaching the above energy in the first half of 2010.

The above arguments can be followed even clearer when one remembers that the strength of interaction is proportional to the density of flux lines. Figure 1 illustrates this behaviour: while the top figure shows the setting with a mass point in the center for one dimension with undiluted flux lines, the lower figure shows (in the magnifying glass) that there are actually two dimensions and the field lines are diluted. Mathematically speaking we find that in the usual three dimensional world the gravitational potential falls-off like 1/distance, however if we have more dimension (e.g. k additional dimensions) to dilute the gravitational flux lines, the potential will fall like 1/distance^{k+1}. Why is the gravitational field then stronger at small distances? This becomes clear, if we remember that the long distance behaviour of gravitation is fixed from daily observation (the Newton's law). If we extrapolate towards smaller distances the additional dilution of the field lines starts, therefore, to get the proper long distance behaviour back, one has to start with more dense flux lines (meaning stronger gravitational interaction) at small distances. Thus, if Newton's law is modified only at small distances when extra dimensions are introduced, it is obvious that the most promising experimental tests for the existence of extra dimensions may simply be measurements of the Newtonian potential at small distances.

3. Mini black hole production

What is the role of black holes in this context? One of the most spectacular consequences of having a low fundamental scale $M_\star$ is the increase of both the gravitational interaction strength and the probability of gravitational collapse, which is nothing but black hole formation. In fact, the increase in gravitational strength is drastic enough that it might allow black hole production at the LHC as a result of high energy hadronic collisions, already within the next couple of months [17, 18, 19]. However, how can one imagine the production of black holes in a particle accelerator? The nuclei of two hydrogen atoms (the protons) are accelerated in opposite directions nearly to the speed of light. When they collide, the subatomic constituents within the protons interact. If the distance between these constituents (the partons) is small enough - i.e. smaller than the corresponding Schwarzschild radius R_H - and the energy is high enough a micro black hole could be formed. The mass of such black holes will be of the order of a TeV

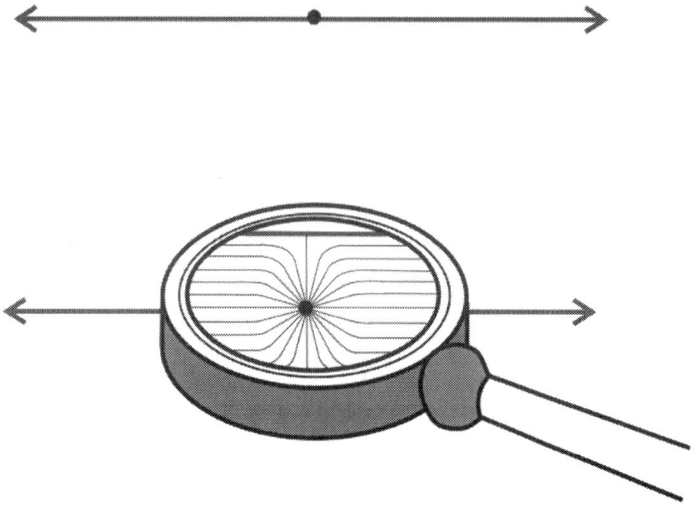

Figure 1. Flux lines emerging from a mass point. Top: Along one dimension in usual space time. Bottom: Spread of the flux lines due to an additional extra dimension. Taken from [16].

and therefore be much lighter than the cosmic black holes we know up to now. This can be strictly derived by using the higher dimensional Schwarzschild-metric [20]

$$ds^2_{(k+4)} = -F(r)dt^2 + F^{-1}(r)dr^2 + r^2 d\Omega^2_{3+k}, \tag{6}$$

where $d\Omega^2_{3+k}$ is the surface element of the $3+k$ - dimensional unit sphere and

$$F(r) = 1 - \left(\frac{R_H}{r}\right)^{k+1}. \tag{7}$$

The constant R_H can be obtained by requiring that the solution reproduces the Newtonian limit for $r \gg R$. This means the derivative of Eq. (7) has to yield the Newtonian potential in $(3+k)$ dimensions:

$$\frac{1}{2}\frac{dF(r)}{dr} = \frac{k+1}{2}\left(\frac{R_H}{r}\right)^{k+1}\frac{1}{r} = \frac{1}{M_\star^{k+2}}\frac{M}{r^{k+2}}. \tag{8}$$

So we have

$$F(r) = 1 - \frac{2}{k+1}\frac{1}{M_\star^{k+2}}\frac{M}{r^{k+1}}, \tag{9}$$

and R_H is

$$R_H^{k+1} = \frac{2}{k+1}\left(\frac{1}{M_\star}\right)^{k+1}\frac{M}{M_\star}. \tag{10}$$

For $M_\star \sim 1$ TeV this radius is $\sim 10^{-4}$ fm. Thus, at high enough energies it might be possible to bring particles closer together than their horizon, which will force them to form a black hole.

Let us now calculate the possible production rate of black holes explicitly. That is, we consider two elementary particles, approaching each other with a very high kinetic energy in the c.o.m. system slightly above the new fundamental scale $M_\star \sim 1$ TeV, as depicted in figure 2. Since the

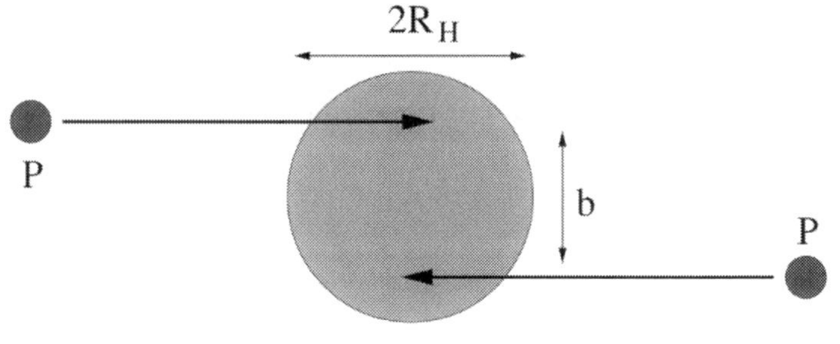

Figure 2. At very high energies, the partons, p, can come closer than the Schwarzschild radius, R_H, associated with their energy. If the impact parameter, b, is sufficiently small, such a collision will inevitably generate a black hole.

black hole is not an ordinary particle of the Standard Model and its correct quantum theoretical treatment is unknown, it is treated as a metastable state, which is produced and decays according to the semi classical formalism of black hole physics. To compute the formation probability, we approximate the cross section of the black holes by their classical geometric area

$$\sigma(M) \approx \pi R_H^2 \Theta(M - M_\star), \tag{11}$$

with Θ being the Heaviside function. Setting $M_\star \sim 1$ TeV and $k = 2$ one finds $\sigma \sim 400$ pb. Using the geometrical cross section formula, it is now possible to compute the differential cross section $d\sigma/dM$ which will tell us how many black holes will be formed with a certain mass M at a c.o.m. energy vs the probability that two colliding particles will form a black hole of mass M in a proton-proton collision at the LHC (at a center of mass energy of 14 TeV) involves the parton distribution functions. These functions, $f_A(x, \hat{s})$, parametrise the probability of finding a constituent A of the proton (quark or gluon) with a momentum fraction x of the total energy of the proton. Here, $\sqrt{\hat{s}}$ is the c.o.m. energy of the parton-parton collision. The differential cross section is then given by summation over all possible parton interactions and integration over the momentum fractions $0 \leq x_1 \leq 1$ and $0 \leq x_2 \leq 1$, where the kinematic relation $x_1 x_2 s = \sqrt{\hat{s}} = M^2$ has to be fulfilled. This yields

$$\frac{d\sigma}{dM} = \sum_{A_1 B_1} \int_0^1 dx_1 \frac{2\sqrt{\hat{s}}}{x_1 s} f_a(x_1, \hat{s}) f_B(x_2, \hat{s}) \sigma(M, k). \tag{12}$$

The particle distribution functions for f_A and f_B are tabulated e.g. in the CTEQ-tables[1]. One might think that the mass distribution diverges when $x_1 = 0$, however, this limit does not fulfill the kinematic constrained $x_1 x_2 = M^2/s$, because it would result in $x_2 > 1$ for finite black hole masses $M = M_\star$. It is now straightforward to compute the total cross section and number by integration over Eq. (12) This also allows us to estimate the total number of black holes, $N_{\rm BH}$, that would be created at the LHC per year. It is

$$N_{\rm BH}/\text{year} = \sigma(pp \to \text{BH}) L, \tag{13}$$

where we insert the estimated luminosity for the LHC, $L = 10^{33}$ cm^{-2}s^{-1}. This yields at a c.o.m. energy of $\sqrt{s} = 14$ TeV the total number of 10 million black holes per year! This means, about one black holes per second would be created. Even if this is only an optimistic upper estimate of the black hole formation rate, the importance of this process led to a high

[1] The parton distribution function can be downloaded from this web page: http://www.phys.psu.edu cteq/

number of publications on the topic of TeV-mass black holes at colliders both in scientific journals [21, 22, 23, 24, 25, 26, 27, 28, 29, 39, 31, 32] and in the popular press[2]. It should be noted that numerical simulation and visualisation packages for the production and decay of black holes in colliders and from cosmic rays are available. These packages are currently used to explore potential signatures for black hole production and go by the names CHARYBDIS [33], CHARYBDIS2 [34], TRUENOIR[35], GROKE [36, 37], CATFISH [38] and, more recently, QBH [39, 40] and BlackMax [41] which employs a richer set of external parameters and allows a certain flexibility among different theoretical models.

4. Mini black hole evaporation

Hawking's revolutionary theoretical discovery about the possibility for black hole to emit thermal radiation like a black body now enters the scene, disclosing new scenarios. Indeed it is very likely that black holes are fated to provide the final answer about our knowledge of Quantum Gravity [43, 42, 44, 45, 46, 47, 48, 49] and close the logical path which they started in 1975, igniting the initial interest in this area. According to theory, black holes would have a temperature

$$T_H = \frac{1}{4\pi} \left. \frac{dF(r)}{dr} \right|_{r=r_H}, \tag{14}$$

and their decay would lead to the sudden emission of a huge quantity of particles independently of their spin, charge, quantum numbers or interaction properties, as long as their rest mass is smaller than the black hole temperature [50, 51]. Therefore mini black holes may be excellent intermediate states for long-sought but yet undiscovered particles, like the Higgs boson or supersymmetric particles, possibly with cross sections much larger than for the direct processes [52]. We could also have key information about the very early universe, as well as solving some basic questions, whose answers are too often taken for granted.

First, do (mini) black holes exist? From velocity measurements for the whirlpool of hot gas surrounding it, astronomers have found convincing evidence for the existence of a supermassive black hole in the center of the giant elliptical galaxy M87. Astronomers have also detected radio emission coming from within 30 million kilometers of the dark object SGR A*, thought to be a colossal black hole, that lies at the center of the Milky Way [53]. Previously, X-ray emission from the binary star system Cygnus X-1 convinced many astronomers that it contains a black hole, which is supported by more precise measurements which have recently become available. To this purpose we also have to mention V404 Cygni, one of the most evident cases for a stellar black hole: observations at X-ray and optical wavelengths have shown that this is a binary system in which a late yellowish G star or maybe an early orange-red K star revolves, every 6.47 days, around a compact companion with a diameter of around $60-90$ km and a probable mass of 8 to 15 solar masses, well above the mass limit at which a collapsed star must become a black hole [54]. In spite of these observations, however, there are still some ranges of mass in which the existence of black holes is unclear, in particular of black holes of less than 3 solar masses. The relevance of these objects is connected with the possibility of observing the Hawking radiation. Indeed, what we know for sure is that for astrophysical black holes the Hawking radiation is negligible because their temperatures can be at most some tens of nK, far below $T_{\rm CMB} \sim 2.7$ K, the temperature of Cosmic Microwave Background (CMB) radiation. On the other hand tidal effects are significant in the case of mini black holes that could be very hot and very bright if their mass is sufficiently small [55].

As second point, we might be able to conclude that (mini) black holes really can evaporate. Indeed for the above reasons the detection of these objects is the unique direct way to have

[2] "Physicists Strive to Build A Black Hole", New York Times, September 11, 2001.

experimental evidence of the Hawking conjecture, one of the most important predictions of Quantum Field Theory in Curved Spacetime and of the associated semiclassical gravity.

Thirdly, we could find out what the fate of a radiating black hole is. If mini black holes can be created in high energy particle collisions, the black holes produced will pass through a number of phases before completing their evaporation. After a loss during the first two phases of their "hair" (i.e. the associated long-range fields) and of their angular momentum (the "balding" and spin-down phases respectively), the picture of the evaporation will be described by the Schwarzschild phase, in which the resulting spherically symmetric black hole loses energy by the emission of Hawking radiation, with a gradual decrease in its mass and an increase in its temperature. Since the Schwarzschild geometry has a curvature singularity at the origin, there would be a divergent behavior of the Hawking temperature if the black hole were to shrink to the origin as a result of losing mass by thermal emission. However, we do not expect that this divergent behavior will, in fact, take place since in the vicinity of the origin, the evaporating black hole will be dramatically disturbed by strong quantum gravitational fluctuations of the spacetime manifold. In other words the black hole will undergo a Plank phase of the evaporation during which a theory of Quantum Gravity must be used. Observations of the final stages of black hole evaporation could provide the profile of the temperature as a function of the mass of the black hole and hence let us pick out, for the first time, the correct quantum gravitational theory.

5. New developments

At this point, it is clear that the the semiclassical approximation given by Quantum Field Theory in Curved Space, turns out to be inadequate when the black hole undergoes the most important phase of the evaporation, the final phase. To this purpose there exists the possibility of improving semiclassical gravity with some arguments coming from String Theory. If we believe in the Brane World scenario and the presence of a 3-brane where all massive objects live, we have to recall that it has been shown that open string end point coordinates do not commute on branes, giving rise to the so called Noncommutative (NC) Geometry [56]

$$[\mathbf{x}^\mu, \mathbf{x}^\nu] = i\Theta^{\mu\nu}, \tag{15}$$

with x^μ, x^ν being the 4-coordinates and Θ being the non-commutativity parameter. Even if we are not fully quantizing gravity, in this approach we obtain an induced model of quantum spacetime. The quantum nature of the manifold can be observed in the consequent loss of resolution coming from the uncertainty relation

$$\Delta x^\mu \Delta x^\nu \geq \theta \tag{16}$$

where $\sqrt{\theta}$ is the universal quantum of length. Since there is no way to determine the exact position of any physical object, we cannot longer speak of pointlike objects because there are no points on such a brane. As a result it can be shown that the conventional Dirac delta distributions $\delta(\mathbf{x})$ switches to a minimal width Gaussian distributions $\rho_\theta(\mathbf{x})$, where the minimal width is $\sqrt{\theta} \sim 1/M_\star$. Therefore if one solves the Einstein equations with an energy momentum tensor written in terms of minimal width Gaussian distributions, one obtains a string improved semiclassical approximation, namely gravity + NC geometry [57, 58, 59, 60, 61, 62, 63, 64, 65, 66, 67, 68, 69, 70]. Looking for spherically symmetric black hole solutions, one assumes a metric of the form

$$ds^2_{(d+1)} = -F(r)\,dt^2 + F^{-1}(r)\,dr^2 + r^2 d\Omega^2_{d-1}, \tag{17}$$

where now $d = 3 + k$ is the total number of spatial dimensions. Plugging the above line element in

$$R^N_M = \frac{1}{M_\star^{d-1}}\left(8\pi\, T^N_M - \frac{8\pi}{d-1}\,\delta^N_M\, T\right), \tag{18}$$

Table 1. Black hole maximal temperatures for different d. See figure (4).

d	3	4	5	6	7	8	9	10
T_H^{max} (GeV)	18×10^{16}	30	43	56	67	78	89	98
T_H^{max} $(10^{15}K)$	$.21\times10^{16}$	.35	.50	.65	.78	.91	1.0	1.1

Table 2. Remnant masses and radii for different values of d, keeping $M_\star \sim 1/\sqrt{\theta}$

	3	4	5	6	7	8	9	10
M_0 (TeV)	2.3×10^{16}	6.7	24	94	3.8×10^2	1.6×10^3	7.3×10^3	3.4×10^4
r_0 $(10^{-4}$ fm)	4.88×10^{-16}	5.29	4.95	4.75	4.62	4.52	4.46	4.40

where T is the trace of the energy momentum tensor, $T = T_M^M$, one finds the noncommutative black hole solution

$$F(r) = 1 - \frac{1}{M_\star^{d-1}} \frac{2M}{r^{d-2}\Gamma(d/2)}\, \gamma\left(\frac{d}{2}\, , \frac{r^2}{4\theta}\right) \tag{19}$$

where

$$\gamma\left(d/2\, , r^2/4\theta\right) \equiv \int_0^{r^2/4\theta} \frac{dt}{t}\, t^{d/2} e^{-t}, \tag{20}$$

is the incomplete gamma function and

$$\Gamma\left(d/2\right) = \left(\tfrac{d}{2} - 1\right)! \qquad d\ even \tag{21}$$

$$\Gamma\left(d/2\right) = \sqrt{\pi}\, \tfrac{(d-2)!!}{2^{(d-1)/2}} \qquad d\ odd. \tag{22}$$

To start commenting the above result we check first of all that at large distances with respect to the quantum of length, stringy effects become negligible and the conventional Schwarzschild solution is reproduced. Indeed for $\sqrt{\theta}/r \to 0$ the incomplete gamma functions $\gamma\left(\frac{d}{2}\, , \frac{r^2}{4\theta}\right)$ match the value of the Euler functions $\Gamma(d/2)$ and the Schwarzschild solution is obtained. The novelty of the solution lies in the following new properties

- the curvature singularity is replaced by a de Sitter core that makes the solution regular and accounts for the quantun gravitational fluctuation, namely for $r \ll \sqrt{\theta}$

$$F(r) = 1 - \frac{1}{d\, M_\star^{d-1}} \frac{4M}{2^{d-1}\pi^{(d-2)/2}\, \theta^{d/2}}\, r^2 + \mathcal{O}\left(r^4\right). \tag{23}$$

- the horizon equation $F(r) = 0$ has three possibilities depending on the value of the mass M, namely two horizons r_+ and r_-, one single degenerate horizon r_0 and no horizon.

The plot in figure 3 shows that the outer horizon radius r_+ decreases as d increases for a given mass M. The minimum of each curve also provides the value of the extremal radius r_0 which decreases with d.

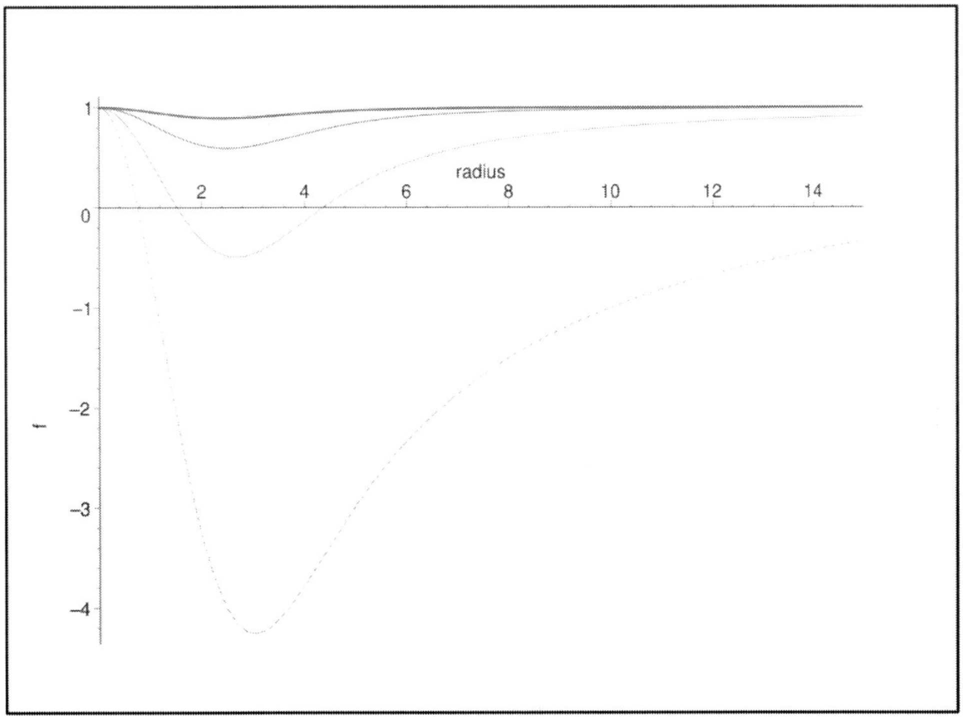

Figure 3. *The noncommutative Schwarzschild solution.* The function F is plotted versus $r/\sqrt{\theta}$, for $M = 10\, M_\star$. We can observe that the curves rise with d and the outer horizon r_+ decreases until $d = 4$. For $d \geq 5$ no black hole can be formed: the mass M is so light that it cannot provide a significant gravitational disturbance, since $F \simeq 1$.

The next step is about the thermodynamical analysis of the noncommutative black hole. Here the temperature is

$$T_H = \frac{d-2}{4\pi r_+}\left[1 - \frac{r_+}{d-2}\frac{\gamma'\left(d/2\,, r_+^2/4\theta\right)}{\gamma\left(d/2\,, r_+^2/4\theta\right)}\right]. \tag{24}$$

From the above formula one obtains again a solution that behaves like the Schwarzschild solution at large distances, thus, reproducing the Hawking result $T_{H\,comm} = (d-2)/4\pi r_+$. Noncommutative effects appear as r_+ approaches $\sim 6\sqrt{\theta}$, depending on d: the manifold is more sensitive to noncommutativity for higher dimensions. Then, from figure 4, we conclude that after an initial Schwarzschild phase, the black hole undergoes a SCRAM[3] phase, namely a cooling down towards a zero temperature extremal black hole remnant. This is the equivalent on the thermodynamics side of the absence of any curvature singularity: no divergent Planck phase occures. One can see also from table 1 that the peak of the temperature increases with d, but even in the case $d = 10$ it is about 98 GeV ($\simeq 10^{15}$ K) which is much lower than $M_\star$.

Remnant masses can be calculated and their values together with remnant radii are summarized in table 2. Since remnant radii are $\sim 4 - 5 \times 10^{-4}$ fm, i.e. smaller than the

[3] The term SCRAM, probably the backronym for "Safety Control Rod Axe Man" or "Super Critical Reactor Axe Man", refers to an emergency shutdown of a thermonuclear reactor. The term has been extended to cover shutdowns of other complex operations or systems in an unstable state, but it also has a "standard" meaning "go away quickly", in particular when we address children or animals.

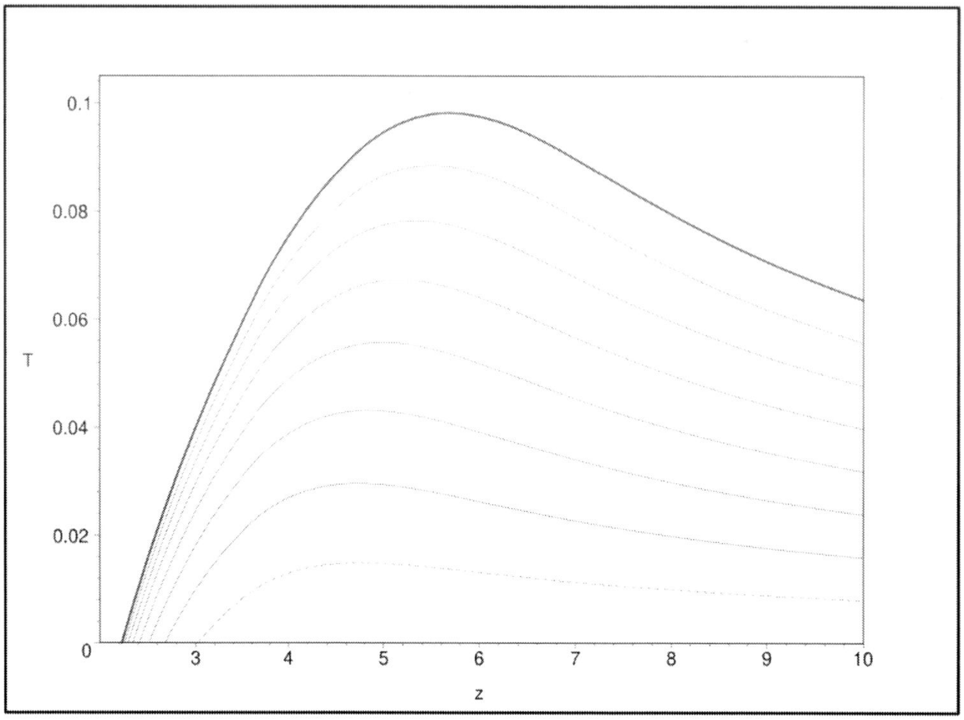

Figure 4. *The noncommutative extra-dimensional Schwarzschild solution.* The temperature T_H as a function of $z = r_+/\sqrt{\theta}$, for various value of d. The lowest curve corresponds to the case $d = 3$, while, from the bottom to top on the right hand side of the figure, the solid lines correspond to $d = 4$ to 10. The temperatures increase with d, while the SCRAM phase leads to smaller remnant for higher d.

size of extra-dimensions, these remnants can be considered, to a good approximation, as totally submerged in a $d+1$ dimensional isotropic spacetime, for any $d = 3 - 10$, confirming the validity of the spherically symmetric noncommutative solution. As a further consequence, estimating the black hole production cross section as the area of the event horizon $\sigma \sim \pi r_+^2$, we find, for every d, an encouraging lower bound $\sigma \sim 10$ nb, almost two orders of magnitude larger than the conventional values. On the other hand, for $d \geq 5$ the remnant is too heavy to be produced at the LHC and could be only detected in Ultra-High-Energy cosmic rays [42].

As a final comment about the String Theory induced noncommutative black holes, we have to say that, starting from different approaches, like Loop Quantum Gravity and Asymptotically Safe Gravity, there exist other black hole solutions which have a SCRAM phase, a regular geometry at the origin and one, two or no horizon [71, 72, 73, 74, 75, 76, 77, 78, 79, 80, 81]. We can conclude that the depicted behavior for noncommutative black holes is universal, once Quantum Gravity is invoked, i.e. independent of the formulation.

6. Final remarks

The production of microscopic black holes at the LHC has recently raised a hot debate about the safety of high energy physics experiment [82]. If, for some reasons, black holes do not evaporate, it was speculated that they may be potentially dangerous, since classically they may start to accrete indefinitely [83]. However, accurate studies have shown that the black hole lifetimes are extremely small. Even in the most pessimistic scenario according to which the black hole has a SCRAM phase and therefore can live longer than expected in conventional theories the lifetimes cannot exceed $\sim 10^{-16}$ s, taking into account all kinds of (bulk/brane) emissions and number of spatial dimensions [84]. Also the destiny of eventual stable black hole remnants is not a source of concern. The probability that a black hole remnant starts to grow inside the Earth is proportional to its volume and therefore negligible [85, 86, 87, 88, 89, 90]. To this purpose, accurate studies performed in [91, 92, 83, 93, 94, 95], addressing the specific case of micro black holes in particle detectors, have excluded any threat from these kind of experiments at the LHC.

Acknowledgments

This work is supported by the Helmholtz International Center for FAIR within the framework of the LOEWE program (Landesoffensive zur EntwicklungWissenschaftlich-Ökonomischer Exzellenz) launched by the State of Hesse. We would like to thank Michelangelo Mangano for valuable comments and Martin Sprenger for discussions.

References

[1] Hawking S W 1975 *Commun. Math. Phys.* **43** 199 [Hawking S W 1976 *Erratum-ibid* **46** 206]
[2] Schwarz J H 1982 *Phys. Rept.* **89** 223
[3] Smolin L 2004 An invitation to loop quantum gravity *Preprint* arXiv:hep-th/0408048
[4] Arkani-Hamed N, Dimopoulos S and Dvali G R 1998 *Phys. Lett.* B **429** 263
[5] Antoniadis I, Arkani-Hamed N, Dimopoulos S and Dvali G R 1998 *Phys. Lett.* B **436** 257
[6] Arkani-Hamed N, Dimopoulos S and Dvali G R 1999 *Phys. Rev.* D **59** 086004
[7] Gogberashvili M 1999 *Mod. Phys. Lett.* A **14** 2025
[8] Gogberashvili M 2000 *Europhys. Lett.* **49** 396
[9] Gogberashvili M 2002 *Int. J. Mod. Phys.* D **11** 1635
[10] Gogberashvili M 2002 *Int. J. Mod. Phys.* D **11** 1639
[11] Kokorelis C 2004 *Nucl. Phys.* B **677** 115
[12] Randall L and Sundrum R 1999 *Phys. Rev. Lett.* **83** 3370
[13] Randall L and Sundrum R 1999 *Phys. Rev. Lett.* **83** 4690
[14] Adelberger E G, Heckel B R, Hoedl S, Hoyle C D, Kapner D J and Upadhye A 2007 *Phys. Rev. Lett.* **98** 131104
[15] Adelberger E G, Gundlach J H, Heckel B R, Hoedl S and Schlamminger S 2009 *Prog. Part. Nucl. Phys* **62** 102
[16] Hossenfelder S 2004 What black holes can teach us *Preprint* arXiv:hep-ph/0412265
[17] Argyres P C, Dimopoulos S and March-Russell J 1998 *Phys. Lett.* B **441** 96
[18] Dimopoulos S and Landsberg G L 2001 *Phys. Rev. Lett.* **87** 161602
[19] Hsu S D H 2003 *Phys. Lett.* B **555** 92
[20] Myers R C and Perry M J 1986 *Annals Phys.* **172** 304
[21] Banks T and Fischler W 1999 A model for high energy scattering in quantum gravity *Preprint* arXiv:hep-th/9906038
[22] Bleicher M, Hofmann S, Hossenfelder S and Stoecker H 2002 *Phys. Lett.* B **548** 73
[23] Hossenfelder S, Hofmann S Bleicher M and Stoecker H 2002 *Phys. Rev.* D **66**, 101502
[24] Giddings S B and Thomas S D 2002 *Phys. Rev.* D **65** 056010
[25] Cavaglia M, Das S and Maartens R 2003, *Class. Quantum Grav.* **20** L205
[26] Cavaglia M and Das S 2004 *Class. Quantum Grav.* **21** 4511
[27] Tanaka J, Yamamura T, Asai S and Kanzaki J 2005 *Eur. Phys. J.* C **41** 19
[28] Harris C M, Palmer M J, Parker M A, Richardson P, Sabetfakhri A and Webber B R 2005 *J. High Energy Phys.* JHEP05(2005)053
[29] Godang R, Bracker S B, Cavaglia M, Cremaldi L, Summers D and Cline D 2005 *Int. J. Mod. Phys.* A **20** 3409
[30] Gingrich D M 2009 Quantum black holes with charge, colour, and spin at the LHC *Preprint* arXiv:0912.0826 [hep-ph]
[31] Aref'eva I Y 2009 Catalysis of Black Holes/Wormholes Formation in High Energy Collisions *Preprint* arXiv:0912.5481 [hep-th]
[32] Gal'tsov D V, Kofinas G, Spirin P and Tomaras T N 2010 *Phys. Lett.* B **683** 331
[33] Harris C M, Richardson P and Webber B R 2003 *J. High Energy Phys.* JHEP08(2003)033 Download page: http://www.ippp.dur.ac.uk/montecarlo/leshouches/generators/charybdis/

[34] Frost J A, Gaunt J R, Sampaio M O P, Casals M, Dolan S R, Parker M A and Webber B R 2009 *J. High Energy Phys.* JHEP10(2009)014 Download page: http://projects.hepforge.org/charybdis2/
[35] Dimopoulos S and Landsberg G L 2001 Black hole production at future colliders in *Proc. of the APS/DPF/DPB Summer Study on the Future of Particle Physics (Snowmass 30 Jun - 21 Jul 2001)*, ed. Graf N (Snowmass Colorado 2001), pp P321
[36] Download page: http://www.phy.olemiss.edu/GR/groke/
[37] Ahn E J and Cavaglia M 2006 *Phys. Rev.* D **73** 042002
[38] Cavaglia M, Godang R, Cremaldi L and Summers D 2007 *Comput. Phys. Commun.* **177** 506 Download page: http://www.phy.olemiss.edu/GR/catfish/
[39] Gingrich D M 2009 Quantum black holes with charge, colour, and spin at the LHC *Preprint* arXiv:0912.0826 [hep-ph]
[40] Gingrich D M 2009 Monte Carlo event generator for quantum black hole production and decay in proton-proton collisions *Preprint* arXiv:0911.5370 [hep-ph] http://projects.hepforge.org/qbh/
[41] Dai D C, Starkman G, Stojkovic D, Issever C, Rizvi E and Tseng J 2008 *Phys. Rev.* D **77** 076007
[42] Cavaglia M 2003 *Int. J. Mod. Phys.* A **18** 1843
[43] Feng J L and Shapere A D 2002 *Phys. Rev. Lett.* **88** 021303
[44] Stoecker H 2006 *J. Phys.* G **32** S429
[45] Casadio R and Harms B 2002 *Int. J. Mod. Phys.* A **17** 4635
[46] Giddings S B 2002 Black hole production in TeV-scale gravity and the future of high energy physics in *Proc. of the APS/DPF/DPB Summer Study on the Future of Particle Physics (Snowmass 2001)* Snowmass Colorado 30 Jun - 21 Jul 2001 pp P328 ed. N. Graf (SLAC, 2001)
[47] Stenmark M 2002 *Chin. J. Phys.* **40** 512
[48] Chamblin A, Cooper F and Nayak G C 2004 *Phys. Rev.* D **70**, 075018
[49] Rizzo T G 2006 *Class. Quantum Grav.* **23** 4263
[50] Casanova A and Spallucci E 2006 *Class. Quantum Grav.* **23** R45
[51] Kanti P 2004 *Int. J. Mod. Phys.* A **19** 4899
[52] Erkoca A E, Nayak G C and Sarcevic I 2009 *Phys. Rev.* D **79** 094011
[53] Ghez A M *et al.* 2004 *Astrophys. J.* **601** L159
[54] Corbel S, Koerding E and Kaaret P 2008 *Mon. Not. Roy. Astron. Soc.* **389** 1697
[55] Page D N and Hawking S W 1976 *Astrophys. J.* **206** 1
[56] Seiberg N and Witten E 1999 *J. High Energy Phys.* JHEP09(1999)032
[57] Nicolini P 2005 *J. Phys. A: Math. Gen.* **38** L631
[58] Nicolini P, Smailagic A and Spallucci E 2005 *ESA Spec. Publ.* **637** 11.1
[59] Nicolini P, Smailagic A and Spallucci E 2006 *Phys. Lett.* B **632** 547
[60] Rizzo T G 2006 *J. High Energy Phys.* JHEP09(2006)021
[61] Spallucci E, Smailagic A and Nicolini P 2006 *Phys. Rev.* D **73** 084004
[62] Ansoldi S, Nicolini P, Smailagic A and Spallucci E 2007 *Phys. Lett.* B **645** 261
[63] Spallucci E, Smailagic A and Nicolini P 2009 *Phys. Lett.* B **670** 449
[64] Nicolini P 2009 *Int. J. Mod. Phys.* A **24** 1229
[65] Arraut I, Batic D and Nowakowski M 2009 *Class. Quantum Grav.* **26** 245006
[66] Nicolini P and Spallucci E 2010 *Class. Quantum Grav.* **27** 015010
[67] Nicolini P and Rinaldi M 2009 A minimal length versus the Unruh effect *Preprint* arXiv:0910.2860 [hep-th]
[68] Banerjee R, Gangopadhyay S and Modak S K 2009 Voros product, Noncommutative Schwarzschild Black Hole and Corrected Area Law *Preprint* arXiv:0911.2123 [hep-th]
[69] Modesto L and Nicolini P 2010 *Phys. Rev.* D **81** 104040
[70] Batic D and Nicolini P 2010 Fuzziness at the horizon *Preprint* arXiv:1001.1158 [gr-qc]
[71] Modesto L 2008 *Adv. High Energy Phys.* **2008** 459290
[72] Modesto L 2008 Space-Time Structure of Loop Quantum Black Hole *Preprint* arXiv:0811.2196 [gr-qc]
[73] Modesto L and Premont-Schwarz I 2009 *Phys. Rev.* D **80** 064041
[74] Hossenfelder S, Modesto L and Premont-Schwarz I 2009 *Preprint* A model for non-singular black hole collapse and evaporation arXiv:0912.1823 [gr-qc].
[75] Bonanno A and M. Reuter M 2000 *Phys. Rev.* D **62** 043008
[76] Bonanno A and M. Reuter M 2006 *Phys. Rev.* D **73** 083005
[77] Burschil T and Koch B 2009 Renormalization group improved black hole space-time in large extra dimensions *Preprint* arXiv:0912.4517 [hep-ph]
[78] Scardigli F 1999 *Phys. Lett.* B **452** 39
[79] Amelino-Camelia G,Arzano M, Ling Y and Mandanici G 2006 *Class. Quantum Grav.* **23** 2585
[80] Husain V and Mann R B 2009 *Class. Quantum Grav.* **26** 075010
[81] Jizba P, Kleinert H and Scardigli F 2009 Uncertainty Relation on World Crystal and its Applications to

Micro Black Holes *Preprint* arXiv:0912.2253 [hep-th]
[82] http://www.foxnews.com/story/0,2933,483477,00.html
[83] Koch B, Bleicher M and Stoecker H 2009 *Phys. Lett.* B **672** 71
[84] Casadio R and Nicolini P 2008 *J. High Energy Phys.* JHEP11(2008)072
[85] Hossenfelder S, Bleicher M, Hofmann S, Stoecker H and Kotwal A V 2003 *Phys. Lett.* B **566** 233
[86] Hossenfelder S, Koch B and Bleicher M 2005 Trapping black hole remnants *Preprint* arXiv:hep-ph/0507140
[87] Koch B, Bleicher M and Hossenfelder S 2005 *J. High Energy Phys.* JHEP10(2005)053
[88] Bleicher M 2007 *Eur. J. Phys.* **28** 509
[89] Koch B, Bleicher M and Stoecker H 2007 *AIP Conf. Proc.* **947** 376
[90] Koch B, Bleicher M and Stoecker H 2007 *J. Phys.* G **34** S535
[91] Giddings S B and Mangano M L 2008 *Phys. Rev.* D **78** 035009
[92] Giddings S B and Mangano M L 2008 Comments on claimed risk from metastable black holes *Preprint* arXiv:0808.4087 [hep-ph]
[93] Casadio R, Fabi S and Harms B 2009 *Phys. Rev.* D **80** 084036
[94] Casadio R, Fabi S, Harms B and Micu O 2009 Theoretical survey of tidal-charged black holes at the LHC *Preprint* arXiv:0911.1884 [hep-th]
[95] Gingrich D M 2010 Production of tidal-charged black holes at the Large Hadron Collider *Preprint* arXiv:1001.0626 [hep-ph]

Coupling coefficients for SO(5) with applications to nuclear physics

M. A. Caprio

Department of Physics, University of Notre Dame, Notre Dame, Indiana 46556-5670, USA

Abstract. The Lie algebra SO(5) has several applications in nuclear theory, involving its various subalgebra chains. In this contribution, the calculation of SO(5) coupling coefficients, reduced with respect to these different subalgebras, is discussed. It is shown that "Racah's method" for extracting coupling coefficients can be formulated generically for a subalgebra chain $G \supset H$, provided the reduced matrix elements of the generators of G and the recoupling coefficients of H are known. The calculation of $\mathrm{SO}(5) \supset \mathrm{SO}(4)$ reduced coupling coefficients, for generic irreps, is considered as an example.

1. Introduction

Continuous symmetries and their associated Lie algebras facilitate the description of many-body systems both directly and indirectly. When a symmetry occurs as a dynamical symmetry of the system, the corresponding algebra immediately gives the spectroscopic properties of the system. However, even when a symmetry is strongly broken, the algebraic structure nonetheless provides a calculational tool, classifying the basis states used in a full computational treatment of the many-body problem and greatly simplifying the underlying calculational machinery. Lie algebras have a long history of application, in both these capacities, to nuclear spectroscopy and related problems [1–3]. The fundamental quantities underlying calculations within a Lie algebraic framework are the coupling coefficients of the algebra, also known as generalized Clebsch-Gordan coefficients or Wigner coefficients. These are needed in order to couple states (or operators) of good symmetry to yield new states (or operators) of good symmetry, and they are required for the calculation of matrix elements through the generalized Wigner-Eckart theorem of the algebra.

The Lie algebra SO(5) has several distinct applications in nuclear theory, involving different physical realizations of the operators, and in which different subalgebra chains are relevant to the symmetry properties. Application as the proton-neutron pairing quasispin algebra [4–8] requires reduction with respect to the $\mathrm{U}(1) \otimes \mathrm{SO}(3)$ algebra of isospin and occupation number operators. For the dynamics of spin-2 bosons (as in the interacting boson model [9, 10]) and for the Bohr collective model [11–15], the appropriate reduction is instead with respect to the physical angular momentum SO(3) subalgebra. Finally, the canonical $\mathrm{SO}(4) \sim \mathrm{SO}(3) \otimes \mathrm{SO}(3)$ subalgebra is the most mathematically natural for computational use.

In this contribution, it is shown that the method of infinitesimal generators ("Racah's method") for extracting coupling coefficients can be formulated generically for a subalgebra chain $G \supset H$, provided the reduced matrix elements of the generators of G and the recoupling coefficients of H are known. After a brief overview of the SO(5) algebra (section 2), the general

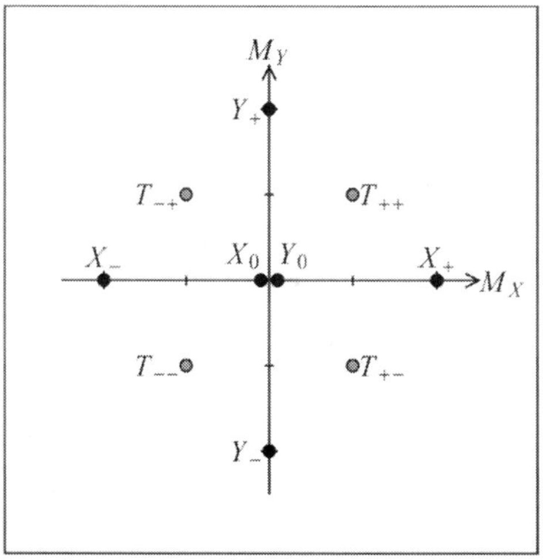

Figure 1. Root vector diagram for SO(5), with the generators labeled by their SO(3) ⊗ SO(3) Cartan weights M_X and M_Y.

problem of calculating reduced coupling coefficients is discussed, and Racah's method is broadly and systematically formulated as a method applicable to the calculation of reduced coupling coefficients for any subalgebra chain (section 3). More specifically, the problem of calculating coupling coefficients for generic irreps of SO(5), reduced with respect to any of the subalgebra chains, is completely resolved by this approach. The calculation of reduced coupling coefficients for the SO(5) ⊃ SO(4) canonical chain is considered in detail (section 4). Coupling coefficients reduced with respect to the noncanonical subalgebra chains of SO(5) may be obtained by a similar application of Racah's method, or they can simply be deduced from the canonical chain coupling coefficients by a straightforward unitary transformation.

2. The SO(5) algebra

The rotation algebra SO(5) in 5-dimensional space is spanned by the infinitesimal generators of rotation,

$$L_{rs} \equiv -i(x_r\partial_s - x_s\partial_r), \tag{1}$$

with r and $s = 1, \ldots, 5$. These operators are Hermitian and antisymmetric in the indices. The structure of the algebra is determined by the commutators

$$[L_{pq}, L_{rs}] = -i(\delta_{qr}L_{ps} + \delta_{ps}L_{qr} + \delta_{sq}L_{rp} + \delta_{rp}L_{sq}). \tag{2}$$

The canonical construction of SO(5) begins with SO(3), with the usual generators

$$\mathrm{SO}(3): \quad J_1 \equiv L_{23} \quad J_2 \equiv L_{31} \quad J_3 \equiv L_{12}. \tag{3}$$

Then, SO(4) is obtained with the addition of generators involving the fourth coordinate,

$$\mathrm{SO}(4): \quad N_1 \equiv L_{14} \quad N_2 \equiv L_{24} \quad N_3 \equiv L_{34}. \tag{4}$$

These generators have commutation relations $[J_r, J_s] = i\varepsilon_{rst}J_t$, $[N_r, N_s] = i\varepsilon_{rst}J_t$, and $[J_r, N_s] = i\varepsilon_{rst}N_t$. However, SO(4) can be recast as the direct sum of two SO(3) algebras, with the definitions

$$\mathrm{SO}_X(3) \otimes \mathrm{SO}_Y(3): \quad X_k \equiv \tfrac{1}{2}(J_k + N_k) \quad Y_k \equiv \tfrac{1}{2}(J_k - N_k), \tag{5}$$

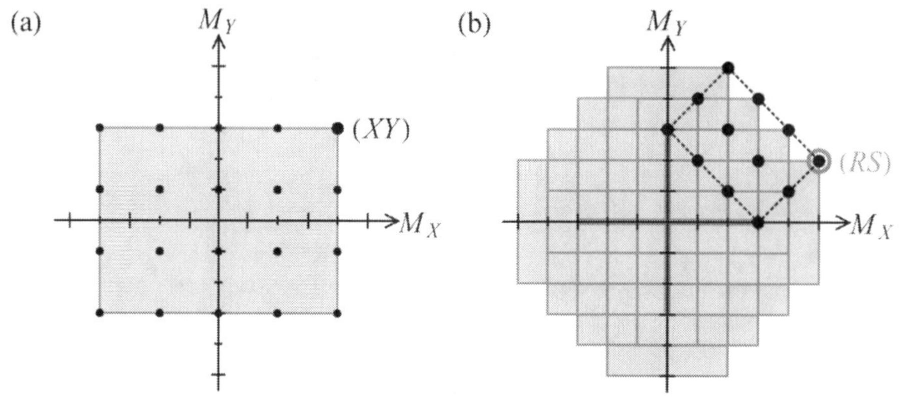

Figure 2. Weight and branching diagrams for SO(4) and SO(5) irreps: (a) Weight points $(M_X M_Y)$ contained within an SO(4) $\sim$ $\mathrm{SO}_X(3) \otimes \mathrm{SO}_Y(3)$ irrep (XY). The irrep $(2\frac{3}{2})$ is used for illustration. (b) Branching of an SO(5) irrep (RS). The irrep $(\frac{5}{2}2)$ is used for illustration. The open circle indicates the highest weight for the SO(5) irrep, and the solid dots indicate SO(4) highest weights, according to branching rule (7). The shaded rectangles are the boundaries of the weight sets for these SO(4) irreps.

so $[X_r, X_s] = i\varepsilon_{rst} X_t$, $[Y_r, Y_s] = i\varepsilon_{rst} Y_t$, and $[X_r, Y_s] = 0$. The ladder operator realizations of these generators ($X_\pm \equiv X_1 \pm iX_2$, $X_0 \equiv X_3$, $Y_\pm \equiv Y_1 \pm iY_2$, and $Y_0 \equiv Y_3$) are indicated in the root vector diagram figure 1, where each operator is placed according to the SO(3) weights M_X and M_Y carried by the operator. The Clebsch-Gordan series and Wigner calculus (coupling and recoupling coefficients) for SO(4) follow immediately from the SO(3)$\otimes$SO(3) structure [16]. Finally, SO(5) is obtained by introducing the four generators L_{r5} ($r = 1, \ldots, 4$), which are most naturally incorporated as the linear combinations

$$\mathrm{SO}(5): \quad \begin{cases} T_{++} = -\frac{1}{2}(L_{15} + iL_{25}) & T_{+-} = \frac{1}{2}(L_{35} + iL_{45}) \\ T_{-+} = \frac{1}{2}(L_{35} - iL_{45}) & T_{--} = \frac{1}{2}(L_{15} - iL_{25}). \end{cases} \tag{6}$$

These together constitute a double spherical tensor $T^{(\frac{1}{2}\frac{1}{2})}_{M_X M_Y}$ ($M_X = \pm\frac{1}{2}$ and $M_Y = \pm\frac{1}{2}$) with respect to the two SO(3) algebras, completing the root vector diagram of figure 1.

The generators of SO(5) as realized in (1) are *differential* operators on $\mathbb{R}^5$, which can only support symmetric irreps. However, the commutation relations (2) provide the fundamental definition of the algebra, and these support irreps of a much more general nature. The weights within a generic SO(4) irrep (XY) follow from the usual angular momentum rules $M_X = -X$, $\ldots$, $X-1$, X and $M_Y = -Y, \ldots, Y-1, Y$, as shown graphically in figure 2(a). The decomposition of an outer product of SO(4) irreps (Clebsch-Gordan series) as $(X_1Y_1)\otimes(X_2Y_2) \to (XY)$ follows by simple application of the triangle inequality to each SO(3) algebra, *i.e.*, $X = |X_1 - X_2|, \ldots,$ $X_1 + X_2 - 1$, $X_1 + X_2$ and $Y = |Y_1 - Y_2|, \ldots, Y_1 + Y_2 - 1, Y_1 + Y_2$. The SO(4) irreps contained within an SO(5) irrep (RS) follow from the branching rule [17, 18]

$$\begin{aligned} X &= R - \tfrac{1}{2}n - \tfrac{1}{2}m \\ Y &= S + \tfrac{1}{2}n - \tfrac{1}{2}m, \end{aligned} \tag{7}$$

with m and n integers such that $0 \le n \le 2(R - S)$ and $0 \le m \le 2S$. The SO(4) irrep highest weights form a lattice bounded by a tilted rectangle, as illustrated in figure 2(b). Note that the $SO_X(3) \otimes SO_Y(3)$ labels for an SO(4) irrep are simply related to the canonical labels by $[pq] = [X+Y, X-Y]$, and the $SO_X(3) \otimes SO_Y(3)$ labels for an SO(5) irrep are related to the canonical labels by $[l_1 l_2] = [R+S, R-S]$.

In addition to the natural subalgebra chains arising from the canonical construction above, involving the $SO(4) \sim SO_X(3) \otimes SO_Y(3)$ subalgebra, two noncanonical subalgebra chains also arise. These involve SO(3) subalgebras [or, equivalently, SU(2) subalgebras] distinct from those already considered. It is the noncanonical chains which appear in physical applications. The chains may be summarized as follows:

$$\begin{aligned}
&\underset{[l_1 l_2]}{SO(5)} \supset \underset{[pq]}{SO(4)} \supset \underset{J}{SO_J(3)} \supset \underset{M_J}{SO_J(2)} && \text{(I)}\\
&\qquad \supset SO(4) \sim \underset{X}{SO_X(3)} \otimes \underset{Y}{SO_Y(3)} \supset \underset{M_X}{SO_X(2)} \otimes \underset{M_Y}{SO_Y(2)} && \text{(I}'\text{)}\\
&\qquad \supset \underset{\kappa\ \ M_S}{U_N(1)} \otimes \underset{T}{SO_T(3)} \supset \underset{M_T}{SO_T(2)} && \text{(II)}\\
&\qquad \supset \underset{\alpha\ \ L}{SO_L(3)} \supset \underset{M_L}{SO_L(2)} && \text{(III)}.
\end{aligned} \tag{8}$$

The irrep labels have been noted beneath each subalgebra, and branching multiplicity labels are required for the noncanonical chains as well.

The isospin algebra $SO_T(3)$ of chain (II) is the relevant subalgebra for the description of proton-neutron pairing [4–8, 19–21]. In this context, the SO(5) generators arise as quasispin operators for pairing of protons and neutrons occupying the same j-shell. The operators

$$X_+ \propto (a_p^\dagger a_p^\dagger)^{(0)} \quad X_0 \propto (a_p^\dagger \tilde{a}_p + \tilde{a}_p a_p^\dagger)^{(0)} \quad X_- \propto (\tilde{a}_p \tilde{a}_p)^{(0)} \tag{9}$$

constitute an SO(3) quasispin algebra for proton pairs (of angular momentum zero), and

$$Y_+ \propto (a_n^\dagger a_n^\dagger)^{(0)} \quad Y_0 \propto (a_n^\dagger \tilde{a}_n + \tilde{a}_n a_n^\dagger)^{(0)} \quad Y_- \propto (\tilde{a}_n \tilde{a}_n)^{(0)} \tag{10}$$

constitute an SO(3) quasispin algebra for neutron pairs. The remaining generators (6) consist of the proton-neutron pair creation and annihilation operators

$$S_+ \propto (a_p^\dagger a_n^\dagger + a_n^\dagger a_p^\dagger)^{(0)} \quad S_- \propto (\tilde{a}_n \tilde{a}_p + \tilde{a}_p \tilde{a}_n)^{(0)} \tag{11}$$

and isospin ladder operators

$$T_+ \propto -(a_p^\dagger \tilde{a}_n)^{(0)} \quad T_- \propto -(a_n^\dagger \tilde{a}_p)^{(0)}, \tag{12}$$

identified as $S_+ = -2T_{++}$, $S_- = 2T_{--}$, $T_+ = 2T_{+-}$, and $T_- = 2T_{-+}$. The natural weight operators, providing the labels $M_S = \frac{1}{2}[N - (2j+1)]$ and $M_T = \frac{1}{2}(N_p - N_n)$ of (II), are $S_0 = X_0 + Y_0$ and $T_0 = X_0 - Y_0$.

On the other hand, the "physical" or "geometric" angular momentum subalgebra $SO_L(3)$ of chain (III) is the relevant subalgebra for application to systems of spin-2 bosons [9, 10] or the nuclear collective model [11–13, 15]. For the bosonic realization, the angular momentum generators are naturally given by the products $L_M = \sqrt{10}(b^\dagger \tilde{b})^{(1)}_M$, which close as the subalgebra $SO_L(3)$. The remaining seven generators are realized as $O_M = \sqrt{10}(b^\dagger \tilde{b})^{(3)}_M$, constituting a spherical tensor of octupole character with respect to $SO_L(3)$. The commutation relations are again found to be those of SO(5), with the L_M and O_M identified as certain linear combinations of the canonical generators, *e.g.*, $L_{+1} = -\sqrt{2}X_+ - \sqrt{6}T_{-+}$.

3. General formulation of Racah's method in reduced form

First, let us establish what is meant by a coupling coefficient for G reduced with respect to a subalgebra H. For a Lie algebra G, with subalgebra H, containing the Cartan subalgebra of G, states which reduce this subalgebra chain may be classified by the corresponding irreps labels Γ and Λ and Cartan weights λ,

$$\underset{\Gamma}{G} \underset{a}{\supset} \underset{\Lambda}{H} \supset \underset{\lambda}{(\text{Cartan subalgebra})}, \tag{13}$$

where multiplicity indices may be required as well. The coupling coefficients for G relate the uncoupled product states of two irreps of G to the coupled states, as

$$\left|\begin{matrix}\Gamma_1\ \Gamma_2 \\ \rho\Gamma \\ a\Lambda \\ \lambda\end{matrix}\right\rangle = \sum_{\substack{a_1\Lambda_1 a_2\Lambda_2 \\ \lambda_1\lambda_2}} \left(\begin{matrix}\Gamma_1 & \Gamma_2 \\ a_1\Lambda_1 & a_2\Lambda_2 \\ \lambda_1 & \lambda_2\end{matrix}\middle|\begin{matrix}\rho\Gamma \\ a\Lambda \\ \lambda\end{matrix}\right) \left|\begin{matrix}\Gamma_1 & \Gamma_2 \\ a_1\Lambda_1 & a_2\Lambda_2 \\ \lambda_1 & \lambda_2\end{matrix}\right\rangle, \tag{14}$$

where ρ is the outer multiplicity index for $G \otimes G$. By Racah's factorization lemma, the coupling coefficient appearing can be decomposed as the product

$$\left(\begin{matrix}\Gamma_1 & \Gamma_2 \\ a_1\Lambda_1 & a_2\Lambda_2 \\ \lambda_1 & \lambda_2\end{matrix}\middle|\begin{matrix}\rho\Gamma \\ a\Lambda \\ \lambda\end{matrix}\right) = \underbrace{\left(\begin{matrix}\Lambda_1 & \Lambda_2 \\ \lambda_1 & \lambda_2\end{matrix}\middle|\begin{matrix}\Lambda \\ \lambda\end{matrix}\right)}_{H} \underbrace{\left(\begin{matrix}\Gamma_1 & \Gamma_2 \\ a_1\Lambda_1 & a_2\Lambda_2\end{matrix}\middle|\begin{matrix}\rho\Gamma \\ a\Lambda\end{matrix}\right)}_{G\supset H \text{ reduced}} \tag{15}$$

of a coupling coefficient of H with a *reduced* coupling coefficient for $G \supset H$. (Here we have taken the subalgebra H and its outer product to be multiplicity free. See [22] for the appropriate generalizations.)

For the *symmetric* irreps of SO(5), one may work with an explicit construction, deriving the five-dimensional spherical harmonics as functions on the four-sphere. Their triple overlap integrals are then proportional to SO(5) coupling coefficients. Such an approach was recently developed for chain (III), where the spherical harmonics $\Psi_{v\alpha LM_L}(\gamma,\Omega)$ are of specific interest as basis functions for nuclear collective model calculations. Further details may be found in refs. [13, 23].

However, calculation of coupling coefficients involving *generic* irreps requires a broader approach. Several methods may be considered, in general, for constructing the reduced coupling coefficients of Lie algebras:

(1) Recurrence relations among coupling coefficients may be obtained by considering the action of an infinitesimal generator $G = G^{(1)} + G^{(2)}$ on uncoupled and coupled states. This approach is broadly termed "Racah's method" [22] and generalizes the classic recurrence method for evaluating SO(3) Clebsch-Gordan coefficients [24].

(2) Recurrence relations may be obtained by considering the action of a "shift tensor", lying outside the algebra [25]. For SO(5), shift tensors can be constructed by vector coherent state methods [26].

(3) Consistency relations among coupling and recoupling coefficients serve as the basis for a "building up" process [8, 27], in which unknown coupling coefficients can be deduced from a few known coefficients.

(4) Explicit realizations of an algebra can be obtained in terms of bosonic or fermionic creation and annihilation operators. Relations among coupling coefficients follow from considering the matrix elements of tensor operators operating on bosonic or fermionic states (*e.g.*, [28]). This approach yields a limited but useful class of coupling coefficients, generally restricted to symmetric irreps, antisymmetric irreps, or irreps which can be obtained as simple combinations thereof.

Indeed, all of these approaches have been applied, in various forms, to the calculation of specific classes of SO(5) coupling coefficients [1, 8, 17, 22, 28–32].

Let us now consider how the first approach, *i.e.*, Racah's method based on the action of infinitesimal generators, can be generally and systematically formulated as a method applicable to the calculation of reduced coupling coefficients involving generic irreps of an arbitrary subalgebra chain. (A comprehensive development of the method will be given in [33].)

Consider the action of a generator $T^{\Lambda_T}_{\lambda_T} = T^{\Lambda_T\,(1)}_{\lambda_T} + T^{\Lambda_T\,(2)}_{\lambda_T}$ of G (written as a tensor operator with respect to H) between coupled and uncoupled states,

$$\left\langle \begin{smallmatrix}\Gamma_1\\ a_1\Lambda_1\\ \lambda_1\end{smallmatrix}\ \begin{smallmatrix}\Gamma_2\\ a_2\Lambda_2\\ \lambda_2\end{smallmatrix}\right| \underset{\longrightarrow}{T^{\Lambda_T}_{\lambda_T}} \left| \begin{smallmatrix}\Gamma_1\,\Gamma_2\\ \rho\Gamma\\ a\Lambda\\ \lambda\end{smallmatrix}\right\rangle = \left\langle \begin{smallmatrix}\Gamma_1\\ a_1\Lambda_1\\ \lambda_1\end{smallmatrix}\ \begin{smallmatrix}\Gamma_2\\ a_2\Lambda_2\\ \lambda_2\end{smallmatrix}\right| \underset{\longleftarrow}{T^{\Lambda_T\,(1)}_{\lambda_T}} \left| \begin{smallmatrix}\Gamma_1\,\Gamma_2\\ \rho\Gamma\\ a\Lambda\\ \lambda\end{smallmatrix}\right\rangle + \left\langle \begin{smallmatrix}\Gamma_1\\ a_1\Lambda_1\\ \lambda_1\end{smallmatrix}\ \begin{smallmatrix}\Gamma_2\\ a_2\Lambda_2\\ \lambda_2\end{smallmatrix}\right| \underset{\longleftarrow}{T^{\Lambda_T\,(2)}_{\lambda_T}} \left| \begin{smallmatrix}\Gamma_1\,\Gamma_2\\ \rho\Gamma\\ a\Lambda\\ \lambda\end{smallmatrix}\right\rangle. \tag{16}$$

Its action (to the right) on the coupled state and (to the left) on each factor of the uncoupled state is then known, and the requirement of equality between the two sides of (16) provides the necessary condition for determination of the coupling coefficients in (14).

In particular, if the right hand side is expanded entirely in terms of uncoupled states, and if the left hand side is expanded entirely in terms of coupled states, by (14) and (15), then (16) constitutes a relation among (unknown) coupling coefficients and (known) matrix elements of the generator T^{Λ_T} between basis states. The Wigner-Eckart theorem for H allows the action of T^{Λ_T} to be expressed in terms of a coupling coefficient of H and a reduced matrix element as

$$\left\langle \begin{smallmatrix}\Gamma'\\ a'\Lambda'\\ \lambda'\end{smallmatrix}\right| T^{\Lambda_T}_{\lambda_T} \left| \begin{smallmatrix}\Gamma\\ a\Lambda\\ \lambda\end{smallmatrix}\right\rangle = \left(\begin{matrix}\Lambda & \Lambda_T\\ \lambda & \lambda_T\end{matrix}\middle| \begin{matrix}\Lambda'\\ \lambda'\end{matrix}\right) \underbrace{\left\langle \begin{smallmatrix}\Gamma'\\ a'\Lambda'\end{smallmatrix}\right\| T^{\Lambda_T} \left\| \begin{smallmatrix}\Gamma\\ a\Lambda\end{smallmatrix}\right\rangle}_{H\text{-reduced matrix element}}. \tag{17}$$

Thus, the basic condition (16) expands to

$$\begin{aligned}
&\sum_{\substack{a'\Lambda'\\(\lambda')}} \left(\begin{matrix}\Lambda & \Lambda_T\\ \lambda & \lambda_T\end{matrix}\middle| \begin{matrix}\Lambda'\\ \lambda'\end{matrix}\right) \left\langle \begin{smallmatrix}\Gamma\\ a'\Lambda'\end{smallmatrix}\right\| T^{\Lambda_T} \left\| \begin{smallmatrix}\Gamma\\ a\Lambda\end{smallmatrix}\right\rangle \left(\begin{matrix}\Lambda_1 & \Lambda_2\\ \lambda_1 & \lambda_2\end{matrix}\middle| \begin{matrix}\Lambda'\\ \lambda'\end{matrix}\right) \left(\begin{matrix}\Gamma_1 & \Gamma_2\\ a_1\Lambda_1 & a_2\Lambda_2\end{matrix}\middle| \begin{matrix}\rho\Gamma\\ a'\Lambda'\end{matrix}\right)\\
&\quad = \sum_{\substack{a_1'\Lambda_1'\\(\lambda_1')}} \left(\begin{matrix}\Lambda_1' & \Lambda_T\\ \lambda_1' & \lambda_T\end{matrix}\middle| \begin{matrix}\Lambda_1\\ \lambda_1\end{matrix}\right) \left\langle \begin{smallmatrix}\Gamma_1\\ a_1\Lambda_1\end{smallmatrix}\right\| T^{\Lambda_T} \left\| \begin{smallmatrix}\Gamma_1\\ a_1'\Lambda_1'\end{smallmatrix}\right\rangle \left(\begin{matrix}\Lambda_1' & \Lambda_2\\ \lambda_1' & \lambda_2\end{matrix}\middle| \begin{matrix}\Lambda\\ \lambda\end{matrix}\right) \left(\begin{matrix}\Gamma_1 & \Gamma_2\\ a_1'\Lambda_1' & a_2\Lambda_2\end{matrix}\middle| \begin{matrix}\rho\Gamma\\ a\Lambda\end{matrix}\right)\\
&\quad + \sum_{\substack{a_2'\Lambda_2'\\(\lambda_2')}} \left(\begin{matrix}\Lambda_2' & \Lambda_T\\ \lambda_2' & \lambda_T\end{matrix}\middle| \begin{matrix}\Lambda_2\\ \lambda_2\end{matrix}\right) \left\langle \begin{smallmatrix}\Gamma_2\\ a_2\Lambda_2\end{smallmatrix}\right\| T^{\Lambda_T} \left\| \begin{smallmatrix}\Gamma_2\\ a_2'\Lambda_2'\end{smallmatrix}\right\rangle \left(\begin{matrix}\Lambda_1 & \Lambda_2'\\ \lambda_1 & \lambda_2'\end{matrix}\middle| \begin{matrix}\Lambda\\ \lambda\end{matrix}\right) \left(\begin{matrix}\Gamma_1 & \Gamma_2\\ a_1\Lambda_1 & a_2'\Lambda_2'\end{matrix}\middle| \begin{matrix}\rho\Gamma\\ a\Lambda\end{matrix}\right).
\end{aligned} \tag{18}$$

However, this relation may be rearranged, by application of the orthonormality conditions on coupling coefficients of H, so that all these coupling coefficients of the lower algebra are bundled into recoupling coefficients, the unitary "6-Λ symbols"

$$\begin{bmatrix}\Lambda_1 & \Lambda_2 & \Lambda_{12}\\ \Lambda_3 & \Lambda & \Lambda_{23}\end{bmatrix} = \sum_{\substack{\lambda_1\lambda_2\lambda_3\\ \lambda_{12}\lambda_{13}}} \left(\begin{matrix}\Lambda_1 & \Lambda_2\\ \lambda_1 & \lambda_2\end{matrix}\middle| \begin{matrix}\Lambda_{12}\\ \lambda_{12}\end{matrix}\right) \left(\begin{matrix}\Lambda_{12} & \Lambda_3\\ \lambda_{12} & \lambda_3\end{matrix}\middle| \begin{matrix}\Lambda\\ \lambda\end{matrix}\right) \left(\begin{matrix}\Lambda_2 & \Lambda_3\\ \lambda_2 & \lambda_3\end{matrix}\middle| \begin{matrix}\Lambda_{23}\\ \lambda_{23}\end{matrix}\right) \left(\begin{matrix}\Lambda_1 & \Lambda_{23}\\ \lambda_1 & \lambda_{23}\end{matrix}\middle| \begin{matrix}\Lambda\\ \lambda\end{matrix}\right). \tag{19}$$

These are the transformation brackets connecting coupling schemes $[(\Lambda_1\Lambda_2)^{\Lambda_{12}}\Lambda_3]^{\Lambda}$ and $[\Lambda_1(\Lambda_2\Lambda_3)^{\Lambda_{23}}]^{\Lambda}$. We can thereby eliminate the coupling coefficients of H, and thus all reference to weights, from the condition (18). Let $\sigma(\Lambda_1\Lambda_2\Lambda)$ be the phase factor incurred when the first and second irreps are interchanged in a coupling coefficient of H. Then (18) simplifies to

$$\sum_{a'}\left\langle \begin{matrix}\Gamma\\ a'\Lambda'\end{matrix}\middle\| T^{\Lambda_T} \middle\| \begin{matrix}\Gamma\\ a\Lambda\end{matrix}\right\rangle \left(\begin{matrix}\Gamma_1\\ a_1\Lambda_1\end{matrix}\quad \begin{matrix}\Gamma_2\\ a_2\Lambda_2\end{matrix}\middle| \begin{matrix}\rho\Gamma\\ a'\Lambda'\end{matrix}\right)$$
$$= \sum_{a_1'\Lambda_1'} \sigma(\Lambda_1\Lambda_2\Lambda')\sigma(\Lambda_1'\Lambda_2\Lambda)\begin{bmatrix}\Lambda_2 & \Lambda_1' & \Lambda\\ \Lambda_T & \Lambda' & \Lambda_1\end{bmatrix}\left\langle \begin{matrix}\Gamma_1\\ a_1\Lambda_1\end{matrix}\middle\| T^{\Lambda_T} \middle\| \begin{matrix}\Gamma_1\\ a_1'\Lambda_1'\end{matrix}\right\rangle \left(\begin{matrix}\Gamma_1\\ a_1'\Lambda_1'\end{matrix}\quad \begin{matrix}\Gamma_2\\ a_2\Lambda_2\end{matrix}\middle| \begin{matrix}\rho\Gamma\\ a\Lambda\end{matrix}\right)$$
$$+ \sum_{a_2'\Lambda_2'} \begin{bmatrix}\Lambda_1 & \Lambda_2' & \Lambda\\ \Lambda_T & \Lambda' & \Lambda_2\end{bmatrix}\left\langle \begin{matrix}\Gamma_2\\ a_2\Lambda_2\end{matrix}\middle\| T^{\Lambda_T} \middle\| \begin{matrix}\Gamma_2\\ a_2'\Lambda_2'\end{matrix}\right\rangle \left(\begin{matrix}\Gamma_1\\ a_1\Lambda_1\end{matrix}\quad \begin{matrix}\Gamma_2\\ a_2'\Lambda_2'\end{matrix}\middle| \begin{matrix}\rho\Gamma\\ a\Lambda\end{matrix}\right), \tag{20}$$

expressed entirely in terms of $G \supset H$ reduced coupling coefficients, H-reduced matrix elements of the generator T^{Λ_T}, and recoupling coefficients of the lower algebra H.

The question remains as to how to *apply* these relations in order to systematically obtain the values of coupling coefficients. Consider the coupling coefficients for $\Gamma_1 \otimes \Gamma_2 \to \rho\Gamma$. Then (20) yields a different relation among specific reduced coupling coefficients for each *quadruplet* of values $(a_1\Lambda_1 a_2\Lambda_2 a\Lambda a'\Lambda')$ for the irrep labels, such that the conditions

$$\begin{gathered}\Lambda\otimes\Lambda_T \to \Lambda' \quad \Lambda_1\otimes\Lambda_2\to\Lambda' \\ \Gamma_1\to\Lambda_1 \quad \Gamma_2\to\Lambda_2 \quad \Gamma\to\Lambda \quad \Gamma\to\Lambda'\end{gathered} \tag{21}$$

are satisfied.

The most familiar and traditional approach to extracting coupling coefficients, after obtaining some set of relations among them, is to proceed by recurrence, both in the familiar case of SO(3) Clebsch-Gordan coefficients [24] as well as in the various approaches noted above for higher algebras. A seed value is given for one coupling coefficient, and further coefficients are deduced inductively from those already obtained.

Since calculation of SO(3) Clebsch-Gordan coefficients provides the archetypical example, let us recall how recurrence relations are applied in that problem [24]. Considering the matrix elements of the generators $J_\pm$ between coupled an uncoupled states, as in (16), *i.e.*,

$$\langle J_1M_1J_2M_2 \,|\, J_\pm \,|\, JM\rangle = \langle J_1M_1J_2M_2 \,|\, J_\pm^{(1)} \,|\, JM\rangle + \langle J_1M_1J_2M_2 \,|\, J_\pm^{(2)} \,|\, JM\rangle, \tag{22}$$

yields a relation among Clebsch-Gordan coefficients only if $M = M_1 + M_2 \mp 1$. Then,

$$K_\pm(JM)\left(\begin{smallmatrix}J_1 & J_2\\ M_1 & M_2\end{smallmatrix}\middle|\begin{smallmatrix}J\\ M\pm1\end{smallmatrix}\right) = K_\pm(J_1M_1\mp1)\left(\begin{smallmatrix}J_1 & J_2\\ M_1\mp1 & M_2\end{smallmatrix}\middle|\begin{smallmatrix}J\\ M\end{smallmatrix}\right) + K_\pm(J_1M_2\mp1)\left(\begin{smallmatrix}J_1 & J_2\\ M_1 & M_2\mp1\end{smallmatrix}\middle|\begin{smallmatrix}J\\ M\end{smallmatrix}\right), \tag{23}$$

where $K_\pm(JM) \equiv \langle JM\pm1 \,|\, J_\pm \,|\, JM\rangle = [(J\mp M)(J\pm M+1)]^{1/2}$ is the generator matrix element. If the Clebsch-Gordan coefficients for $J_1 \otimes J_2 \to J$ are represented as a lattice of points (M_1, M_2), as in figure 3(a), these relations connect neighboring points at the vertices of an upward-oriented triangle (for J_+) or downward-oriented triangle (for J_-). A natural order for traversing the coefficients can therefore easily be chosen, such that only one unknown arises at each step. It is necessary to involve certain known-zero or "forbidden" Clebsch-Gordan coefficients, represented by the triangle vertices without dots in figure 3(a), in the relations. However, from figure 3(b), it is apparent that all nonvanishing coefficients may also be directly connected by the relations (23), without involving any forbidden coefficients. This yields a system of equations which fully determines the coefficients, to within an overall phase and normalization, without the need for choosing a specific seed coefficient and recurrence order.

Returning to the general problem for $G \supset H$ reduced coupling coefficents, if there are N coupling coefficients for the coupling $\Gamma_1 \otimes \Gamma_2 \to \rho\Gamma$, then the relations obtained from (20)

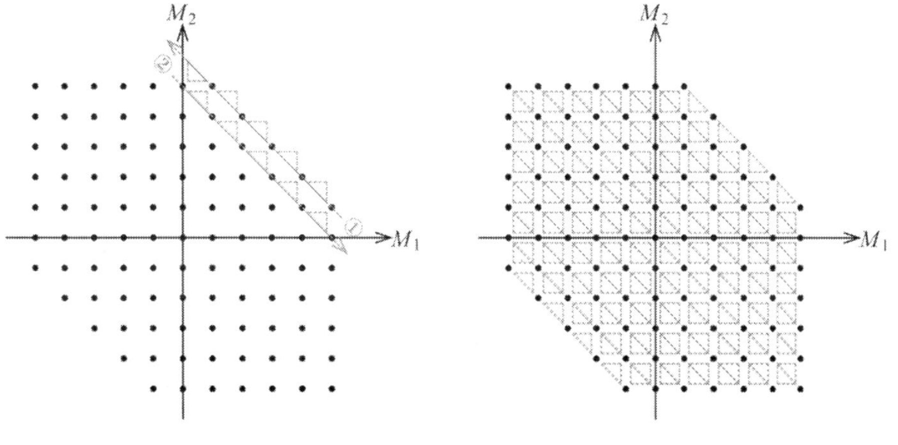

Figure 3. The classic problem of constructing the SO(3) Clebsch-Gordan coefficients $(J_1 M_1 J_2 M_2 | J, M_1 + M_2)$ by use of the relations (23). Dots indicate allowed non-zero coefficients. Coefficients at the vertices of a dashed triangle are connected by (23). (a) The conventional recurrence approach, in which coefficients are calculated inductively from a seed coefficient, making use of relations which in some cases invoke known-zero Clebsch-Gordan coefficients. (b) A full set of relations among allowed Clebsch-Gordan coefficients, yielding a linear, homogeneous system of equations in the Clebsch-Gordan coefficients.

constitute a system of equations in N unknowns for these coupling coefficients. In general, many irreps of H will be connected by the generator $T^{(\Lambda_T)}$, and therefore each relation obtained from (20) will involve many unknown coupling coefficients. A simple recurrence pattern, as in figure 3(a), may be impractical to devise.

Instead, it is possible to solve the full system of equations derived from (20), as in figure 3(b), simultaneously. This is a linear, homogeneous system of equations in the unknown coupling coefficients. Let us label the N unknown coupling coefficients for $\Gamma_1 \otimes \Gamma_2 \to \rho\Gamma$ with a single counting index, as $C_i \equiv \left(\begin{smallmatrix} \Gamma_1 & \Gamma_2 \\ a_1\Lambda_1 & a_2\Lambda_2 \end{smallmatrix} \middle| \begin{smallmatrix} \rho\Gamma \\ a\Lambda \end{smallmatrix}\right)$, where $i = 1, \ldots, N$. Then, for each generator $T^{(\Lambda_T)}$ in G but not in H, and for each quadruplet of values $(a_1\Lambda_1 a_2\Lambda_2 a\Lambda a'\Lambda')$ allowed by (21), the relation (20) yields an equation, which we label by an index k, of the form $\sum_{i=1}^{N} a_{ki} C_i = 0$. In matrix form, the system of equations for the unknown C_i is then

Coupling coefficient $(a_1\Lambda_1 a_2\Lambda_2 a\Lambda) \longrightarrow$

Racah relation $(a_1\Lambda_1 a_2\Lambda_2 a\Lambda a'\Lambda') \downarrow$

$$\begin{bmatrix} & \vdots & \\ \cdots & a_{ki} & \cdots \\ & \vdots & \end{bmatrix} \begin{bmatrix} C_1 \\ \vdots \\ C_N \end{bmatrix} = \begin{bmatrix} 0 \\ \vdots \\ 0 \end{bmatrix}. \tag{24}$$

The problem of solving this linear, homogeneous system of equations (24) is equivalent to finding the null vector (or vectors)

$$\begin{matrix} [\ C_{11} & C_{12} & \cdots & C_{1N}\ ] & \rho = 1 \\ & & \cdots & & \\ [\ C_{D1} & C_{D2} & \cdots & C_{DN}\ ] & \rho = D \end{matrix} \tag{25}$$

of the matrix on the left hand side. These may be obtained most simply by Euler row reduction.

The entries of the null vectors $[C_{\rho 1}C_{\rho 2}\cdots C_{\rho N}]$ are the reduced coupling coefficients, to within (ortho)normalization and a possible overall phase, to be determined by convention. The appropriate orthonormality condition for *reduced* coupling coefficients is

$$\sum_{a_1\Lambda_1 a_2\Lambda_2}\begin{pmatrix}\Gamma_1 & \Gamma_2 & \rho\Gamma\\ a_1\Lambda_1 & a_2\Lambda_2 & a\Lambda\end{pmatrix}\begin{pmatrix}\Gamma_1 & \Gamma_2 & \rho'\Gamma'\\ a_1\Lambda_1 & a_2\Lambda_2 & a'\Lambda\end{pmatrix}=\delta_{(\rho\Gamma)(\rho'\Gamma')}\delta_{aa'}. \tag{26}$$

In the case where $\Gamma_1\otimes\Gamma_2\rightarrow\Gamma$ is free of outer multiplicity, the matrix in (24) is of rank $N-1$, yielding a unique null vector. More generally, if the coupling $\Gamma_1\otimes\Gamma_2\rightarrow\Gamma$ has outer multiplicity D ($\rho=1,\ldots,D$), the matrix is of rank $N-D$, yielding D linearly independent null vectors. The coupling coefficients in the presence of multiplicity are only defined to within a unitary transformation, which in the present approach is manifested as the freedom in choosing the orthonormal basis vectors spanning the D-dimensional null space.

4. Calculation of SO(5) ⊃ SO(4) coupling coefficients

Let us now consider the application of Racah's method in "reduced" form, as described in section 3, to the canonical chain of SO(5), *i.e.*, for the case in which $G\supset H$ is SO(5) ⊃ SO(4). The necessary ingredients for constructing relations of the type (20) are:

(1) matrix elements for generators of G, reduced with respect to H, and

(2) the recoupling coefficients of H.

Both components are readily available, as simple closed-form expressions, for SO(5) ⊃ SO(4). The operators $T^{(\frac{1}{2}\frac{1}{2})}$ of (6) are generators of SO(5) but not SO(4). They constitute a double spinor with respect to $SO_X(3)\otimes SO_Y(3)$ and thus a tensor with respect to SO(4). The reduced matrix elements were evaluated by Hecht [17] and by Kemmer, Pursey, and Williams [18], *e.g.*,

$$\left\langle\begin{matrix}(RS)\\(X+\frac{1}{2}Y+\frac{1}{2})\end{matrix}\middle\|T^{(\frac{1}{2}\frac{1}{2})}\middle\|\begin{matrix}(RS)\\(XY)\end{matrix}\right\rangle=\frac{\begin{matrix}[(R+S-X-Y)(R+S+X+Y+3)\\ \times(-R+S+X+Y+1)(R-S+X+Y+2)]^{1/2}\end{matrix}}{2\widehat{X+\frac{1}{2}}\widehat{Y+\frac{1}{2}}}, \tag{27}$$

where $\hat{J}\equiv(2J+1)^{1/2}$, with similar expressions obtained for the other reduced matrix elements. The recoupling coefficients for SO(4) factorize into ordinary SO(3) recoupling coefficients, by the SO(3) ⊗ SO(3) structure, as

$$\begin{aligned}\begin{bmatrix}(X_1Y_1) & (X_2Y_2) & (X_{12}Y_{12})\\(X_3Y_3) & (XY) & (X_{23}Y_{23})\end{bmatrix}&=\begin{bmatrix}X_1 & X_2 & X_{12}\\X_3 & X & X_{23}\end{bmatrix}\begin{bmatrix}Y_1 & Y_2 & Y_{12}\\Y_3 & Y & Y_{23}\end{bmatrix}\\&=(-)^{X_1+X_2+X_3+X}(-)^{Y_1+Y_2+Y_3+Y}\hat{X}_{12}\hat{X}_{23}\hat{Y}_{12}\hat{Y}_{23}\begin{Bmatrix}X_1 & X_2 & X_{12}\\X_3 & X & X_{23}\end{Bmatrix}\begin{Bmatrix}Y_1 & Y_2 & Y_{12}\\Y_3 & Y & Y_{23}\end{Bmatrix},\end{aligned} \tag{28}$$

in terms of the conventional (nonunitary) 6-J symbols.

The system of equations for the coupling coefficients is assembled as described in section 3. A different condition is obtained from (20) for each quadruplet of SO(4) irreps (X_1Y_1), (X_2Y_2), (XY), and $(X'Y')$, chosen from the branchings $(R_1S_1)\rightarrow(X_1Y_1)$, $(R_2S_2)\rightarrow(X_2Y_2)$, $(RS)\rightarrow(XY)$, and $(RS)\rightarrow(X'Y')$. Furthermore, a nonnull condition on the coupling coefficients is obtained only if the SO(4) outer product conditions $(XY)\otimes(\frac{1}{2}\frac{1}{2})\rightarrow(X'Y')$ and $(X_1Y_1)\otimes(X_2Y_2)\rightarrow(X'Y')$ are satisfied (the relevant branching and Clebsch-Gordan series relations are given in section 2). Since the matrix elements (27) and recoupling coefficients (28) defining the system of equations are known exactly, as (signed) square roots of rational numbers,

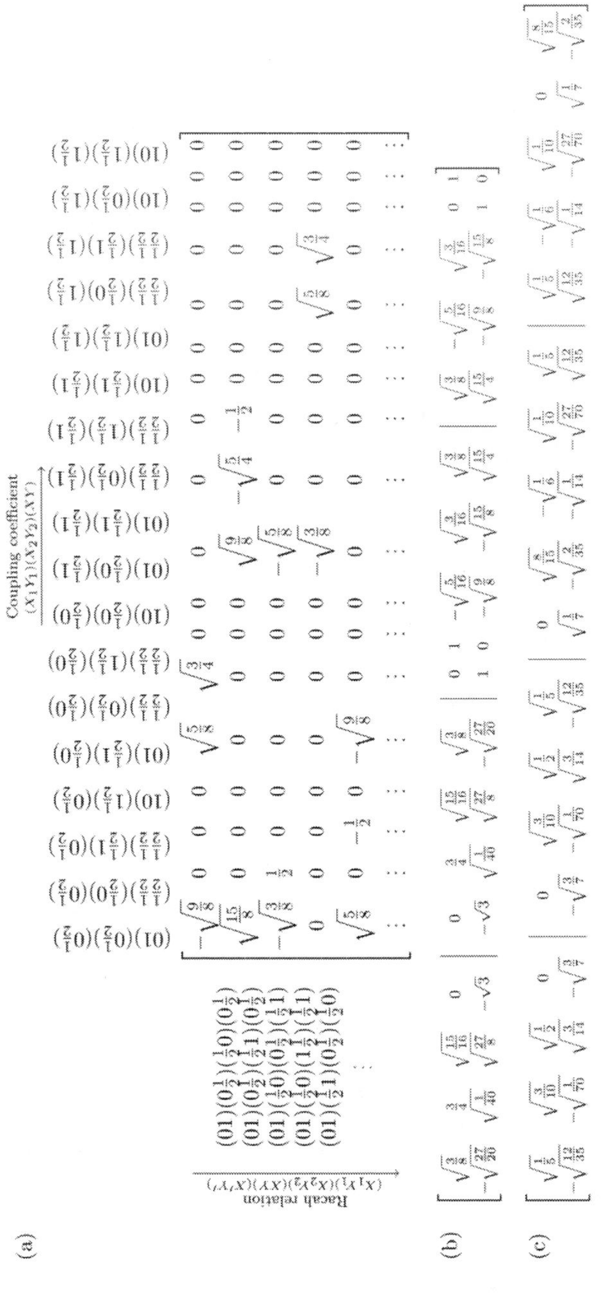

Figure 4. Calculation of SO(5) $\supset$ SO(4) reduced coupling coefficients for $(10) \otimes (1\frac{1}{2}) \to (1\frac{1}{2})$: (a) Coefficient matrix for the linear, homogenous system of equations determining the coupling coefficients. The full matrix has dimension 36×18. (b) Basis vectors for the null space, which is of dimension 2. (c) Orthonormal vectors, containing the desired coupling coefficients as entries. The orthonormality condition (26) interrelates groups of coupling coefficients sharing the same (XY) label, which are delimited by the vertical lines in parts (b) and (c).

the SO(5) $\supset$ SO(4) reduced coupling coefficients can be obtained exactly by Euler row reduction, again as (signed) square roots of rational numbers.

A concrete numerical example is provided by the SO(5) coupling $(10) \otimes (1\frac{1}{2}) \rightarrow (1\frac{1}{2})$. (In the canonical labeling scheme, this coupling is $[11] \otimes [\frac{3}{2}\frac{1}{2}] \rightarrow [\frac{3}{2}\frac{1}{2}]$.) There are 18 SO(4)-reduced coupling coefficients to be determined, related by a system of 36 equations. The first few rows of the coefficient matrix are shown in figure 4(a). The coupling has outer multiplicity $D = 2$. Correspondingly, the coefficient matrix is of rank 16, admitting two null vectors, given in figure 4(b). Orthonormal null vectors, with respect to the condition (26), are obtained by the Gram-Schmidt procedure, with the result given in figure 4(c). The entries of the upper and lower rows may thus be taken as the SO(5) $\supset$ SO(4) reduced coupling coefficients for $\rho = 1$ and 2, respectively.

5. Conclusions

It has been shown that Racah's method of infinitesimal generators can be systematically generalized to the calculation of reduced coupling coefficients for an arbitrary subalgebra chain, provided the matrix elements of the generators (reduced with respect to the lower algebra) and the recoupling coefficients of the lower algebra are known. For the algebra SO(5), the problem of calculating coupling coefficients for generic irreps, reduced with respect to the canonical or noncanonical chains, is thereby completely resolved.

The necessary ingredients take on a particularly simple form for the canonical chain of SO(5). However, the generator matrix elements reduced with respect to the noncanonical isospin subalgebra [chain (II)] and physical angular momentum algebra [chain (III)] are also known, from vector coherent state theory [34, 35]. Hence, Racah's method may be applied directly to the calculation of chain (II) and chain (III) reduced coupling coefficients.

Alternatively, once coupling coefficients for the canonical chain have been obtained, the coupling coefficients for the noncanonical chains can readily be deduced by a unitary transformation. Transformation brackets between basis states reducing the canonical and noncanonical chains are obtained by diagonalizing the appropriate Casimir operator, *i.e.*, $\mathbf{T}^2$ or $\mathbf{L}^2$, in the canonical basis (as in [14]). Then the transformed reduced coupling coefficients follow, *e.g.*, for chain (II), as

$$\begin{aligned}
&\underbrace{\left(\begin{array}{cc|c} (R_1S_1) & (R_2S_2) & \rho(RS) \\ M_{S1}\kappa_1T_1 & M_{S2}\kappa_2T_2 & M_S\kappa T \end{array}\right)}_{\mathrm{SO}(5)\supset\mathrm{U}_N(1)\otimes\mathrm{SO}_T(3)} \\
&\quad = \sum_{\substack{(X_1Y_1)(X_2Y_2)(XY)\\ M_{T1}(M_{T2})}} \underbrace{\left(\begin{array}{cc|c} T_1 & T_2 & T \\ M_{T1} & M_{T2} & M_T \end{array}\right)}_{\mathrm{SO}(3)} \underbrace{\left(\begin{array}{cc|c} (X_1Y_1) & (X_2Y_2) & (XY) \\ M_{X1}M_{Y1} & M_{X2}M_{Y2} & M_XM_Y \end{array}\right)}_{\mathrm{SO}(4)} \\
&\quad\times \underbrace{\left\langle \begin{array}{c} (R_1S_1) \\ M_{S1}\kappa_1T_1 \\ M_{T1} \end{array} \middle| \begin{array}{c} (R_1S_1) \\ (X_1Y_1) \\ M_{X1}M_{Y1} \end{array} \right\rangle \left\langle \begin{array}{c} (R_2S_2) \\ M_{S2}\kappa_2T_2 \\ M_{T2} \end{array} \middle| \begin{array}{c} (R_2S_2) \\ (X_2Y_2) \\ M_{X2}M_{Y2} \end{array} \right\rangle \left\langle \begin{array}{c} (RS) \\ M_S\kappa T \\ M_T \end{array} \middle| \begin{array}{c} (RS) \\ (XY) \\ M_XM_Y \end{array} \right\rangle}_{\mathrm{SO}(5)\supset[\mathrm{U}_N(1)\otimes\mathrm{SO}_T(3)\leftrightarrow\mathrm{SO}(4)]} \\
&\quad\times \underbrace{\left(\begin{array}{cc|c} (R_1S_1) & (R_2S_2) & \rho(RS) \\ (X_1Y_1) & (X_2Y_2) & (XY) \end{array}\right)}_{\mathrm{SO}(5)\supset\mathrm{SO}(4)}, \qquad (29)
\end{aligned}$$

where $M_X = \frac{1}{2}(M_S + M_T)$, $M_Y = \frac{1}{2}(M_S - M_T)$, and similarly for M_{X1}, M_{Y1}, M_{X2}, and M_{Y2}.

The specific example of SO(5) coupling coefficients may be considered as a prototype for the systematic calculation of coupling coefficients for other higher algebras. For instance, the computational machinery for SU(3) is well established [36, 37] and may therefore be used as the starting point for calculation of Sp(6) $\supset$ U(3) reduced coupling coefficients, for the fermion dynamical symmetry model [38], or Sp(6, $\mathbb{R}$) $\supset$ U(3) reduced coupling coefficients, for the symplectic shell model [39]. The requisite generator matrix elements for Sp(6) and Sp(6, $\mathbb{R}$) may again be calculated from vector coherent state realizations [40, 41]. The Sp(6, $\mathbb{R}$) $\supset$ U(3) coupling coefficients are required, for instance, if large-scale calculations are to be carried out in the *ab initio* symplectic scheme of Dytrych *et al.* [42, 43].

Acknowledgments

The author is grateful to K. D. Sviratcheva, A. E. McCoy, D. J. Rowe, T. A. Welsh, and F. Iachello for collaboration on various problems outlined in this contribution, and to C. Bahri, T. Dytrych, J. P. Draayer, and S. De Baerdemacker for valuable discussions. This work was supported by the US DOE under grant DE-FG02-95ER-40934.

References

[1] Racah G 1965 *Springer Tracts Mod. Phys.* **37** 28
[2] Hecht K T 1983 *Annu. Rev. Nucl. Sci.* **23** 123
[3] Iachello F 2006 *Lie Algebras and Applications* (*Lecture Notes in Physics* vol 708) (Berlin: Springer)
[4] Helmers K 1961 *Nucl. Phys.* **23** 594
[5] Flowers B H and Szpikowski S 1964 *Proc. Phys. Soc. (London)* **84** 193
[6] Ichimura M 1965 *Prog. Theor. Phys.* **33** 215
[7] Ginocchio J N 1965 *Nucl. Phys.* **74** 321
[8] Hecht K T 1967 *Nucl. Phys.* A **102** 11
[9] Arima A and Iachello F 1976 *Ann. Phys. (N.Y.)* **99** 253
[10] Iachello F and Arima A 1987 *The Interacting Boson Model* (Cambridge: Cambridge University Press)
[11] Bohr A and Mottelson B R 1998 *Nuclear Structure* vol 2 *Nuclear Deformations* (Singapore: World Scientific)
[12] Eisenberg J M and Greiner W 1987 *Nuclear Theory* 3rd ed vol 1 *Nuclear Models: Collective and Single-Particle Phenomena* (Amsterdam: North-Holland)
[13] Rowe D J, Turner P S and Repka J 2004 *J. Math. Phys.* **45** 2761
[14] De Baerdemacker S, Heyde K and Hellemans V 2008 *J. Phys.* A **41** 304039
[15] Rowe D J, Welsh T A and Caprio M A 2009 *Phys. Rev.* C **79** 054304
[16] Biedenharn L C 1961 *J. Math. Phys.* **2** 433
[17] Hecht K T 1965 *Nucl. Phys.* **63** 177
[18] Kemmer N, Pursey D L and Williams S A 1968 *J. Math. Phys.* **9** 1224
[19] Engel J, Langanke K and Vogel P 1996 *Phys. Lett.* B **389** 211
[20] Sviratcheva K D, Georgieva A I and Draayer J P 2005 *Phys. Rev.* C **72** 054302
[21] Sviratcheva K D, Draayer J P and Vary J P 2006 *Phys. Rev.* C **73** 034324
[22] Wybourne B G 1974 *Classical Groups for Physicists* (New York: Wiley)
[23] Caprio M A, Rowe D J and Welsh T A 2009 *Comput. Phys. Commun.* **180** 1150
[24] Edmonds A R 1960 *Angular Momentum in Quantum Mechanics* 2nd ed (*Investigations in Physics* no 4) (Princeton, New Jersey: Princeton University Press)
[25] Rowe D J and Repka J 1997 *J. Math. Phys.* **38** 4363
[26] Hecht K T 1993 *J. Phys.* A **26** 329
[27] Hecht K T and Pang S C 1969 *J. Math. Phys.* **10** 1571
[28] Van Isacker P 1987 *J. Math. Phys.* **28** 957
[29] Iachello F and Kuyucak S 1981 *Ann. Phys. (N.Y.)* **136** 19
[30] Hecht K T 1989 *Nucl. Phys.* A **493** 29
[31] Han Q Z, Sun H Z and Wang J J 1993 *Commun. Theor. Phys.* **20** 313
[32] Caprio M A and Iachello F 2007 *Nucl. Phys.* A **781** 26
[33] Caprio M A, Sviratcheva K D and McCoy A E in preparation
[34] Hecht K T and Elliott J P 1985 *Nucl. Phys.* A **438** 29
[35] Turner P S, Rowe D J and Repka J 2006 *J. Math. Phys.* **47** 023507
[36] Draayer J P and Akiyama Y 1973 *J. Math. Phys.* **14** 1904
[37] Rowe D J and Bahri C 2000 *J. Math. Phys.* **41** 6544
[38] Wu C L, Feng D H, Chen X G, Chen J Q and Guidry M W 1986 *Phys. Lett.* B **168** 313
[39] Rosensteel G and Rowe D J 1980 *Ann. Phys. (N.Y.)* **126** 343

[40] Rowe D J 1984 *J. Math. Phys.* **25** 2662
[41] Hecht K T and Chen J Q 1990 *Nucl. Phys.* A **512** 365
[42] Dytrych T, Sviratcheva K D, Bahri C, Draayer J P and Vary J P 2007 *Phys. Rev. Lett.* **98** 162503
[43] Dytrych T, Sviratcheva K D, Draayer J P, Bahri C and Vary J P 2008 *J. Phys.* G **35** 123101

Attosecond neutron scattering from open quantum systems: entanglement and decoherence phenomena of protons in condensed matter and molecules

C A Chatzidimitriou-Dreismann[1] and I C Tietje[2]

[1] Institute of Chemistry (Sekr. C2), Technical University of Berlin, D-10623 Berlin, Germany
[2] Institutes of Physics, Technical University of Berlin, D-10623 Berlin, Germany

E-mail: `dreismann@chem.tu-berlin.de`

Abstract. In general, nuclei and electrons in condensed matter and/or molecules are entangled, due to the prevailing (electromagnetic) interactions. As a matter of fact, the "environment" interacting with a microscopic system of interest (e.g. a proton) causes the destruction of the entanglement. This process, called decoherence, has until now prevented experimenters from directly accessing atomic and/or nuclear entanglement effects in real experiments. However, our neutron and electron Compton scattering experiments from protons (H-atoms) in several condensed systems and molecules at ambient conditions demonstrated a new striking effect, i.e. an "anomalous" decrease of scattering intensity from protons, which seem to become partially "invisible" to the neutrons. This effect, which has no interpretation within conventional neutron scattering theory, is proposed to be caused by the *non-unitary* time evolution (due to decoherence) during the ultrashort, but finite, time-window of the neutron-proton scattering process. Due to the large energy (several eV) and momentum (20-200 Å^{-1}) transfers of these experiments the collisional time between the probe particle and a struck proton is 100–1000 attoseconds long. It is shown that, due to this short timescale, the scattering process must be theoretically treated within quantum dynamics of *open* quantum systems. New experimental neutron Compton scattering results from a single crystal $KHCO_3$ are presented. They provide the first direct evidence for a connection between the momentum distribution and the "anomalous" scattering intensity of H. Additionally, recent experiments with electron-atom Compton scattering at momentum transfers are mentioned, which reveal the presence of this effect even in scattering from single H_2 molecules in the gas phase. Theoretical discussions "from first principles" are presented, also in relation to the quantum Zeno effect, which underline the crucial role of decoherence in the considered experiments. The experimental results and their qualitative interpretation show that epithermal neutrons being available at spallation sources (e.g. ISIS/U.K., SNS/U.S.A. or ESS/Sweden), and electron spectrometers with large scattering angles, provide tools for investigating new physical and chemical phenomena in the sub-femtosecond timescale.

1. Introduction

The counter-intuitive phenomenon of quantum entanglement (QE) — also called Einstein-Podolsky-Rosen correlations or Schrödinger cat states — between two or more quantum systems has emerged as the most emblematic feature of quantum mechanics [1]. QE is at the heart of the intriguing complexity of describing quantum many-body systems in physics, chemistry and biology.

Nuclei and electrons in condensed matter and/or molecules are usually entangled, due to the prevailing (electromagnetic) interactions. Usually, the "environment" of a microscopic scattering system (e.g. a proton) causes an ultrafast decoherence (see, e.g. [2]), thus making atomic and/or nuclear entanglement effects not directly accessible to experiments. It may be noted that the phenomenon of decoherence represents the main obstacle preventing the construction of a quantum computer [3]. In this article, the term entanglement may refer to various dynamical variables (e.g. position, momentum and/or spin degrees of freedom).

However, as shown below, neutron Compton scattering (NCS) is an ultrafast scattering technique and thus it may open up a way for investigation of short-lived quantum entanglement even in condensed matter at ambient conditions. In this respect it is important to emphasize that, under the physical conditions of NCS, each impinging neutron collides with *one* nucleus (e.g. one proton of H_2) only, and not with the "entire molecule". This is clearly demonstrated with the aid of the measured time-of-flight (TOF) spectra: E.g., the position of the measured H-recoil peak of molecular hydrogen, H_2, lies always in the TOF-position corresponding to the collision with a particle of one atomic mass unit (i.e., the proton), and not two atomic mass units (i.e., the whole hydrogen molecule).

This fact also shows that the NCS process represents scattering from *open quantum systems* [2], e.g. one proton of H_2 molecules. Consequently, the remaining adjacent particles (e.g., the second proton and the two electrons of H_2) represent the *environment* which strongly interacts with the scattering system under investigation (i.e., a single proton, in the considered example). These strong interactions are the cause of ultrafast decoherence. The conceptual connection and/or interrelation with the physical field of *quantum optics* is obvious.

Starting in 1995, our NCS experiments from protons (H-atoms) in several condensed systems and molecules [4] demonstrated a new striking effect, i.e. an "anomalous" decrease of scattering intensity from protons, which seem to become partially "invisible" to the neutrons [4, 5]. Several experiments were done at ambient conditions. Due to the large energy (several eV) and momentum transfers of these experiments, the collisional time between the probe particle and a struck proton is about 100–1000 attoseconds long (see below).

Many NCS experiments in various systems (molecular hydrogen, metallic hydrides, polymers, inorganic crystals, gases, etc.) have been carried out successfully and demonstrated this attosecond quantum effect; see e.g. [4, 5] and references cited therein.

This effect has no interpretation within conventional scattering theory of NCS (also called deep-inelastic neutron scattering, DINS) [6, 7]. According to our theoretical predictions and/or investigation, it is caused by the non-unitary time evolution (due to decoherence) during the ultrashort, but finite, time-window of the neutron-proton scattering process; see theoretical discussions below.

For clarity, and to be specific, in the following we constrain our considerations to the physical context of NCS only. The following sections present:

(A) a short description of the NCS technique and a novel experimental result;

(B) a discussion of the characteristic ultashort (attosecond) timescale of NCS from protons (H-atoms), and the necessity to apply the theory of *open* quantum systems to the scattering process;

(C) a qualitative theoretical understanding "from first principles" of the new effect of "anomalous" decrease of scattering intensity, which is based on the quantum *Zeno* effect [8, 9, 10] and/or a fundamental theoretical result by Schulman and Gaveau [11, 12];

(D) a theoretical model of scattering which "extends" conventional theory through the inclusion of entanglement and decoherence [13];

(E) a "first principles" theoretical model of scattering from *open* quantum systems subject to a Lindblad-type dynamics [14]. In particular we consider the case that the duration of the scattering process is of similar order as the decoherence time of the scatterer.

The theoretical models (C)-(E), which have a common conceptual basis, show that a reduction of the system's transition rate due to scattering is derived. This is tantamount to a shortfall of scattering intensity.

The results and their qualitative interpretation show that epithermal (say, 1–100 eV) neutrons being available at spallation sources (presently at ISIS/U.K.), and electron spectrometers with large scattering angles and associated large momentum transfers, provide novel tools for investigation of new attosecond physical and chemical phenomena in condensed matter and molecules.

2. Neutron Compton scattering

The experimental method utilized here is neutron Compton scattering (NCS), also called deep inelastic neutron scattering (DINS), provided by the electron-volt spectrometer Vesuvio (formerly eVS) at the ISIS neutron spallation source, Rutherford Appleton Laboratory, UK. This is an inverse-geometry spectrometer, meaning that the final neutron energy is fixed, see e.g. [7, 15].

Neutron scattering at high energy transfers($\hbar\omega \gtrsim$ 1eV) and momentum transfers ($\hbar\mathbf{q}$, with $q \gtrsim 20$ Å^{-1}) can be used to directly measure the distribution of momentum $\hbar\mathbf{p}$ of light atoms in condensed matter systems[16]. This experimental method (NCS) is analogous to the measurement of the electronic momentum distribution through Compton scattering of X- and γ-rays from electrons [17], or the nucleon momentum distribution via quasi-elastic electron scattering from nuclei. Here it is assumed that the scattering is essentially incoherent and the so-called *impulse approximation* (IA) applies [16, 18], and thus single-particle properties are probed [6, 7].

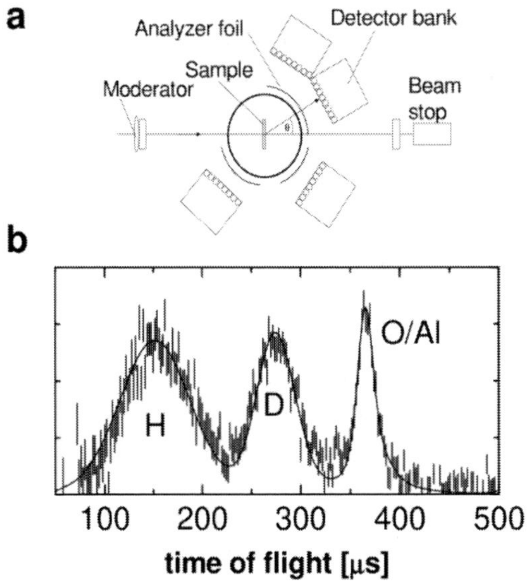

Figure 1. (a) Schematic representation of time-of-flight (TOF) spectrometer eVS/Vesuvio of ISIS; adapted from [15]. (b) A TOF-spectrum of a 20:80 H_2O-D_2O mixture in an Al can together with the corresponding fit (full line). The H and D recoil peaks are well separated from each other and from the joint O/Al peak; adapted from [4]. The peak area of atom X (i.e., H, D, etc.) gives the neutron-X scattering intensity.

A well founded brief review of the theory of NCS was given by Sears [18], which is as follows. In a time-of-flight (TOF) scattering experiment, cf. figure 1a, one determines from the measured spectra the partial (or: double) differential cross section and the associated dynamic structure factor $S(\mathbf{q}, \omega)$. For the scattering of a neutron by a system of N identical atoms it holds

$$\frac{d^2\sigma}{d\omega d\Omega} = Nb^2 \frac{k_1}{k_0} S(\mathbf{q}, \omega) \tag{1}$$

(b: bound scattering length of atom; k_0, k_1: absolute values of wavevectors of incident and scattered neutron; $d\Omega$: small solid angle subtended by the neutron detector in direction of k_1 at the target). $\hbar\mathbf{q}$ and $\hbar\omega$ are the momentum and energy transfers from the neutron to a scattering nucleus, respectively.; i.e, $\hbar\mathbf{q} = \hbar\mathbf{k}_0 - \hbar\mathbf{k}_1$, $\hbar\omega = E_0 - E_1$. The subscripts "0" and "1" refer to neutron quantities before and after collision, respectively.

The introduced potential of the neutron-nucleus interaction effectuating the neutron scattering is known as Fermi's pseudo-potential [19, 20, 21]

$$V(\mathbf{r}) = \frac{2\pi\hbar^2}{m}\, b\,\delta(\mathbf{r}) \tag{2}$$

(m: neutron mass) which approximates the short-range strong interaction. According to the basic van Hove theory [19] $S(\mathbf{q},\omega)$ is given by

$$S(\mathbf{q},\omega) = \frac{1}{2\pi}\int_{-\infty}^{\infty} \exp((-i\omega t)F(\mathbf{q},t)\, dt \tag{3}$$

where

$$F(\mathbf{q},t) = \frac{1}{N}\sum_{j,k}^{N}\langle \exp(-i\mathbf{q}\cdot\mathbf{r}_j(0))\exp(-i\mathbf{q}\cdot\mathbf{r}_k(t))\rangle. \tag{4}$$

is the so-called intermediate correlation function. $\langle ... \rangle$ denotes a thermodynamics average, and $\mathbf{r}_j(t)$ denotes the position operator of atom j at time t in the *Heisenberg* representation.

Here it should be pointed out that the whole derivations of the above formulas are based on Fermi's golden rule, which, for scattering processes, is equivalent to the first Born approximation [19, 20, 21]. Recall also that both these approximations are essentially equivalent to first-order perturbation theory. However, this theory does not hold for a singular potential like Eq. (2). The justification for the use of Fermi's golden rule in these circumstances is that, in combination with Eq. (2), it gives the required result of isotropic (s-wave) scattering for a single fixed nucleus; cf. [20, 21] and references therein.

In NCS $q = |\mathbf{q}| \gg 2\pi/d$, where d is the nearest-neighbor distance, and therefore one may neglect the $j \neq k$ terms in the sum and obtains

$$F(\mathbf{q},t) = \langle \exp(-i\mathbf{q}\cdot\mathbf{r}(0))\exp(+i\mathbf{q}\cdot\mathbf{r}(t))\rangle \tag{5}$$

where $\mathbf{r}(t)$ denotes the position operator of a scattering atom at time t in the Heisenberg representation. However, it should be pointed out that $\mathbf{r}(t)$ is a Heisenberg-operator, and thus it is an N-body operator, despite the (probably misleading) notation of "position operator of a scattering atom".

The last equation is referred to as the *incoherent approximation*. Here one neglects coherent scattering contributions, because the condition $q \gg 2\pi/d$ implies that the coherent superposition of scattering amplitudes from different atoms averages to zero over a small solid angle $d\Omega$ associated with the detector (neutron counter).

To introduce the *impulse approximation*, IA, one may proceed as follows. In classical mechanics one has $\mathbf{r}(t) - \mathbf{r}(0) = \int_0^t \mathbf{v}(t')dt'$, where $\mathbf{v}$ is the velocity of the scattering atom. In quantum mechanics, however, the Heisenberg operators $v(t')$ at different times do not commute. Nevertheless, as shown by Sears, it holds

$$F(\mathbf{q},t) = \exp(i\omega_r t)\left\langle \mathcal{T}\exp\left[i\mathbf{q}\cdot\int_0^t \mathbf{v}(t')dt'\right]\right\rangle \tag{6}$$

where $\mathcal{T}$ is the time-ordering operator and

$$\hbar\omega_r = \hbar^2 q^2/2M \tag{7}$$

represents the energy of recoil of a stationary nucleus with mass M due to the collision. For sufficiently large q, formally $q \to \infty$, the right-hand-side (rhs) of Eq. (4) is appreciably different from zero only if $t \to 0$, and thus $\mathbf{v}(t') \approx \mathbf{v}(0) \equiv \mathbf{v}$. Thus, for $q \to \infty$ one obtains

$$F(\mathbf{q},t) \to \langle \exp[i(\omega_r + \mathbf{q}\cdot\mathbf{v})t)]\rangle \equiv F_{IA}(\mathbf{q},t) \tag{8}$$

and therefore

$$S(\mathbf{q},\omega) \rightarrow \langle \delta(\omega - \omega_r + \mathbf{q}\cdot\mathbf{v})\rangle \equiv S_{IA}(\mathbf{q},\omega)\,. \tag{9}$$

The delta function expresses conservation of energy and momentum in the collision of a neutron with an atom having initial velocity $\mathbf{v}$, and follows from the assumed validity of Fermi's golden rule [20]. This formula represents the *impulse approximation* (IA). As pointed out by Sears, this approximation neglects interatomic forces in the final state, but not in the initial state, since the thermodynamic average $\langle ... \rangle$ refers to the interacting system.

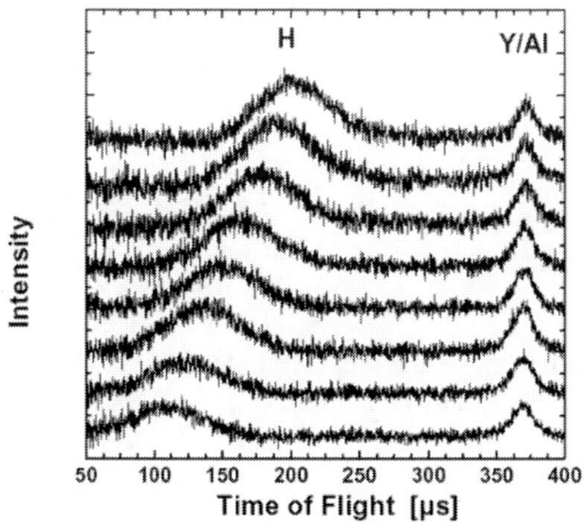

Figure 2. Time-of-flight (TOF) spectra of YH_3 at various scattering angles θ. (For clarity, the spectra are parallel shifted along the ordinate.) The peak separation between H and the heavier atom (here: Y and Al) increases with θ. The position of the H peak shows that the mass of the recoiling particle is 1 a.m.u.

It should be emphasized that the measured TOF-positions of the recoil peaks correspond to scattering from single nuclei. E.g., the peaks marked with "H" in figure 2 correspond to a recoiling particle of 1.0 atomic mass unit (a.m.u.), that is, to a single proton. The same holds also for NCS from molecular H_2 [22], that is, the scattering particle is a "single proton", and not a H_2 molecule "as a whole" (since the latter has the mass of 2 a.m.u.).

To simplify notations, and in accordance with standard notation used in the literature, the subscript IA is dropped in the following.

The dynamic structure factor in the IA can be expressed equivalently as

$$\begin{aligned} S(\mathbf{q},\omega) &= \int n(\mathbf{p})\,\delta\!\left(\hbar\omega + \frac{\hbar^2 p^2}{2M} - \frac{\hbar^2(\mathbf{p}+\mathbf{q})^2}{2M}\right) d\mathbf{p} \\ &= \int n(\mathbf{p})\,\delta(\hbar\omega - \hbar\omega_r - \hbar\mathbf{q}\cdot\mathbf{p}/M)\,d\mathbf{p}\,. \end{aligned} \tag{10}$$

M and $n(\mathbf{p})$ are the mass and momentum distribution of the scattering atom (nucleus) before collision, respectively. A useful feature of NCS is that the recoil peaks of light atoms (e.g. H, D, He) are well resolved from those of heavier atoms in the TOF-spectra (compare figure 1). In a typical NCS experiment, the characteristic time τ_q of a neutron-nucleus collision, termed "scattering time", is very short. For the neutron-proton collision τ_q is of the order $10^{-16} - 10^{-15}$; see below.

As the term $\mathbf{q}\cdot\mathbf{p}$ in Eq. (1) shows, the energy distribution of the scattered neutrons are directly related to the distribution of particle momenta $\hbar\mathbf{p}$ (before collision) parallel to the wavevector transfer $\mathbf{q}$. In fact, the nucleus is in a quantum state having a distribution of initial momenta, and the recoil peak will therefore be Doppler broadened.

In the IA can be shown [23] that, for each kind of struck particles of mass M, the two variables ω and $q = |\mathbf{q}|$ are explicitly coupled through the scaling variable y defined as[18, 23]

$$y = \frac{M}{\hbar^2 q}\left(\hbar\omega - \hbar\omega_r\right)\,, \tag{11}$$

which reduces Eq. (10) to

$$S_{\hat{q}}(q,\omega) = \frac{M}{\hbar q} J(y, \hat{\mathbf{q}}) \,, \tag{12}$$

with

$$J(y, \hat{\mathbf{q}}) = \int n(\mathbf{p}) \, \delta(\mathbf{p} \cdot \hat{\mathbf{q}} - \hbar y) \, d\mathbf{p} \tag{13}$$

where the neutron Compton profile (also termed "longitudinal momentum distribution"), $J(y, \hat{\mathbf{q}})$, is the distribution of the one-dimensional projection $\hbar y$ of atomic momentum $\mathbf{p}$ along the $\hat{\mathbf{q}}$-direction indicated by the dimensionless unit vector $\hat{\mathbf{q}}$ or, equivalently, by the scattering angle θ. This property is known as y- or West-scaling[18, 7, 23] and implies that all detectors yield the same $J(y)$, as pointed out above.

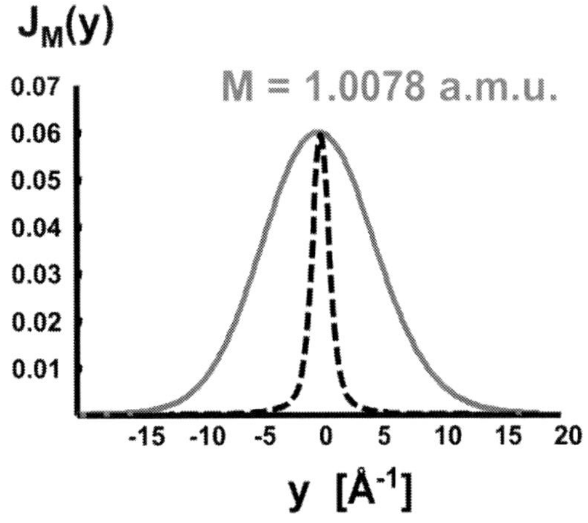

Figure 3. A typical Compton profile $J(y)$ of H at scattering angle $\approx 45°$. The broken line represents the experimental resolution function using the Au-foil analyzer, see [15].

For isotropic systems, mostly studied by NCS, the direction $\hat{\mathbf{q}}$ becomes immaterial and one writes $J(y)$. Thus, for large momentum transfers, the Compton profiles obtained from all detectors (which are set at different scattering angles θ) are expected to "collapse" onto one $\hat{\mathbf{q}}$-independent Compton profile $J(y)$. According to standard theory, this quantity, and the associated $n(p)$, depend on the effective Born-Oppenheimer (BO) potential $V(r)$ of the struck particle. For a harmonic $V(r)$, the associated $J(y)$ is Gaussian [6, 7], i.e.,

$$J(y) \propto \exp(-y^2/2\sigma^2), \tag{14}$$

where σ is the width of the longitudinal momentum distribution.

The standard theoretical results described above all presuppose that $J(y)$ can be described by a single-particle in a potential, that is, that the BO is valid [6, 7]. It is believed that NCS is the only technique capable of measuring momentum distributions and BO potentials directly. The explicit relations between a Compton profile $J(y)$, a momentum distribution $n(p)$ and the associated effective BO potential $V(r)$ are well known in standard NCS theory [6, 7].

2.1. The "anomalous" scattering effect

The first direct experimental evidence of short-lived (that is, attosecond) QE involving protons in condensed matter was provided by means of the novel neutron Compton scattering (NCS) method. Starting in 1995, our experiments on liquid water and H_2O-D_2O mixtures [4] have revealed the following striking effect: The intensity of neutrons scattered from protons exhibits a considerable "anomalous" shortfall which can be as much as 30%. This finding unequivocally contradicts conventional theory [4]. Further NCS experiments confirmed the existence of this effect in various condensed matter systems, e.g., metallic hydrides [24, 25], polymers [26, 27], "soft" condensed matter [28], and liquid benzene

[26]. Experimental evidence of the effect under consideration has been provided even in liquid H_2 and HD [22, 26].

We now shortly discuss the striking decrease of NCS-intensity from H, which is represented by the violation of the basic equation $R_{exp} = R_{conv}$, see below. As an example, let us consider the scattering results [4] from H_2O-D_2O mixtures. From a measured TOF spectrum, the data analysis procedure [15] determines the partial differential cross section $d^2\sigma/d\omega d\Omega$, Eq. (1). The relevant peak areas A_X (with X=H, D, O) are then extracted from $d^2\sigma/d\omega d\Omega$. Thus one can determine the ratio

$$R_{exp} \equiv A_H/A_X . \qquad (15)$$

According to standard NCS theory [6, 18] the conventionally expected value R_{conv} of this ratio is calculated with (cf. [4, 15])

$$R_{conv} = \frac{N_H b_H^2}{N_X b_X^2} . \qquad (16)$$

N_X is the number density of atom X. Instead of the conventionally expected equality $R_{exp} = R_{conv}$, however, the experimental results presented below (amd various others reported in the literature) show that this equation is strongly violated, i.e.,

$$R_{exp} < R_{conv} . \qquad (17)$$

This surprising effect has no interpretation in the frame of conventional theory [6, 7].

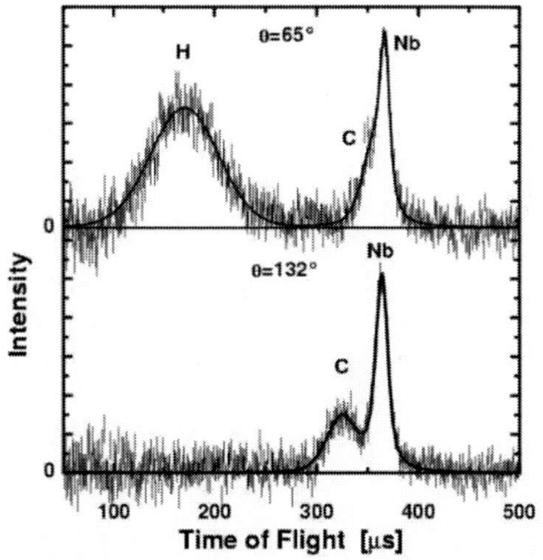

Figure 4. TOF spectra from liquid C_6H_6 in an Nb can, at T=295 K. The error bars are due to counting statistics only. The full lines denote the fitted TOF spectra. The positions of the peaks are not fitting parameters. For the scattering angle $\theta = 65^o$ the C- and Nb-recoil peaks overlap. For $\theta = 132^o$, however, the maxima of these two peaks are well resolved, thus facilitating the reliable determination of the peak intensities.

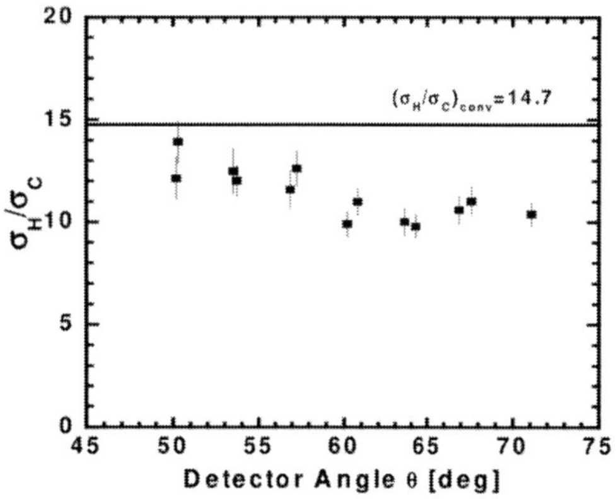

Figure 5. The experimentally determined ratios of neutron cross-sections of H and C, from the NCS spectra from liquid C_6H_6, for different scattering angles (positions of the detectors). The vertical line at 14.7 denotes the ratio of cross-sections $\sigma_H/\sigma_C = b_H^2/b_C^2$ according to conventional theory and tabulated values of these quantities.

2.2. NCS results from liquid benzene

For illustration of this "anomalous" intensity effect, earlier results obtained with the eVS-spectrometer from liquid benzene [26] are presented; see figures 4 and 5.

The expected ratio of scattering intensities from H and C, R_{conv} is equal to the tabulated values of corresponding cross-sections $\sigma_X = 4\pi b_X^2$ (with $X = H, C$). The measured ratio R_{exp} is about 25% smaller than R_{conv}.

3. NCS from a single crystal $KHCO_3$: momentum distributions and intensities

We present a striking new result of our current analysis of NCS data [29] obtained from a single crystal of $KHCO_3$ at room temperature with the eVS/Vesuvio spectrometer. $KHCO_3$ is a prototypical system for proton transfer dynamics. In a single crystal, all H atoms belong to OH bonds and all these OH bonds are virtually parallel to each other throughout the crystal. Therefore, with a single crystal properly oriented with respect to the neutron beam it is possible to probe the proton dynamics specifically along each degree of freedom of H vibrational motion. Figure 6 shows results of current investigations, which compare for the first time:

(a) the measured kinetic energy of H (which is proportional to σ_H^2, where σ_H is the width of H-Compton profile $J_H(y)$) at a given scattering angle, which is determined from the measured Compton profile $J_H(y)$ by standard theory [6, 7],
with

(b) the associated measured intensity of the H-recoil peak at given scattering angle (normalized by the sum of the scattering intensities of all heavier peaks, i.e. C, O and K.

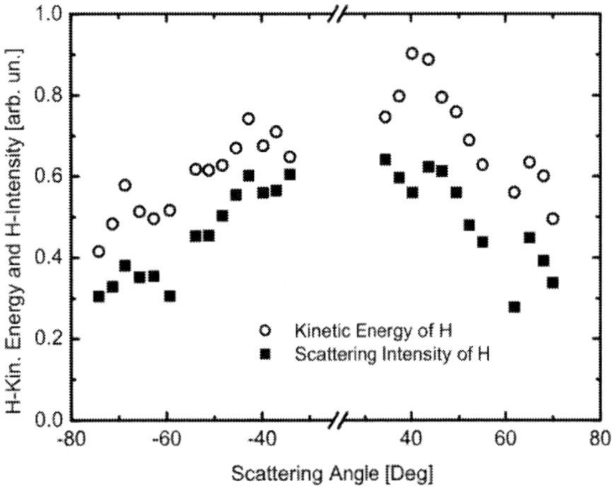

Figure 6. NCS from a single crystal $KHCO_3$. Measured NCS intensities from H (normalized with those of all heavier peaks) and measured kinetic energies of H, at various scattering angles. Negative (positive) angles: left (right) to beam direction. The H recoil peaks are well separated from the "common" recoil peak of C, O and K. Experimental data from [29].

These new results are quite remarkable for several reasons:

(c) The scattering angle (and thus momentum transfer q) dependences of both quantities are very strong. The higher kinetic energies at small (positive and negative) scattering angles are due to the simple fact that the momentum transfer q — which varies over the TOF-H-peak — is chosen to be roughly parallel to the OH bonds, which have a "stronger" local BO potential (and thus also higher vibrational energy, which is well known from conventional vibrational spectroscopy). At larger (positive and negative) scattering angles, q and the OH-direction build a large angle, and thus q probes another (and much weaker) local BO potential. Thus the strong scattering-angle dependence of H-kinetic energy is qualitatively well understood.

(d) However, the strong scattering-angle dependence of the H-peak intensity I_H contradicts the standard NCS-theory [6, 7], according to which I_H must remain constant for all angles. This result represents the novel "anomalous" scattering intensity effect under consideration.

(e) Moreover, and most surprisingly, the comparison of the two quantities shown in figure 2 shows a clearly visible, strong correlation between them, which remained unnoticed until now. Qualitatively speaking, the "anomalous" intensity deficit from H seems to become stronger for "broader" BO potentials and the associated more delocalized protons. In other terms, it obviously exists a physical connection between H-Compton profile ($J(y)$) and H-intensity (I_H). This finding appears to be in blatant contrast to the associated statement of Ref. [7], see p. 463.

4. Remarks on electron-atom Compton scattering experiments

The "anomalous" scattering effect under consideration has also been validated by electron-atom Compton scattering (ECS) at large momentum transfers. The experiments have been taking place in Australia [5, 30, 31], and most recently in Canada [32]. The experiment [32] may be of particular interest to theoreticians: H_2 is the best theoretically investigated molecule, and the scattering protons are those of isolated H_2 molecules in the gas phase. A short description of the results is as follows.

ECS from gaseous H_2, D_2, a 50:50 mixture of H_2 and D_2, and HD was investigated with 2.25 keV impact energy and $100°$ scattering angle, corresponding to a momentum transfer with ca. 40 $Å^{-1}$. In contrast to the typical conditions of NCS, the energy transfers applied here were less than the dissociation energies of the molecules. The spectral positions of the H and D recoil peaks follow standard Rutherford scattering theory. However, the scattering intensities contradict standard theory. Namely, in the spectrum of the 50:50 H_2-D_2 mixture, the integrated intensity of the H peak was about 30% lower (as compared to that of D) than predicted by Rutherford scattering, despite equal screening of nuclear charges by the electrons in all molecules. In contrast, the ratio of scattering intensities from HD was found to agree with Rutherford scattering. (This represents an interesting difference to the NCS experiment [22], and is currently under theoretical investigation.) Recently, a detailed theoretical analysis by Bonham et al. demonstrated that this striking effect cannot be explained by scattering theory calculations based on the first Born and Born-Oppenheimer approximations [33].

Very recently, these novel experimental findings have been criticized to be instrumental artifacts, and it was speculated that the anomaly is caused by differences in the molecular speeds of H_2 and D_2 [34]. However, these criticisms have been refuted by further experimental investigations [35]. Especially, the measured decrease of H-intensity of about 30% could be expected if the sample were an effusive gas jet which allowed free escape of the sample gas molecules. This is not the case in the ECS experiments since a gas cell was used [32, 35]. Moreover, if such a molecular speed effect were the cause of the observed effect, then the simultaneously performed $(e, e + ion)$ mass spectra would show the same anomaly since the $(e, e + ion)$ coincidence experiment involves the same gas cell collision region, the same electron transfer and focusing lenses, and the same electron detection hardware as the energy loss function of the spectrometer [35]. Based on these facts, the conclusion of [35] was that the arguments advanced in [34] to "explain" the experimental quasi-elastic H_2-D_2 relative intensity anomaly under consideration, are erroneous.

5. Criticisms and their experimental tests

Considerable efforts to identify possible sources of NCS-experimental errors have been made during the last 10 years; for an account in detail of data analysis and experimental procedures and tests, see [15].

Senesi et al. [36] investigated all the criticisms raised in [37] concerning the effects of instrumental resolution and filter absorption profile, by applying for the first time the so-called "exact method" of data analysis proposed [37]. Data obtained from solid HCl at $T = 4$ K were analyzed. The results demonstrated unequivocally the presence of the same anomaly in the scattering from protons, in agreement with the results of the standard data analysis procedure of ISIS [36].

The standard method of NCS-data processing [7, 15] was recently compared with a newly proposed (by B. Dorner, ILL) model-free data-analysis procedure. This procedure is independent of the form of the momentum distribution and the resolution function [38]. In [38], the original results from the metallic hydride $NbH_{0.8}$ [24] were analyzed. The comparison of results obtained with the mentioned two qualitatively independent methods underline the importance of the effect under consideration.

The differential cross-section $d\sigma/d\Omega$ of neutrons in the 24–150 keV incident energy range scattered from H_2O relative to that of D_2O and H_2O–D_2O mixtures was reported recently in Ref. [39], claiming that the measured neutron scattering intensity ratios in the keV range exhibit no anomalous behavior. It should be stressed that this experiment in not equivalent to the standard NCS experiments which measure the *partial* differential cross-section. Despite this fact, however, a detailed scrutiny of [39], for single [40] and multiple [41] neutron-nucleus scattering events, demonstrated that the data analysis procedure in [39] was erroneous: We find that an anomalous reduction of H_2O scattering intensity (relative to that of D_2O) of about 20% is still present in the original experimental data.

Most recently, Stock et al. [42] presented a different approach to our "anomalous" scattering effect by investigating the hydrogen cross section in polyethylene (PE) using the direct geometry time-of-flight spectrometer MARI of ISIS with the incident energy fixed at a series of values ranging from $E_0 = 0.5$ eV to 100 eV. They claim that their measurements do not show any anomaly and thus contradict the results obtained at eVS/Vesuvio. However, the experimental method and data processing of [42] were scrutinized in great detail by Mayers et al. [43], who showed that, contrary to the claims of the authors of Ref. [42], the MARI energy resolution at eV energies is greatly inferior (i.e., by 1-2 orders of magnitude) to that on eVS/Vesuvio. This crude resolution of MARI in the aforementioned energy range invalidates the claim in [42] to have a unique demonstration that the "anomalous" intensity effect (or: intensity deficit) in the Compton profile measured on eVS/Vesuvio is an experimental artifact.

6. Collisional time window: attosecond scattering time

In the context of NCS, in which the effectuated momentum and energy transfers are large,

$$\Delta E \approx 1 - 100\,\text{eV}\,, \quad |\mathbf{q}| \sim 20 - 200\ \text{Å}^{-1}\,, \tag{18}$$

the Impulse Approximation (IA) is valid [18, 6] and the characteristic time scale — often termed "scattering time", τ_{sc} — of the neutron-proton scattering process is very short [5, 24],

$$\tau_{sc} \sim 100 - 1000\ \text{as} \tag{19}$$

(as: attosecond). These sub-femtosecond scattering times are a consequence of the large energy and momentum transfers attained with the Vesuvio instrument, Eqs. (18), and they follow from the theoretical result valid in the IA [18, 6]

$$\tau_{sc}\,|\mathbf{q}|\,v_0 \approx 1\,, \tag{20}$$

where v_0 is the root-mean-square (rms) velocity of the nucleus and $\hbar\mathbf{q}$ is the momentum transfer from the neutron to the proton. The time τ_{sc} is given by the t-width of the intermediate correlation function $F(q,t)$, which is connected to the dynamic structure factor $S(q,\omega)$ by Fourier transform [18]

$$S(q,\omega) = \frac{1}{2\pi}\int_{-\infty}^{\infty} \exp\left(-i\omega t\right) F(q,t)\,dt\,. \tag{21}$$

Additionally, as shown by Gidopoulos within first-order time-dependent perturbation theory[46], another physical meaning for τ_q is the inverse of the energy spread ΔE_{Ψ_q} of the nuclear wavepacket Ψ_q after collision, i.e. $\tau_q = \hbar/\Delta E_{\Psi_q}$. Also this physical view provides a similar scattering time (19).

To shed more light upon the issue of "relevant scattering time", one may also refer to the Margolus-Levitin theorem [44]. Let us consider the neutron-proton system during the collisional process. Obviously, the initial and final states of it are very different and thus orthogonal to each other. The theorem asserts that in takes at least a time $T_\perp \geq (\pi\hbar)/(2E_s)$ for the system to evolve from its initial to any orthogonal final state. E_s is the system's average energy minus its ground state energy. $T_\perp$ provides a strict bound for the considered dynamical process [44]. Note that in NCS one has $E_s \approx E_0$ and thus E_s is larger than, but of similar order as, the aforementioned energy spread ΔE. Thus it is revealing that also this characteristic time $T_\perp$ is very similar to the aforementioned scattering time, i.e. $T_\perp \sim \tau_{sc}$.

A numerically similar estimate can be obtained with the relation

$$\Delta_E \tau_0 \geq \pi\hbar/2 \tag{22}$$

by Ballentine [45]. τ_0 denotes the shortest time required for the state vector $|\psi(t)\rangle$ to become orthogonal to the initial state, and Δ_E is a standard measure of the statistical spread of energy in the state (e.g. the square root of the variance). Thus we have $\tau_{sc} \approx \tau_0$, which is satisfactory.

It is interesting to note that the intuitively expected "actual duration" of a neutron-proton interaction via the strong interaction may be even shorter, as a classical estimate indicates. E.g. a neutron with kinetic energy $E_0 \approx 10$ eV (and associated velocity v_0) will pass a distance $R_{strong} \approx 10^{-5}$ Å (i.e. the range of the strong interaction) in a much shorter time, i.e.

$$\Delta t_{strong} = R_{strong}/v_0 \approx 10^{-19} - 10^{-20} \text{ s.}$$

However, this is not in conflict with the above estimate, for the following reason. As standard theory shows [6], $S(q,\omega)$ is peaked around the nuclear recoil energy $E_q = \hbar^2 q^2/2m$. The scattering time τ_{sc} is the inverse of the width ΔE of $S(q,\omega)$, and $S(q,\omega)$ plays the role of the probability density distribution for transferring energy $\hbar\omega$ from the neutron to the proton, when the momentum transfer is $\hbar q$. It then holds $\tau_{sc} \approx \hbar/\Delta E$. For a typical value $\Delta E \approx 10$ eV, one gets $\tau_{sc} \approx 10^{-16} - 10^{-17}$ s. In other words [46](b):

> The scattering time τ_{sc} gives a statistical measure of the length of the time interval during which an elementary neutron-proton collision may occur, in the same way that the spatial extent of a particle wavefunction (or wavepacket) gives a statistical measure of the extent of the region in which the particle may be found.

7. Causality: open quantum systems

In the light of Relativity Theory, this has a crucial consequence: The effective scattering system — i.e., a recoiling proton and some of the adjacent electrons — must have a linear dimension smaller than

$$\Delta s_{caus} = c \cdot \Delta t_{strong} \lesssim 0.3\,\text{Å}$$

(for the above mentioned example with $E_0 = 10$ eV), where c is the velocity of light, since the neutron-nucleus scattering dynamics during Δt_{strong} cannot be causally influenced by other particles being more than R_{strong} (i.e. the range of the strong force) apart from the colliding nucleus. Consequently, the scattering system must necessarily be of "microscopic" dimensions; it contains the scattering nucleus and a part of the adjacent electron density.

In other terms, causality implies that any cause of a single neutron-proton Compton scattering event can not be situated farther than Δs_{caus} from the struck nucleus. In this context it should be noted that a typical covalent bond H-X (with X= C, O, N, etc.) is about 1 Å and therefore the atom X is beyond the causality horizon defined by Δs_{caus}, i.e. the X-nucleus has no influence on the quantum dynamics of the neutron-proton Compton scattering event.

For example, the internuclear distance of the H_2 molecule is about 0.75 Å. Consequently, it is only one proton and a part of the electrons' degrees-of-freedom that is/are causally participating in the collision process; it is not the whole molecule that is interacting with the neutron. This thus implies that the usual approximation of a H_2 molecule to be a *closed* system interacting with a neutron contradicts the principle of causality.

Summarizing, the systems we examine in NCS experiments must be treated — for fundamental reasons — as open quantum systems.

8. Origins of entanglement: theoretical remarks

Since conventional theory cannot explain the observed anomalies, it has been suggested to extend the existing theory to allow for a treatment of neutron scattering by quantum entangled particles [4]. However, the theoretical models existing until now appear to differ in important theoretical aspects:

(i) the relevance of quantum exchange correlations ([47] and earlier references cited therein), and

(ii) the relevance of decoherence of quantum entangled states during the scattering process [13, 14].

(iii) In this context one should also mention the suggested interpretation based on excitation of electronic degrees of freedom [46, 49, 50].

For details about the possibility of excitation of electronic degrees of freedom in a neutron-nuclear collision due to non-adiabatic effects, i.e. the breakdown in the Born-Oppenheimer (BO) approximation, see [46, 49]. Obviously, in this case it would be impossible to determine one-particle effective BO potentials, as standard NCS theory claims [6, 7]. It should also be noted that the predicted additional peaks due to these excitations have not been observed in the experimental TOF spectra of any sample investigated thus far.

Quantum exchange correlations, mentioned in (i), represent a well known effect of standard quantum mechanics, which applies to two (or more) identical particles and leads to quantum entanglement between spatial and spin degrees of freedom, as e.g. in the well-known case of ortho- and para-H_2. These two states of molecular hydrogen have different total neutron scattering cross-sections for slow (thermal and/or cold) neutrons [20]. It was suggested by Karlsson [47, 48] that exchange correlations between proton pairs – even the two H atoms belong to different molecules, e.g. in liquid HD – can be the cause of the "anomalous" scattering effect under consideration, under the condition that they fall within the coherence volume of the neutron wavepacket; see [47] and papers cited therein for details.

However, the variety of physical conditions characterizing protons in the investigated condensed systems indicate that exchange correlations cannot play the dominant role in the context of NCS. E.g., the considered effect appears, with roughly similar magnitude, even in systems where the protons occupy non-equivalent (and thus distinguishable) sites, like polystyrene [26] and amphiphiles [28]. Furthermore, our NCS investigations of the heteronuclear hydrogen molecule HD (with no exchange correlations) and of H_2 (which exhibits such correlations) reveal that the considered QE-effect does appear in both samples even with the same magnitude [22].

Moreover, our current analysis of NCS-data from a single-crystal sample of $Rb_3H(SO_4)_2$ [work in progress] reveals a strong "anomalous" intensity effect, although the minimal distance between two H atoms in the crystal is 5.07 Å. This distance is larger than the coherence length of the neutron wavepacket (estimated to be about 2.5 Å in the applied instrumental setup, see [48]).

Alternatively, QE in condensed matter may be caused by many-body interactions. This follows from first principles of quantum mechanics. Let Ψ_A and Ψ_B be the state vectors of two systems A and B, which may be assumed either as indistinguishable (thus exhibiting exchange correlations) or not. The two systems with Hamiltonians H_A and H_B shall interact either directly or indirectly (via a third subsystem, e.g. environmental electronic charges), the corresponding Hamiltonian being V_{AB}. As usually assumed, let the state of the composite system 'A and B', in short AB, at time $t = 0$ be not entangled, i.e. $\Psi_{AB}(0) = \Psi_A \cdot \Psi_B$. In most cases, V_{AB} does not commute with H_A and H_B. Then it holds for $t > 0$ that the complete evolution operator of the composite system AB,

$$U_{AB}(t) = \mathrm{e}^{-\frac{i}{\hbar}(H_A + H_B + V_{AB})t} \tag{23}$$

does not factorize into a product of two 'individual' evolution operators, say U_A and U_B, representing the dynamics of the individual (isolated) systems A and B. Consequently, the wave function $\Psi_{AB}(t)$ at time t, $\Psi_{AB}(t) = U_{AB}(t)\Psi_{AB}(0)$, does not factorize into a product of two wave functions (say, Ψ'_A and Ψ'_B), each of them representing the state of one of the two systems A and B; i.e.

$$\Psi_{AB}(t) = U_{AB}(t)\Psi_{AB}(0) \neq \Psi'_A(t) \cdot \Psi'_B(t)\,. \tag{24}$$

In other words, $\Psi_{AB}(t)$ represents an entangled state. This indicates that *spatial* QE is of particular interest, because the strongest interactions in the systems under investigation are given by Coulombic interactions, which depend on the spatial degrees of freedom of the participating particles (i.e., nuclei and electrons).

8.1. Theoretical models

We consider this "dynamical origin" of quantum entanglement to play the dominant role in the context of our NCS scattering experiments. It should be noted that this type of (creation, control and/or manipulation of) QE is also used and/or addressed in all experimental and theoretical fields related with quantum computing and/or quantum information [3].

The following sections of the paper deal with the theoretical treatment and interpretation of the striking attosecond effect under consideration. Various theoretical approaches are presented, which however are based on a common physical context, i.e. entanglement and its fast decoherence in open quantum systems (like molecules and/or condensed matter systems) at ambient experimental conditions.

9. Quantum Zeno effect and reduced scattering intensity

The so-called "Zeno paradox", or quantum Zeno effect, is a well known quantum mechanical phenomenon, theoretically introduced by Misra and Sudarshan [8, 9, 10]. A short outline is as follows: Let $|\phi\rangle$ be the state vector of a quantum system at time $t = 0$ and H the relevant Hamiltonian. Putting $\hbar = 1$, the time evolution of the state vector, $|\phi(t)\rangle$ is given by $\exp(-iHt)||\phi\rangle$, and the survival probability $S(t)$ for the system to "remain" in the initial state is then

$$S(t) = |\langle\phi|\exp(-iHt)|\phi\rangle|^2$$

If now one considers small times t, a power series expansion is possible:

$$\exp(-iHt) \approx 1 - iHt - \frac{1}{2}H^2t^2 + ...$$

so that the survival probability becomes

$$S(t) = |\langle\phi|\exp(-iHt)|\phi\rangle|^2 \approx 1 - (\Delta H)^2t^2$$

where

$$(\Delta H)^2 \equiv \langle\phi|H^2|\phi\rangle - \langle\phi|H|\phi\rangle^2 .$$

That is, $S(t) \to 1$ for sufficiently small times $t \to 0$.

This leads to the following considerations. The time evolution leads from $|\phi(t)\rangle$ to a linear superposition of available quantum states. For example, an excited atomic state may evolve into a superposition of itself and the (stable) unexcited states, emitting a photon.) A measurement to determine whether the initial state survives can be formally described as an instantaneous projection of the superposition back onto the initial state $|\phi(t)\rangle$. (This is usually termed projective or von Neumann-type measurement.) Therefore, in this process, quantum phase relations between the involved states are destroyed, i.e. decoherence takes place. From these considerations follows that, if such projective measurements are done repeatedly at times 0, Δt, $2\Delta t$, $3\Delta t$, ... $n\Delta t$, ..., then the initial quantum state $|\phi\rangle$ is hindered and evolves slower, and in the limit $\Delta t \to 0$ the time evolution becomes completely interrupted.

There exists a vast literature on the quantum Zeno effect, in which various scenarios of the mentioned projective measurements are proposed and/or analyzed. In the present NCS experimental context, of particular interest are *continuous* measurements which may also lead to the Zeno effect [9, 10]. Here, the time evolution of the neutron-proton collision process is continuously "observed", or continuously "measured" by the adjacent particles (electrons and nuclei), i.e. the "environment".

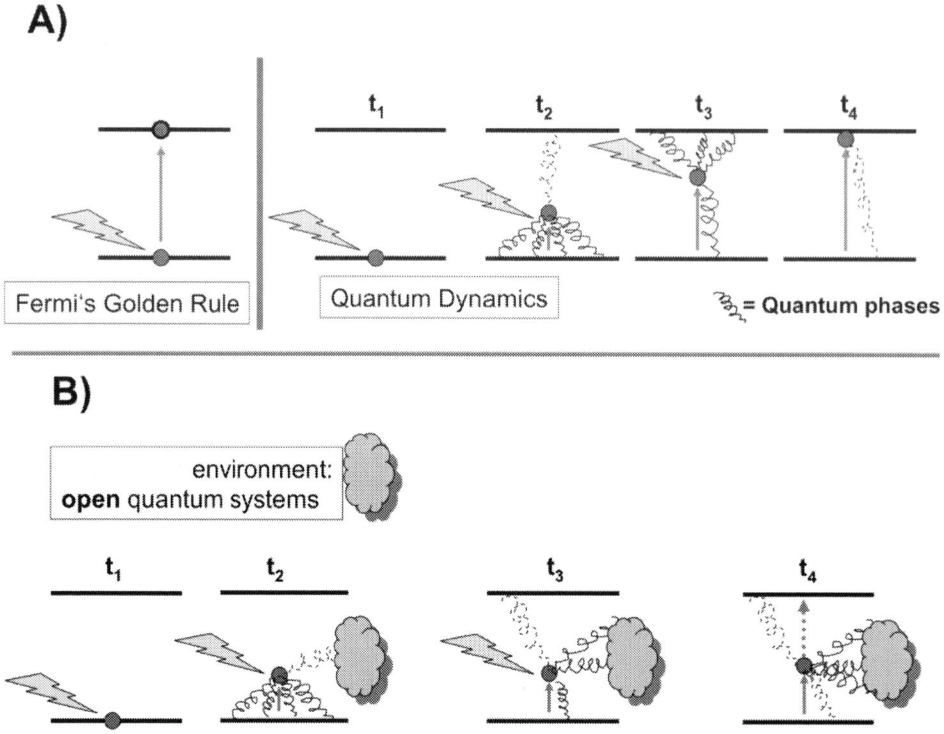

Figure 7. Illustrative description of the Zeno effect. Closed system, left panel A): Fermis' golden rule illustrated as a jump from initial (Ψ_i) to final Ψ_f state of the impinging neutron and the recoiling atom. Closed system, right panel A): The dynamical process of the transition creates quantum phases between Ψ_i and Ψ_f which develop in the course of time. Open system, panel B): The system's interaction with the environment destroys (or hinders) the dynamical development of quantum phase relations between Ψ_i and Ψ_f, which is tantamount to a reduced transition rate (and associated scattering neutron intensity).

Moreover, the continuous observation can be incorporated in the Hamiltonian, and thus one dispenses with instantaneous quantum jumps [9]. E.g., a neutron-proton scattering process from H_2 becomes continuously observed by the "environment", i.e. the two electrons and the second proton which interact with the struck proton during the whole collisional process. In qualitative terms, this leads to the aforementioned destruction of quantum phases and the associated hindered transition to the final scattering state(s). Obviously, this is tantamount with a reduced neutron-proton scattering intensity and/or cross-section.

This physical picture also shows that, in the considered physical context, the "time window" τ_{sc} of scattering must be sufficiently short, of the order of the fast evolving quantum phase relations (or: coherences) of the collisional system "neutron plus proton plus environment".

The cartoon of figure 7 may help to illustrate this process. The "environment" of the neutron-proton collisional system is represented as a "cloud" interacting with the system.

10. Schulman–Gaveau effect and reduced scattering intensity

Further insight into the quantum-phase destruction mechanism described above can be provided with the aid of the Schulman–Gaveau effect [11, 12]. This work permits a first principles interpretation of, and provides new physical insight into, the reduced scattering intensity phenomenon [4, 5], which may be outlined as follows.

A probe particle (neutron, with free Hamiltonian H_n) of sufficiently high energy impinges on a molecule (condensed system, etc., with Hamiltonian H_{Sys}). We shall consider scattering from a proton p (with free Hamiltonian H_p) of this molecule (system) and thus write its Hamiltonian as $H_{Sys} = H_p + H_{Rest}$. Let the neutron-proton interaction of this scattering be $g\widehat{V_{pn}}$, g being a c-number representing the coupling strength. The full Hamiltonian is then

$$H = H_n + H_p + H_{Rest} + g\widehat{V_{pn}} \,. \tag{25}$$

[In the notation of [11] we have: $H_A = H_n$ and $H_B = H_p + H_{Rest}$. To make contact with [11], the coupling Hamiltonian can be formally considered to be of the form $\widehat{V_{pn}} = V_n \otimes V_p$, with V_n and V_p some operators of the neutron and the proton, respectively. In specific cases, the coupling strength g may be explicitly time-dependent. To make contact with [12], Sect. V, Eqs. (70,74), the coupling Hamiltonian may be a function of the $p - n$ distance $|r_p - r_n|$, i.e. $\widehat{V_{pn}} = W(|r_p - r_n|)$; r_X: position of particle X.]

As discussed above, the scattering (interaction, collisional) time, τ_{sc}, of the neutron-proton interaction in the experiments [4, 5] lies in the sub-femtosecond time scale. This supports the present physical assumption (or, in other words, the working hypothesis) that the effect of Schulman and Gaveau [11] can be applicable here.

The following remarks may be helpful. The standard (i.e. well established and widely applied) neutron scattering theory, which is based on the first Born approximation, is known as the van Hove theory; see e.g. [20]. This is the only theory applied to thermal and cold neutron scattering, which are "slow" scattering processes as compared to NCS. As first-order perturbation theory is used, here there is no notion at all of entanglement between the neutron and the scatterer (proton, system) after the collision. Thus this general theory corresponds to the case of a factorizing final-state density operator, i.e.

$$\rho(t) = \rho_n(t) \otimes \rho_{Sys}(t) \quad \text{for} \quad t \geq \tau_{sc}. \tag{26}$$

Let us assume here that in the sub-femtosecond (NCS) experiments [4, 5] the above assumed factorization is physically incorrect. This is due to the fact that even the electronic rearrangements and/or relaxation, the fastest relaxation processes in condensed and/or molecular matter, lie in the sub-femtosecond timescale too. That is, we assume here that the exact total density operator $\rho(t)$ is entangled and denote the difference as

$$\Delta\rho(t) \equiv \rho_n(t) \otimes \rho_{Sys}(t) - \rho(t) \tag{27}$$

where the partial trace $\rho_n(t) = Tr'\rho(t)$ is introduced, with Tr' the trace over all other variables up to n; and correspondingly $\rho_p(t) = Tr'\rho(t)$.

It is shown in [11, 12] that

$$\Delta H = Tr[\Delta\rho(t)H] = Tr[\Delta\rho(t)(g\widehat{V_{pn}})] > 0, \tag{28}$$

under rather general conditions for the interaction Hamiltonian. That is, if one would evaluate the energy using the exact density operator $\rho(t)$ it would be unchanged from its time-0 value. However, if $\rho(t)$ is replaced by the uncorrelated density operator $\rho_n(t)\otimes\rho_{Sys}(t)$, then the energy could change, and for short times it in fact increases. This result is shown [12] to be more general than was quoted in the original paper [11]. These results demonstrate that operations that seem quite innocent, in particular the erasure of certain entanglement, can have significant effects.

Proposed connection with NCS experiments. — This result is the crucial point in the following comparison between results from conventional "slow" neutron scattering experiments (i.e. thermal and cold neutron scattering) and from the new "fast" NCS experiments. The "anomalous" NCS effect can be understood by noting that the scattering intensity I_H from protons (and the associated cross-section σ_H)

may be assumed to be a momotonous function of the expectation value of the interaction Hamiltonian squared:

$$I_H \sim |\langle \widehat{gV_{pn}} \rangle|^2 \,, \tag{29}$$

or the time average of this quantity over the duration of the scattering process, for time-dependent $g(t)$ and/or time-dependent density operator.

In the conventional "slow" scattering experiments the density operator is assumed to be not entangled, $\rho_n(t) \otimes \rho_{Sys}(t)$. This should be a good approximation, because decoherence is an extremely fast process in strongly interacting systems (like condensed matter or molecules).

In the "fast" NCS experiments we may assume that this decoherence mechanism does not destroy the quantum entanglement (QE) completely, and therefore let us denote the density operator by ρ_{QE}.

Thus using the Schulman–Gaveau result we conclude that

$$Tr[\rho_n(t) \otimes \rho_{Sys}(t)\, \widehat{gV_{pn}}] > Tr[\rho_{QE}\, \widehat{gV_{pn}}] \tag{30}$$

and equivalently

$$I_H(\text{slow scattering}) > I_H(\text{NCS}) \,. \tag{31}$$

This difference is a natural consequence of the decoherence, as revealed by the Schulman–Gaveau effect, and does not depend on any specific property of the interaction Hamiltonian. This point was already noted in [5], in which results of neutron and electron scattering from protons were compared. Obviously, if one denies the existence of decoherence in the NCS physical frame [51], the last inequality becomes an equality and, therefore, the "anomalous" scattering effect should "disappear". However, this difference can be considerable; cf. the novel experimental results presented in figure 6.

11. Introducing entanglement and decoherence into conventional formalism

In this section, we "extend" the standard expression Eq. (1) of the partial differential cross-section of conventional scattering theory [19, 20, 6, 7] by introducing ultrafast entanglement and decoherence of final and initial states of the scattering system. Recall that decoherence is an unknown phenomenon in the frame of conventional theory. Here we follow the presentation of [13].

To simplify notations, let us temporarily assume that the condensed system consists of N atoms of the same kind only, say, hydrogen atoms. The starting point is the expression for the partial differential cross section, Eq. (1), which we shall consider here in more detail.

The number of neutrons (I_{sc}) with fixed initial energy E_0, scattered per second into a small solid angle $d\Omega$ (in some given direction) with final energy between E_1 and $E_1 + dE_1$ (and thus $dE_1 = d\omega$),

$$I_{sc}(E_1, \Omega) \propto \frac{d^2\sigma}{d\Omega dE_1} \,, \tag{32}$$

i.e., it is proportional to the partial differential cross section [20]

$$\frac{d^2\sigma}{d\Omega dE_1} = \frac{k_1}{k_0} \sum_{\nu\nu'} W_\nu \Big| \sum_j b_j \langle \nu' | \exp(i\mathbf{q}\cdot\mathbf{r}_j) | \nu \rangle \Big|^2 \delta(E_\nu - E_{\nu'} + E_0 - E_1) \,. \tag{33}$$

b_j is the bound scattering length of atom j $(= 1, ...N)$. ν and ν' refer to the initial and final stationary states of the whole N-body scattering system (with energies E_ν and $E_{\nu'}$, respectively), corresponding to the transition $|\nu\rangle \to |\nu'\rangle$.

Recall that this formula is derived from Fermi's golden rule, after introducing the additional delta-function, which represents strict energy conservation [20, 21]. (See also additional remarks in the Discussion section below.)

The sum $\sum_j$ contains N terms. So the square of its absolute value, $|\sum_j|^2$, is the sum of N^2 terms of which a typical member is

$$\begin{aligned} M_{jj'(\nu\nu')} &\equiv b_j b_{j'} \langle\nu'|\exp(i\mathbf{q}\cdot\mathbf{r}_j)|\nu\rangle\langle\nu|\exp(-i\mathbf{q}\cdot\mathbf{r}_{j'})|\nu'\rangle \\ &= b_j b_{j'}\, Tr\left[\rho_{\nu'}\{\exp(i\mathbf{q}\cdot\mathbf{r}_j)|\nu\rangle\langle\nu|\exp(-i\mathbf{q}\cdot\mathbf{r}_{j'})\}\right] \\ &= b_j b_{j'}\, Tr\left[\rho_{\nu'}\exp(i\mathbf{q}\cdot\mathbf{r}_j)\,\rho_\nu\exp(-i\mathbf{q}\cdot\mathbf{r}_{j'})\right] \\ &= b_j b_{j'}\, Tr\left[\rho_{\nu'}\,V_j\,\rho_\nu\,V_{j'}^\dagger\right] \end{aligned} \tag{34}$$

where $V_j \equiv \exp(i\mathbf{q}\cdot\mathbf{r}_j)$, $V_j^\dagger \equiv \exp(-i\mathbf{q}\cdot\mathbf{r}_j)$ and $\rho_{\nu'} \equiv |\nu'\rangle\langle\nu'|$, $\rho_\nu \equiv |\nu\rangle\langle\nu|$. $Tr[...]$ denotes the trace operation.

In the physical context of NCS one has $q \gg 2\pi/d$, where d is the nearest-neighbor distance of two scattering nuclei. As a consequence, the spatial scale of the scattering event, represented by $1/q$ (where $q = |\mathbf{q}|$), is too small for one to detect interference effects due to scattering from pairs of different nuclei, $j \neq j'$. This is an important point, because sometimes one confuses the limiting case of large momentum transfers — which, unfortunately, is commonly called "incoherent approximation" [6, 18]— with the well-known "incoherent scattering component" (which constitutes only one part of the total scattering intensity) appearing in scattering of thermal [20] or cold neutrons.

Thus all terms with $j \neq j'$ do not contribute to the measured scattering intensity, and therefore one obtains from Eq. (34)

$$M_{jj(\nu\nu')} = b_j^2\, Tr\left[\rho_{\nu'}\,V_j\,\rho_\nu V_j^\dagger\right] \quad \text{or} \quad M_{(\nu\nu')} = b^2\, Tr\left[\rho_{\nu'}\,V\,\rho_\nu V^\dagger\right]\,. \tag{35}$$

The index j may be dropped, since all particles are by assumption equivalent. Since the quantity $V\rho_\nu V^\dagger = |V\nu\rangle\langle\nu V|$ is a projector (thus having eingenvalues 0 or 1) and $\rho_{\nu'}$ is a density operator (thus having non-negative eigenvalues), the product $(\rho_{\nu'}\,V\,\rho_\nu\,V^\dagger)$ must be hermitian and positive semidefinite. Consequently, $Tr\left[\rho_{\nu'}\,V\,\rho_\nu\,V^\dagger\right] \geq 0$ and $M_{(\nu\nu')} \geq 0$. Thus, in the conventional theoretical expression given by Eq. (33), and for sufficiently large momentum transfers (so that $j = j'$), occur only non-negative terms.

The initial and final states, ν and ν', refer to the whole (closed) N-body system and are assumed to be eigenstates of the system's complete Hamiltonian H_c [20]. Each individual NCS process, however, is expected to involve only a much smaller number of "relevant" degrees of freedom, i.e. those belonging to the relevant microscopic *open quantum system*; see above. The latter can be described by reduced density operators $\widetilde{\rho_{\nu'}}$ and $\widetilde{\rho_\nu}$ obtained from $\rho_{\nu'}$ and ρ_ν through partial trace $Tr_{(env)}$ over the huge number of the remaining degrees of freedom of the N-body system (the "environment"):

$$\widetilde{\rho_{\nu'}} = Tr_{(env)}\left[\rho_{\nu'}\right]\,, \quad \widetilde{\rho_\nu} = Tr_{(env)}\left[\rho_\nu\right]\,. \tag{36}$$

Thus one easily obtains the relation

$$0 \leq M_{(\nu\nu')} = b^2\, Tr\left[\rho_{\nu'}\,V\,\rho_\nu\,V^\dagger\right] = b^2\, Tr_{(rel)}\left[\widetilde{\rho_{\nu'}}\,V\,\widetilde{\rho_\nu}\,V^\dagger\right]\,, \tag{37}$$

where $Tr_{(rel)}$ denotes the trace over the degrees of freedom of the relevant system.

In the following we investigate possible implications of QE and its decoherence for the scattering process. The physical context is associated with the relation

$$\tau_{dec} \sim \tau_{sc}\,. \tag{38}$$

Let $\{|\xi\rangle\}$ be the *preferred* representation [52] being selected by the decoherence accompanying the dynamics of the relevant scattering system. Using as an explicit example a Lindblad-type decoherence model [2], we have

$$\langle\xi|\widetilde{\rho_{\nu'}}(t)|\xi'\rangle = \langle\xi|\widetilde{\rho_{\nu'}}(0)|\xi'\rangle\, \mathrm{e}^{-\Lambda|\xi-\xi'|^2 t}\,. \tag{39}$$

In general, the positive constant Λ should strongly depend on the state ν', but we drop a related subscript for simplicity of notations. In order to simplify the following derivations, we may temporarily assume $\widetilde{\rho_\nu}$ to be time-independent.

Performing now the trace in Eq. (37) with respect to the basis $\{|\xi\rangle\}$ and noting the closure relation $\int d\xi'|\xi'\rangle\langle\xi'| = \hat{1}$, Eq. (37) yields

$$\begin{aligned} M_{(\nu\nu')}(t) &= b^2 \iint d\xi d\xi' \, \langle\xi|\widetilde{\rho_{\nu'}}(t)|\xi'\rangle\langle\xi'|V\,\widetilde{\rho_\nu}\,V^\dagger|\xi\rangle \\ &= b^2 \iint d\xi d\xi' \, \langle\xi|\widetilde{\rho_{\nu'}}(0)|\xi'\rangle\langle\xi'|V\,\widetilde{\rho_\nu}\,V^\dagger|\xi\rangle \, \mathrm{e}^{-\Lambda|\xi-\xi'|^2 t} \, . \end{aligned} \tag{40}$$

This quantity contains a t-dependence due to decoherence, which concerns the non-diagonal elements of the reduced (or: relevant) density operator $\widetilde{\rho_{\nu'}}(t)$. Clearly, for $\Lambda = 0$ this equation is identical with Eq. (37), that is: $M_{(\nu\nu')} \equiv M_{(\nu\nu')}(0) = M_{(\nu\nu')}(t)$, for all t.

Since time-resolved details of this very fast t-dependence are not accessible experimentally, comparison with measurements should be made after taking the time average over the duration of the scattering process, given by the scattering time τ_{sc}. Thus, the associated experimentally relevant quantity is

$$\overline{M}_{(\nu\nu')} = \frac{1}{\tau_{sc}} \int_0^{\tau_{sc}} dt \, M_{(\nu\nu')}(t) \, . \tag{41}$$

As expected, the last formula includes also the cases in which decoherence is absent, i.e. $\Lambda = 0$, since then it trivially holds $\overline{M}_{(\nu\nu')} = M_{(\nu\nu')}(0)$. But for the general case $\Lambda > 0$, all (absolute values of the) contributions to the rhs of Eq. (40) are reduced due to the exponentials $\exp(-\Lambda|\xi - \xi'|^2 t)$.

Based on physical considerations, one can conclude that this causes a decrease of the quantity $\overline{M}_{\nu\nu'}$. For this purpose, note first that the "diagonal" terms with $\xi = \xi'$ in Eq. (40) are t-independent and positive, as easily seen:

$$\langle\xi|\widetilde{\rho_{\nu'}}(0)|\xi\rangle\langle\xi|V\,\widetilde{\rho_\nu}\,V^\dagger|\xi\rangle = \langle\xi|\widetilde{\rho_{\nu'}}(0)|\xi\rangle\langle\xi V^\dagger|\widetilde{\rho_\nu}|V^\dagger\xi\rangle \geq 0. \tag{42}$$

By continuity, all associated terms with $\xi \approx \xi'$ should be positive, too, and they contribute most significantly to the process. This point has been emphasized by Mensky [2](b). Thus, we finally obtain the desired result $\overline{M}_{(\nu\nu')} < M_{(\nu\nu')}(0)$, which is tantamount to a decrease of scattering intensity.

Now we can drop the preceding assumption of $\widetilde{\rho_\nu}$ being decoherence-free. Then the above procedure of Eqs. (39)–(42) can be straightforwardly extended. The localization rates Λ will then depend on both states ν and ν'. The main conclusion remains unchanged.

Moreover, investigations of the limiting cases (i) $\tau_{dec} \gg \tau_{sc}$, and (ii) $\tau_{dec} \ll \tau_{sc}$ (applicable to all other neutron scattering techniques than NCS), showed that the associated theoretical results are completely in line with the predictions of conventional theory [19, 20, 6, 7] — simply because in these cases the decoherence dynamics becomes irrelevant.

In summary, it has been shown that the exponentially decaying factors due to decoherence (in a microscopic open quantum system) may lead to a reduction of the time averages $\overline{M}_{(\nu\nu')}$ and of the scattering intensity, Eq. (32), thus providing a qualitative theoretical interpretation of the new NCS effect under consideration.

12. A "first principles" model of fast decoherence in scattering dynamics

Here we present a simple theoretical model [14] for the theoretical interpretation of the considered NCS phenomenon, which is more explicit that the aforementioned qualitative considerations. The basic physical idea remains however the same, i.e. the crucial role that quantum entanglement and ultrafast decoherence play in the physical context under consideration; for more details see [14].

Let us consider neutron scattering from a system consisting of N particles with the same scattering length b, and the N-body Hamiltonian $H_{total} = H_0 + V$ with the interaction

$$V(\mathbf{r}) = \lambda\, n(\mathbf{r}), \quad \lambda = \frac{2\pi\hbar^2}{m}\, b\,. \tag{43}$$

m is the neutron mass, $n(\mathbf{r})$ is the particle density operator

$$n(\mathbf{r}) = \frac{1}{V}\sum_{j=1}^{N} \delta(\mathbf{r} - \mathbf{R}_j) \tag{44}$$

where V is the volume, and $\mathbf{R}_j$ is the spatial position of the j-th particle; cf. the textbook [20].

In the *interaction* picture, the Schrödinger equation is now (setting for simplicity $\hbar = 1$) $i\partial_t \Psi = \lambda\, n(\mathbf{r},t)\Psi$, with the perturbative solution

$$\Psi(t) = \Psi(0) - i\lambda \int_0^t n(\mathbf{r}, t')\, dt' \Psi(0). \tag{45}$$

We write the transition probability $W(t)$ between initial states ψ_i (with probability P_i that the scattering system is in the state ψ_i) and final states ψ_f of the scattering system to be given by

$$W(t) = \sum_{i,f} \mid \langle \psi_f \mid \lambda \int_0^t n(\mathbf{r}, t')\, dt' \mid \psi_i \rangle \mid^2 P_i. \tag{46}$$

It should be noted that ψ_i and ψ_f are eigenstates of the N-body Hamiltonian H_0 omitting the probe system [20]. This allows us to write the transition probability in the form

$$W(t) = \lambda^2 \int_0^t dt' \int_0^t dt'' \sum_f \langle \psi_f \mid n(\mathbf{r},t')\, \rho\, n(\mathbf{r},t'') \mid \psi_f \rangle, \tag{47}$$

with $\rho = \sum_i \mid \psi_i \rangle P_i \langle \psi_i \mid$, where we have noted that $n^\dagger(\mathbf{r}, t) = n(\mathbf{r}, t)$.

To describe an actual scattering experiment from condensed matter, one takes an appropriate average over all these states [20], as done in Eq. (46).

Furthermore, the initial ($\mathbf{k}_0$) and final ($\mathbf{k}_1$) momenta of an impinging probe particle (neutron) may be assumed to be well defined [20]. Introducing the momentum transfer $\mathbf{q} = \mathbf{k}_0 - \mathbf{k}_1$ from the probe particle to the scattering system, the Fourier transform of the particle density reads

$$n(\mathbf{r}, t) = \frac{1}{(2\pi)^3} \int d\mathbf{q}\, n(\mathbf{q}, t) \exp(i\, \mathbf{q} \cdot \mathbf{r}) \tag{48}$$

where, in the case of neutron scattering, cf. Eq.(44),

$$n(\mathbf{q}, t) = \sum_j \exp\left[-i\mathbf{q} \cdot \mathbf{R}_j(t)\right]\,. \tag{49}$$

Since $n(\mathbf{r}, t)$ is Hermitian, we have $n^\dagger(\mathbf{q}, t) = n(-\mathbf{q}, t)$ and one obtains from Eq. (46)

$$W(t) = \lambda^2 \int_0^t dt' \int_0^t dt'' \sum_f \langle \psi_f | n(\mathbf{q}, t')\, \rho\, n(-\mathbf{q}, t'') | \psi_f \rangle. \tag{50}$$

The sum over ψ_f runs over all possible eigenstates of H_0 which constitute a complete set, i.e. $\Sigma_f|\psi_f\rangle\langle\psi_f| = \mathbf{1}$. Hence

$$\sum_f \langle\psi_f|n(\mathbf{q},t')\,\rho\,n(-\mathbf{q},t'')|\psi_f\rangle = Tr\left[n(\mathbf{q},t')\,\rho\,n(-\mathbf{q},t'')\right] \tag{51}$$

where $Tr[...]$ denotes the trace operation. As done in standard theory, in Eq. (51) one first sums over all final states, keeping the initial state ψ_i fixed, and then averages over all ψ_i (see e.g. [20], p. 19). The right-hand-side of Eq. (51) contains the density operator ρ of the system before collision. Also, if the integration in Eq. (50) is extended over all times (i.e., $t \to \infty$), there ensues over-all energy conservation. This reproduces the well known result of standard neutron scattering theory.

Here, however, it is important to retain the finite duration of the scattering time. This introduces an additional freedom into the theory, because we may be able to observe the influence of the decoherence on the scattering yield. The result will be expressed in terms of the correlation function

$$C(\mathbf{q},\tau) = Tr[n(\mathbf{q},t)\,\rho\,n(-\mathbf{q},t+\tau)] = Tr[n(\mathbf{q},0)\,\rho\,n(-\mathbf{q},\tau)], \tag{52}$$

where we have utilized the fact that the scattering system is stationary.

By introducing a measurement time (the so-called scattering time) τ_{sc}, representing the time interval within the scattering process may happen, we find

$$W(\tau_{sc}) = \lambda^2 \int_0^{\tau_{sc}} dt' \int_0^{\tau_{sc}} dt'' C(q,t''-t') = \lambda^2 \int_0^{\tau_{sc}} dt' \int_0^{t'} d\eta\,[C(q,\eta)+C(q,-\eta)]\,. \tag{53}$$

Here we use the stationarity of the correlation function [20]. If we assume this function to be real, and that $C(q,\eta) \approx 0$ for $\eta \gtrsim \tau_{sc}$, we obtain the result

$$W(\tau_{sc}) \approx 2\lambda^2\tau_{sc} \int_0^{\tau_{sc}} d\eta\, C(q,\eta). \tag{54}$$

Now we can introduce the transition rate, $\dot{W}$ say, which is defined as

$$\dot{W} \equiv \frac{W(\tau_{sc})}{\tau_{sc}} = 2\lambda^2 \int_0^{\tau_{sc}} d\eta\, C(q,\eta). \tag{55}$$

Here the correlation function is analogous to the so-called intermediate function of neutron scattering theory [20]. This result for the scattering yield is analogous to that of standard theory.

We now introduce a set of preferred coordinates $\{\,|\xi\rangle\}$, cf. [2, 52]. These represent the relevant degrees of freedom coupled to the neutron probe, thus representing and/or defining an *open quantum system*. The density matrix needed in (52) is then the *reduced* one in the space spanned by these states, and it is obtained by tracing out the (huge number of the) remaining degrees of freedom belonging to the "environment". To simplify notations, we denote this reduced density matrix by ρ too.

In the *subspace* spanned by the preferred coordinates (also termed "pointer basis"), we may assume the relevant density matrix to obey a Lindblad-type equation of the form [2]

$$\partial_t\rho = -i\,[H,\rho] + \mathcal{R}\rho \equiv \mathcal{L}\rho \tag{56}$$

with the formal solution: $\rho(t) = e^{\mathcal{L}t}\rho(0)$. Let us look at a time-dependent expectation value

$$\langle A(t)\rangle \equiv Tr\,(\rho(t)A) = Tr\left(e^{\mathcal{L}t}\rho(0)A\right) = Tr\left(\rho(0)e^{\mathcal{L}^{\dagger}t}A\right)\,, \tag{57}$$

where we define $\mathcal{L}^\dagger$ by setting $Tr\left((\mathcal{L}X)Y\right) = Tr\left(X\left(\mathcal{L}^\dagger Y\right)\right)$. Thus we obtain a Lindblad time evolution for the operators too by writing

$$\partial_t A(t) = \mathcal{L}^\dagger A(t). \tag{58}$$

This form was actually the original Lindblad result as introduced in the quantum algebraic formulation of the theory of semigroups. Note that this assumes that $\mathcal{L}$ does not depend on time.

Now we find that we may use this formalism to calculate correlation functions like the one in (52). We write

$$\langle A(t)B\rangle = Tr\left[\rho(0)\left(e^{\mathcal{L}^\dagger t}A\right)B\right] = Tr\left[Ae^{\mathcal{L}t}\left(B\rho(0)\right)\right] \equiv Tr\left(A\rho_B(t)\right), \tag{59}$$

where $\rho_B(t)$, as defined in Eqs. (59), obeys the equation

$$\partial_t \rho_B(t) = \mathcal{L}\rho_B(t) \tag{60}$$

and the initial condition $\rho_B(0) = B\rho(0)$. Thus, except for the initial condition, we have to solve the same equation of motion as for the density matrix (56).

We here assume a simple Lindblad-type ansatz for the master equation in the relevant subspace. In order to show the effect of dephasing, we simply assume one Lindblad variable X; in the real system we would have a multitude of such variables. We set

$$\partial_t \rho = -i\left[H,\rho\right] - K\left[X,\left[X,\rho\right]\right] = \mathcal{L}\rho, \tag{61}$$

where $K > 0$ and H is the reduced (or relevant Hamiltonian) of a microscopic or mesoscopic scattering system and the double commutator term describes dephasing (and/or decoherence). For simplicity of the further calculations, we here assume that we can take the preferred coordinates to commute with the total Hamiltonian

$$H \mid \xi\rangle = \mathcal{E}_\xi \mid \xi\rangle\,, \qquad X \mid \xi\rangle = \xi \mid \xi\rangle. \tag{62}$$

This time evolution is now introduced into the correlation function $C(\mathbf{q},\tau)$, Eq. (52). A short straightforward calculation (see Ref. [14] for full details) yields for the transition rate the result

$$\dot{W} = 2\lambda^2 \int_0^{\tau_{sc}} \sum_{\xi,\xi'} \exp\left[-i\left(\mathcal{E}_{\xi'} - \mathcal{E}_\xi\right)\tau\right]\, \exp\left[-K\left(\xi' - \xi\right)^2 \tau\right] \times$$
$$\times \langle \xi \mid n(-\mathbf{q},0) \mid \xi'\rangle\langle \xi' \mid n(\mathbf{q},0)\,\rho(0) \mid \xi\rangle\, d\tau\,. \tag{63}$$

Obviously, the decoherence-free limit of this result, i.e. with $K = 0$, corresponds to the conventional result of scattering theory. The oscillating factors $\exp\left[-i\left(\mathcal{E}_{\xi'} - \mathcal{E}_\xi\right)\tau\right]$ are characteristic for the "unitary-type" dynamics caused by the commutator part $-i\left[H,\rho\right]$ of the master equation (61) for the reduced (or relevant) density matrix ρ. These factors have the absolute value 1 and are well known from standard theory [20]. On the other hand, decoherence (with a characteristic time τ_{dec}) is present in the case $\tau_{dec} = K^{-1} \sim \tau_{sc}$, leading to the additional restrictive factors $\exp(-K\left(\xi' - \xi\right)^2\tau) \leq 1$, which can be seen to cause a decrease of the transition rate and thus of the associated cross-section. This can be illustrated in physical terms as follows.

Let us first assume that the reduced density operator $\rho(0)$ can be chosen to be *diagonal* in the preferred ξ−representation (which corresponds to the usual random phase approximation at $t = 0$). Then each term of Eq. (63) contains the factor

$$\langle\xi|n(-\mathbf{q},0)|\xi'\rangle\langle\xi'|n(\mathbf{q},0)\,\rho(0)|\xi\rangle = |\langle\xi \mid n(-\mathbf{q},0) \mid \xi'\rangle|^2\langle\xi|\rho(0)|\xi\rangle \geq 0\,. \tag{64}$$

If the assumption of diagonal form for $\rho(0)$ would be considered as being "too strong", one may note the following: The decoherence factors $\exp(-K\left(\xi' - \xi\right)^2\tau)$ imply that only terms with $\xi \approx \xi'$

contribute significantly to the transition rate. Thus we may conclude that, by continuity, all associated terms with $\xi \approx \xi'$ in Eq. (63) should be positive, too. The further terms with ξ being much different from ξ' can be positive or negative. But they may be approximately neglected, since they decay very fast and thus contribute less significantly to $\dot{W}$; cf. [2](b).

The main conclusion from the preceding considerations is that the time average in Eq. (63) always decreases the value of $\dot{W} \equiv W(\tau_{sc})/\tau_{sc}$, due to the presence of the exponential factors $\exp(-K(\xi' - \xi)^2 \tau) \leq 1$. In other words, the effect of decoherence (and/or dephasing) during τ_{sc} plays a crucial role and may lead to an "anomalous" decrease of the transition rate and the associated scattering intensity.

Interestingly, in the limit of "very slow" decoherence, $K \to 0$, this "anomaly" disappears, i.e. the scattering results are predicted to agree with conventional theoretical expectations.

In the opposite limiting case of "very fast" decoherence, $K \to \infty$, one immediately sees that only the "diagonal" terms with $\xi = \xi'$ survive in Eq. (63) — which is due to the action of the factors $\exp(-K(\xi' - \xi)^2 \tau) \leq 1$. At the same time, for all oscillating factors holds: $\exp\left[-i\left(\mathcal{E}_{\xi'} - \mathcal{E}_{\xi}\right)\tau\right] \to 1$. Consequently, Eq. (63) tends to the standard expression Eq. (55) in the limit of short scattering times. Also this result is according to conventional expectations.

Additionally, the above theoretical remarks and derivations clearly show that Fermis' golden rule (GR) cannot be strictly valid for the quantum dynamical description of a microscopic open quantum system in the ultrashort timescale of NCS; see also section 13(B-C) below. E.g., for an open quantum system, the assumed stationary states for the derivation of GR are totally fictitious. Confusion related with this fact can lead one to conclude that the decoherence phenomenon is irrelevant and/or nonexistent in the physical context of NCS [51].

13. Additional remarks and discussion

(A) Common in all the theoretical considerations and models of the theoretical sections 9–12 is the physical insight that short-lived quantum entanglement and decoherence between a struck quantum system (say, proton) and its adjacent particles play a central role in the dynamical process of scattering. The phenomenon of decoherence is unknown in conventional neutron scattering theory [20, 21, 6, 7]. The discussion of the NCS physical parameters showed that the characteristic timescale of NCS lies in the attosecond range. A similar conclusion holds also for electron-atom Compton scattering.

(B) Considering scattering from single molecules in the gas phase, the following question may arise. E.g., for the case of H_2 and D_2 scattering experiments, the "environment" consists of only 3 other particles (the second proton/deuteron and two electrons). Such a "mini" environment seems to offer the unique chance to investigate the possibility to go beyond the aforementioned decoherence-type theoretical approaches. That is, instead of treating the impinging neutron and the scattering nucleus (proton) as an open system, one could as well consider the closed system "neutron plus H_2" as a few body system. Scattering of one electron or neutron from a fully entangled four body system — that is, considering a five body closed system — seems to be in reach of present days theory. Consequently, the experimental results could potentially be explained by a five body theory (for the H_2 case) without talking about open quantum systems.

Although a definitive answer to this question is presently not known, we believe that the arguments presented in section 7 deny the suggested possibility. Namely, the ultrafast attosecond neutron-proton collisional time was shown to imply that the second proton of H_2 has no causal influence on the dynamics of the neutron-proton collisional system.

(C) General neutron scattering theory (cf. [20, 21]), as well as the particular NCS theory [6, 7] are based on the quantum theory of systems with well defined Hamiltonians and the associated *unitary* time evolutions (in the frame of first-order perturbation theory). This is the case for *closed* quantum systems. In this theoretical framework, of course, the phenomenon of decoherence plays no role "by definition". Therefore it should be stressed that the physical conditions of the real NCS process are consistent with the theoretical frame of scattering from *open* quantum systems. E.g. for NCS from molecular hydrogen (gas phase) the relevant quantum system consists of an impinging neutron and one proton of H_2 only — as the H-peak position in the TOF spectrum proves beyond doubt — probably being "dressed" with some electronic degrees of freedom of the electrons of the chemical bond. The remaining adjacent particles and/or degrees of freedom (i.e., the second proton and a part of the two electrons of H_2) represent the "environment". The latter strongly interacts with the scattering system (dressed proton) through electromagnetic forces, thus causing ultrafast decoherence. The conceptual connection with the physical fields of "quantum optics" and/or "quantum computing" should be obvious.

In this context it cannot be overemphasized that in the general theory [20, 21], as well as in the more specific standard NCS theory [6, 7], the time evolution is assumed to be *unitary*. Let us discuss this in some detail. The transition from Eq. (33),

$$\frac{d^2\sigma}{d\Omega dE_1} = \frac{k_1}{k_0}\sum_{\nu\nu'} W_\nu \Big| \sum_j b_j \langle \nu' | \exp{(i\mathbf{q}\cdot\mathbf{r}_j)} |\nu\rangle \Big|^2 \delta(E_\nu - E_{\nu'} + E_0 - E_1)\,, \tag{65}$$

— which represents Fermi's golden rule [20, 21] with the additional inclusion of the shown delta-function (in order to guarantee strict energy conservation) — to

$$\frac{d^2\sigma}{d\omega d\Omega} = b^2 \frac{k_1}{k_0}\frac{1}{2\pi}\int_{-\infty}^{\infty} dt\, \exp((-i\omega t)\sum_{j,k}^{N} \langle \exp(-i\mathbf{q}\cdot\mathbf{r}_j(0))\exp(-i\mathbf{q}\cdot\mathbf{r}_k(t))\rangle\,, \tag{66}$$

— that is, Eqs. (1) and (4) — is effectuated by the spectral decomposition of the delta function

$$\delta(E_\nu - E_{\nu'} + E_0 - E_1) = \frac{1}{2\pi\hbar}\int_{-\infty}^{\infty} dt\, \exp[i(E_{\nu'} - E_\nu)t/\hbar]\exp(-i\omega t) \tag{67}$$

and the introduction of the unitary time evolution operator $U(t)$ through the relation

$$U(t)|\nu\rangle = \exp(-iH_c t)|\nu\rangle = \exp(-iE_\nu t)|\nu\rangle \tag{68}$$

because $|\nu\rangle$ and $|\nu'\rangle$ are assumed to be eigenfunctions of the complete Hamiltonian H_c of the N-body system. Furthermore, application of the (assumed validity of the) closure relation over the final states

$$\sum_{\nu'} |\nu'\rangle\langle\nu'| = \hat{1}$$

leads to the introduction of the Heisenberg representation of the position operators

$$\mathbf{r}_j(t)) = U^\dagger(t)\mathbf{r}_j U(t)$$

in Eq. (4), i.e.

$$F(\mathbf{q},t) = \frac{1}{N}\sum_{j,k}^{N} \langle \exp(-i\mathbf{q}\cdot\mathbf{r}_j(0))\exp(-i\mathbf{q}\cdot\mathbf{r}_k(t))\rangle$$

(see e.g. the textbooks [20, 21] for a detailed derivation). These results are due to the assumption that the closed N-body scattering system is considered to have a well defined Hamiltonian, H_c, which determines the complete quantum dynamics with the aid of $U(t)$.

Of course, this is not true for *open* quantum systems which must exhibit non-unitary time evolutions and decoherence [2]. In the context of NCS, rejection of these facts leads to confusion, in which Fermi's golden rule is claimed to be strictly valid and/or non-stationary states of the scattering system are claimed to be totally fictitious.

(D) Concerning the time range of validity of first-order time-dependent perturbation theory, and thus also of validity of Fermi's golden rule, one should keep in mind that one necessary condition by the derivation is the assumed "long time",

$$t \gg \hbar/\Delta E$$

with ΔE being the energy difference of final and initial states; see textbooks of quantum mechanics. The same holds for the delta-function in Eq. (65), which is justified only if $t \gg \hbar/\Delta E$. This condition, however, is in conflict with the estimated characteristic ultrashort collisional time of NCS; see Section 6.

(E) As discussed in Section 7 — and illustrated with the example of NCS from a single H_2 molecule — the physical reason which requires the scattering particle (e.g. a proton) to be an *open* quantum system is the causality principle and the finite velocity ($\leq c$) of any physical signal. Thus, in the physical context of NCS, this has a far-reaching consequence: Not a wavefunction (state vector) Ψ, but a (reduced) density operator ρ is the physical entity of basic importance. In other terms, ρ is not a mere "approximation" related with "missing knowledge" of an "exact state" Ψ, but rather the opposite holds true: Ψ ought to be a "bad" approximation, because it is not consistent with causality (in the NCS case under consideration).

This conclusion appears to be consistent with a theoretical result by Beretta [53], which should be mentioned here. The theoretical work [53] deals with model evolution equations (applicable not only to open systems but also to closed isolated systems) capable of describing – simultaneously and in competition with the usual Hamiltonian unitary evolution – the natural tendency of any initial nonequilibrium state to relax towards thermodynamic equilibrium. The author also makes the following interesting point:

> "The central conceptual difference between the proposed approach, and the approaches based on attempting to derive the KSGL [Kossakowski-Sudarshan-Gorini-Lindblad] equation, is that this approach regards a non-pure density matrix as representing a real ontological object, the actual state of the world, and is not understood as just an epistemic ignorance of which particular pure state the world is 'really' in." (See [53], p. 4.)

Furthermore, the characteristic time parameter τ_D in Beretta's theory (see [53], Eq. (35)) called "intrinsic dissipative time" seems to be conceptually connected with the characteristic decoherence time τ_{dec} in the preceding theoretical models of Sections 11 and 12.

In this context it should be pointed out that the description of open quantum systems is possible not only in terms of a reduced density matrix, but also in terms of wave functions, but these are then usually solutions of effective nonlinear Schrödinger equations (see, e.g. [54]).

(F) The novel NCS-experimental findings from a single crystal (presented in Section 3) provide strong evidence for the following effect: For protons, variations of the widths of Compton profiles $J_H(y)$ are strongly correlated with the variations of NCS-intensity I_H, and therefore also with the "anomalous" decrease of scattering intensity. This possible correlation and/or connection is novel and, until now, it has been categorically denied (see e.g. last paragraph on p. 463 of [7]). Moreover, current high-precision NCS-experiments on H_2 (gas phase, various temperatures) provide further evidence for this possible connection [E. Gray et al., work in progress].

(G) As mentioned in Section 4, the "anomalous" scattering intensity effect has been independently confirmed using electron-proton Compton scattering (ECS) [5, 31, 32, 35]. ECS-investigations from a solid polymer showed exactly the same shortfall in scattered electrons (with energy about 15-30 keV) from hydrogen nuclei, comparable to the shortfall of scattered neutrons in accompanying NCS experiments on the same polymer. Thus this effect appears to be independent of the two fundamental interactions involved (i.e., the electromagnetic and strong interactions) [5, 55].

(H) It may be interesting to note that the theoretical approaches to explain a reduced scattering signal (see the preceding sections) are sensitive to H/D isotope effects as they are seen in some NCS experiments. One reason for this is that the scattering time (τ_{sc}, see section 6) varies with the momentum transfer, which strongly depends on the mass of the struck nucleus.

(I) The experimental results discussed and/or presented above and their qualitative theoretical interpretation show that "fast" (say, with 1–100 eV kinetic energy) neutrons provided at spallation sources (e.g., ISIS/U.K., SNS/U.S.A., ESS/Sweden), and electron spectrometers with large scattering angles and associated large momentum transfers, represent novel tools for investigation of attosecond physical and chemical phenomena in condensed matter and molecules, even at ambient conditions. The experimental results represent a challenge to conventional (time-independent) theories of scattering and molecular spectroscopy, which may need to be supplemented by time-dependent theories (also beyond the Born–Oppenheimer approximation) to describe phenomena of Attosecond Physics.

Acknowledgments
We are indebted to F. Fillaux (LADIR-CNRS, Thiais, France) for making available his experimental NCS-data from a single crystal $KHCO_3$.

[1] (a) Einstein A, Podolsky B and Rosen N 1935 *Phys. Rev.* **47**, 777; (b) Bell J S 1964 *Physics* **1** 195
[2] (a) Breuer H P and Petruccione F 2002 *The Theory of Open Quantum Systems* (Oxford: Oxford University Press); (b) Mensky M B 2000 *Quantum Measurements and Decoherence* (Dordrecht: Kluwer); (c) Schlosshauer M 2007 *Decoherence and the Quantum-to-Classical Transition* (Berlin: Springer)
[3] Nielsen M A and Chuang I L 2000 *Quantum Computation and Quantum Information* (Cambridge: Cambridge University Press)
[4] Chatzidimitriou-Dreismann C A, Abdul-Redah T, Streffer R M F and Mayers J 1997 *Phys. Rev. Lett.* **79** 2839
[5] Chatzidimitriou-Dreismann C A, Vos M, Kleiner C and Abdul-Redah T 2003 *Phys. Rev. Lett.* **91** 057403
[6] Watson G I 1996 *J. Phys.: Condens. Matter* **8** 5955
[7] Andreani C, Colognesi D, Mayers J , Reiter G F and Senesi R 2005 *Adv. Phys.* **54** 377
[8] Misra B and Sudarshan E C G 1977 *J. Math. Phys.* **18** 756
[9] Schulman L S 1997 *Phys. Rev.* A **57** 1509
[10] Kofman A G and Kurizki G 2000 *Nature* **405** 546
[11] Schulman L S and Gaveau B 2006 *Phys. Rev. Lett.* **97** 240405
[12] Gaveau B and Schulman L S 2008 *preprint*
[13] Chatzidimitriou-Dreismann C A 2005 Laser Physics **15** 780
[14] Chatzidimitriou-Dreismann C A and Stenholm S 2005, in: *Decoherence, Entanglement and Information Protection in Complex Quantum Systems*, NATO Science Series II – Vol 189, ed V M Akulin, A Sarfati, G Kurizki and S Pellegrin (Dordrecht: Springer) pp 555-562; also available at arXiv:quant-ph/0702038v1
[15] Mayers J and Abdul-Redah T 2004 *J. Phys: Condens. Matter* **16** 4811
[16] Hohenberg P C and Platzman P M 1966 *Phys. Rev.* **152** 198
[17] Compton A H 1923 *Phys. Rev.* **21** 483
[18] Sears V F 1984 *Phys. Rev.* B **30** 44
[19] van Hove L 1954 *Phys. Rev.* **95** 249
[20] Squires G L 1996 *Introduction to the Theory of Thermal Neutron Scattering* (Mineola: Dover)
[21] (a) Marshall W and Lovesey S W 1971 *Theory of Thermal Neutron Scattering* (Oxford: Oxford University Press); (b) Lovesey S W 1984 *Theory of Neutron Scattering form Condensed Matter* (Oxford: Oxford University Press)
[22] Chatzidimitriou-Dreismann C A, Abdul-Redah T and Krzystyniak M 2005 *Phys. Rev.* B **72** 054123
[23] West G B 1975 *Phys. Rep.* **18** 263
[24] Karlsson E B, Chatzidimitriou-Dreismann C A, Abdul-Redah T, Streffer R M F, Hjörvarsson B, Öhrmalm J and Mayers J 1999 *Europhys. Lett.* **46** 617
[25] Karlsson E B Abdul-Redah T, Streffer R M F, Hjörvarsson B, Mayers J and Chatzidimitriou-Dreismann C A 2003 *Phys. Rev.* B **67** 184108
[26] Chatzidimitriou-Dreismann C A, Abdul-Redah T Streffer R M F and Mayers J 2002 *J. Chem. Phys.* **116** 1511
[27] Chatzidimitriou-Dreismann C A, Abdul-Redah T and Sperling J 2000 *J. Chem. Phys.* **113** 2784
[28] Chatzidimitriou-Dreismann C A, Abdul-Redah T and Kolarić B 2001 *J. Am. Chem. Soc.* **123** 11945
[29] (a) Fillaux F 1999 "Proton Dynamics in $KHCO_3$" ISIS Experimental Report; (b) Postorino P, Fillaux F, Mayers J, Tomkinson J and Holt R S 1991 *J. Chem. Phys.* **94** 4411
[30] Vos M 2001 *Phys. Rev.* A **65** 012703
[31] Vos M, Chatzidimitriou-Dreismann C A, Abdul-Redah T and Mayers J *Nucl. Instr. Meth.* B 2005 **227** 233
[32] Cooper G, Hitchcock A P and Chatzidimitriou-Dreismann C A 2008 *Phys. Rev. Lett.* **100** 043204
[33] Bonham R A, Cooper G and Hitchcock A P 2009 *J. Chem. Phys.* **130** 144303
[34] Moreh R and Nemirovsky D 2009 *J. Chem. Phys.* **131** 054305

[35] Hitchcock A P, Cooper G, Bonham R A and Chatzidimitriou-Dreismann C A 2009 Quasi-elastic electron scattering from atoms and molecules (in press), *J. Electron Spectrosc. Relat. Phenom.* doi:10.1016/j.elspec.2009.12.002
[36] Senesi R, Colognesi D, Pietropaolo A and Abdul-Redah T 2005 *Phys. Rev.* B **72** 054119
[37] Blostein J J, Dawidowski J and Granada J R 2001 *Physica* B **304** 357; 2005 *Phys. Rev.* B **71** 054105
[38] (a) Dorner B 2005 *J. Neutron Research* **13** 267; (b) Dorner B 2006 *Nucl. Instr. Meth.* B **247** 390; (c) Krzystyniak M and Chatzidimitriou-Dreismann C A 2005 *Phys. Rev.* B **72** 174117
[39] Moreh R, Block R C, Danon Y and Neumann M 2005 *Phys. Rev. Lett.* **94** 185301
[40] Chatzidimitriou-Dreismann C A and Krzystyniak M 2006 *J. Phys.: Condens. Matter* **18** 4741
[41] Chatzidimitriou-Dreismann C A 2006 *Physica* B **385-386** 1
[42] Stock C, Cowley R A, Taylor J W and Bennington S M, *preprint* arXiv:0907.1945v2 [Cond-Mat.Stat-Mech Arxiv, 2009]
[43] Mayers J, Adams M and Reiter G *preprint* arXiv:0909.2633v1 [Cond-Mat.Other Arxiv, 2009]
[44] Margolus N and Levitin L B 1998 *Physica* D **120** 188
[45] Ballentine L E 1998 *Quantum Mechanics – A Modern Development* (Singapore: World Scientific)
[46] (a) Gidopoulos N I 2005 *Phys. Rev.* B **71** 054106; (b) 2004 "Non-adiabatic electronic excitations in neutron Compton scattering", ISIS Science Highlights
[47] Karlsson E B 2008 *Phys. Scr.* **66** 065301
[48] Karlsson E B 2003 *Phys. Rev. Lett.* **90** 95301
[49] Reiter G F and Platzman P M 2005 *Phys. Rev.* B **71** 054107
[50] Colognesi D, Pietropaolo A and Senesi R 2008 *J. Phys.: Condens. Matter* **20** 445225
[51] Colognesi D 2007 *Physica* B **398** 89
[52] Zurek W H 2003 *Rev. Mod. Phys.* **75** 715
[53] Beretta G P (2009) Nonlinear Quantum Evolution Equations to Model Irreversible Adiabatic Relaxation with Maximal Entropy Production and Other Nonunitary Processes *preprint* arXiv:0907.1977v1
[54] Strunz W T 2002, in: *Coherent Evolution in Noisy Environments* ed A Buchleitner and K Hornberger (Berlin: Springer) paragraph 5.3.5
[55] 2003 *Physics Today* Physics Update p 9

Quantum theory – essential from cosmos to consciousness

T Görnitz
Institut für Didaktik der Physik, FB Physik
J. W. Goethe-Universität Frankfurt/Main
Mail: Karl-Mangold-Str. 13, D-81245 München, Germany

E-mail: goernitz@em.uni-frankfurt.de

Abstract. Quantum theory is the most successful physical theory. About one third of the gross national product in the developed countries results from its applications. But very often quantum theory is still declared as "crazy" or "not understandable". However, quantum theory has a clear mathematical structure that expresses well-known experiences from every day life: A whole is often more than the sum of its parts, and not only the facts also the possibilities can act. If such structures become important then the consequences differ from the models of classical physics which rests on the fundamental differences between matter and motion, material and force, localization and extension, fullness and emptiness. From quantum theory one can learn that all these differences are useful in many cases but are not fundamental. There are equivalences between them, and these can be extended even to the equivalence between matter, energy and abstract quantum information. It is cosmological funded and is denominated as "Protyposis" to avoid the connotation of information and meaning. Protyposis enables a fundamentally new understanding of matter which can be seen as "formed", "condensed" or "designed" abstract quantum information. One result of the Protyposis is a derivation of Einstein's equations from the abstract quantum information. Another consequence is the ontological reality of the mind and its connection to a brain which can be explained without any dualistic model.

1. Introducing quantum theory

Quantum theory is the most successful physical theory. About one third of the gross national product in the developed countries results from its applications. The applications of this theory range from nuclear power to all tools for computing, laser, solar cells and so on. Up to now no border for its range of validity has been found. Quantum theory has a clear mathematical structure, so a student can it learn in a short time.

What are quanta? "Quantum" is a genus, like "fruit". Precisely, there is never "fruit" in a bag, because seen in a detailed way there are apples, bananas, grapes etc. The same applies to quantum theory.

It seems to be useful to divide the quantum objects into three classes. Material quantum objects, like electrons, protons and so on and also macroscopic quantum systems have a rest mass, and therefore they can be at rest somewhere. Energetic objects like photons have no rest mass, so they have to move always with the velocity of light. Very interesting are structural quanta. Phonons, the quanta of vibrations of sound in solids, cannot appear outside of the solid body; nevertheless they are real in a certain sense. They interact with other quantum objects and are indispensable for an understanding of the structure of solid bodies. Also the quarks and gluons belong to the structural quanta. They also are indispensable for an understanding of the structure and properties of hadrons, even if they do not exist as free particles outside the hadrons. Also quantum bits represent essential structural quanta and also cannot exist in a localized form without a carrier.

Quantum theory has developed in an enormous manner in the last century. Today it comprehends much more than only good old quantum mechanics.

There are some essential steps in the evolution of this theory. The year 1900 is considered as the birth of quantum theory because Planck discovered the quantum of action. Einstein was the first one who discloses in 1905 the basic structure of quantum theory with his hypothesis of light quanta. In 1925 Heisenberg invented quantum mechanics. It was soon developed further by Schrödinger, Dirac and many other people. Already in 1930 Heisenberg and Pauli made some first approaches towards a quantum theory of force fields. The last essential step further can be seen in the fifties. In 1955 Weizsäcker started to think of a universal theory of abstract quantum information. Later on Finkelstein, Wheeler and others followed.

There exist some popular but insufficient ideas about the distinctions between quantum physics and classical physics. One misleading distinction is on the difference of microphysics and macrophysics. It is true that in microphysics only quantum theory is applicable; nevertheless there are also many macroscopic quantum phenomena. The difference between discontinuous and continuous effects should be another one. This neglects that there are also many operators in quantum physics that have continuous spectra. Popular but false is also the distinction between a fuzzy quantum theory and a sharp classical physics. It neglects that quantum theory does supply us with the most accurate description of nature we ever had. Classical physics lives with the illusion of exactness. Its mathematical basis stands on the assumption of an "arbitrarily smooth change" for any variable. This is a precondition for calculus, but it is by no means always fulfilled in nature. At very high precision in any case the quantum structure will become important because quantum physics is the physics of preciseness.

Often there is no need for the precision of quantum theory. At first sight, most of the processes in nature do appear smooth. But strictly speaking, all actions are quantized; they appear only in "numbers" or "steps". Therefore one can say that all changes are strictly speaking quantum jumps. So the quantum jump is the smallest nonzero change in nature. Perhaps that makes this concept so attractive in politics and economics.

2. What is the central structure for an understanding of quantum theory?

To explain the central structure of quantum theory one has to look on the way in which composite systems are constructed.

For classical physics the composition of a many body system happens in an additive way. The state space of the composed system is the direct sum of the state spaces of the single particle systems. This results in a "Lego world view" of smallest building blocks – one or another kind of "atoms". The world has to be decomposed into ultimately elementary objects – which never change – and the forces between such objects. This picture about the structure of reality was generally accepted for more than two millennia.

Composite systems in quantum physics are constructed in a fundamentally different way. The state space of a composed quantum system is the tensor product of the state spaces of the single particle systems.

To explain quantum theory we have to start with this structural difference. But "tensor product" is a very technical concept. Is it possible to relate it to something known in everyday life?

Here is to remind that "relations" create a product structure. One can say that the new states of a composed object are the relational structures between the states of its parts. Therefore quantum theory can be characterized as the physics of relations and appears as a clear mathematical implementation of a familiar life experience: A whole is often more than the sum of its parts.

Relational structures create networks, in the essence they are plurivalent. In such a net many different connections between two goals are possible. This explains a further characterization of quantum theory as "the physics of possibilities". Quantum theory deals with possibilities only. As one consequence this explains that also possibilities can act – and not only the facts. If in a double slit experiment quantum particles have the possibility of going through both holes without controlling them then one will find behind the holes more than two bunches. If there is a controlling of the slits and thereby a change does happen from possibilities into facts then only two bunches will result. This is in the same manner as in every day life where controlling restricts possibilities and thereby influences human behavior.

3. The indissoluble relation between classical physics and quantum physics: The dynamic layering-process

Since the first days of quantum mechanics Bohr has insisted that classical physics is a precondition for speaking about quantum results. It is impossible to ignore that for mankind, there exist not only possibilities but also facts. Therefore for a good description of nature we need both parts of physics, classical physics and quantum physics. Its connection can be described as a "dynamic layering-process". By going to the classical limit a quantum theoretical description changes into a classical one, by the process of quantization classical physics goes over to quantum physics.

About quantization there exist many concepts (a good overview can be found in [1]), so one may ask whether there is a hope to find a simple fundamental structure for it?

If we ignore, meanwhile, the canonical quantization then a general structure does appear. From a bit with its two states {0,1} to a qubit the quantization happens by constructing all the complex valued functions over two points, resulting in a two-dimensional complex space C^2. Path integral quantization can also be interpreted as constructing all functions over the classical pathways. At a first look second quantization seems to be different. But the construction of a state of the quantum field by states of quantum particles has the same structure as the construction of an analytical function from powers of the variable itself. The analytical functions are dense e.g. in the continuous or the measurable functions, even distributions can be explained as boundary values of analytical functions. So the analytical functions represent in some sense "all functions", and a quantum field can be interpreted as "the functions over the quantum particles".

Therefore quantization can be distinguished by the sentence: Quantization is the transition from the manifold of the facts to the possibilities over the facts – that are the functions over the facts. It seems possible to formulate: "Quantization is (actually) second quantization." In this sense Einstein's invention of photons was the first demonstration of quantization.

But how to interpret the canonical quantization from mechanics to quantum mechanics which is characterized by one or the other form of polarization, of a bisection of variables?

The explanation can be given as follows: Mechanics is a limit theory of quantum mechanics in such a way that all operators commute. In quantum mechanics the position operator acts multiplicatively on wave functions over position space and the momentum operator acts multiplicatively on wave functions over momentum space. A duplication from $\Phi(x)$ or $\Psi(p)$ to $R[x, p]$ enables the commutation of position and momentum. Therefore classical mechanics acts over positions and momenta, polarization reverses this duplication.

The core of quantization can be explained by the sentence: Quantization is the transition from the facts to the relational network of possibilities which could develop from the facts – or in a more mathematical language – which span a linear space of functions over the facts.

A further characterization is: The quantization of a system is the transition from its nonlinear description in a low dimensional space having many or infinitely many degrees of freedom (e.g. classical mechanics or electromagnetic fields) to a linear description in an infinite dimensional space of many or infinitely many systems with few degrees of freedom (e.g. quantum bits respectively photons or other field quanta). This reminds of the exponential map and its conversion of products, which are nonlinear, into sums, which are linear.

It seems to be evident that quantum theory is the foundation for classical physics. For all of the objects that are handled so successfully by classical physics its existence can be understand only by quantum theory. It may be recalled that the existence of an atom is forbidden by classical electrodynamics. On the other hand classical physics is a precondition for the appearance of quantum properties. The quantum properties of a system become clear only if its entanglement with the environment is cut of. Such a cut can be modeled in classical physics only.

4. Quantum theory relativized distinctions

Quantum theory is based on every day experiences indeed, but in non-living nature they become essential only at a very high precision. At such a precision, however, effects may appear that are not so evident in every day experiences of the world around us.

Already in school the so-called wave-particle-duality is mentioned. In quantum theory depending on the circumstances one and the same quantum object can act as a wave or as a particle, as a localized or as an extended object. Quantum theory shows up that transformation between matter and motion or between force and material is possible. Of course, Einstein's famous formula $E = mc^2$ was found in special relativity, but in any experiment the pure quantum concept of antimatter is included. The distinction of force and material appears only as the difference between quanta of integer or halfinteger spin. Dirac's model shows that even the distinction between emptiness and plenitude is relativized by quantum theory. The central philosophical aspect of second quantization is that the distinction between object and attribute depends on the context. A quantum particle is an object in quantum mechanics and is an attribute of a quantum field.

5. The quantum theoretical equivalence of matter, energy and quantum information

In addition, quantum theory discloses a completely new perspective on the three entities matter, energy and abstract quantum information. Already since 1955 C. F. v. Weizsäcker has speculated on the possibility of founding physics on abstract quantum information. His "Ur-Theory" grows up from the intention [2] „Physik ist eine Erweiterung der Logik" (Physics is an extension of logics). As the basis for a new reduction he has proposed quantized binary alternatives – Ur-Alternativen or Urs. Werner Heisenberg wrote on Weizsäcker's concept [3] „…daß die Durchführung dieses Programmes »ein Denken von so hoher Abstraktheit erfordert, wie sie bisher, wenigstens in der Physik, nie vorgekommen ist. « Ihm, Heisenberg, »wäre das sicher zu schwer«, aber Weizsäcker solle es mit seinen Mitarbeitern unbedingt versuchen. (… that the realization of this program "requires thinking at such a high degree of abstraction that up to now – at least in physics – has never happened." For him, Heisenberg, "it would be too difficult", but Weizsäcker and his coworkers should definitely do so.)

Why was for a long time no better resonance on this project? The first reason may be that it was far too abstract, and further that there were too little relations to experience. At this times the values that Weizsäcker has proposed have overburden the imagination of the physicists. That one proton is 10^{40} bit is till now hardly to communicate in physics. But a serious problem was that, at this time, the models were inconsistent with general relativity.

An essential step was to go beyond the Urs. Weizsäcker proposes [4] „Ein »absoluter« Begriff der Information hat keinen Sinn" (An »absolute« value of information is meaningless). But this is a contradiction to his claim [5]: "Materie ist Information" (Matter is information). Because matter has an absolute value for equivalence the information has to have one as well.

Therefore, there was a need to make the concept of »information« so abstract that neither emitter nor receiver and – of capital importance – also no meaning or knowledge is primarily included. This is a basic precondition for the equivalence of matter and information. To fulfill this it was necessary to show up the connection to modern theoretical und empirical structures of physics, especially to Bekensteins und Hawkins's theory of the entropy of black holes and to cosmology. Physics is more than an »extension of logics« and in physics information is different from destination or meaning or knowledge. Meaning has always a subjective aspect as well, so meaning cannot be a basis for science and objectivity.

If quantum information will become a basis for science it must be introduced as abstract quantum information, free of meaning. This is the Protyposis. It provides a base for a new understanding of the world that ranges from matter to consciousness. The Protyposis adds to $E=mc^2$, the equivalence of matter and energy, a further formula:

$$N = m\, c^2\, t_{cosmos}\, 6\pi/\hbar.$$

A mass m or an energy mc^2 is equivalent to a number N of qubits. The proportionality factor contains t_{cosmos}, the age of the universe. Today a proton is 10^{41} qubits.

The Protyposis opens new insights into the relations between quantum information and group theory. Two examples will be given. Weizsäcker [4, 5] and Drieschner [6] proposed that the three dimensional position space is a consequence of quantum theory – but their models disagreed with general relativity. This can be repaired by group theoretical considerations. Relativistic particles were represented by irreducible representations of the Poincaré group. Such representations can be constructed in a process of second quantization from quantum bits.

6. Cosmology and quantum bits

That the position space is a consequence of the symmetry for quantum bits was first proposed by Weizsäcker. An object is represented by all the decisions that are possible about it. Any decision that can be scientifically decided can be reduced to quantum bits. The states of a quantum bit are represented by its symmetry group. The symmetry group for a quantum bit is spanned by the groups *SU (2)* and *U (1)* and the complex conjugation. The essential part of the symmetry group for a quantum bit is the *SU (2)*, a three parameter compact group. Any number of quantum bits can be represented in the Hilbert space of measurable functions over the *SU (2)*, which is a S^3. This three dimensional space S^3 is identified with the three dimensional position space. This was already explained by Weizsäcker and Drieschner, but the interpretation of the qubits as "the smallest generally thinkable objects" ("die kleinsten überhaupt denkbaren Objekte") [7] has set the qubits in a strong connection to spacelike atomistic associations. This has hindered the developing of the theory into a structure compatible to general relativity.

The connection to general relativity was established in [8, 9 and 10].

On a first sight it is not so obvious that a qubit is primarily a cosmic structure, but all localization properties arose only in connection with a material carrier and without carrier no localization. A qubit corresponds to a two-dimensional representation of *SU(2)*. Such a representation can behold as a subrepresentation of the regular representation. Then any state of a qubit is given by a function having only one knot surface on the *SU(2)*, the three-dimensional sphere S^3. Such a function divides the S^3 into two parts. Many qubits are represented by the tensor product of the two-dimensional representations of the *SU(2)*. Such a product representation can be reduced into irreducible representations. In this reduction scheme more localized functions appear. Therefore only many qubits can represent something that is localized on the S^3.

For an easier understanding we remind to the sinus. It divides the circle into two halves. With products of many sin-functions strongly localized functions can be constructed. The same is valid for the qubits.

Suppose there are N qubits in the cosmos. The tensor product of the N two-dimensional representations $D_{1/2}$ can be decomposed. The Clebsch-Gordan-series of the reduction into irreducible representations has the form

$$({}^2D_{1/2})^{\otimes N} = \bigoplus_{j=0}^{|N/2|} \frac{N!(N+1-2j)}{(N+1-j)!\,j!}\left({}^{2|N/2|-2j+1}D_{|N/2|-j}\right).$$

The factors of multiplicities

$$f(j) = \frac{N!(N+1-2j)}{(N+1-j)!\,j!}$$

grow almost linearly from $f\,(N/2) = O\,(2^N N^{3/2})$ for the largest wavelength D_0 respectively $D_{1/2}$ to its maximum at $D_{\sqrt{N}}$ with a value

$$f[(1/2)(N - \sqrt{N})] = O(2N\,N^{1}) = (\sqrt{N})\;O(2N\,N^{3/2}) = (\sqrt{N})\,f(N/2).$$

Afterwards an exponential degree follows until the value 1. For large N the maximum is very sharp and the functions with a wavelength smaller than $\sqrt{N}$ almost never occur and can be ignored. By this group theoretical definition a smallest length on the S^3 is introduced, which will be identified with the Planck-length λ_0.

7. A cosmological model from the Protyposis

If it is acknowledged that the S^3 can be identified with position space, then with three physical plausible assumptions a cosmological model follows. [11]

- The energy of a quantum system is inversely to its extension.

- The first law of thermodynamics is correct: $dU + p\, dV = 0$.
- There exists a universal and distinguished velocity: c.

Because one qubit has a wavelength proportional to the radius R of the S^3

$$R = \lambda_0 \sqrt{N}$$

its energy must be proportional to

$$1/R = 1/\sqrt{N}\,.$$

Then the N qubits have an energy proportional to $N/R = \sqrt{N} = R$. This means that the total energy U is proportional to R.

The volume V of a S^3, the "surface" of a four dimensional sphere, is $V = 2\pi^2 R^3$. If one chooses the units for energy in such a way that $U = 2\pi^2 R$, then the energy density behaves like $\mu = 1/R^2$.

From the first law of thermodynamics $dU + p\, dV = 0$ it follows

$$2\pi^2 (dR + p\, 3 R^2 dR) = 0$$

or

$$p = -1/(3R^2)$$

and the state equation for qubits results as

$$p = -\mu/3.$$

The energy-momentum-tensor is $T_i^k = diag(\mu, \mu/3, \mu/3, \mu/3)$.

If there is – as supposed – a distinguished velocity c, then it is natural that the cosmic radius R increases proportional to time t_{cosmos}

$$R = c\, t_{cosmos}\,.$$

So for the Hubble-parameter $H = (dR/dt)/R$ it follows

$$H = 1/t_{cosmos}\,.$$

The resulting cosmological model is a Friedman-Robertson-Walker-cosmos with the line element

$$ds^2 = dt^2 - R^2(t)\,[(1 - r^2)^{-1} dr^2 + r^2 d\Omega^2)]$$

with

$$R(t) = c\,t.$$

The S^3-cosmos expands with velocity of light.

From this metric the Einstein-tensor G_i^k can be computed, then

$$G_i^k \sim T_i^k = diag(\mu, \mu/3, \mu/3, \mu/3).$$

If it is demanded that the proportionality between G_i^k and T_i^k should be conserved also for local variations of the energy then Einstein's equations of general relativity appear as a consequence of the abstract quantum information.

The Protyposis can appear as matter and/or energy and/or vacuum and/or as a no-particle substance, that today is declared as dark energy.

$$T^i_{k\,(Protyposis)} = T^i_{k\,(matter)} + T^i_{k\,(light)} + T^i_{k\,(vacuum)} + T^i_{k\,(no\ particles/dark\ energy)}.$$

The energy density of the Protyposis is denoted by μ, the energy density of matter and light by $\mu_{(matter)}$, $\mu_{(light)}$ and the energy density of the vacuum by λ. The energy density of the vacuum behaves like an effective cosmological term. Because of $\mu>0$, $\mu_{(matter)}>0$, $\mu_{(light)}>0$ it follows for the relation between matter and vacuum $2\lambda \geq \mu_{(matter)} + \mu_{(light)} \geq \lambda$. It has the right order of magnitude. There is no restriction for the size of the no-particle form of the Protyposis.

$$
\left[T^{\,i}_{\,k}\right]_{\text{Protyposis}} = \begin{bmatrix} \mu & & & \\ & \frac{\mu}{3} & & \\ & & \frac{\mu}{3} & \\ & & & \frac{\mu}{3} \end{bmatrix} =
$$

$$
\begin{bmatrix} \mu_M & & & \\ & 0 & & \\ & & 0 & \\ & & & 0 \end{bmatrix} + \begin{bmatrix} \mu_L & & & \\ & -\frac{\mu_L}{3} & & \\ & & -\frac{\mu_L}{3} & \\ & & & -\frac{\mu_L}{3} \end{bmatrix} + \begin{bmatrix} \lambda & & & \\ & \lambda & & \\ & & \lambda & \\ & & & \lambda \end{bmatrix} + \begin{bmatrix} \mu_{NP} & & & \\ & \frac{\mu_{NP}}{3} & & \\ & & \frac{\mu_{NP}}{3} & \\ & & & \frac{\mu_{NP}}{3} \end{bmatrix}
$$

A comparison with data shows that this model is not rejected by the observations. The cosmological model implicates $H\, t_{cosmos} = 1$. The probability distribution for $H_0 t_0$ given by the *SN Ia* observations is tightly constrained to 0.96 ± 0.04, and is an approximating Gaussian curve [12]. And the supernova Data for higher z move towards the Protyposis-cosmology.

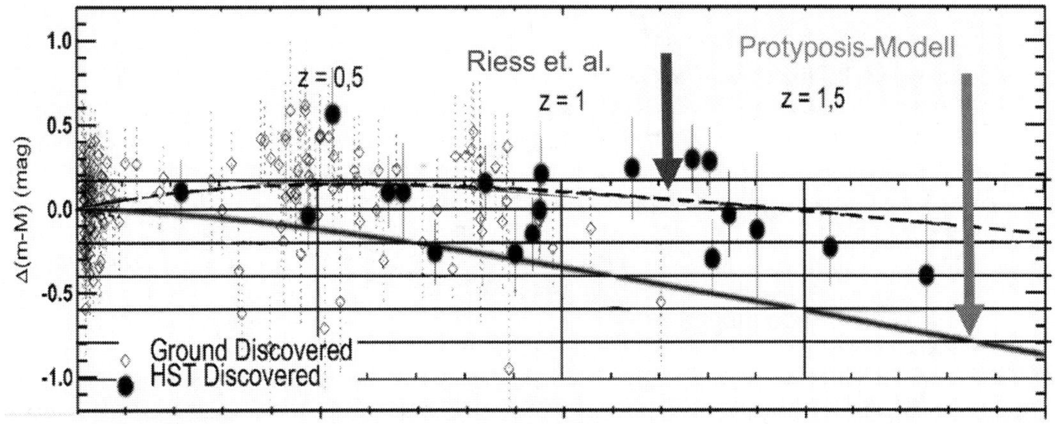

Figure 1. Type Ia Supernova Discoveries at z > 1 From the Hubble Space Telescope: Evidence for Past Deceleration and Constraints on Dark Energy Evolution [13]

8. Relativistic particles from quantum bits

For a precise definition of a particle one has to work in Minkowski space. A relativistic particle is then represented by an irreducible representation of the Poincaré group. Such a representation can be constructed from quantum information by Parabose creation and destruction operators for qubits and anti-qubits (Urs and Anti-Urs) with states running from 1 to 4.

Let be r,s,t $\in\{1,2,3,4\}$, $|\Omega\rangle$ the vacuum for qubits and p the order of Parabose statistics. The commutation relations for Parabose are

$$
[\hat{a}_t, \{\hat{a}_r^+, \hat{a}_s^+\}] = -2\delta_{rt}\hat{a}_s^+ - 2\delta_{st}\hat{a}_r^+ \qquad [\hat{a}_t^+, \{\hat{a}_r, \hat{a}_s\}] = -2\delta_{rt}\hat{a}_s - 2\delta_{st}\hat{a}_r \qquad [\hat{a}_t^+, \{\hat{a}_r^+, \hat{a}_s\}] = -2\delta_{st}\hat{a}_r^+
$$

$$
[\hat{a}_s, \{\hat{a}_r, \hat{a}_s\}] = 0 \qquad [a_s^+, \{a_r^+, a_s^+\}] = 0
$$

$$
\hat{a}_s \hat{a}_r^+ |\Omega\rangle = \delta_{rs}\, p\, |\Omega\rangle.
$$

The vacuum of Minkowski space $|0\rangle$ is an eigenstate of the Poincaré group with mass, energy, momentum and spin equal to zero. It can be constructed over the vacuum of the qubits $|\Omega\rangle$: [14]

$$|0\rangle = \sum_{n_1=0}^{\infty} \sum_{n_2=0}^{\infty} \frac{(-1)^{n_1+n_2}}{n_1! \cdot n_2!} \left(\frac{\hat{a}_1^+ \hat{a}_3^+ + \hat{a}_3^+ \hat{a}_1^+}{2} \right)^{n_1} \left(\frac{\hat{a}_2^+ \hat{a}_4^+ + \hat{a}_4^+ \hat{a}_2^+}{2} \right)^{n_2} |\Omega\rangle.$$

Another example [15, 16] is a massive fermion at rest with mass $m=P_0\neq 0$, momentum $P_1=P_2=P_3=0$, $spin_z=-½$, Parabose-order $p>1$:

$$\Phi(m) = \sum_{n_1=0}^{\infty} \sum_{n_2=0}^{\infty} \sum_{n_3=0}^{\infty} \frac{(-1)^{(n_1+n_2+n_3)} m^{(2\cdot n_1+n_2+n_3)} (p-1+n_1+n_2+n_3)!}{(p+2\cdot n_1+n_2+n_3)! \cdot (p-2+n_1+n_3)! \cdot (p-1+n_1+n_2)!} *$$
$$* \left[\hat{a}_1^+ + m\,\hat{a}_2^+ \left(\frac{\hat{a}_4^+ \hat{a}_1^+ + \hat{a}_1^+ \hat{a}_4^+}{2} \right) \right] \frac{\left(\frac{\hat{a}_4^+ \hat{a}_2^+ + \hat{a}_2^+ \hat{a}_4^+}{2} \right)^{n_3} \left(\frac{\hat{a}_4^+ \hat{a}_1^+ + \hat{a}_1^+ \hat{a}_4^+}{2} \right)^{n_1} \left(\frac{\hat{a}_3^+ \hat{a}_2^+ + \hat{a}_2^+ \hat{a}_3^+}{2} \right)^{n_1} \left(\frac{\hat{a}_3^+ \hat{a}_1^+ + \hat{a}_1^+ \hat{a}_3^+}{2} \right)^{n_2}}{n_1! \cdot n_2! \cdot n_3! \cdot (p+1+2n_1+n_2+n_3) \cdot (p-1+n_1+n_3)} |0\rangle.$$

Matter can be seen indeed as a special form of abstract quantum information. For every day life an appropriate differentiation may be:

- Matter is inactive, it offers resistance against change.
- Energy can move matter.
- Information can trigger Energy.

9. Conclusions for life and consciousness

The Protyposis enables clear and accessible perceptions for quantum phenomena. Suitable imaginations for quantum behaviour cannot be found among the facts of the objects and the forces, but rather in circumstances, where quantum physics acts on the macroscopic world, where possibilities exist as reality, e.g. in quantum controlling processes of/on something unstable, like our body.

Life is characterized by controlling and timing via quantum information. Only unstable systems can be controlled. Living systems are unstable because they are far from the thermodynamical equilibrium. In the process of self-regulation – which in the progression of biological evolution extend even to consciousness – at organisms quantum effects can become operative macroscopically. Consciousness is quantum information with a living brain as its carrier, quantum information that experiences and knows himself. This is intended not as an analogy but as a physical characterization. Quantum information is of equal reality like e.g. the quarks. Of course, a human personality as we know from daily life needs a living brain as its carrier, but the consciousness of a person is much more than only "a function" of his brain. That consciousness is widely based on quantum physics becomes clear also in many psychic phenomena. We can perceive many possibilities at the same time, we can feel ambivalence, and we can have contradictory thoughts about the same object at the same time, and so on. Controlling and timing of the body happen by mental information. The unconscious is physically seen a form of quantum information just like the consciousness. For the unconscious does not exist – like for any unknown quantum system – no time, no clear and certain „NO". The ambivalence, a characteristic of any quantum system so to say, is much more expressed in the unconscious than in our conscious thoughts.

At last it should be mentioned that all the experimental tryouts about the information processing at living brains, like fNMR or PET, are based on quantum physics.

Quantum theory offers by the Protyposis an understandable new view of the world, mankind and consciousness. More can be found in the Literature. [17, 18]

References

[1] Ali S T, Engliš M 2005 *Rev. Math. Phys.* **17** 391
[2] Weizsäcker C F v 1958 *Weltbild der Physik*, (Stuttgart: Hirzel) p.357
[3] Heisenberg W 1969 *Der Teil und das Ganze* (München: Piper) p 332
[4] Weizsäcker C F v 1982 *Aufbau der Physik*, (München: Hanser) p 172
[5] Weizsäcker C F v 1971 *Einheit der Natur*, (München: Hanser) p 361
[6] Drieschner M 1979 *Voraussage – Wahrscheinlichkeit – Objekt* (Springer: Berlin)

[7] [6] p 183
[8] Görnitz T 1988 *Int. J. Theor. Phys.* **27** 527
[9] Görnitz T 1988 *Int. J. Theor. Phys.* **27** 659
[10] Görnitz T, Ruhnau E 1989 *Int. J. Theor. Phys.* **28** 651
[11] Görnitz T 2009 arXiv:0904.1784
[12] Tonry J L et al 2003 *Astrophys. J.* **594** 1, Fig. 15
[13] Riess A G et al 2004 Astrophys. J. **607** 665 (*Preprint* arXiv:astro-ph/0402512)
[14] Görnitz T, Graudenz D, Weizsäcker C F v 1992 *Intern. J. Theoret. Phys.* **31** 1929
[15] Görnitz T and Schomäcker U 1997 Group theoretical aspects of a charge operator in an urtheoretical framework *Proc. XXI Int. Colloq. On Group Theoretical Methods in Physics (Group 21)* pp 334-338 (Singapore: World Scientific).
[16] Görnitz T, Görnitz B 2002 *Der kreative Kosmos - Geist und Materie aus Information* (Heidelberg: Spektrum Akadem. Verl.)
[17] Görnitz T 1999 *Quanten sind anders - Die verborgene Einheit der Welt* (Heidelberg: Spektrum Akadem. Verl.)
[18] Görnitz T, Görnitz B 2008 *Die Evolution des Geistigen/ Quantenphysik – Bewusstsein – Religion* (Göttingen: Vandenhoeck&Ruprecht)

Journal of Physics: Conference Series **237** (2010) 012012 doi:10.1088/1742-6596/237/1/012012

A group-theoretical derivation of the S-matrix for the Pöschl-Teller potentials

J. Guerrero

Departamento de Matemática Aplicada, Universidad de Murcia, Campus de Espinardo, 30100 Murcia, Spain

E-mail: juguerre@um.es

Abstract. A derivation of the S-matrix for the Pöschl-Teller and related potentials is done by group-theoretical methods using the group $SU(1,1)$, the dynamical symmetry for these systems. Light is shed on the representation of $SU(1,1)$ which must be used to this end, and the nature of bound (and anti-bound or virtual) states viewed as poles of the S-matrix in the complex plane.

1. Introduction

In this work we shall study the scattering states of the modified Pöschl-Teller potential:

$$V(x) = -\frac{D}{\cosh^2(\alpha x)}\,. \tag{1}$$

For that purpose, we shall determine the classical symmetries of the system, derive the Schrödinger equation from the Casimir of the symmetry group, for both bound and scattering states, find its solutions and derive the S matrix for scattering states. Finally, we shall determine the poles of the S matrix and interpret the results in a group-theoretical setting.

To find the classical symmetries of the system, we follow the procedure designed in [1] for the modified Pöschl-Teller potential and further developed in [2] for the Trigonometric Pöschl-Teller and Morse potentials[1] where a transformation is performed relating an open subset of the phase space of the system (that with a definite sign of the energy, either positive or negative) with (an open subset of) coadjoint orbits of the Lie groups $SO(2,1)$ or $SO(3)$. This transformation is not symplectic and in some cases it fails to be differentiable at some points ($H = 0$), although when restricted to the open subsets of positive or negative energy it is differentiable. This has important implications at the quantum level, where the system is realized as certain representations of these groups.

Although we are interested in the scattering states of the Modified Pöschl-Teller potential (MPT), for completeness we shall also discuss the Trigonometric Pöschl-Teller potential (TPT), having only bound states, and the bound states of the MPT, in order to compare with the results

[1] See also [3] for a similar construction, viewed as a classsical factorization, applied sistematicaly to a larger family of potentials. However, the symmetry algebras $SO(2,1)$ and $SO(3)$ are only obtained for the same cases as those considered in [1, 2].

obtained for the poles of the S matrix, which are related with bound and anti-bound or virtual states, as well as with resonant states (although in the present case there are no resonances).

It should be stressed that the symmetries we are considering are dynamical symmetries, in the sense of spectrum generation, as opposed to degeneracy symmetry which considers states with the same energy. Dynamical symmetry groups are also distinct from potential groups, were the action of the group keeps the energy but changes the potential parameters, therefore mixing states belonging to different physical systems. Despite of this lack of physical interpretation, the potential group approach has been very successful in finding the analytical solutions for many solvable potentials [4, 5, 6].

The paper is organized as follows. Sec. 2 introduces the main categories of Pöschl-Teller potentials, Sec. 3 is devoted to the Trigonometric Pöschl-Teller potential, finding its classical and quantum symmetries, Sec. 4 is focused on the Modified Pöschl-Teller potential, studying its classical symmetries, Sec. 5 is devoted to the bound states of Modified Pöschl-Teller potential, Sec. 6 to the scattering states, paying attention to the properties of the S matrix. Finally, Sec. 7 comments on some possible extensions of the work.

2. The family of Pöschl-Teller potentials

The Pöschl-Teller potentials are a family of potentials which can be classified in two types:

- Trigonometric Pöschl-Teller potentials (TPT):

$$V(x) = D\left(\frac{\lambda}{\cos^2(\alpha x)} + \frac{\kappa}{\sin^2(\alpha x)}\right) \tag{2}$$

- Hyperbolic (Modified) Pöschl-Teller potentials (MPT)

$$V(x) = -D\left(\frac{\lambda}{\cosh^2(\alpha x)} + \frac{\kappa}{\sinh^2(\alpha x)}\right) . \tag{3}$$

They have many applications in different branches of Molecular Physics, where they describe out-of-plane bending vibrations, with some variations like the Rosen-Morse Potential [7], and in Solid State Physics, where they model 1-D crystals (Scarf Potential) [8]. The PT potentials also appear in generalizations of the harmonic oscillator potential to spaces of constant curvature [9, 10, 11]. The PT potentials have the same solutions as an harmonic oscillator with position dependent mass, or with energy dependent frequency [12, 1]. They are integrable and related to a $SO(2,1)$ (or SO(3)) dynamical symmetry. In fact, they can be derived in a unified way, together with the Morse potential, from free motion in homogeneous spaces under the $SU(2)$ and $SU(1,1)$ Lie groups [13, 14], assuring in this way their integrability.

They can be solved by different approaches: dynamical group [1], potential group [4, 5, 6], SUSY QM and shape invariance [15], etc.

Let us study the phase space associated with these systems, in particular the cases of the symmetric TPT and MPT potentials.

3. Trigonometric Pöschl-Teller potential

We shall concentrate on the Trigonometric PT of the form:

$$V(x) = \frac{D}{\cos^2(\alpha x)} , \tag{4}$$

which is symmetric and has the form given in figure 1a. The parameter $D > 0$ is the minimum of the potential (the value at the origin), and α is related to the width of the potential.

3.1. Phase space

The trajectories in phase space for the TPT potential can be derived from the equation of motion, either in Lagrangian or Hamiltonian form, and are given by:

$$\begin{aligned} x(t) &= \frac{1}{\alpha}\arcsin\left[\sin(\alpha x_0)\cos(\omega(E)t) + \frac{\alpha}{m\omega(E)}p_0\cos(\alpha x_0)\sin(\omega(E)t)\right] \\ p(t) &= m\dot{x}(t)\,, \end{aligned} \tag{5}$$

where (x_0, p_0) are the initial coordinate and momentum parameterizing each solution and $E = \frac{p_0^2}{2m} + \frac{D}{\cos(\alpha x_0)^2}$ is the energy of this trajectory. In the last expression $\omega(E) \equiv \sqrt{\frac{2\alpha^2 E}{m}}$ is an energy dependent frequency[2].

These trajectories are shown in figure 1b. If we represent these trajectories in a 3D-phase space (x, p, H), we obtain the graphic shown in figure 1c. Note that in this case $E \geq D > 0$.

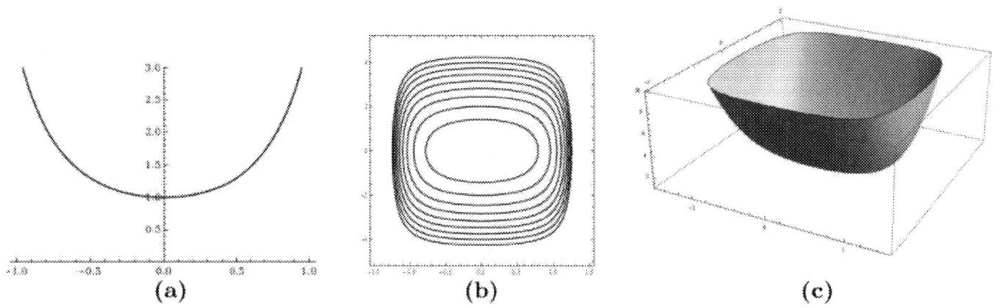

Figure 1. *Symmetric PT potential (a) and trajectories in 2D (b) and 3D phase space (c).*

3.2. Closing a subalgebra of the Poisson algebra

To obtain the dynamical group of the system, we are now interested in closing a Lie algebra with functions of the energy, the momentum and the position. We start with the basic Poisson bracket, derived from the Poincaré-Cartan 1-form taking the quotient by the solutions of the equation of motion (5), which has the canonical form:

$$\{x, p\} = 1\,, \tag{6}$$

with $p = \frac{\partial L}{\partial \dot{x}}$. Although $\{H, x, p\}$ do not close a Poisson subalgebra, we can find the functions closing the algebra $SO(2,1)$:

$$\mathcal{E} \equiv 2\sqrt{D}\sqrt{H}\,, \quad \mathcal{X} \equiv \frac{2}{\sqrt{m\Omega}}\sqrt{H}\sin(\alpha x)\,, \quad \mathcal{P} \equiv \sqrt{2}p\cos(\alpha x)\,, \tag{7}$$

with $\Omega = \omega(D)$, i.e. the frequency at the bottom of the potential, corresponding to the limit of small oscillations.

In fact, we have:

$$\{\mathcal{E}, \mathcal{P}\} = m\Omega^2\mathcal{X}\,, \quad \{\mathcal{E}, \mathcal{X}\} = -\frac{1}{m}\mathcal{P}\,, \quad \{\mathcal{X}, \mathcal{P}\} = \frac{1}{D}\mathcal{E}\,. \tag{8}$$

[2] The solutions of the classical equation of motion correspond to that of an harmonic oscillator with energy dependent frequency (see [1] and also [16]).

This is an $SO(2,1)$ algebra where the compact generator is $\mathcal{E}$ and the Casimir is:

$$\frac{1}{D}\mathcal{E}^2 - m\Omega^2\mathcal{X}^2 - \frac{1}{m}\mathcal{P}^2 = 4D > 0\,. \tag{9}$$

The coadjoint orbit of $SO(2,1)$ associated with this system is the upper sheet of a two-sheet hyperboloid, since the Casimir is positive and $\mathcal{E} > 0$.

3.3. *The quantum TPT oscillator*

The time-independent Schrödinger equation for a particle of mass m in a TPT potential with depth D and width $1/\alpha$ is:

$$\frac{\hbar^2}{2m}\frac{d^2\Psi}{dx^2} + \left(E - \frac{D}{\cos^2(\alpha x)}\right)\Psi = 0\,. \tag{10}$$

The solutions to this equation are given in terms of Gegenbauer polynomials:

$$\Psi_n^k(x) = (\cos(\alpha x))^k C_n^k(\sin(\alpha x))\,, \tag{11}$$

where $k(k-1) = \frac{2mD}{\hbar^2\alpha^2}$.

The energy eigenvalues are given by:

$$E_n = \frac{\hbar^2\alpha^2}{2m}(n+k)^2 = -\frac{\hbar^2\Omega^2}{4D}(n+k)^2 = -\frac{D}{k(k-1)}(n+k)^2\,,\ n = 0, 1, \ldots\,. \tag{12}$$

The scalar product is the usual one in $L^2(\mathbb{R}, dx)$, and with this scalar product Ψ_n^k, , $n = 0, \ldots$, are orthogonal and normalizable, the orthonormal basis being:

$$\begin{aligned} \bar{\Psi}_n^k(x) &= N_n^k\,\Psi_n^k(x) \\ N_n^k &= \sqrt{\frac{\alpha n!(n+k)\Gamma(k)\Gamma(2k)}{\pi^{\frac{1}{2}}\Gamma(k+\frac{1}{2})\Gamma(n+2k)}} \end{aligned} \tag{13}$$

for $n = 0, 1, \ldots$.

3.3.1. *The dynamical group for the TPT.*

It is possible to find ladder operators for the TPT wavefunctions:

$$\begin{aligned} \hat{Z} &= \frac{1}{\sqrt{2k}}\sqrt{\frac{n+k-1}{n+k}}\left[\frac{1}{\alpha}\cos(\alpha x)\frac{d}{dx} + (n+k)\sin(\alpha x)\right] \\ \hat{Z}^\dagger &= \frac{1}{\sqrt{2k}}\sqrt{\frac{n+k+1}{n+k}}\left[-\frac{1}{\alpha}\cos(\alpha x)\frac{d}{dx} + (n+k)\sin(\alpha x)\right]\,, \end{aligned} \tag{14}$$

which, on normalized solutions, act as:

$$\begin{aligned} \hat{Z}\bar{\Psi}_n^k &= \sqrt{n\frac{2k+n-1}{2k}}\,\bar{\Psi}_{n-1}^k \\ \hat{Z}^\dagger\bar{\Psi}_n^k &= \sqrt{(n+1)\frac{2k+n}{2k}}\,\bar{\Psi}_{n+1}^k\,. \end{aligned} \tag{15}$$

These operators, together with the number of quanta operator, $\hat{n}$, close the Lie algebra of $SU(1,1)$. Therefore, we corroborate that $SU(1,1)$ is the dynamical algebra for the TPT potential ([17]).

3.3.2. Deriving the Schrödinger equation for the TPT. It should be stressed that the transformation (7) is not symplectic since $\{\mathcal{X},\mathcal{P}\} = \frac{1}{D}\mathcal{E} \neq 1$. This is not a drawback, however, and the dynamics (both classical and quantum) can be derived entirely from $SO(2,1)$. In this case the transformation (7) is differentiable, and the phase space shown in figure 1c is mapped to a full coadjoint orbit of $SO(2,1)$. This implies that at the quantum level this system will be realized as a unitary irreducible representation of the discrete series of representations of $SO(2,1)$ (see, for instance, [17]), the one associated with the upper sheet of a two-sheet hyperboloid in the sense of the Coadjoint Orbit Method of Kirillov [18] (see also [19]).

The time independent Schrödinger equation for the TPT potential can be derived from the Casimir of $SO(2,1)$ in a representation of the Discrete positive series $\mathcal{D}_k^+$, realized as a free particle in Anti-de Sitter space-time (which can be seen as a relativistic oscillator):

$$\hat{C}\varphi \equiv -\frac{c^2}{\omega^2}\Box\varphi = k(k-1)\varphi\,, \tag{16}$$

where $k = \frac{mc^2}{\hbar\omega}$ and $\Box$ is the D'Alambertian in an Anti-de Sitter space-time:

$$\Box \equiv \frac{1}{c^2(1+\frac{\omega^2 y^2}{c^2})}\frac{\partial^2}{\partial\tau^2} - \frac{2\omega^2 y}{c^2}\frac{\partial}{\partial y} - (1+\frac{\omega^2 y^2}{c^2})\frac{\partial^2}{\partial y^2}\,. \tag{17}$$

Defining $x \equiv \frac{1}{\alpha}\mathrm{arctanh}(\sqrt{\frac{m\omega}{\hbar k}}y)$, we arrive at

$$-\frac{\hbar^2\alpha^2}{2m}\frac{1}{\omega^2}\frac{\partial^2}{\partial\tau^2}\varphi = \left[\frac{\hbar^2}{2m}\frac{\partial^2}{\partial x^2}\varphi - \frac{D}{\cos^2(\alpha x)}\varphi\right]\,, \tag{18}$$

where $k(k-1) = \frac{2mD}{\alpha^2\hbar^2} = (\frac{2D}{\hbar\Omega})^2$.

Denoting $t \equiv \frac{2\omega}{\Omega}\tau$, $\hat{E} \equiv i\hbar\frac{\partial}{\partial t}$, and $\chi \equiv e^{-\frac{i}{\hbar}\sqrt{D}\sqrt{E}t}\varphi$ we recover the time-independent Schrödinger equation for a particle of mass m in a MPT potential with depth D and width $1/\alpha$:

$$\frac{\hbar^2}{2m}\frac{d^2\chi}{dx^2} + \left(E - \frac{D}{\cos^2(\alpha x)}\right)\chi = 0\,. \tag{19}$$

3.3.3. Deriving the ladder operators for the TPT. The ladder operators can also be derived from ladder operators for the discrete series representation $\mathcal{D}_k^+$ of $SU(1,1)$. In the same parametrization of the free particle in Anti-de Sitter spacetime, they are given by (see [1]):

$$\begin{aligned}\hat{Z} &= \sqrt{\frac{\hbar}{2m\omega}}e^{i\omega t}\sqrt{1+\frac{\omega^2y^2}{c^2}}\left[\frac{\partial}{\partial y} + i\frac{\omega y}{c^2(1+\frac{\omega^2 y^2}{c^2})}\frac{\partial}{\partial\tau}\right]\\ \hat{Z}^\dagger &= \sqrt{\frac{\hbar}{2m\omega}}e^{-i\omega t}\sqrt{1+\frac{\omega^2y^2}{c^2}}\left[-\frac{\partial}{\partial y} + i\frac{\omega y}{c^2(1+\frac{\omega^2 y^2}{c^2})}\frac{\partial}{\partial\tau}\right]\,.\end{aligned} \tag{20}$$

Performing the same change of variables as in the derivation of the Schrödinger equation, the resulting ladder operators acting on energy eigenfunctions are the ones given in Eq. (14), except for the global factor depending on the state n. This factor must be added to restore the unitarity of the representation when the time variable is factored out (see [1] and references therein for a detailed discussion).

Journal of Physics: Conference Series **237** (2010) 012012 doi:10.1088/1742-6596/237/1/012012

4. The modified Pöschl-Teller potential

We shall concentrate on the MPT of the form:

$$V(x) = -\frac{D}{\cosh^2(\alpha x)}\,, \tag{21}$$

which is also symmetric and has the form given in figure 2a. The parameter $D > 0$ is the potential depth and α is related to the width of the potential.

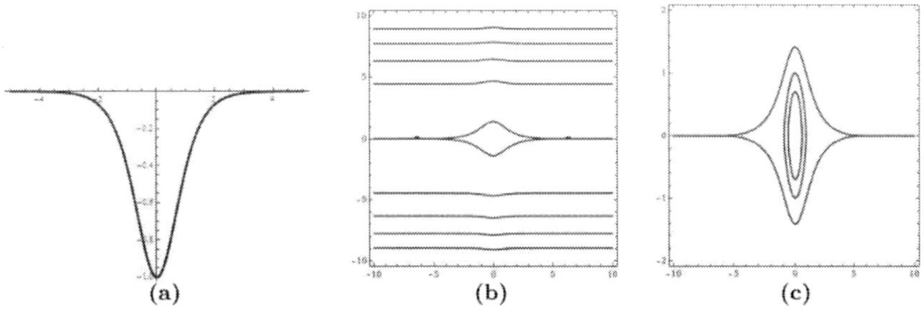

Figure 2. *Symmetric MPT potential (a) and trajectories in 2D phase space for $E > 0$ (b) and $E < 0$ (c).*

4.1. Phase space

The trajectories in phase space for the MPT potential are given by:

$$\begin{aligned} x(t) &= \frac{1}{\alpha}\text{arcsinh}\left[\sinh(\alpha x_0)\cosh(\omega(E)t) + \frac{\alpha}{m\omega(E)}p_0\cosh(\alpha x_0)\sinh(\omega(E)t)\right] \\ p(t) &= m\dot{x}(t)\,, \end{aligned} \tag{22}$$

where (x_0, p_0) are the initial coordinate and momentum parameterizing each solution and $E = \frac{p_0^2}{2m} - \frac{D}{\cosh^2(\alpha x_0)}$ is the energy of this trajectory. Here $\omega(E)$ has the same meaning as in the TPT case. Note that when $E < 0$ the time-dependent hyperbolic functions change to their trigonometric counterparts. These trajectories are shown in figure 2b for $E > 0$ and figure 2c for $E < 0$. For $E = 0$ they degenerate to $x(t) = \frac{1}{\alpha}\text{arcsinh}\left[\sinh(\alpha x_0) + \cosh(\alpha x_0)\frac{\alpha p_0}{m}t\right]$ which is a free motion in the coordinate $\xi = \sinh(\alpha x)$, $\xi = \xi_0 + \dot{\xi}_0\, t$ with constant velocity $\dot{\xi}_0 = \frac{\alpha}{m}p_0\sqrt{1+\xi_0^2}$, and p_0 related to x_0 through the constraint $E = 0$.

The trajectories in a 3D-phase space (x, p, H) are show in figure 3a for $E > 0$, figure 3b for $E < 0$ and figure 3c for the union of both cases.

4.2. Closing a subalgebra of the Poisson algebra

The basic Poisson bracket, derived as in the case of the TPT, has the canonical form:

$$\{x, p\} = 1\,. \tag{23}$$

Although $\{H, x, p\}$ do not close a Poisson subalgebra, we can find functions closing an algebra, but we must distinguish between the cases $E > 0$ and $E < 0$.

- $E > 0$

  The following functions close a $SO(2,1)$ algebra:

$$\mathcal{E} \equiv 2\sqrt{D}\sqrt{H}\,, \quad \mathcal{X} \equiv \frac{2}{\sqrt{m\Omega}}\sqrt{H}\sinh(\alpha x)\,, \quad \mathcal{P} \equiv \sqrt{2}p\cosh(\alpha x). \tag{24}$$

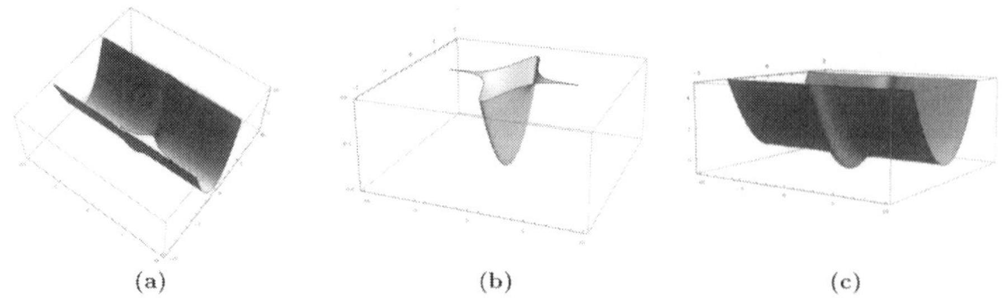

(a) (b) (c)

Figure 3. *Trajectories in 3D phase space for $E > 0$ (a), $E < 0$ (b) and the union of both cases (c), for the MPT Potential*

In fact, we find:

$$\{\mathcal{E},\mathcal{P}\} = -m\Omega^2\mathcal{X}\,, \quad \{\mathcal{E},\mathcal{X}\} = -\frac{1}{m}\mathcal{P}\,, \quad \{\mathcal{X},\mathcal{P}\} = \frac{1}{D}\mathcal{E}\,, \tag{25}$$

where again $\Omega = \omega(D)$. This is a $SO(2,1)$ algebra where the compact generator is $\mathcal{P}$ and the Casimir is:

$$\frac{1}{m}\mathcal{P}^2 - \frac{1}{D}\mathcal{E}^2 - m\Omega^2\mathcal{X}^2 = 4D > 0\,. \tag{26}$$

The coadjoint orbit of $SO(2,1)$ associated with this system is a half of a two-sheet hyperboloid, shown in figure 4a.

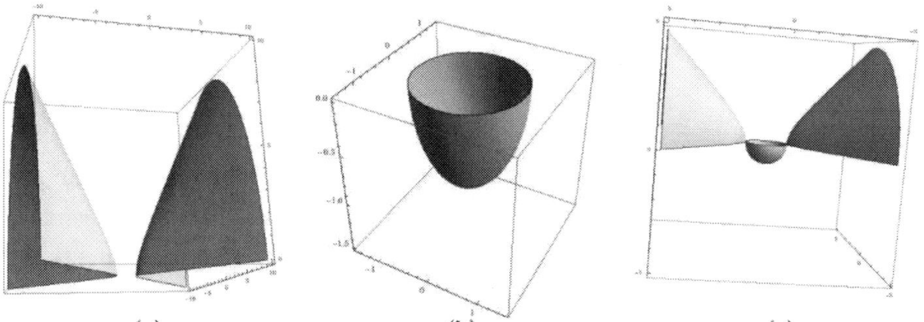

(a) (b) (c)

Figure 4. *Coadjoint orbits of the group $SO(2,1)$ for $E > 0$ (a), $SO(3)$ for $E < 0$ (b) and the union of both cases (c).*

- $E < 0$
  The following functions close a $SO(3)$ algebra[3]:

$$\mathcal{E} \equiv -2\sqrt{D}\sqrt{-H}\,, \quad \mathcal{X} \equiv \frac{2}{\sqrt{m\Omega}}\sqrt{-H}\sinh(\alpha x)\,, \quad \mathcal{P} \equiv \sqrt{2}p\cosh(\alpha x). \tag{27}$$

In fact, we find:

$$\{\mathcal{E},\mathcal{P}\} = m\Omega^2\mathcal{X}\,, \quad \{\mathcal{E},\mathcal{X}\} = -\frac{1}{m}\mathcal{P}\,, \quad \{\mathcal{X},\mathcal{P}\} = \frac{1}{D}\mathcal{E}\,. \tag{28}$$

[3] Note the minus sign in $\mathcal{E}$ in order to have the same sign as H.

This is an $SO(3)$ algebra with the Casimir function given by:

$$\frac{1}{D}\mathcal{E}^2 + m\Omega^2\mathcal{X}^2 + \frac{1}{m}\mathcal{P}^2 = 4D > 0\,. \tag{29}$$

The coadjoint orbit of $SO(3)$ associated with this system is a hemisphere, shown in figure 4b.

Putting the two orbits together we obtain a phase space, shown in figure 4c, resembling that of figure 3c.

As in the TPT case, the transformations (24) and (27) are not symplectic. In this case they are not even differentiable at $H = 0$. As shown in figure 4c, the whole $E = 0$ closed subset of phase space in (x, p, H), which in turn is formed by two disconnected curves, is mapped to two points in $(\mathcal{X}, \mathcal{P}, \mathcal{E})$, namely $(0, \pm 2\sqrt{mD}, 0)$. This means that the two cases, $E > 0$ and $E < 0$ are disconnected, and there is no way to connect them in the framework of the groups $SO(2,1)$ and $SO(3)$ (see the last section for a way to connect them).

4.3. The quantum MPT oscillator

At the quantum level, this system has special features, since we must distinguish between the cases $E > 0$ and $E < 0$, and in neither cases the phase space is a coadjoint orbit, but a half of it.

- $E > 0$
  The system will be realized as part of the sum of two unitary and irreducible representations of the discrete series (positive and negative) of (the universal covering group of) $SO(2,1)$. The Hamiltonian is a non-compact operator with positive continuum spectrum. This result seems to be in agreement with [20], where the discrete series of $SO(2,1)$ is used to describe the continuum spectrum of the MPT potential.
- $E < 0$
  The system will be realized as part of an irreducible representation of $SU(2)$. The Hamiltonian is a compact operator with discrete spectrum, but only one half of the $2s+1$ states of the $SU(2)$ representations are realized.

In the last case, there is a better interpretation as a non-unitary, finite-dimensional representation of (the universal covering group of) $SO(2,1)$ (see [1]). The reason is that in the context of $SU(2)$ there is no explanation for the fact that half of the states are missing, and that the potential depth should have only definite values, since it is related to the values of the Casimir, which in turns depends on the discrete values of the spin.

Here $SO(3)$ appears as contained in the complexification of $SO(2,1)$, but then it should be realized in a non-unitary way. But $SO(2,1)$ admits non-unitary, finite dimensional representations, which in many respects behave as those of $SU(2)$. The lack of unitarity manifest itself in that half of the states are non-normalizable, and therefore are outside of the Hilbert space of the physical (normalizable) states. These non-normalizable states are precisely the anti-bound states (ABS), which represent outgoing states growing at $\pm\infty$ (see [21]).

We shall study separately the bound states and the scattering states.

5. Bound states for the MPT potential

The time-independent Schrödinger equation for a particle of mass m in a MPT potential with depth D and width $1/\alpha$ is:

$$\frac{\hbar^2}{2m}\frac{d^2\Psi}{dx^2} + \left(E + \frac{D}{\cosh^2(\alpha x)}\right)\Psi = 0\,. \tag{30}$$

The bounded solutions to this equation are given in terms of Gegenbauer polynomials:

$$\Psi_n^q(x) = (\cosh(\alpha x))^{n-q} C_n^{q-n+\frac{1}{2}}(\tanh(\alpha x)), \tag{31}$$

where $q(q+1) = \frac{2mD}{\hbar^2\alpha^2}$.

The energy eigenvalues are given by ($n = 0, 1, \ldots, [q]$):

$$E_n = -\frac{\hbar^2\alpha^2}{2m}(q-n)^2 = -\frac{\hbar^2\Omega^2}{4D}(q-n)^2 = -\frac{D}{q(q+1)}(q-n)^2 . \tag{32}$$

The scalar product is the usual one, and with this scalar product Ψ_n^q, , $n = 0, \ldots, [q]$, are orthogonal and normalizable, the orthonormal basis being:

$$\begin{aligned} \bar{\Psi}_n^q(x) &= N_n^q \Psi_n^q(ux) \\ N_n^q &= \sqrt{\frac{\alpha n! \Gamma(q-n+\frac{1}{2})}{\pi^{\frac{1}{2}}\Gamma(q-n)(2q-2n+1)_n}} \end{aligned} \tag{33}$$

for $n = 0, \ldots, [q]$.

5.1. *The dynamical group for bound states*

It is possible to find ladder operators (see, for instance [22]) for the MPT wavefunctions:

$$\begin{aligned} \hat{Z} &= \frac{1}{\sqrt{2q}}\sqrt{\frac{q-n+1}{q-n}}\left[\frac{1}{\alpha}\cosh(\alpha x)\frac{d}{dx} + (q-n)\sinh(\alpha x)\right] \\ \hat{Z}^\dagger &= \frac{1}{\sqrt{2q}}\sqrt{\frac{q-n-1}{q-n}}\left[-\frac{1}{\alpha}\cosh(\alpha x)\frac{d}{dx} + (q-n)\sinh(\alpha x)\right], \end{aligned} \tag{34}$$

which, on normalized solutions, act as:

$$\begin{aligned} \hat{Z}\bar{\Psi}_n^q &= \sqrt{n\frac{2q-n+1}{2q}}\bar{\Psi}_{n-1}^q \\ \hat{Z}^\dagger\bar{\Psi}_n^q &= \sqrt{(n+1)\frac{2q-n}{2q}}\bar{\Psi}_{n+1}^q . \end{aligned} \tag{35}$$

These operators, together with the number of quanta operator, $\hat{n}$, close a Lie algebra which can be identified with $SU(2)$ in a representation of index $j = q$. Therefore, we could think of $SU(2)$ as the dynamical algebra for the MPT potential.

However, there are some inconsistencies in this scheme:

- The parameter q can take any positive real value, and this is not possible in $SU(2)$, where q must be integer or half-integer.
- There are only $[q] + 1$ normalizable states, instead of $2q + 1$ (for q integer or half-integer).
- Although the Schrödinger equation for the MPT oscillator admits more solutions, with $n > [q]$, these are either identical to the previous ones or are non-normalizable.
- In particular, if q is an integer, there exist states Ψ_n^q with $n = q, q+1, \ldots, 2q$, but the state with $n = q$ is non-normalizable, and $\Psi_{2q-n}^q = \Psi_n^q$, for $n = 0, 1, \ldots, q-1$.
- If q is not an integer (half-integer or real), there is an infinite number of states Ψ_n^q with $n = [q]+1, [q]+2, \ldots$, but all these states are non-normalizable.

Therefore, $SU(2)$ is not a good candidate to be the dynamical algebra for the MPT oscillator.

5.2. Deriving the Schrödinger equation for bound states.
As in the case of the TPT potential, the time independent Schrödinger equation for the MPT potential can be derived from the Casimir of $SO(2,1)$ in a representation of the discrete positive series $\mathcal{D}_k^+$, realized as a free particle in Anti-de Sitter space-time. If in the Casimir given by Eq. (16) we consider a negative Bargmann index $k = -q < 0$, we arrive at (see [1] for details):

$$-\frac{\hbar^2\alpha^2}{2m}\frac{1}{\omega^2}\frac{\partial^2}{\partial\tau^2}\varphi = \left[\frac{\hbar^2}{2m}\frac{\partial^2}{\partial x^2}\varphi + \frac{D}{\cosh^2(\alpha x)}\varphi\right], \tag{36}$$

where $x = \frac{1}{\alpha}\text{arctanh}(\sqrt{\frac{m\omega}{\hbar q}}y)$ and $q(q+1) \equiv k(k-1) = \frac{2mD}{\alpha^2\hbar^2} = (\frac{2D}{\hbar\Omega})^2$.

Denoting $t \equiv \frac{2\omega}{\Omega}\tau$, $\hat{E} \equiv i\hbar\frac{\partial}{\partial t}$, and $\chi \equiv e^{-\frac{i}{\hbar}\sqrt{D}\sqrt{-E}t}\varphi$ we recover at the time-independent Schrödinger equation for a particle of mass m in a MPT potential with depth D and width $1/\alpha$:

$$\frac{\hbar^2}{2m}\frac{d^2\chi}{dx^2} + \left(E + \frac{D}{\cosh^2(\alpha x)}\right)\chi = 0\,. \tag{37}$$

5.2.1. Deriving the ladder operators for the MPT potential. Similarly to the TPT case, the ladder operators (34) can be derived from the ladder operators for the Discrete series representations $\mathcal{D}_k^+$ of $SU(1,1)$, given in Eq. (20) performing the corresponding change of variables and replacing $k = -q$. Also, a global factor depending on the state n must be introduced to restore the unitarity.

As was commented previously, the assumption that $SU(2)$ is the dynamical group for the bound states of the MPT potential leads to inconsistences, as was shown in [23]. The question that now arises is, which is then the dynamical group?. A possible candidate for a dynamical group for the bound states of the MPT potential is $SU(1,1)$. Although $SU(1,1)$ is non-compact and therefore all its unitary representations are infinite-dimensional, it admits finite dimensional irreducible representations which are non-unitary (besides infinite-dimensional irreducible non-unitary representations). These can be obtained simply by changing the sign $k = -q\,, q > 0$ of the Bargmann index in the positive discrete series of $SU(1,1)$, D_k^+. If $2q \in \mathbb{Z}$, the representation has $2q+1$ states (although it can be a factor representation). If $2q \notin \mathbb{Z}$, the representation is infinite-dimensional (and non-unitary).

This procedure of changing the sign $k = -q\,, q > 0$ is precisely what we have done to derive the Schrödinger equation for the MPT potential from the Casimir of $SO(2,1)$ in a particular realization of the Discrete Series D_k^+, corresponding to a model of a relativistic oscillator. The wave functions and a pair of ladder operators are also obtained, with the same features mentioned before (non-normalizable states, etc).

The non-normalizability of some of the states is a consequence of the non-unitarity of the representation. By the restriction to the subspace of normalizable states, a pair of ladder operators belonging to the enveloping algebra of $SU(1,1)$ can be constructed, coinciding with (34) on normalized energy eigenstates (see [1]).

6. Scattering states for the MPT potential

For this section we shall denote the Bargmann index by N, instead of k, to avoid confusion with the momentum, which is generally denoted by k. The solution of the Schrödinger equation of the MPT potential for scattering states of energy $E = \frac{k^2}{2m}$ are given in terms of Hypergeometric functions as:

$$\Psi(x) = (1-u^2)^{-i\kappa}F(N - i\kappa, -N+1-i\kappa, 1-i\kappa, \frac{1-u}{2}), \tag{38}$$

where $u = \tanh(\alpha x)$, $\kappa = \frac{k}{\hbar\alpha}$ and $N(N-1) = \frac{2mD}{\alpha^2\hbar^2}$.

6.1. Deriving Schrödinger equation for scattering states.
After the change of variables:

$$
\begin{aligned}
x &= \frac{c}{\omega}\text{arccosh}\left(\frac{1}{\sqrt{\sin^2(\omega\tau)-\frac{\omega^2}{c^2}y^2\cos^2(\omega\tau)}}\right)\\
s &= mc\,\text{arcsinh}\left(\frac{\omega y}{c\sqrt{\sin^2(\omega\tau)-\frac{\omega^2}{c^2}y^2\cos^2(\omega\tau)}}\right),
\end{aligned}
\tag{39}
$$

and the identifications $t \equiv \frac{2\omega}{mc\Omega}s$, $\hat{E} \equiv i\hbar\frac{\partial}{\partial t}$, and $\chi \equiv e^{-\frac{i}{\hbar}\sqrt{D}\sqrt{E}t}\varphi$, the Casimir equation (16) turns into the time-independent Schrödinger equation for the MPT potential.

6.2. The S matrix for the MPT potential
To determine the S matrix, whose components are the transmission T and reflection R amplitudes, we study the asymptotic behaviour of the solutions at $x \to \pm\infty$:

- For $x \to +\infty$, $\Psi(x) \sim 4^{-i\kappa/2}e^{ikx/\hbar}$
- For $x \to -\infty$

$$
\Psi(x) \sim 4^{-i\kappa/2}\left[\frac{\Gamma(1-i\kappa)\Gamma(i\kappa)}{\Gamma(1-N)\Gamma(N)}e^{-ikx/\hbar} + \frac{\Gamma(1-i\kappa)\Gamma(-i\kappa)}{\Gamma(1-N-i\kappa)\Gamma(N-i\kappa)}e^{ikx/\hbar}\right], \tag{40}
$$

where use of the transformation properties of the hypergeometric functions has been made:

$$
\begin{aligned}
F(a,b,c;z) &= \frac{\Gamma(c)\Gamma(c-a-b)}{\Gamma(c-a)\Gamma(c-b)}F(a,b,a+b-c+1,1-z)\\
&+ (1-z)^{c-a-b}\frac{\Gamma(c)\Gamma(a+b-c)}{\Gamma(a)\Gamma(b)}\\
&\times F(c-a,c-b,c-a-b+1,1-z)\,.
\end{aligned}
\tag{41}
$$

From these expressions the transmission and reflection amplitudes are readily obtained (see [4, 5, 6]:

$$
\begin{aligned}
T(k) &= \frac{\Gamma(N-i\kappa)\Gamma(1-N-i\kappa)}{\Gamma(1-i\kappa)\Gamma(-i\kappa)}\\
R(k) &= \frac{\Gamma(N-i\kappa)\Gamma(1-N-i\kappa)\Gamma(i\kappa)}{\Gamma(N)\Gamma(1-N)\Gamma(-i\kappa)}\,.
\end{aligned}
\tag{42}
$$

The S matrix, defined as

$$
S = \begin{pmatrix} T & R \\ R & T \end{pmatrix} \tag{43}
$$

is unitary: $|T|^2 + |R|^2 = 1$. See figure 5 for a plot of $|T|^2$ and $|R|^2$ for different values of N.

For integer N, $R = 0$ and $T = 1$, see figure 5b. Therefore there is no reflection and the quantum system behaves as the classical one. The Hilbert space splits into two invariant subspaces, corresponding to (a half of) $\mathcal{D}_N^+$ and $\mathcal{D}_N^-$.

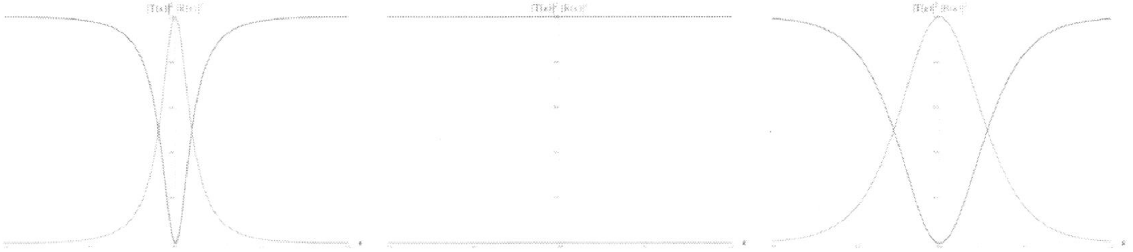

(a) (b) (c)

Figure 5. *Transmission and reflection coefficients: (a)* $N = 2.9$*, (b)* $N = 3$ *and (c)* $N = 3.5$.

6.3. Poles of the S matrix

The S matrix admits an analytical extension to the complex k plane, since the potential is rapidly decreasing at spatial infinity (see, for instance [24] and references therein). From the poles of T, which happen at the imaginary k axis we recover the bound states and the virtual or anti-bound states (ABS). In this case there are no resonances.

To study them, define $\kappa = i\nu$ and $q = -N$, then:

$$\begin{aligned} T(\nu) &= \frac{\Gamma(-q+\nu)\Gamma(1+q+\nu)}{\Gamma(1+\nu)\Gamma(\nu)} \\ R(\nu) &= \frac{\Gamma(-q+\nu)\Gamma(1+q+\nu)\Gamma(-\nu)}{\Gamma(-q)\Gamma(1+q)\Gamma(\nu)} \,. \end{aligned} \tag{44}$$

Bound states occur as poles at $\nu > 0$, and there are $[q]+1$: $\nu = q - n$ $n = 0, 1, \ldots, [q]$, at the energies corresponding with bound states.

ABS occur at $\nu < 0$, and their number depend on the character of q:

- For integer q, there is only one ABS state for $\nu = 0$ or $n = q$ ($E = 0$). But here $|T| = 1$ (no pole) !!
  There are zeros of T at symmetrical positions to bound states: $\nu = n - q < 0$. See figure 6b.
- For half-integer q, there is an infinite number of ABS, located at positions $\nu = q - n < 0$, $n = [q]+1, \ldots$.
  There are zeros of T at $\nu \in \mathbb{Z}^-$. See figure 6c.
- For real q, there is an infinite number of ABS, located at positions $\nu = q - n$, $n = [q]+1, \ldots$. There are also poles of T at $\nu = -q - n$, $n = 1, \ldots$. There are also zeros of T at $\nu \in \mathbb{Z}^-$. See figure 6a.

Although ABS are non-normalizable states and therefore are unphysical, those with a small value of ν have an effect on the transmission and reflection coefficients for real momentum. In particular, for integer q the ABS state at $\nu = 0$ causes that $|T(\kappa)|^2 = 1, \forall\kappa \in \mathbb{R}$. Even more, for $q = 2.9$, the presence of an ABS state at $\nu = -0.1$ (see figure 6a) causes the transmission coefficient to increase faster to 1, see figure 5a, than in other cases, as in figure 5c, where the ABS state is at $\nu = -0.5$.

It should be noted that the S matrix has a symmetry $N \to 1 - N$ (or $-q \to 1 + q$), which reflects the symmetry of the value of the Casimir under this transformation. The transmission and reflection coefficients are also periodic in N for k real, with period 1, as can be seen in figure 7. This property does not hold in the imaginary k axis. The reason is that, using the properties

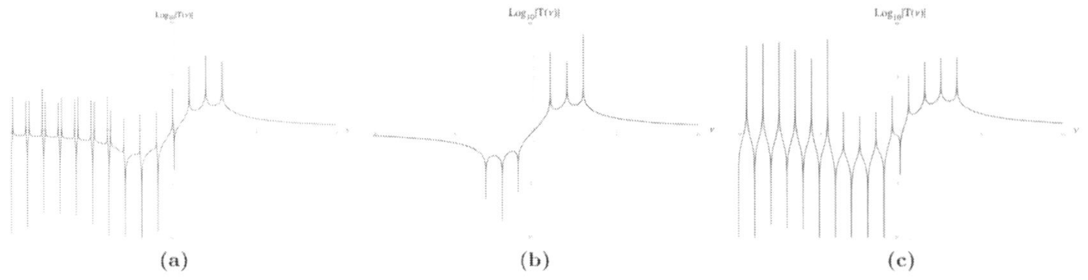

Figure 6. *Transmission coefficient in the imaginary axis, $\kappa = i\nu$, in a logaritmic scale, for different values of q: (a) $q = 2.9$, (b) $q = 3$ and (c) $q = 3.5$.*

of Gamma functions for complex arguments, they can be written as:

$$
\begin{aligned}
|T(k)|^2 &= \frac{\sinh^2(\pi\kappa)}{\sinh^2(\pi\kappa) - \sin^2(\pi N)} \\
|R(k)|^2 &= \frac{\sin^2(\pi N)}{\sin^2(\pi N) - \sinh^2(\pi\kappa)} \,.
\end{aligned}
\qquad (45)
$$

This periodic behaviour is shown in figure 7.

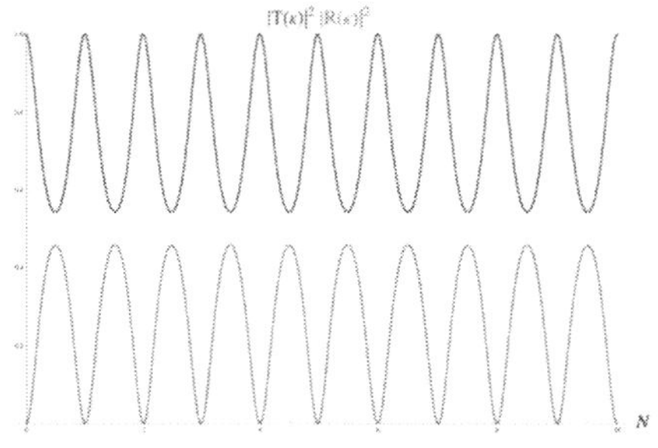

Figure 7. *Periodicity of the transmission and reflection coefficients in the real axis, as a function of N, for $\kappa = 0.3$.*

7. Comments and outlook

In this paper a group theoretical description of the modified Pöschl-Teller potential and related potentials, both for positive and negative energies, has been presented. Each case must be studied separately, since the dynamical symmetries and the representations associated are different for each case.

This behaviour is rather unpleasant, and a unified description, into a single Lie group, would be desirable. Probably this would require an infinite-dimensional group able to accommodate the Hamiltonian, which has a mixed spectrum (both positive continuum and negative discrete spectra), as a single generator. However, an intermediate step can be done including both groups into a finite-dimensional group, the price we must pay being that the Hamiltonian is associated with different generators for the positive and negative energy cases. The minimal algebra including both $SO(2,1)$ and $SO(3)$ algebras is the complexification of any of them, the $SL(2,\mathbb{C})$ algebra, which is isomorphic to $SO(3,1)$, the Lorentz algebra. The complexification appears here in a natural way due to the presence of the square root of the Hamiltonian in the transformation (24). Thus, $SL(2,\mathbb{C})$ seems the natural framework to study this problem from a group-theoretical point of view (this was already anticipated in [1]).

Although at this point this asumption is simply a conjecture, there are clues indicating that this can be the right choice. For instance, if in the MPT potential for positive energy, besides the functions (24) we introduce $\mathcal{E}' \equiv i\mathcal{E}$, $\mathcal{X}' \equiv i\mathcal{X}$ and $\mathcal{P}' \equiv i\mathcal{P}$, the set $\{\mathcal{E},\mathcal{X},\mathcal{P},\mathcal{E}',\mathcal{X}',\mathcal{P}'\}$ closes a $SL(2,\mathbb{C})$ algebra where the $SO(2,1)$ subalgebra is the original $\{\mathcal{E},\mathcal{X},\mathcal{P}\}$ and the $SO(3)$ subalgebra is $\{\mathcal{E}',\mathcal{X}',\mathcal{P}\}$, which coincides with the one given in (27) and that are real for negative energies.

What remains to be done is to relate the phase spaces of these systems to the coadjoint orbits (or open subsets of them) of $SL(2,\mathbb{C})$, and to derive the Hilbert space of quantum states from the unitary (or non-unitary) representations of $SL(2,\mathbb{C})$. This is work in progress.

The complexification of $SO(2,1)$ to $SL(2,\mathbb{C})$ seems to be related to the analytical extension of the S matrix to the complex plane. The fact that the S matrix provides all the information for a quantum system, not only the scattering states and bound states, but also the ABS states and resonant states, suggest that a detailed study of the S matrix and its symmetries could provide a unified description of the system.

This study can be generalized to other potential, like the Morse potential and their generalization (Ginnochio potentials, Natanzon potentials), and to the 3 dimensional version of the potentials considered here.

Acknowledgments
Work partially supported by the Spanish MCYT, Junta de Andalucía and Fundación Séneca under projects FIS2008-06078-C03-01/FIS, P06-FQM-01951 and 08816/PI/08.

References
[1] Aldaya V and Guerrero J 2005 *J. Phys. A* **38** 6939-6953
[2] Guerrero J, López-Ruiz F F, Calixto M and Aldaya V 2009 *Rep. Math. Phys.* **64** 329–340
[3] Kuru S and Negro J 2008 *Ann. Phys.* **323** 413–431
[4] Alhassid Y, Gürsey F and Iachello F 1983 *Phys. Rev. Lett.* **50** 873
[5] Alhassid Y, Gürsey F and Iachello F 1983 *Ann. Phys.* **148** 346
[6] Frank A and Wolf K B 1984 *Phys. Rev. Lett.* **52**, 1737
[7] Rosen N and Morse P M 1932 *Phys. Rev.* **42** 210
[8] Scarf F 1958 *Phys. Rev.* **112** 1137
[9] Kozlov V V and Harin A O 1992 *Celestial Mech. Dynam. Astron.* **54** 393-399
[10] Slawianowski J J 2000 *Rep. Math. Phys.* **46** 429
[11] Cariñena J F, Rañada M F and Santander M 2005 *J. Math. Phys.* **46** 052702
[12] Cariñena J F, Rañada M F and Santander M 2004 *Rep. Math. Phys.* **54** 285
[13] Calzada J A, del Olmo M A and Rodríguez M A 1999 *J. Math. Phys.* **40** 188–209
[14] Calzada J A, Negro J, del Olmo M A and Rodríguez M A 2000 *J. Math. Phys.* **41** 317–336
[15] Gendenshtein L 1983 *JETP Lett.* **38** 356
[16] Nieto M M and Simmons L M Jr 1979 *Phys. Rev. D* **20** 1342-1350
[17] Antoine J P, Gazeau J P, Monceau P, Klauder J R and Penson K 2001 *J. Math. Phys.* **42** 2349,
[18] Kirillov A A 1974 *Elements of the Theory of Representations* Moscow: MIR
[19] Aldaya V and Guerrero J 2000 *J. Math. Phys.* **41** 6747-6765
[20] Arias J M, Gómez-Camacho J and Lemus R 2004 *J. Phys. A* **37** 877-893
[21] Berrondo M and Recamier J 1997 *Int. J. Quant. Chem.* **62** 239-244
[22] Dong S and Lemus R 2002 *Int. J. Quant. Chem.* **86** 265–272
[23] Guerrero J and Aldaya V 2006 *J. Phys. A* **39** L267-L276
[24] Galindo A and Pascual P 1990 *Quantum Mechanics vol. I* Berlin: Springer-Verlag

Symmetry in music

O F Herrero

Conservatorio Superior de Música "Eduardo Martínez Torner"
Corrada del Obispo s/n 33003 – Oviedo – Asturias, Spain

E-mail: o.f.herrero@hotmail.com

Abstract. Music and Physics are very close because of the symmetry that appears in music. A periodic wave is what music really is, and there is a field of Physics devoted to waves researching. The different musical scales are the base of all kind of music. This article tries to show how this musical scales are made, how the consonance is the base of many of them and how symmetric they are.

1. Introduction

At first sight connection between music and physics could seem to be strange, fields so different like art and science, but in the end, music is sound, sound is a wave and physics studies waves. Both subjects can research the same thing but from a different point of view.

This article, tries to show, how Physics can contribute to the study of Music, and how we can see some elements of this art from a scientific point of view.

2. Theoretical introduction of musical Acoustics

First of all, let's talk about acoustics, just the items that will appear later.

The first item to talk about is sound. *Sound* is a travelling wave, a pressure oscillation transmitted through a solid, a liquid, or a gas, composed of frequencies within the range of hearing and of a level strong enough to be heard, or the sensation stimulated in organs of hearing by such vibrations. So to produce a sound, you need a sender, a transmitter and a receiver. In the experiment referred to in this article, it is used: as a sender, musical instruments (violin, bowed string; clarinet, semi-open tube; pipe organ, open tube, monochord); as a transmitter, the air; and as a receiver, the human ear has been replaced by a spectrum analyzer, with characteristics similar to the human ear which allows us to obtain numerical results and graphs of what the ear would receive.

Another item needed to understand this experiment is the *Principle of Superposition*. This Principle says that if two sounds are emitted, with the same amplitude but with different frequencies f_1 and f_2, a new sound is produced, its amplitude changes in time and whose frequency is the semi-sum of both frequencies.

$$
\begin{aligned}
y_1 &= A\cos(\varpi_1 t + \varphi) = A\cos(2\pi f_1 t + \varphi) \\
y_2 &= A\cos(\varpi_2 t + \varphi) = A\cos(2\pi f_2 t + \varphi) \\
y_1 + y_2 &= 2A\cos\left(2\pi \frac{f_1 - f_2}{2} t\right)\cos\left(2\pi \frac{f_1 + f_2}{2} t + \varphi\right)
\end{aligned}
\qquad (1)
$$

The *Fourier theorem* is necessary for this experiment. In our context, this theorem says: the sound from a musical instrument is usually not a pure sine wave, but a "fundamental frequency accompanied by a number of overtones of frequencies 2, 3, 4, etc., times the frequency of the fundamental. The combination of the fundamental and the overtones is a complicated periodic function with the period of the fundamental. Given the complicated function, we could ask how to write it as a sum of terms corresponding to the various harmonics. In general it might require all the harmonics, that is, an infinite series of terms. This is called Fourier series. Expanding a function into a Fourier series then amounts to breaking it down into its various harmonics. In fact, this process is sometimes called harmonic analysis [1], and it represented musically:

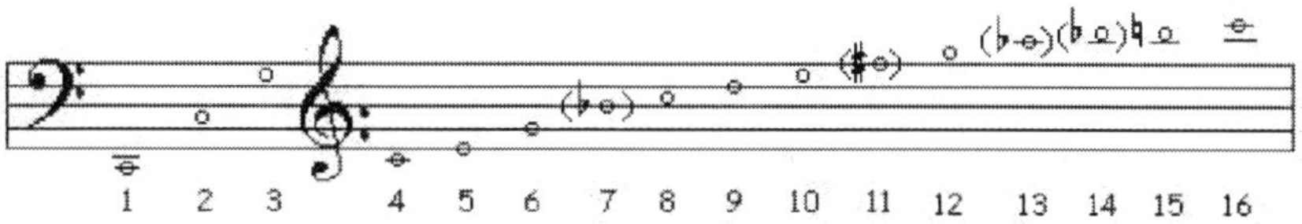

Figure 1. Fourier Theorem

In this graph one can see the fundamental frequency with the number 1 and its representation on a staff, and the musical representation of spectral analysis.

All of these items are theoretical knowledge needed in order to interpret the results but to give them a real sense it is necessary to define *the critical bandwidth* discovered by Békésy, it is the range of frequencies where the brain knows that the received sound is not just one but it cannot recognise two. When the Fourier Theorem is applied to these sounds, the spectrum shows how the peaks of frequency that the main cannot separate fuse in one, with a frequency that is the semi-sum of harmonics that fuse, and its intensity changes in time with a beat frequency similar to the difference in frequency between peaks. The hard sound for quick beats can be perceived in the critical bandwidth and the nice sound for slow beats.

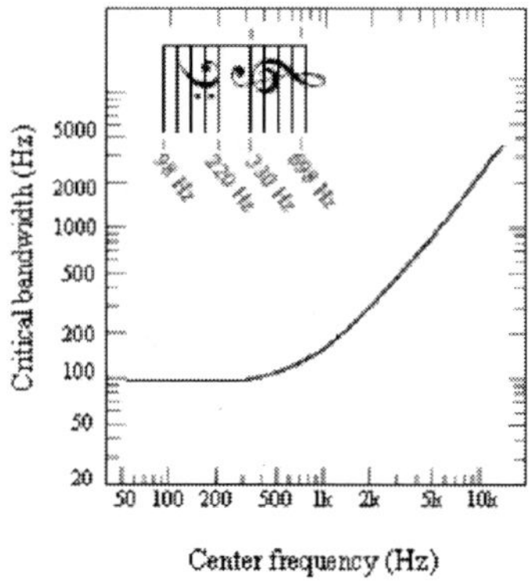

Figure 2. Critical Bandwidth

3. Theoretical introduction of musical Dissonance

The relationship between Music, Mathematics and Physics has been known since s.VI b.C. when Pythagoras and his pupils studied vibrant string. These studies set the beginning of the study of Harmony in Music.

The interest of the consonance and dissonance come from this time to nowadays. Throughout this time many people worked on it trying to set a definition. Some of the last theories about it are exposed in this article.

John Tyndall, for example, said in 1893 [2], that an interval would be more consonant if the ratios are more simple.

Hermann von Helmholtz [3] was the first person who introduced the sense of beat.

Georg von Békésy focused his studies on acoustic hardness. The existence of the beats can create an unpleasant sensation, but not on all occasions. If the frequency of a beat is low, simply a tremolo is

appreciated, not dissonance. By contrast, two sounds whose frequencies differ enough to produce no audible pulsation, you can create, when they sound together, a feeling of roughness.
Carl Stumpf said [4], in 1898, neither harmonics nor difference tones are essential to discriminate between consonant and dissonant intervals, whereas he rejected the frequency-ratio theory as more speculation. Stumpf called attention to the fact, investigated by him before [5] and onfirmed by many others after him, that the degree of fusion of intervals depends on simple frequency ratio in the same order as consonance does. By fusion, he meant the tendency of two simultaneous tones to be perceived as a unity.[6]
Plomp and Levelt add [7], that they assume that the total dissonance of such an interval is equal to the sum of the dissonances of each pair of adjacent partials.
Kameoka and Kuriyagawa [8] based their studies about consonance and dissonance on the harmonic structure of the emitted sounds. They said that the consonance between two frequencies without harmonics only depends on the distance. And the intensity has an important role, because it can mask one of the sounds and have influence on the consonance.
Ernst Terhardt thinks that musical dissonance has to be studied in two ways, one of them is sensorial dissonance, the concept of sounding well, on the other hand, the harmony or the influence of the culture.

4. Experiment

The objective of the experiment is to prove experimentally the results that Plomp and Levelt obtained in 1965 [6] statistically and numerically about the consonance and dissonance, and the application of these results to the construction of musical scales (western and eastern).
This experiment tries to show the dissonance as something objective. All the theories until now show it as something subjective, something that depends on culture, taste or statistics and numerical results, but this experiment tries to give a value of dissonance. Obviously, the concept of dissonance is different for everyone, but with this experiment a scale can be made, with values, and it can be known from what value something dissonant is listened.
Some intervals produced by different instruments have been recorded and then, a Fourier analysis with different precision has been done to obtain a very precise measurement or numerical results, very similar to the ear reception capacity. A spectrum has been used with a bandwidth as the human ear.
The same experiment has been done with the violin and the clarinet. With the first one, a violinist played two sounds at the same time. The violinist tried to find a specific tune, the tune that allows an integer ratio between both sounds. The same procedure was done with the clarinet, but with two clarinettists, one of the sounds stays constant and the other rises from the unison to the octave chromatically.
Analyzing the results with two different types of spectrum, one of them with a very thin bandwidth, to obtain very exact results, and the other one with a bandwidth similar to the human ear, in order to obtain numerical results about the human ear reception capacity. With the thin bandwidth, we obtained the exact frequencies of each peak. In the analysis with wide bandwidth, a spectrum was done every millisecond during the second second that was chosen, making a movie with them. The second second, because in the first one there would be imprecise tunes. In that movie we can see how the intensity of some peaks of frequency (the peaks which are very closed in the spectrum with thin bandwidth) change in time, the beats. Representing the results on a graph, intensity to time:

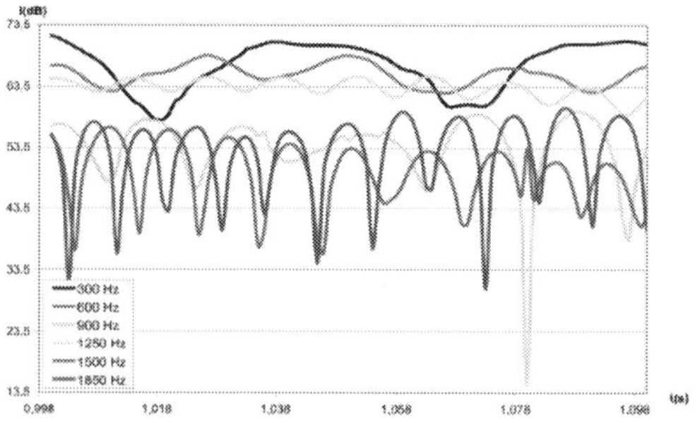

Figure 3. Variation of intensity vs. time for minor second

In this graph it can be seen how intensity changes in time for most of the harmonics, that is to say the beats.
Numerical results for this interval are shown in the next table:

Table 1. Results for minor second

i	fi	fi_1	fi_2	fi_2-fi_1	$(fi_1+fi_2)/2$	fb	Dissonance
1	300	294	313	19	303.5	17.79	0.96
2	600	589	623	34	606.0	34.72	0.99
3	900	880	936	56	908.0	55.17	0.87
4	1250	1173	1247	74	1210.0	72.80	0.83
5	1500	1467	1559	92	1513.0	87.79	0.79
6	1850	1761	1872	111	1816.5	114.58	0.72

This table shows how the beat frequency is very close to the difference of original frequencies and how the semisum of these frequencies is very similar to the fundamental frequency of each harmonic.
The value of the dissonance was calculated taking into account the percentage of bandwidth that corresponds to the beat frequency and applying the calculation of Plomp and Levelt. The total value for the dissonance of this interval is the addition of the dissonance of all of harmonics for this interval.
If the beat of frequency is outside the bandwidth, this beat doesn't contribute to the dissonance.
These results correspond to the minor second from Violin. The same procedure was followed with the rest of intervals and with both instruments. The experiment started with the unison, one of the sounds stays in the same frequency and the other goes up chromatically to the octave. When the results are represented in a similar graph to Plomp and Levelt's, we see the following:

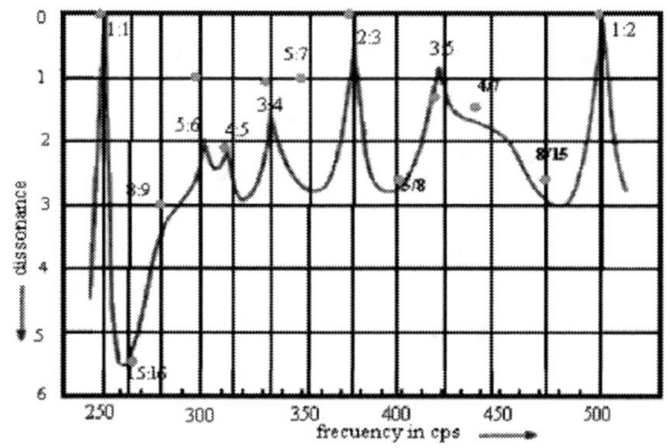

Figure 4. Comparison between Plomp and Levelt results and our results for violin

The red points correspond to the results of the experiment. A continuous graph couldn't be done, like Plomp and Levelt, since all the rank of frequencies were not swept, they only go up chromatically. They only work with harmonics until 6, so they can only obtain a relationship from frequency in natural numbers up to 6 (1:1, 5:6, 4:5, 3:4, 2:3, 3:5, 1:2) that is why frequency relationship for numbers greater than 6 (15: 16, 8:9, 5:7, 5:8, 4:7, 8:15) don't appear, their sounds are artificial (for that reason they can control the number of overtones), whereas the results of the experiment are natural. That is the reason why there are more overtones.
It can also be observed that the frequency ratios that agree with that are in the graph. In the experiment there are more consonant relationship than for Plomp and Levelt.
Although at first sight it would be possible to hope that since the greater number of overtones used, respecting to Plomp and Levelt, would contribute to a greater dissonance. There are a series of factors that

Journal of Physics: Conference Series **237** (2010) 012013 doi:10.1088/1742-6596/237/1/012013

influence it, for example, the intensity. For that reason, the intensity of the overtones that Plomp and Levelt handle can produce greater dissonances than the experiment (still being less harmonic). Also it is necessary to consider that the point of greater dissonance obtained in the experiment has been done in accordance with the point of greater dissonance obtained by Plomp and Levelt. Relative values are used at any moment, the sounds are more or less consonant compared to others, then to be able to establish a comparison, the most dissonant value and the minus have been taken as a reference, comparing the rest based on that scale, i.e., if the most dissonant value for the experiment were more dissonant than the one of Plomp and Levelt, the differences of the other tips (between our results and those of Plomp and Levelt) would be different.

If the dots are connected, a graph that is very similar to the one of Plomp and Levelt can be obtained.

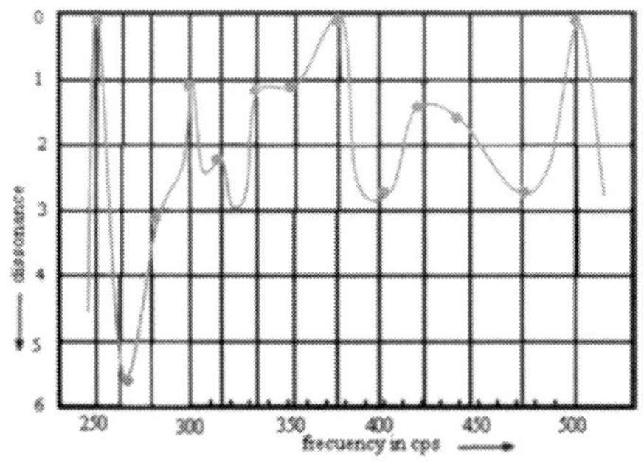

Figure 5. Dots connected

The same was done with the clarinet and the pipe organ. The difference with these instruments is that they are tempered instruments, so the ratio between the frequencies is not of natural numbers.

The next graph shows the results of these three instruments.

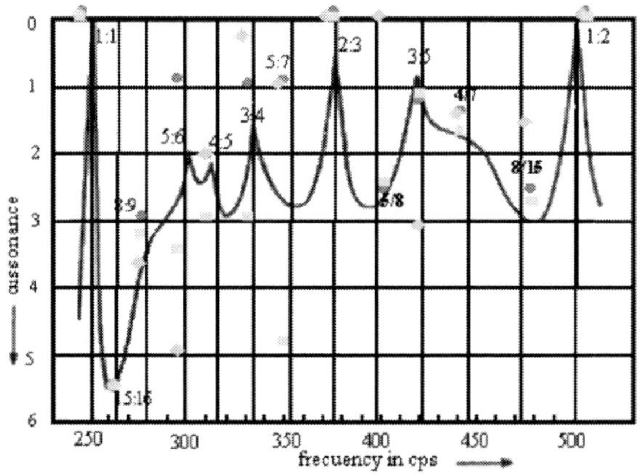

Figure 6. Plot with Plomp and Levelt results, and our results for violin, clarinet and pipe organ

The red points correspond to the violin, blue squares to the clarinet, and green rhombus to the pipe organ. The plot shows how the most consonant dots are the same for the three instruments and how for the ratio 5:7 the triton it can be seen that it is a consonant interval for no tempered instrument (when the ratio is simple) and very dissonant for the organ, with tempered tuning.

Most of the ratios haven't got the same consonance or dissonance, for many reasons. If the intensity of one of the overtones is different to another, the frequency ratio would be different. For the clarinet there are only the pair overtones, so there are less than for the rest of the instruments, and this, influences in the consonance and dissonance.

5. Scales

These results try to prove how eastern and western scales are based on the sense of consonance.

5.1. Western scales

It is well-known, that western music is based on the superposition of fifth and third. Some theoreticians will use only one of the intervals using therefore the second and third overtones, and others use both, that's the reason why they would use overtones 2, 3 and 5 (remembering that the fifth corresponds with the ratio $\frac{3}{2}$, Mayor third $\frac{5}{4}$, minor third $\frac{5}{3}$ and the octave $\frac{2}{1}$).

Now some of the western scales will be shown, in first place the harmonic structure in which they are constructed and secondly the ordered frequencies ascending indicating the intervals, that in this case formed between consecutive frequencies. a-Figures show the way to make the scales, how all of them are made in the same way, and b-Figures show the periodicity or the symmetry in the scales, or in many of them.

5.1.1. Pythagoras

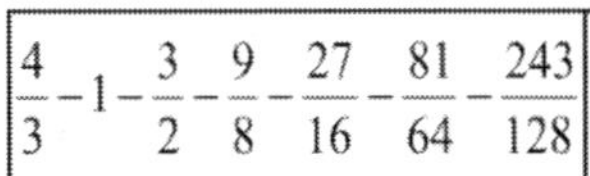

Figure 7.a Pythagoras

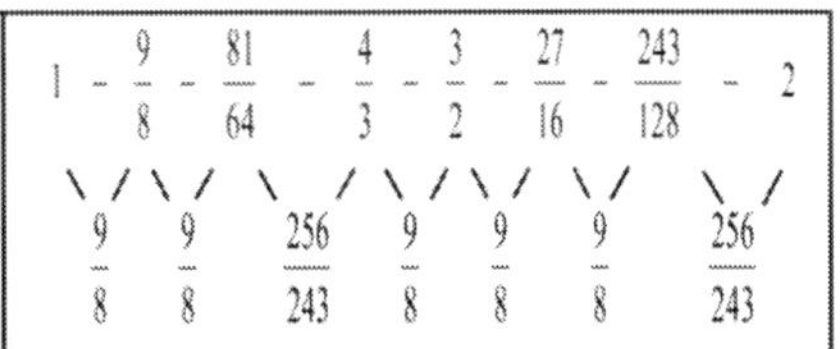

Figure 7.b Pythagoras

5.1.2. Ramos de Pareja

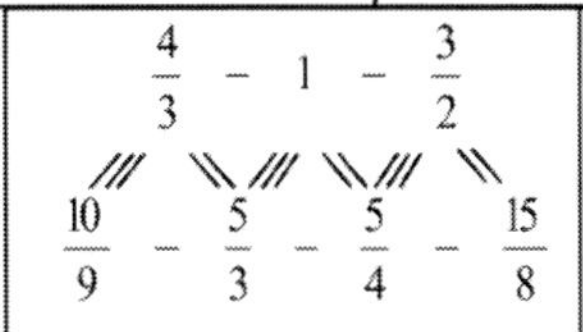

Figure 8.a Ramos de Pareja

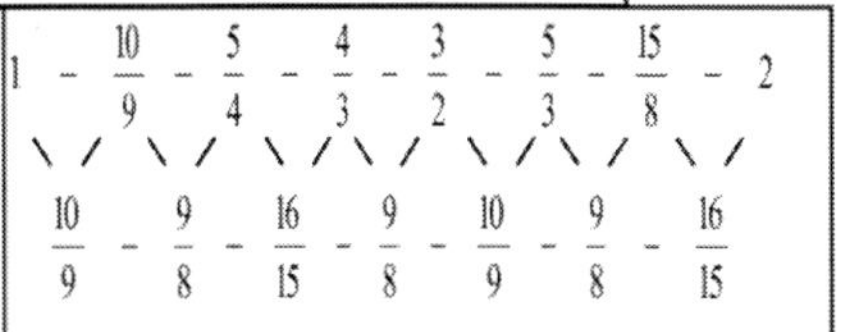

Figure 8.b Ramos de Pareja

5.1.3. Salinas

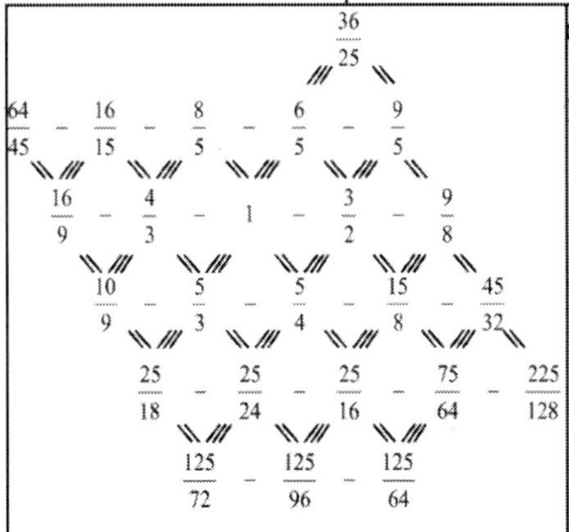

Figure 9.a Salinas

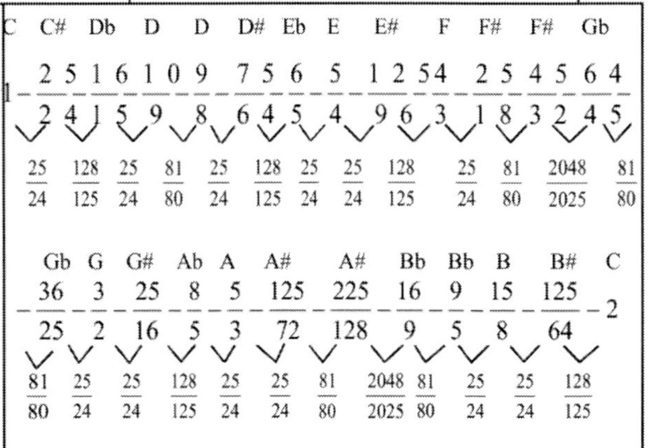

Figure 9.b Salinas

5.1.4. Zarlino

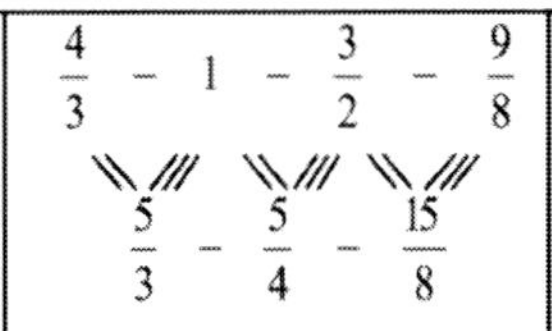

Figure 10.a Zarlino

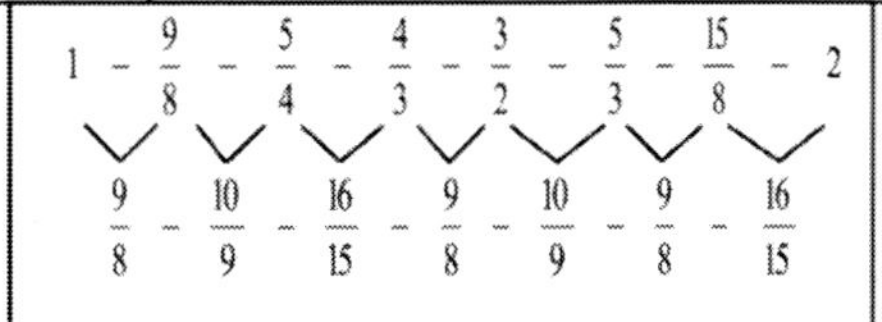

Figure 10.b Zarlino

All these scales are constructed, as it can be seen, from fifth $\frac{3}{2}$, Mayor third $\frac{5}{4}$ and minor third $\frac{5}{3}$, i.e. using overtones 2, 3 and 5, first of the series, therefore, simple ratios, which implies little dissonant intervals. Once the frequencies have been placed in ascending sequence, intervals are formed between each consecutive sound, but they are always powers of these same overtones. It can be said that western music is constructed with consonant intervals, where consonant means sound with few beatsin the critical bandwidth [9].

In the case of Eastern music it will also be verified that the scales are made with consonant intervals from the first overtones of the series. The values for the frequency ratios have been taken from the works of Gruber [10], Sanchez González [11] and Barkechli [12].

5.2. Indian scales

5.2.1. 14 Tones

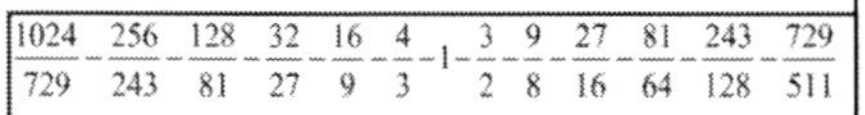

Figure 11.a 14-tone scale

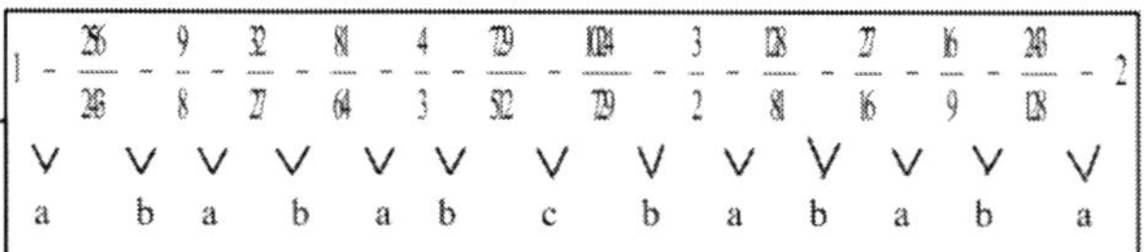

Figure 11.b 14-tone scale

As it can be seen how this 14 tone scale is made only with the fifth, as Pythagoras scale. But these scales show that, once the frequencies are placed in ascending sequence, the intervals between two consecutive sounds are always the same, and make system of symmetry.

5.2.2. 23 Tones

Figure 12.a 23-tone scale

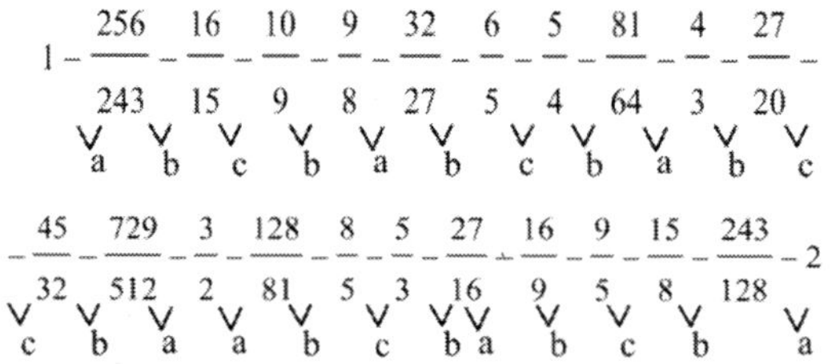

Figure 12.b 23-tone scale

This is the most common scale for Indian people. When they play their music, they never reach this frequency exactly, because when Indian people talk about a sound, they never talk about a specific sound, they just talk about an area around that sound, but the center of this area is this frequency.

Between two consecutive sounds, in this scale there are three different intervals, the first of them, $a = \frac{256}{243}$, was used by Pythagoras, and the others, $b = \frac{81}{80}$ and $c = \frac{25}{24}$ are used by Salinas. This fact shows how eastern and western music are linked.

The next figure shows the name that Indian musicians give to these frequencies.

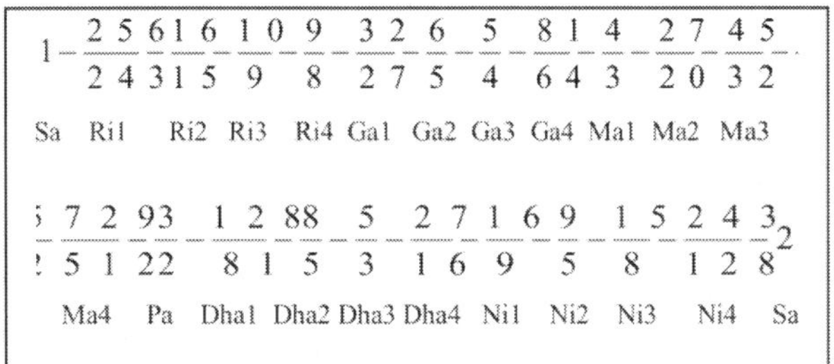

Figure 13. 23-tone scale

5.2.3. 26 Tones

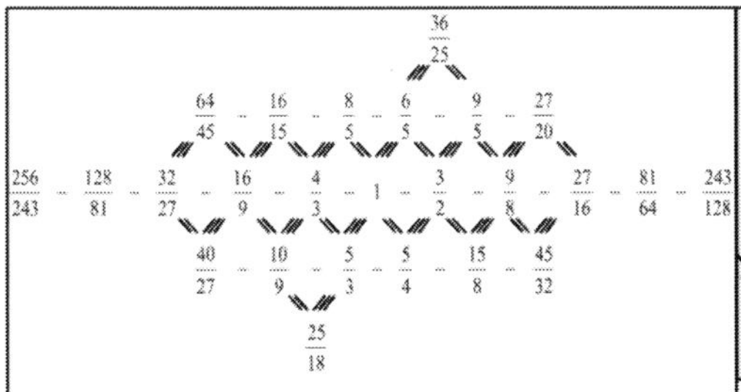

Figure 14.a 26-tone scale

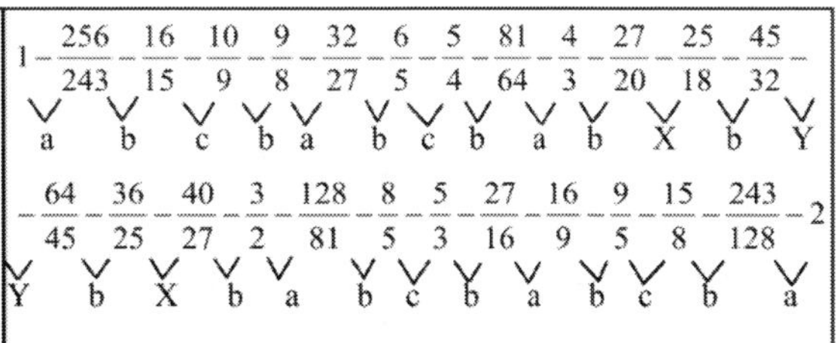

Figure 14.b 26-tone scale

5.2.4. 30 Tones

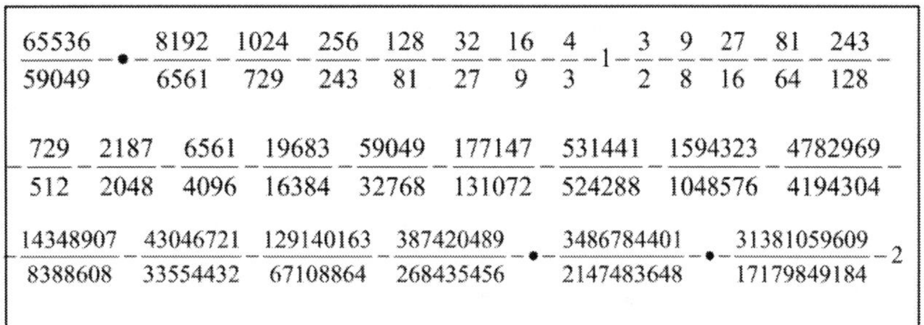

Figure 15.a 30-tone scale

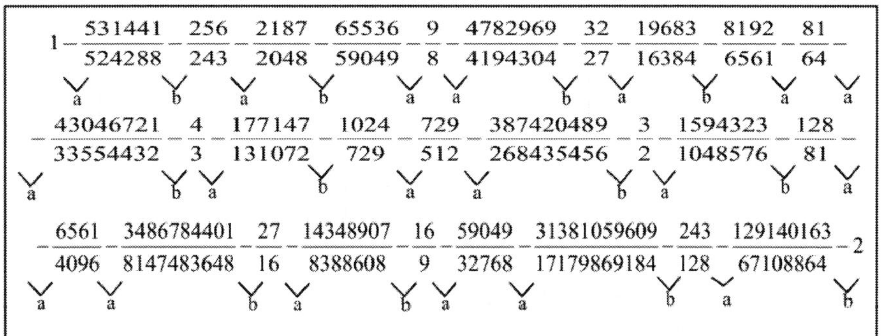

Figure 15.b 30-tone scale

5.2.5. 32 Tones

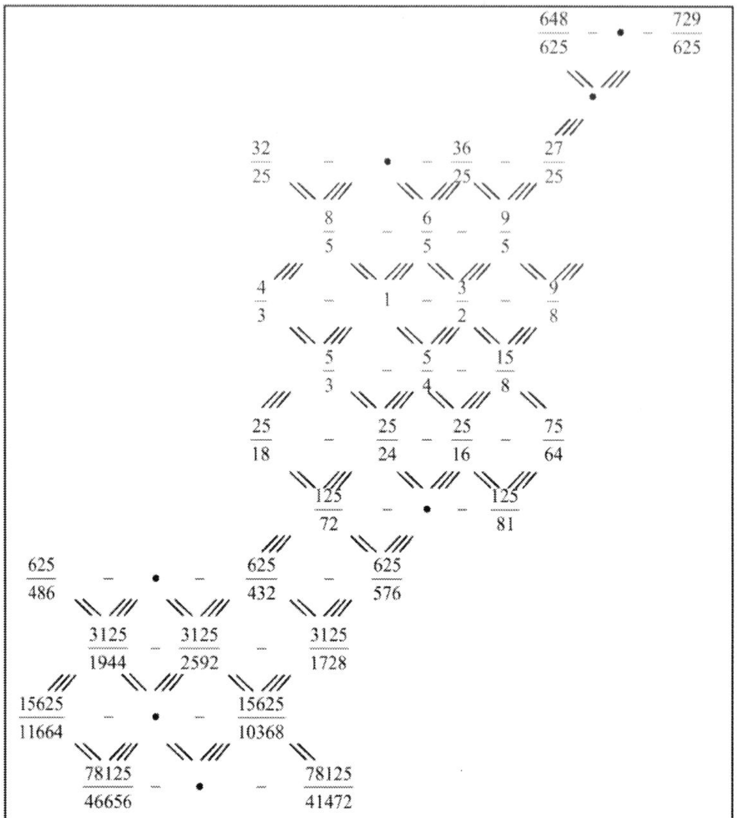

Figure. 16.a 32-tone scale

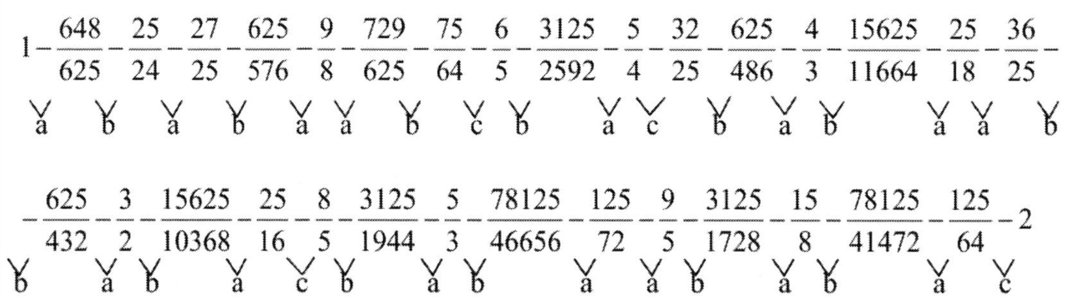

Figure 16.b 32-tone scale

All Indian scales are made using the same intervals as western scales, $\frac{3}{2}, \frac{5}{4}, \frac{5}{3}$, i.e. using the harmonics 2, 3 and 5, the most consonant intervals.

5.3. Arabian scales

These scales are made overlapping tetrachords. For this music, the experiment obtained very similar schemes to the ones for Indian music.

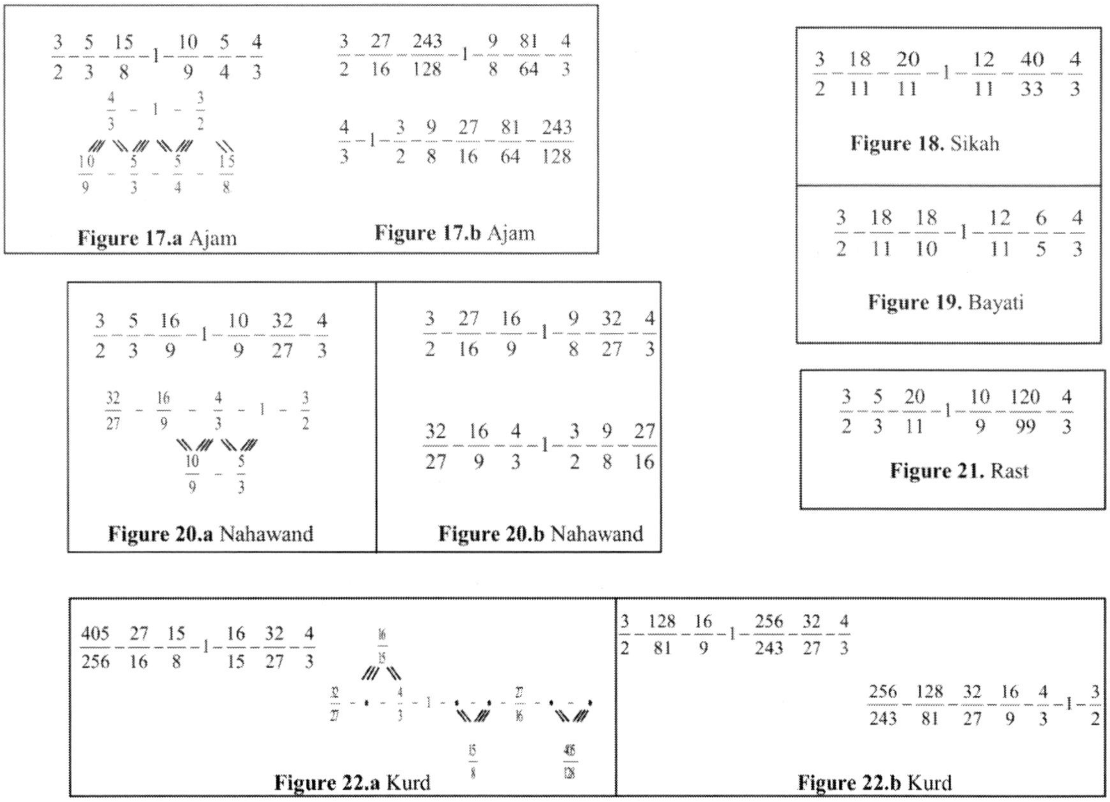

Figure 17.a Ajam

Figure 17.b Ajam

Figure 18. Sikah

Figure 19. Bayati

Figure 20.a Nahawand

Figure 20.b Nahawand

Figure 21. Rast

Figure 22.a Kurd

Figure 22.b Kurd

This music used two more overtones to make their scales, the 7 and the 11, but remember that the interval $\frac{7}{5}$ is consonant [9].

6. Conclusions

This experiment tries to demonstrate experimentally the Helmholtz beats theory, seeing them, and obtaining similar results to Plomp and Levelt's, although they obtain statistical and numerical results and the experiment just obtains experimental results.

It demonstrates as well than triton is consonant if it is played with exact tuning $\frac{7}{5}$, and how eastern and western scales are made using consonant intervals.

References

[1] Boas M L 1960 *Mathematical Methods in the Physical Sciences* (New York: John Wiley & Sons) p 286

[2] Tyndall J 1893 *Sound* (London: Longmans, Green & Co.)

[3] Helmholtz H 1913 *Die Lehre von den Tonempfindungen als Physiologische Grundlage für die Theorie der Musik* (Braunschweig: Verlag F. Vieweg & Sohn) 1st ed. 1863

[4] Stumpf C 1898 *Konsonanz und Dissonanz* (Beitr. Akust. Musikwiss. 1) p 1

[5] Stumpf C 1890 *Tonpsychologie* **2** (Leipzig: Verlag S. Hirzel) p 127

[6] Plomp R and Levelt W J M 1965 Tonal consonance and critical bandwidth *Journal of the Acoustical Society of America*, **38** 883

[7] Kameoka A and Kuriyagawa M 1969 Consonance theory, part I: Consonance of dyads *Journal of the Acoustical Society of America* **45** 1451

[8] Kameoka A and Kuriyagawa M 1969 Consonance theory, part II: Consonance of complex tones

and its computation method *Journal of the Acoustical Society of America* **45** 1460
[9] Herrero O F and Lorente M 2006 Comprobación experimental de la teoría de la consonancia y disonancia musical *Revista de Acústica* **37** 5
[10] Gruber B J 2006 Mathematical-Physical Properties of Musical Tone Systems II (Wien: Sitzungsber. Österr. Akad. Wiss. math.-nat. Kl.) p 45
[11] Sanchez Gonzalez J 1989 *The great tone (8/7) interval in Arab and Related Musics* Symposium on oriental music, Bagdad (unpublished) CSIC
[12] Barkechli M M 1959 L'Évolution de la gamme dans la musique orientale *Accoustique musicale* (Paris: Editions du centre national de la recherché scientifique)

Journal of Physics: Conference Series **237** (2010) 012014 doi:10.1088/1742-6596/237/1/012014

Symmetry and phase transitions: Quantum phase transitions in algebraic models

Francesco Iachello
Center for Theoretical Physics, Sloane Laboratory, Yale University, New Haven, CT 06520-8120, USA

Abstract. A brief description of the role of symmetry in phase transitions is given. Quantum phase transitions (QPT) in algebraic models are discussed, both at the semi-classical and quantal level. The newly introduced concept of excited states quantum phase transitions (ESQPT) is presented.

1. Introduction

Quantum phase transitions (QPT) are phase transitions that occur as a fucntion of coupling constants, $\xi_1, \xi_2, ...$, the so-called *control parameters*, that appear in the quantum Hamiltonian, H, that describes the system

$$H = \varepsilon \left(H_1 + \xi_1 H_2 + \xi_2 H_3 + ... \right). \tag{1}$$

Associated with phase transitions there are *order parameters*, the expectation values of suitable chosen operators that describe the state of the system $\langle P \rangle$. A simple example is the Ising model in a transverse magnetic field

$$H = -V \left(\xi \sum_i \hat{\sigma}_i^x - \sum_{\langle ij \rangle} \hat{\sigma}_i^z \hat{\sigma}_j^z \right) \tag{2}$$

which has a second order QPT at $\xi = \xi_c = 1/2$. The order parameter is the magnetization, $M \propto \langle \hat{\sigma}^z \rangle$. The two phases are: (i) ferromagnet for $\xi < \xi_c$ and (ii) paramagnet for $\xi > \xi_c$. QPTs are also called ground state phase transitions [1] and/or zero temperature phase transitions.

An algebraic model of a physical many-body system is a mapping of all operators of the system onto an *algebraic space*, g, usually taken as the unitary (compact) algebra in n dimensions, u(n). The representations of u(n) are labelled by the total number of particles N. They are totally symmetric for bosons and totally antisymmetric for fermions with a one-row (boson) or one-column (fermions) Young tableau. Since all physical operators are expanded onto elements of g, the algebra g is called the *spectum generating algebra* (SGA). Many-body systems composed of a mixture of bosons and fermions can also be treated by algebraic models by mapping onto an *algebraic superspace*, g*, usually taken as the unitary (compact) superalgebra in n bosonic and m fermionic dimensions, u(n/m). States are characterized by the totally supersymmetric representations of u(n/m), labeled by the total number of bosons plus fermions, $N = N_B + N_F$. The algebra g* is called the *spectrum generating superalgebra* (SGSA).

In this contribution a study of quantum phase transitions in algebraic models will be presented.

2. QPT in algebraic models

The generic Hamiltonian H is written in terms of elements $G_{\alpha\alpha'}$ of g,

$$H = \sum_{\alpha} \varepsilon_\alpha G_{\alpha\alpha'} + \sum_{\alpha\alpha'\beta\beta'} u_{\alpha\alpha'\beta\beta'} G_{\alpha\alpha'} G_{\beta\beta'} + ... \quad ; \quad G_{\alpha\alpha'} \in \mathrm{g} \, . \tag{3}$$

In most applications, only terms up to quadratic in the elements of g are kept. In this presentation, bosonic systems will be considered for which the elements of g are bilinear products of boson creation and annihilation operators,

$$G_{\alpha\alpha'} = b_\alpha^\dagger b_{\alpha'} \qquad (\alpha, \alpha' = 1, ..., n). \tag{4}$$

The algebra g is $\mathrm{U}(n)$. The basis states are totally symmetric

$$[N] \equiv \square\square ... \square \tag{5}$$

and can be constructed by acting with boson operators on the vacuum

$$|N\rangle = \frac{1}{\sqrt{N!}} b_\alpha^\dagger b_{\alpha'}^\dagger ... |0\rangle \, . \tag{6}$$

2.1. Phases

Phases in algebraic models are all possible breakings of the algebra g into its subalgebras (subject to some conditions, for example, rotational invariance, i.e. the chain must contain so(3)). The subalgebras $\mathrm{g}_1, \mathrm{g}_2, ..., \mathrm{g}_\phi$ are often called dynamic symmetries

$$\begin{aligned} g &\supset g_1 \supset g_1' \supset ... \\ g &\supset g_2 \supset g_2' \supset ... \\ &... \\ g &\supset g_\phi \supset g_\phi' \supset ... \end{aligned} \tag{7}$$

A classification of phases of interacting boson models in arbitrary number of dimensions n is given in [2]. Whether or not there is a phase transition between two phases will be determined by further analysis as shown in the following sections. Throughout this article, two examples will be discussed, the Interacting Boson Model of nuclei and the Vibron Model of molecules.

2.1.1. The Interacting Boson Model (IBM) The Interacting Boson Model [3] is a model of atomic nuclei in terms of correlated pairs of nucleons with angular momentum $J = 0$ and $J = 2$, treated as bosons and called s and d bosons [4]. The model can also be seen as a treatment of S and D-wave pairing in 3-dimensions. There are six boson operators $b_\alpha(\alpha = 1, ..., 6) \equiv s, d_\mu(\mu = 0, \pm 1, \pm 2)$, spanning a six-dimensional space. The algebraic structure is u(6), with 6×6=36 elements. The phases are:

$$\begin{aligned} u(6) &\supset u(5) \supset so(5) \supset so(3) \supset so(2) \quad \text{(I)} \\ u(6) &\supset su(3) \supset so(3) \supset so(2) \quad \text{(II)} \\ u(6) &\supset so(6) \supset so(5) \supset so(3) \supset so(2) \quad \text{(III)}. \end{aligned} \tag{8}$$

2.1.2. The Vibron Model (VM) The Vibron Model [5] is a model of vibrations of molecules in terms of quanta of vibrations, called vibrons [6]. It can be formulated in 1, 2 or 3 dimensions. In this contribution, the vibron model in 2-dimensions will be considered [7]. The vibron operators are a singlet σ and a doublet τ_+, τ_-. They span a three-dimensional space. The algebraic structure is u(3) with 3×3=9 elements. The phases are:

$$\begin{aligned} u(3) &\supset u(2) \supset so(2) \quad (\mathrm{I}) \\ u(3) &\supset so(3) \supset so(2) \quad (\mathrm{II}) \end{aligned} \tag{9}$$

The vibron model in 2-dimensions is isomorphic to, i.e. has the same algebraic structure of, a model of high-T_c superconductors in terms of correlated electron pairs in two dimensions, called s, d_+, d_- bosons which can also be viewed as a treatment of S and D-wave pairing in two dimensions [8]. [A discrete version of this model on a square lattice, with symmetry D_{4h} is given in [9]. The boson operators that transform as representations of D_{4h} are $\Gamma_1^\dagger \equiv s^\dagger$, $\Gamma_3^\dagger \equiv \left(d_-^\dagger + d_+^\dagger\right)$, $\Gamma_4^\dagger \equiv \left(d_-^\dagger - d_+^\dagger\right)$.]

2.2. Phase diagram

Quantum phase transitions in algebraic models are best studied by writing the Hamiltonian in terms of (invariant) Casimir operators of the algebras $g_i (i = 1, 2, ..., \phi)$ that define the phases

$$H = \alpha_1 C(g_1) + \alpha_2 C(g_2) + ... + \alpha_\phi C(g_\phi). \tag{10}$$

Taking out a scale factor this can be rewritten as

$$H = \alpha_1 \left[C(g_1) + \frac{\alpha_2}{\alpha_1} C(g_2) + ... + \frac{\alpha_\phi}{\alpha_1} C(g_\phi) \right]. \tag{11}$$

The control parameters are the ratios

$$\xi_1 = \frac{\alpha_2}{\alpha_1},\ \xi_2 = \frac{\alpha_3}{\alpha_1},\ ...,\ \xi_{\phi-1} = \frac{\alpha_\phi}{\alpha_1}. \tag{12}$$

The number of control parameters is equal to the number of phases ϕ minus one. It is sufficient to take one Casimir operator for each g_i. The corresponding Hamiltonian is called the "essential" Hamiltonian.

In the two cases introduced in the previous subsection, the essential Hamiltonian is, for the interacting boson model, g≡u(6)

$$H = \varepsilon \left[C_1(u(5)) + \xi_1 C_2(su(3)) + \xi_2 C_2(so(6)) \right] \tag{13}$$

and for the two-dimensional vibron model, g≡u(3),

$$H = \varepsilon \left[C_1(u(2)) + \xi C_2(so(3)) \right]. \tag{14}$$

Here ε is a scale factor, and $C_p(g)$ denotes the Casimir operator of order p of the algebra g.

The essential Hamiltonian is often parametrized in such a way that the control parameters vary in a finite interval, usually $0 - 1$ instead of $0 - \infty$, and the Casimir operators are scaled by appropriate powers of the number of particles N to guarantee that the location of the critical points is independent of N in the large N-limit. Since the eigenvalues of the a Casimir operator of order p are in the large N-limit $\propto N^p$, this scaling is often adopted. For example, for the two-dimensional vibron model, the scaled essential Hamitlonian is

$$H = \varepsilon \left[\frac{(1-\xi)}{N} C_1(u(2)) + \frac{\xi}{N^2} C_2(so(3)) \right] \quad ; \quad 0 \leq \xi \leq 1. \tag{15}$$

In practice, combinations of Casimir operators are often used.

2.2.1. Essential Hamiltonian of the interacting boson model In most studies, the essential Hamiltonian of this model is taken as

$$\begin{aligned} H &= \varepsilon\left[(1-\xi)\,\hat{n}_d - \frac{\xi}{4N}\hat{Q}^\chi\cdot\hat{Q}^\chi\right] \\ \hat{n}_d &= \left(d^\dagger\cdot\tilde{d}\right) \\ \hat{Q}^\chi &= \left(d^\dagger\times\tilde{s}+s^\dagger\times\tilde{d}\right)^{(2)}+\chi\left(d^\dagger\times\tilde{d}\right)^{(2)}. \end{aligned} \tag{16}$$

There are three phases, characterized by the first algebra in the subalgebra chains of Eq.(8), $g_1 \equiv u(5)$, $g_2 \equiv su(3)$, $g_3 \equiv so(6)$, and two control parameters, ξ_1, ξ_2, in the parametrization of Eq.(13) or ξ, χ in the parametrization of Eq.(16). The phase diagram of this model is thus two-dimensional and usually represented as a triangle.

2.2.2. Essential Hamiltonian of the two-dimensional vibron model In most studies, the Hamiltonian of the two-dimensional vibron model is parametrized as

$$\begin{aligned} H &= \varepsilon\left[(1-\xi)\,\hat{n}+\frac{\xi}{N-1}\hat{P}\right] \\ \hat{n} &= \tau_+^\dagger\tau_+ + \tau_-^\dagger\tau_- \quad ; \quad \hat{\ell}=\tau_+^\dagger\tau_+ - \tau_-^\dagger\tau_- \\ \hat{P} &= N(N+1)-\left[\frac{1}{2}\left(\hat{D}_+\hat{D}_- + \hat{D}_-\hat{D}_+\right)+\hat{\ell}^2\right] \\ \hat{D}_+ &= \sqrt{2}\left(\tau_+^\dagger\sigma-\sigma^\dagger\tau_-\right) \\ \hat{D}_- &= \sqrt{2}\left(-\tau_-^\dagger\sigma+\sigma^\dagger\tau_+\right). \end{aligned} \tag{17}$$

Similarly, in the s-d model of high-T_c superconductivity, the essential Hamiltonian is parametrized as

$$\begin{aligned} H &= \varepsilon\left[(1-\xi)\,\hat{n}_d+\frac{\xi}{2N}\left(\hat{T}_+\hat{T}_- + \hat{T}_-\hat{T}_+\right)\right] \\ \hat{n}_d &= d_+^\dagger d_+ + d_-^\dagger d_- \\ \hat{T}_+ &= \sqrt{2}\left(d_+^\dagger s - s^\dagger d_-\right) \\ \hat{T}_- &= \sqrt{2}\left(-d_-^\dagger s + s^\dagger d_-\right). \end{aligned} \tag{18}$$

In either parametrization there are two phases, characterized by the algebras, $g_1 \equiv u(2)$ and $g_2 \equiv so(3)$, and one control parameter, ξ. The phase diagram is one dimensional (line).

2.3. Construction of the phase diagram. "Semiclassical" mean field method
The construction of the the phase diagram of algebraic models can be done by a semiclassical method using an algorithm developed by Gilmore [1].

2.3.1. Coset spaces The first step is the introduction of a geometry of algebraic models through coset spaces. A review of coset spaces is given in [10]. The algebra g is split into a stability algebra, h, and a remainder, p,

$$g = h\oplus p, \qquad g\supset h \tag{19}$$

with factor algebra g/h. For bosonic systems described by u(n), a convenient coset is u(n)/u($n-1$)$\oplus$u(1), with

$$\begin{aligned} h &\doteq b_1^\dagger b_1, b_\alpha^\dagger b_\beta \qquad (\alpha, \beta = 2, ..., n) \\ p &\doteq b_1^\dagger b_\alpha, b_\alpha^\dagger b_1 \qquad (\alpha = 2, ..., n). \end{aligned} \tag{20}$$

The coset space u(n)/u($n-1$)$\oplus$u(1) has dimension $2\,(n-1)$. There are $(n-1)$ complex variables in this space.

2.3.2. Coherent states The second step is the introduction of coherent or intrinsic states

$$|\eta_i\rangle = \exp\left[\sum_i p_i \eta_i\right] |\Lambda_{ext}\rangle\,. \tag{21}$$

For bosonic systems a convenient extremal state is

$$|\Lambda_{ext}\rangle = \frac{1}{\sqrt{N!}} \left(b_1^\dagger\right)^N |0\rangle \tag{22}$$

and the coherent states are written as

$$|N, \eta_\alpha\rangle = \exp\left[\sum_\alpha \eta_\alpha b_\alpha^\dagger b_1 + \sum_\alpha \eta_\alpha^* b_1^\dagger b_\alpha\right] \frac{1}{\sqrt{N!}} \left(b_1^\dagger\right)^N |0\rangle\,. \tag{23}$$

It is convenient to use projective coherent states with a fixed number of particles, N. For the interacting boson model, the coset space is five dimensional, and the projective coherent states are

$$|N; \alpha_\mu\rangle = \left(s^\dagger + \sum_\mu \alpha_\mu d_\mu^\dagger\right)^N |0\rangle \qquad (\mu = 0, \pm 1, \pm 2). \tag{24}$$

The variables α_μ are in general complex, the real part representing the coordinates themselves, α_μ, and the imaginary part the associated momenta, π_μ . However, for the construction of the phase diagram discussed here it is sufficent to take them real. The variables α_μ are then transformed to the so-called Bohr variables $\beta, \gamma, \theta_1, \theta_2, \theta_3$, where β, γ are called intrinsic variables and $\theta_1, \theta_2, \theta_3$ are Euler angles characterizing the orientation of the intrinsic frame in the laboratory frame.

For the two-dimensional vibron model and the model of high-T_c superconductors the space is two dimensional, and

$$|N; \eta_\mu\rangle = \left(s^\dagger + \sum_\mu \eta_\mu d_\mu^\dagger\right)^N |0\rangle \qquad (\mu = \pm). \tag{25}$$

The real variables η_μ can be transformed into polar coordinates r, θ with r the intrinsic variable and θ the polar angle.

2.3.3. Ground state energy functional The ground state energy functional, sometimes called the Landau potential, is the expectation value of the Hamiltonian in the coherent state of Eq.(23)

$$E(\eta_i) = \frac{\langle \eta_i | H | \eta_i \rangle}{\langle \eta_i \mid \eta_i \rangle} \tag{26}$$

By a rotation to the intrinsic frame the angles θs can be eliminated and the energy functional can be written in terms only of the intrinsic variables. For the interacting boson model,

$$E(N;\beta,\gamma) = \frac{\langle N;\beta,\gamma\,|H|\,N;\beta,\gamma\rangle}{\langle N;\beta,\gamma\mid N;\beta,\gamma\rangle}. \tag{27}$$

Explicit evaluation of the expectation value of the Hamiltonian Eq.(16) gives the Landau potential, or scaled energy per particle, $V(\beta,\gamma) \equiv E(\beta,\gamma)/\varepsilon N$

$$\begin{aligned} V(\beta,\gamma) &= \frac{\beta^2}{1+\beta^2}\left[(1-\xi) - (\chi^2+1)\frac{\xi}{4N}\right] \\ &\quad - \frac{5\xi}{4N(1+\beta^2)} - \frac{\xi(N-1)}{4N(1+\beta^2)^2} \\ &\quad \times \left[4\beta^2 - 4\sqrt{\frac{2}{7}}\chi\beta^3\cos 3\gamma + \frac{2}{7}\chi^2\beta^4\right]. \end{aligned} \tag{28}$$

For the two-dimensional vibron model

$$E(N;r) = \frac{\langle N;r\,|H|\,N;r\rangle}{\langle N;r\mid N;r\rangle}. \tag{29}$$

and the Landau potential is

$$V(r) = \left[(1-\xi)\frac{r^2}{1+r^2} + \xi\left(\frac{1-r^2}{1+r^2}\right)^2\right]. \tag{30}$$

2.3.4. Classical order parameters A minimization of E with respect to the η_is gives the equilibrium values η_{ie}. These are the *classical order parameters.* The value of E at equilibrium $E_{\min}$ is the ground state energy

$$E_{\min} = E(N;\eta_{ie}). \tag{31}$$

For the interacting boson model,

$$E_{\min} = E(N;\beta_e,\gamma_e). \tag{32}$$

The equilibrium value is $\gamma_e = 0°$ for $\chi < 0$ and $\gamma_e = 60°$ for $\chi > 0$. For $\chi = 0$, E does not depend on γ.

For the two-dimensional vibron model

$$E_{\min} = E(N;r_e). \tag{33}$$

2.3.5. Order of the phase transition The ground state energy $E_{\min}$ depends on the number of particles N and the control parameter(s) ξ, $E_{\min}(N;\xi)$. When $N \to \infty$, discontinuities (phase transitions) may appear in $E_{\min}$ and/or its derivatives with respect to ξ for some value of the control parameter(s) ξ, called critical values, ξ_c. It has become customary to classify phase transitions according to Ehrenfest classification

$$\begin{array}{ll} E_{\min} & \text{0-th order} \\ \dfrac{\partial E_{\min}}{\partial \xi} & \text{1-st order} \\ \ldots & \\ \dfrac{\partial^n E_{\min}}{\partial \xi^n} & \text{n-th order} \\ \text{No discontinuity} & \text{crossover} \end{array} \tag{34}$$

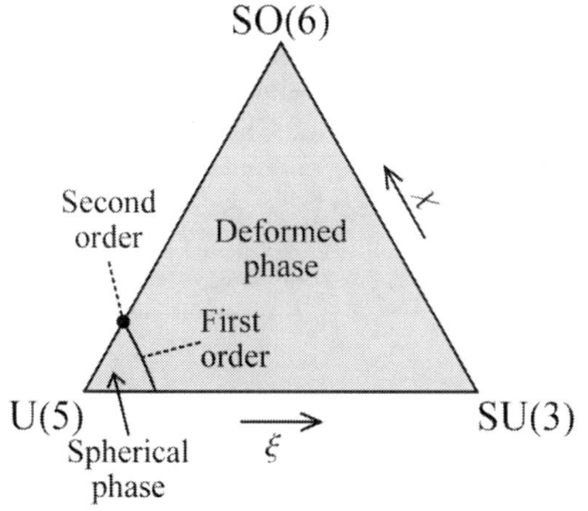

Figure 1. Phase diagram of IBM. A curve of first-order transitions and point of second-order phase transitions separate spherical and deformed phases.

2.3.6. Phase diagrams Phase diagrams of bosonic models of the type described in the subsections above (called s-b boson models) in an arbitrary number of dimensions have been constructed [2]. For the interacting boson model the phase diagram is a triangle [11], shown in figure 1. In this triangle, the three dynamic symmetries are placed at the vertices and the parameters ξ and χ are as shown. There is a line of first order transitions ending in a point of second order transitions. For the two-dimensional vibron model the phase diagram is a line, as shown in the insert to figure 2. In this case there is only a point of second order transitions.

2.3.7. Behavior of the order parameters The behavior of the order parameters as a function of the control parameters is of great interest in the theory of phase transitions. For second order transitions this behavior has been extensively investigated. Around the critical point, the order parameter, generically denoted by r_e here, behaves as

$$\begin{aligned} r_e &\propto (\xi - \xi_c)^{\mu} & \xi \geq \xi_c \\ r_e &= 0 & \xi < \xi_c \end{aligned} \tag{35}$$

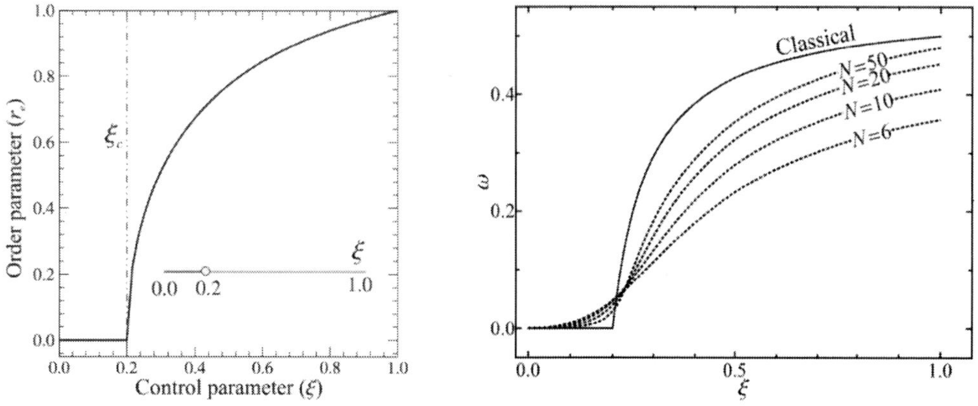

Figure 2. Classical and quantum order parameters for the u($n-1$)-so(n) phase transition. (Left) Evolution of classical equilibrium coordinate r_e. (Right) Quantum calculations of order parameter ω at finite N and the classical limit ($N \to \infty$). Left part adapted from Ref. [16].

The exponent μ is called the critical exponent. This behavior is shown in the left panel of figure 2 for the two-dimensional vibron model. The critical exponent is $\mu = \frac{1}{2}$. All second order phase transitions of the type u($n-1$)-so(n) in s-b boson models have the same behavior and the same critical exponent. For first order transitions, the situation is much more complex. The energy E as a function of the intrinsic variables may have more than one minimun. The energy functional of the interacting boson model, Eq.(28), has, as a function of β, one minimum (at $\beta_e = 0$) for $\xi \leq \xi^*$ (ξ^* is called spinodal point), two minima for $\xi^* < \xi < \xi^{**}$ (ξ^{**} is called antispinodal point) and one minimum for $\xi \geq \xi^{**}$. Within the coexistence region, $\xi^* < \xi < \xi^{**}$, the potential at the minima has equal value for $\xi = \xi^c$ (called critical point). It has ben suggested to define a spinodal exponent that characterizes the behavior of the order parameter around the spinodal point as

$$\begin{aligned} (\beta_e - \beta_e^*) &\propto (\xi - \xi_c)^{\mu^*} & \xi \geq \xi^* \\ \beta_e &= 0 & \xi < \xi^*. \end{aligned} \tag{36}$$

The interacting boson model with up to quadratic interaction does not have phase transitions in the order parameters γ_e and therefore critical exponents for this order parameter will not be discussed here.

Critical exponents have been evaluated for u(n) models with at most quadratic Hamiltonians. For the two-dimensional vibron model, as discussed above, the critical exponent is $\mu = \frac{1}{2}$. For the interacting boson model, $\mu = \frac{1}{2}$ for the u(5)-so(6) transition and $\mu^* = \frac{1}{2}$ for the u(5)-su(3) transition. To understand this behavior, it is instructive to consider a simple, analytically solvable case [12], that of the Landau potential

$$V(\beta) = \beta^2 + \xi\left[\left(2-\beta^2\right)^2 - \eta\beta^3\right], \qquad \beta \geq 0, \tag{37}$$

with two control parameters ξ and η. Up to a change of variables, this is the generic form of the potential for s-b boson models with at most two-body interactions. Using the method discussed above, one finds that for $\eta \neq 0$, a phase transition occurs as a function of ξ. The spinodal point, ξ^*, the critical point, ξ^c, and the antispinodal point, ξ^{**}, can be calculated to be at

$$\xi^* = \frac{32}{128+9\eta^2}; \; \xi^c \cong \frac{32}{128+8\eta^2}; \; \xi^{**} = \frac{1}{4}. \tag{38}$$

The upper branch of the order parameter is given by

$$\beta_e = \frac{3}{8}\eta + \frac{1}{2}\sqrt{\frac{9}{16}\eta^2 + 8\left(1-\frac{1}{4\xi}\right)}, \qquad \xi \geq \xi^*, \tag{39}$$

with spinodal exponent $\mu^* = \frac{1}{2}$. For $\eta = 0$, the transition becomes second order. The spinodal, critical and antispinodal points coincide and the order parameter is given by

$$\beta_e = \sqrt{2\left(1-\frac{1}{4\xi}\right)}, \qquad \xi \geq \xi^c, \tag{40}$$

with critical exponent $\mu = \frac{1}{2}$. The ground state energy is

$$\begin{aligned} E_0 &= 4\xi, & \xi \leq \xi^c = \frac{1}{4} \\ E_0 &= 2 - \frac{1}{4\xi} & \xi \geq \xi^c = \frac{1}{4}. \end{aligned} \tag{41}$$

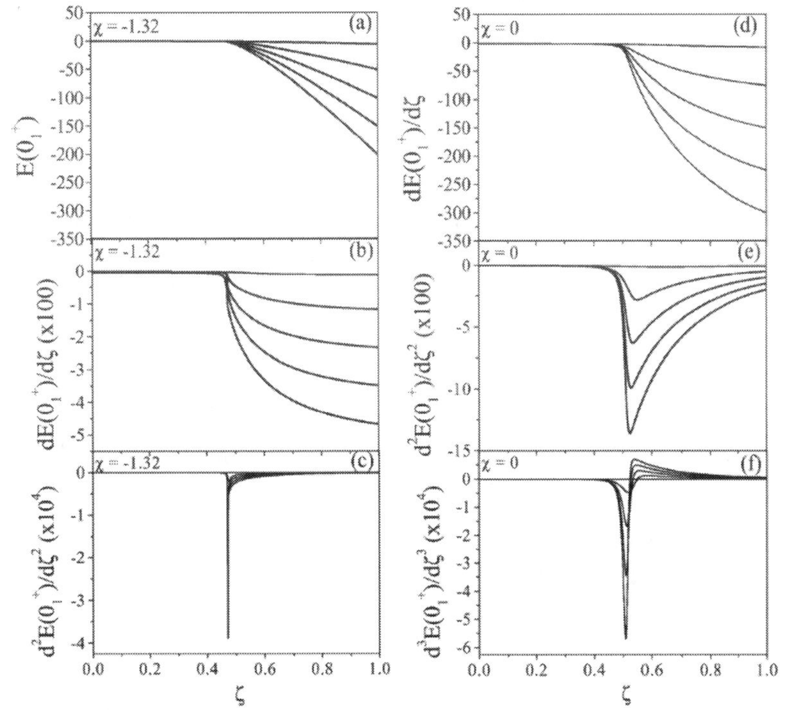

Figure 3. The ground state energy $E(0_1^+)$ and its derivatives across first-order (left) and second-order QPTs for ther IBM at finite N . The control parameter used for these plots is $\zeta \equiv 4\xi/(1+3\xi)$. Figure from Ref. [13].

2.4. Construction of the phase diagram. Quantal method

2.4.1. Diagonalization of H for finite N In this method, the essential Hamiltonian is diagonalized numerically as a function of the control parameters. The ground state energy and its derivatives are computed [13]. If the diagonalization is done for sufficiently large N, one obtains indications of phase transitional behavior and its order. An example is shown in figure 3 where the ground state energy of the interacting boson model and its derivatives are studied as a function of the control parameter ξ for the transition u(5)-su(3), $\chi = -\frac{\sqrt{7}}{2}$, and u(5)-so(6), $\chi = 0$. From these plots, one can see that there are indications of a first order transition for u(5)-su(3) and of a second order transition for u(5)-so(6).

2.4.2. Order parameters. Computation for finite N The *quantal order parameters* are the expectation values of some suitable chosen operators constructed from the elements $G_{\alpha\alpha'}$ of the algebra g. In general, for u(n) models, the number of independent order parameters is $n-1$. If constraints are imposed, the actual number is less than $n-1$. In s-b boson models, one of the quantal order parameters is always taken as the expectation value in the ground state of the number operator for b-bosons, $\hat{n}_b$,

$$\omega = \frac{\langle \hat{n}_b \rangle}{N}. \tag{42}$$

Several choices are discussed in the literature for the other order parameters, if needed.

For the interacting boson model, there are in general five order parameters. However, the constraint of rotational invariance reduces this number to two. The classical order parameters are β_e, γ_e. The quantal order parameters can be chosen as

$$\begin{aligned} \omega_1 &= \frac{\langle \hat{n}_d \rangle}{N} \rightarrow \frac{\beta_e^2}{1+\beta_e^2} \\ \omega_2 &= \frac{1}{N^2} \left\langle \left(d^\dagger \times \tilde{s} + s^\dagger \times \tilde{d} \right)^{(2)} \cdot \left(d^\dagger \times \tilde{d} \right)^{(2)} + h.c. \right\rangle \rightarrow \frac{\beta_e^3 \cos 3\gamma_e}{(1+\beta_e^2)^2}. \end{aligned} \tag{43}$$

(The arrow indicates the classical order parameters that correspond to the quantal order parameters.) For the two-dimensional vibron model, there are in general two order parameters. However, two-dimensional rotational invariance reduces this number to one

$$\omega = \frac{\langle \hat{n} \rangle}{N} \rightarrow \frac{r_e^2}{1+r_e^2}. \tag{44}$$

The behavior of the order parameter with ξ for this model obtained numerically for finite N is shown in the right part of figure 2 where it is compared with the classical behavior. This behavior is common to all second order phase transitions of the type, u$(n-1)$-so(n). A similar study has been done for the first order transition u(5)-su(3) in the interacting boson model. Qualitatively, in finite systems, the discontinuities are smoothed out, a zeroth order transition looks like a first order, a first like a second and a second like a crossover.

2.4.3. Finite size scaling The scaling behavior of various quantities of physical interest as a function of the number of particles, N, has been investigated both numerically and analytically. For second order transitions, all physical quantities Φ are expected to scale as a power law [14]

$$\Phi(\xi^c) \sim N^{-(A_\Phi + 2B_\Phi/3)}. \tag{45}$$

Particularly important is the behavior of the ground state energy, E_0, and the order parameter ω. The scaling exponents $S_\Phi = -(A_\Phi + 2B_\Phi/3)$ for the u$(n-1)$-so(n) phase transition have been determined to be [15], for E_0, $S_\Phi = -\frac{4}{3}$ and, for ω, $S_\Phi = -\frac{2}{3}$. The numerical determination [16] of the scaling exponent for ω for the two-dimensional vibron model gives $-0.623(5)$.

3. Excited state quantum phase transitions (ESQPT)

In recent years, interest has grown in the study of excited states QPT. Discontinuities in certain quantities related to excited states are called excited state quantum phase transitions (ESQPT) [17]. ESQPT are best studied by constructing the so-called correlation diagram, that is how energies of excited states evolve as a function of the control parameter. In figure 4, the correlation diagram for a second order phase transition in a two-level pairing model is shown. One can see from this figure that levels accumulate along a curve separating the two phases. The level density diverges along this curve. Phase transitions thus appear to occur as one traverses the diagram both in the horizontal direction, as a function of the control parameter ξ, and in the vertical direction as a function of the excitation energy, E, for fixed value of ξ. One can also see that, at the critical value $\xi = \xi_c$, all levels accumulate at zero energy. For finite N, the divergence in the level density is smoothed out and a gap Γ appears. Finite size scaling for the gap in second order transitions has been investigated both analytically and numerically. The scaling exponent for the gap Γ in u$(n-1)$-so(n) transitions is $S_\Phi = -\frac{1}{3}$. A numerical investigation [16] of the scaling exponent for the gap in the two-dimensional vibron model gives $-0.335(1)$.

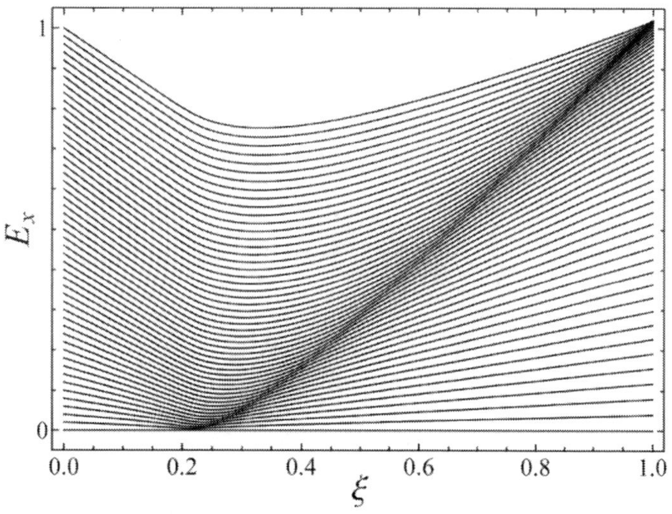

Figure 4. Excitation energies for a two-level fermionic pairing model, illustrating the compression of energy levels associated with a ESQPT. Figure from Ref. [17].

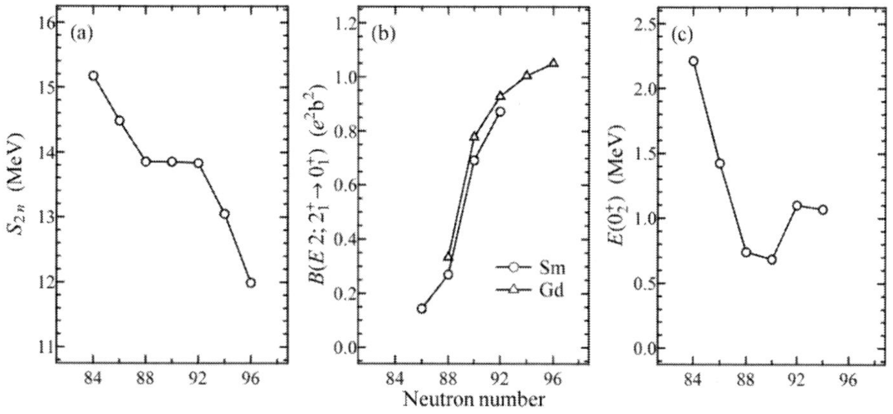

Figure 5. Experimental data for Sm isotopes (circles) and Gd isotopes (triangles) (a) two-neutron separation energy S_{2n}, (b) electromagnetic transition strength $B(E2; 2_1^+ \to 0_1^+)$, and (c) energy of the lowest $J = 0$ excitation $E(0_2^+)$.

4. Experimental evidence for QPT in nuclei

Quantum phase transitions have been observed in nuclei. Medium-mass and heavy nuclei can be well described by the interacting boson model with Hamiltonian

$$H = \varepsilon \hat{n}_d - \kappa \hat{Q}^\chi \cdot \hat{Q}^\chi. \tag{46}$$

The interactions in (46) are effective interactions that depend on proton and neutron number. The "control parameters" thus vary from nucleus to nucleus and can thus be swept by varying neutron and proton number. The Hamiltonian can be converted to the form (16) with a control parameter ξ that depends linearly (to first approximation) on neutron and proton number. Figure 5 shows the behavior of several quantities as a function of neutron number in the Sm isotopes. Panel (a) shows the two-neutron separation energies, $S_{2n}(N) = E_0(N+1) - E_0(N) \propto$

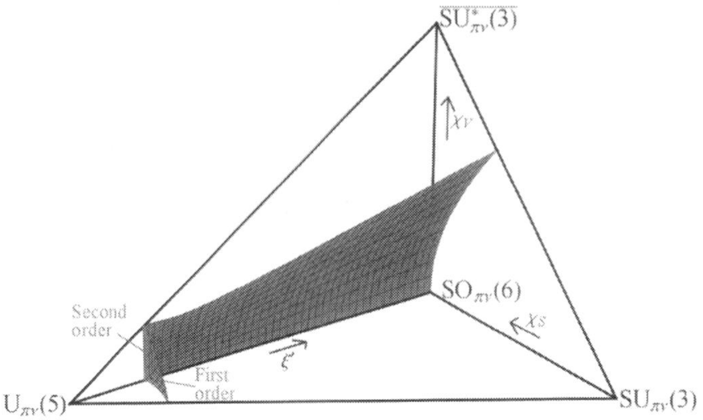

Figure 6. Phase diagram of the proton-neutron interacting boson model (IBM-2). Figure from Ref. [18].

$\frac{\partial E_0}{\partial \xi}$. Panel (b) shows $B(E2; 2_1 \rightarrow 0_1) \propto \langle \hat{n}_d \rangle^2$ and panel (c) $E(0_2) - E(0_1) \propto \Gamma$. The discontinuity in S_{2n} is clearly visible as well as the sudden increase in the order parameter $\langle \hat{n}_d \rangle$.

5. Symmetry and phase transitions

Dynamic symmetries

$$\begin{array}{ccc} & & g_1 \supset g_1 \supset \dots \\ & \nearrow & \\ g & & \dots \\ & \searrow & \\ & & g_\varphi \supset g'_\phi \supset \dots \end{array} \tag{47}$$

determine the phases of the system and the structure of the corresponding phase diagram. In this presentation, two examples have been discussed, the interacting boson model with algebraic structure

$$\begin{array}{ccc} & & u(5) \supset so(5) \supset so(3) \supset so(2) \\ & \nearrow & \\ u(6) & \longrightarrow & su(3) \supset so(3) \supset so(2) \\ & \searrow & \\ & & so(6) \supset so(5) \supset so(3) \supset so(2) \end{array} \tag{48}$$

and the two-dimensional vibron model with algebraic structure

$$u(3) \begin{array}{c} \nearrow \;\; u(2) \supset so(2) \\ \\ \searrow \;\; so(3) \supset so(2) \end{array} . \tag{49}$$

In the last few years, several other cases have been investigated. Results are available for:
(A) Quantum phase transitions in two-fluid bosonic systems [18, 19].
(B) Quantum phase transitions in two-level pairing models (bosonic and fermionic) [20].
The algebra g is, in these cases,

$$g \equiv u(n_1) \oplus u(n_2). \tag{50}$$

The phase diagrams for these systems are more complicated, as there are more phases and more control parameters. The phase structure of the proton-neutron interacting boson model (IBM-2) is shown in figure 6. The phase diagram is here three dimensional, represented as a tetrahedron. There is a plane of first order transitions and a sheet of second order transitions.

Investigations have started for:

(C) Quantum phase transitions in mixed Bose-Fermi systems [21, 22].

The algebra g is here

$$g \equiv u_B(n_1) \oplus u_F(n_2). \tag{51}$$

It has been found that supersymmetry (Bose-Fermi symmetry) plays a role in this case, if g is embedded onto $g^* \equiv u(n_1/n_2)$.

6. Conclusions

Major progress has been made in recent years in the study of QPT in algebraic models, both in the ground state (QPT) and in excited states (ESQPT). In this presentation the general framework for QPT in algebraic models has been given, including both semiclassical and quantal analysis. The phase diagrams of two bosonic models, the interacting boson model with algebraic structure U(6) and the two-dimensional vibron model U(3) have been constructed and their properties (control parameters, order parameters, critical exponents and scaling exponents) discussed. The analysis has been completed for second order transitions. For first order transitions it is to a large extent completed although some problems still remain (finite size scaling behavior). QPT have been verified experimentally to occur in nuclei and in molecules. Evidence for QPT in nuclei (interacting boson model) has been presented. ESQPT have been introduced and briefly presented. In particular the correlation diagrams and gaps of the interacting boson model and of the two-dimensional vibron model have been discussed. Experimental evidence for ESQPT is not yet available, although some indications show that they occur in molecules.

Dynamic symmetry plays a crucial role in the study of QPT in algebraic models, since it determines the phases of the system and the corresponding phase transitions between them.

Acknowledgments

This work was supported in part by the US DOE Grant DE-FG02-91ER-40608. I thank M. A. Caprio for providing figures 1, 2(b), and 5.

References

[1] Gilmore J 1979 *J. Math. Phys.* **20** 891
[2] Cejnar P and Iachello F 2007 *J. Phys. A: Math. Theor.* **40** 581
[3] Arima A and Iachello F 1975 *Phys. Rev. Lett.* **35** 1069
[4] Iachello F and Arima A 1987 *The Interacting Boson Model* (Cambridge: Cambridge University Press)
[5] Iachello F 1981 *Chem. Phys. Lett.* **78** 581
[6] Iachello F and Levine R D 1991 *Algebraic Theory of Molecules* (Oxford: Oxford University Press)
[7] Iachello F and Oss S 1996 *J. Chem. Phys.* **104** 6956
[8] Iachello F 2002 *Phil. Mag. Lett.* **82** 289
[9] Iachello F 2006 *Symmetry and Heterogeneity in High Temperature Superconductors*, ed A Bianconi (Berlin: Springer) p. 165
[10] Gilmore R 1974 *Lie Groups, Lie Algebras and Applications* (New York: J. Wiley and Sons)
[11] Feng D H, Gilmore R and Deans S R 1981 *Phys Rev C* **23** 1254
[12] Iachello F and Zamfir N V 2004 *Phys. Rev. Lett.* **92** 212501
[13] Williams E 2009 Ph.D. Thesis Yale University
[14] Fisher M E and Barber M N 1972 *Phys. Rev. Lett.* **28** 1516
[15] Dusuel S, Vidal J, Arias J M, Dukelsky J and Garcia-Ramos J E 2005 *Phys. Rev. C* **62** 011301(R)
[16] Perez-Bernal F and Iachello F 2008 *Phys. Rev. A* **77** 032115
[17] Caprio M A, Cejnar P and Iachello F 2008 *Ann. Phys. (NY)* **323** 1106

[18] Caprio M A and Iachello F 2004 *Phys. Rev. Lett.* **93** 242502
[19] Arias J M, Dukelsky J, Garcia-Ramos J E 2004 *Phys. Rev. Lett.* **93** 212501
[20] Caprio M A, Skrabacz J H and Iachello F 2009 in preparation
[21] Jolie J, Heinze S, Van Isacker P and Casten R F 2004 *Phys. Rev. C* **70** 011305(R)
[22] Alonso C E, Arias J M, Fortunato L and Vitturi A 2005 *Phys. Rev. C* **72** 061302(R)

Journal of Physics: Conference Series **237** (2010) 012015 doi:10.1088/1742-6596/237/1/012015

In memoriam of two distinguished participants of the Bregenz Symmetries in Science Symposia: Marcos Moshinsky and Yurii Fedorovich Smirnov

Maurice R Kibler[a,b,c]

[a] Université de Lyon, F–69622, Lyon, France
[b] Université Claude Bernard Lyon 1, Villeurbanne, France
[c] CNRS/IN2P3, Institut de Physique Nucléaire de Lyon, France

E-mail: m.kibler@ipnl.in2p3.fr

Abstract. Some particular facets of the numerous works by Marcos Moshinsky and Yurii Fedorovich Smirnov are presented in these notes. The accent is put on some of the common interests of Yurii and Marcos in physics, theoretical chemistry, and mathematical physics. These notes also contain some more personal memories of Yurii Smirnov.

1. Introduction

Yurii Fedorovich Smirnov passed away in 2008 and Marcos Moshinsky in 2009. They were two famous physicists with common interests in nuclear physics, atomic and molecular physics, and mathematical physics. More generally, both of them were at the origin of significant achievements in symmetry methods in physics. They actively participated in several Symmetries in Science Symposia in Bregenz. These two giants had parallel centers of interest in the sense that they developed separately some complementary works in nuclear physics (that led in particular to the concept of Moshinsky-Smirnov coefficients), dealt with some related problems in atomic, molecular and mathematical physics, and, finally, combined efforts to produce a beautiful and very useful book [1] on the applications of the harmonic oscillator system to various areas of physics and chemistry.

It is not the purpose of these notes to extensively list and analyse the numerous papers by Marcos and Yurii. I shall focus on some particular facets of their works. I had the opportunity to meet Marcos and Yurii several times in Bregenz and on several other occasions, and to discuss with them about Wigner-Racah algebras for finite groups, Lie groups and quantum groups. I had also a chance to collaborate with Yurii Smirnov. Therefore, I shall devote the main part of these notes to some specific domains of importance to Yurii and Marcos and to some more personal reminiscences on Yurii.

2. Marcos Moshinsky

Marcos Moshinsky was a Mexican physicist. He was born in Kiev (Ukraine) in 1921. He arrived as a refugee in Mexico when he was three years old and obtained Mexican citizenship in 1942. He received a Bachelor's degree in physics from the *Universidad Nacional Autónoma de México* (U.N.A.M.) and a Ph.D. degree in theoretical physics, under the guidance of Eugene P. Wigner, from Princeton University. Marcos was also the recipient of a post-doctoral fellowship at the *Institut Henri Poincaré* in Paris. Afterwards, he returned to Mexico and pursued a brillant career at the U.N.A.M. in Mexico City.

Professor Marcos Moshinsky had important responsibilities as the President of the *Sociedad Mexicana de Física*, as a member of *El Colegio Nacional*, and as a member of the editorial board of several international scientific reviews. He produced and/or co-produced more than 200 scientific papers and four books among which the most well-known are the one written in collaboration with Thomas A. Brody on transformation brackets for nuclear shell-model calculations [2] and the one with Yurii F. Smirnov on the applications of the harmonic oscillator in various fields of physics and quantum chemistry [1]. He received several prizes, namely the *Premio Nacional de Ciencias y Artes* in 1968, *Premio Luis Elizondo* in 1971, *Premio U.N.A.M. de Ciencias Exactas* in 1985, *Premio Príncipe de Asturias de Investigación Científica y Técnica* in 1988, and the prestigious UNESCO Science Prize in 1997 for his work in nuclear physics. He also received the Wigner medal in 1998.

3. Marcos and Yurii

A first seminal paper by Marcos concerned the transient dynamics of particle wavefunctions, a phenomenon that gives rise to diffraction in time [3]. However, most of his scientific work dealt with collective models of the nucleus, canonical transformations in quantum mechanics, and group theoretical methods in physics, with a special emphasis on symplectic symmetry in nuclear, atomic and molecular physics. These themes were of interest to Yurii too. Following the pioneering work of Talmi (who prepared his Ms.S. thesis with Guilio Racah, his Doctorate thesis with Wolfgang Pauli and who was a post-doctoral fellow with Eugene P. Wigner) [4], both Marcos and Yurii were interested in the description of pairs of nucleons in a harmonic-oscillator potential. In 1959, Moshinsky developed a formalism to connect wavefunctions in two different coordinate systems for two particles (with identical masses) in a harmonic-oscillator potential [5]. In this formalism, any two-particle wavefunction $|n_1\ell_1, n_2\ell_2, \lambda\mu\rangle$, expressed in coordinates with respect to the origin of the harmonic-oscillator potential, is a linear combination of wavefunctions $|n\ell, NL, \lambda\mu\rangle$, expressed in relative and centre-of-mass coordinates of the two particles. The so-called transformation brackets $\langle n\ell, NL, \lambda | n_1\ell_1, n_2\ell_2, \lambda\rangle$ make it possible to pass from one coordinate system to the other. Moshinsky gave an explicit expression of these coefficients in the case $n_1 = n_2 = 0$ and derived recurrence relations that can be used to obtain the coefficients for $n_1 \neq 0$ and $n_2 \neq 0$ from those for $n_1 = n_2 = 0$ [5]. Along this vein, Brody and Moshinsky published extensive tables of transformation brackets [2]. At the end of the fifties, Smirnov worked out a similar problem, viz. the calculation of the Talmi coefficients for unequal mass nucleons, and gave solution for the case $n_1 \neq 0$ and $n_2 \neq 0$ [6, 7]. (Indeed, the transformation brackets and the Talmi coefficients are connected via a double Clebsch-Gordan transformation.) The coefficients $\langle n\ell, NL, \lambda | n_1\ell_1, n_2\ell_2, \lambda\rangle$, called *transformation brackets* by Moshinsky and *total Talmi coefficients* by Smirnov, are now referred to as Moshinsky-Smirnov coefficients. Both the Moshinsky-Smirnov coefficients and the Talmi coefficients were revisited at the end of the seventies in terms of generating functions in the framework of the approaches of Julian S. Schwinger and Valentine Bargmann to the harmonic-oscillator bases [8]. (The work by Mehdi Hage Hassan [8], who prepared his State Doctorate thesis at the *Institut de Physique Nucléaire de Lyon* and conducted his career in Beyrouth, constitutes a very deep and original approach to the Talmi coefficients and Moshinsky-Smirnov coefficients.) It should be noted that the transformation brackets or Moshinsky-Smirnov coefficients are also of importance for atoms and molecules as shown by Marcos and Yurii in their book [1] written during the time Yurii was a visiting professor at the *Instituto de Física* of the *Universidad Nacional Autónoma de México*.

A second area of common interest to both Marcos and Yurii concerns the many-body problem considered from the point of view of unitary and symplectic groups and the use of nonlinear and nonbijective canonical transformations. In this vein, Moshinsky and some of his collaborators introduced the concept of an ambiguity group, a group required for taking into account the nonbijectivity of certain canonical transformations [9]. Indeed, this concept is closely related to the one of Lie algebra under constraint [10], which in turn is connected to nonbijective transformations like Hopf fibrations on spheres and Hopf fibrations on hyperbolids [11]. Among the nonbijective transformations, one may mention the $\mathbf{R}^4 \to \mathbf{R}^3$ Kustaanheimo-Stiefel transformation and the $\mathbf{R}^2 \to \mathbf{R}^2$ Levi-Civita transformation as well as their various extensions [11, 12]. In particular, the Kustaanheimo-Stiefel transformation allows one to pass from a four-dimensional isotropic harmonic oscillator subjected to a constraint to the three-dimensional Kepler-Coulomb system (see for instance [13]). This subject was of interest to Yurii and he revisited the hydrogen-oscillator connection with Corrado Campigotto [14].

The harmonic oscillator is a central ingredient in numerous studies by Smirnov and Moshinsky. Many applications of the nonrelativistic and relativistic harmonic oscillators to modern physics (from molecules, atoms and nuclei to quarks) were pedagogically exposed in the book by Marcos and Yurii [1] with a special attention paid to the n-body problem (in the Hartree-Fock approximation), the nuclear collective motion and the interacting boson model.

But their common interests were not limited to transformation brackets and harmonic oscillators. Let us briefly mention that both of them were also interested in group theoretical methods and symmetry methods in physics and also contributed to several fields of mathematical physics including, for instance, the state labelling problem, special functions, and generating functions (see, for example, [15, 16]).

4. Yurii Fedorovich Smirnov

Yurii Fedorovich Smirnov was a Russian physicist. He was born in the city of Il'inskoe (Yaroslavl' region, Russia) in 1935. He graduated from Moscow State University. Subsequently, he completed his Doctorate thesis at the same university under the guidance of Yurii M. Shirokov and benefited from fruitful contacts with other distinguished physicists including Yakov A. Smorodinsky. He pursued his career in the (Skobeltsyn) Institute of Nuclear Physics and in the Physics Department of (Lomonosov) Moscow State University with many stays abroad. The last fifteen years of his life were shared between Moscow and Mexico City where he was a visiting professor and later a professor at the *Instituto de Física* and at the *Instituto de Ciencias Nucleares* of the U.N.A.M. (he spent almost 11 years in Mexico). He received prestigious awards: the K.D. Sinel'nikov Prize of the Ukrainian Academy of Sciences in 1982 and the M.V. Lomonosov Prize in 2002. He was also a member of the Academy of Sciences of Mexico.

Yurii Smirnov authored and/or co-authored eleven books and more than 250 scientific articles. He also translated several scientific books into Russian. He translated, for example, a book on the harmonic oscillator written by Marcos Moshinsky in 1969, precisely the book that was a starting point for their common book on the same subject, published in 1996 [1]. He was a member of the editorial board of several journals and a councillor of the scientific councils of the Skobeltsyn Institute of Nuclear Physics and of the Chemistry Department of Moscow State University, as well as of the Institute for Theoretical and Experimental Physics (ITEP) in Moscow.

5. Some personal reminiscences on Yurii

My first contact with the work of Yurii Smirnov dates back to 1978 when my colleague J. Patera showed me, on the occasion of a NATO Advanced Study Institute organised in Canada by J.C. Donini, a beautiful book written by D.T. Sviridov and Yu.F. Smirnov [17]. This book dealt with the spectroscopy of d^N ions in inhomogeneous electric fields (part of a disciplinary domain known as crystal- and ligand-field theory in condensed matter physics and explored via the theory of level splitting from a theoretical point of view). In 1979, B.I. Zhilinskiĭ, while visiting Dijon and

Lyon in France in the context of an exchange programme between the USSR and France, gave me another interesting book, on f^N ions in crystalline fields, written by D.T. Sviridov, Yu.F. Smirnov and V.N. Tolstoy [18]. At that time, the references for mathematical aspects of crystal- and ligand-field theory were based on works by Y. Tanabe, S. Sugano and H. Kamimura from Japan [19], J.S. Griffith from England [20], and Tang Au-chin and his collaborators from China [21] (see also some contributions by the present author [22]). The two above-mentioned books by Smirnov and his colleagues shed some new light on the mathematical analysis of spectroscopic and magnetic properties of partly filled shell ions in molecular and crystal surroundings. In particular, special emphasis was put on the derivation of the Wigner-Racah algebra of a finite group of molecular and crystallographic interest from that of the group $SO(3) \sim SU(2)/Z_2$.

My second (indirect) contact with Yurii is related to an invitation to participate in the fifth workshop on *Symmetry Methods in Physics* in Obninsk in July 1991. Unfortunately, I did not get my visa on time reducing my participation to a paper in the proceedings of the workshop edited by Yu.F. Smirnov and R.M. Asherova [23].

In the beginning of the 1990's, I had a chance to discover another facet of Yurii's work. In 1989, a Russian speaking student from Switzerland, C. Campigotto, spent one year in the group of Prof. Smirnov. He started working on the above-mentioned Kustaanheimo-Stiefel transformation, an $\mathbf{R}^4 \rightarrow \mathbf{R}^3$ transformation associated with the Hopf fibration $S^3 \rightarrow S^2$ with compact fiber S^1. Then, Campigotto (well-prepared by Smirnov and his team, especially Andrey M. Shirokov and Valeriy N. Tolstoy) came to Lyon to prepare a Doctorate thesis (partly published in [24]). He defended his thesis in 1993 with George S. Pogosyan (representing Yu.F. Smirnov) as a member of the jury.

A fourth opportunity to work with Yurii stemmed from our mutual interest in quantum groups and in nuclear and atomic spectroscopy. I met him for the first time in Dubna in 1992. We then started a collaboration (partly with R.M. Asherova) on q- and qp-boson calculus in the framework of Hopf algebras associated with the Lie algebras $su(2)$ and $su(1,1)$ [25]. In addition, we pursued a group-theoretical study of the Coulomb energy averaged over the $n\ell^N$-atomic states with a definite spin [26]. We also had fruitful exchanges in nuclear physics. Indeed, Prof. Smirnov and his colleagues D. Bonatsos (from Greece), S.B. Drenska, P.P. Raychev and R.P. Roussev (all from Bulgaria) developed a model based on a one-parameter deformation of $SU(2)$ for dealing with rotational bands of deformed nuclei and rotational spectra of molecules [27] (see also [28]). Along the same line, a student of mine, R. Barbier, developed in his thesis a two-parameter deformation of $SU(2)$ with application to superdeformed nuclei in mass region $A \sim 130-150$ and $A \sim 190$ (partly published in [29]). It was a real pleasure to receive Yurii in Lyon on the occasion of the defence of the Barbier thesis in 1995. Indeed, from 1992 to 1995, Yurii made four stays in Lyon (one with his wife Rita and one with his daughter Tatyana) and we jointly participated in several meetings, one in Clausthal in Germany (organised by H.-D. Doebner, V.K. Dobrev and A.G. Ushveridze) and two in Bregenz in Austria (organised by B. Gruber and M. Ramek).

I cannot do justice to all of the fields in which Yurii was recognized as a superb researcher. It is enough to say that he contributed to many domains of mathematical physics (e.g., finite groups embedded in compact or locally compact groups, Lie groups and Lie algebras, quantum groups, special functions) and theoretical physics (e.g., nuclear, atomic and molecular physics, crystal- and ligand-field theory). Let me mention, among other fields, that he achieved alone and with collaborators significant advances in the theory of clustering of shell-model (nuclear) systems [30], in projection operator techniques for simple Lie groups [31], in the theory of heavy ion collisions [32], and in the so-called J-matrix formalism for quantum scattering theory (see [33] and references therein). The J-matrix formalism requires the solution of three-term recurrence relations (or second-order difference equations); along this line, Yurii and some of his collaborators published several works (see for instance [34]). As another major contribution, at the end of the sixties he proposed in collaboration with Vladimir G. Neudatchin a method, the so-called (e,2e) method (an analog of the (p,2p) method used in nuclear physics), for the experimental investigation of the electronic structure of atoms, molecules and solids; this method was successfully tested in many laboratories around the world (see [35] and references therein).

Yurii was also an exceptional teacher. It was very pleasant, profitable and inspiring to be taught by him. I personally greatly benefited from discussions with Yurii Smirnov.

Closing

Yurii and Marcos had many students who are now famous physicists. They interacted with many collaborators in their countries and abroad, and had an influence on many scientists. Marcos Moshinsky and Yurii Fedorovich Smirnov will remain examples for many of us. We shall not forget them.

References

[1] Moshinsky M and Smirnov Yu F 1996 *The Harmonic Oscillator in Modern Physics* (Amsterdam: Harwood Academic Publishers)
[2] Brody T A and Moshinsky M 1960 *Tables of Transformation Brackets for Nuclear Shell-Model Calculations* (Mexico City: Universidad Nacional Autónoma de México)
[3] Moshinsky M 1952 *Phys. Rev.* **88** 625
[4] Talmi I 1952 *Helv. Phys. Acta* **25** 185
[5] Moshinsky M 1959 *Nucl. Phys.* **13** 104
[6] Smirnov Yu F 1961 *Nucl. Phys.* **27** 177
[7] Smirnov Yu F 1962 *Nucl. Phys.* **39** 346
[8] Hage Hassan M 1980 *J. Phys. A: Math. Gen.* **13** 1903
[9] Moshinsky M and Seligman T H 1978 *Ann. Phys., NY* **114** 243
Moshinsky M and Seligman T H 1979 *Ann. Phys., NY* **120** 402
Deenen J, Moshinsky M and Seligman T H 1980 *Ann. Phys., NY* **127** 458
[10] Kibler M and Winternitz P 1988 *J. Phys. A: Math. Gen.* **21** 1787
[11] Lambert D and Kibler M 1987 Levi-Civita, Kustaanheimo-Stiefel and other transformations *Group Theoretical Methods in Physics* ed R Gilmore (Singapore: World Scientific)
Lambert D and Kibler M 1988 *J. Phys. A: Math. Gen.* **21** 307
Kibler M 1988 *Lecture Notes in Physics* **313** 238
Kibler M and Labastie P 1989 On some transformations generalizing the Levi-Civita, Kustaanheimo-Stiefel, and Fock transformations *Group Theoretical Methods in Physics* eds Y Saint-Aubin and L Vinet (Singapore: World Scientific)
[12] Hage Hassan M and Kibler M 1990 Non-bijective quadratic transformations and theory of angular momentum *Selected Topics in Statistical Mechanics* eds A A Logunov, N N Bogolubov, Jr, V G Kadyshevsky and A S Shumovsky (Singapore: World Scientific)
Hage Hassan M and Kibler M 1991 On Hurwitz Transformations *Le Problème de Factorisation de Hurwitz : Approche Historique, Solutions, Applications en Physique* eds A Ronveaux and D Lambert (Namur: FUNDP) (*Preprint* hepth/9409051)
[13] Kibler M and Négadi T 1983 *Lett. Nuovo Cimento* **37** 225
Kibler M and Négadi T 1983 *J. Phys. A: Math. Gen.* **16** 4265
Kibler M and Négadi T 1984 *Phys. Rev.* A **29** 2891
[14] Campigotto C and Smirnov Yu F 1991 *Helv. Phys. Acta* **64** 48
[15] Moshinsky M, Patera J, Sharp R T and Winternitz P 1975 *Ann. Phys., NY* **95** 139
[16] Raychev P P, Roussev R P and Smirnov Y F 1991 *J. Phys. A: Math. Gen.* **24** 2943
[17] Sviridov D T and Smirnov Yu F 1977 *Teoriya Opticheskikh Spektrov Perekhodnykh Metallov* (Moscow: Izd. Nauka)
Sviridov D T and Smirnov Yu F 1968 *Soviet. Phys. Doklady* **13** 565
Vonsovski C V, Grimailov C V, Tcherepanov V I, Meng A N, Sviridov D T, Smirnov Yu F and Nikiforov A E 1969 *Crystal-Field Theory and Optical Spectra of Partly Filled d Shell Transition Ions* (Moscow: Nauka)
Sviridov D T, Sviridova R K and Smirnov Yu F 1976 *Optical Spectra of Transition-Metal Ions in Crystal* (Moscow: Nauka)
[18] Sviridov D T, Smirnov Yu F and Tolstoy V N 1975 *Spektroskopiya Kristallov* (Moscow: Akad. Nauk SSSR)
[19] Tanabe Y and Sugano S 1954 *J. Phys. Soc. Japan* **9** 753

Tanabe Y and Sugano S 1954 *J. Phys. Soc. Japan* **9** 766
Tanabe Y and Sugano S 1956 *J. Phys. Soc. Japan* **11** 864
Tanabe Y and Kamimura H 1958 *J. Phys. Soc. Japan* **13** 394
Sugano S and Tanabe Y 1958 *J. Phys. Soc. Japan* **13** 880
Sugano S, Tanabe Y and Kamimura H 1970 *Multiplets of Transition-Metal Ions in Crystals* (New York: Academic Press)
[20] Griffith J S 1960 *Molec. Phys.* **3** 79
Griffith J S 1960 *Molec. Phys.* **3** 285
Griffith J S 1960 *Molec. Phys.* **3** 457
Griffith J S 1960 *Molec. Phys.* **3** 477
Griffith J S 1961 *The Theory of Transition-Metal Ions* (Cambridge: Cambridge Univ. Press)
Griffith J S 1962 *The Irreducible Tensor Method for Molecular Symmetry Groups* (Englewood Cliffs: Prentice-Hall)
[21] Tang Au-chin, Sun Chia-chung, Kiang Yuan-sun, Deng Zung-hau, Liu Jo-chuang, Chang Chian-er, Yan Go-sen, Goo Zien and Tai Shu-shan 1966 *Sci. Sinica (Peking)* **15** 610
Tang Au-chin, Sun Chia-chung, Kiang Yuan-sun, Deng Zung-hau, Liu Jo-chuang, Chang Chain-er, Yan Guo-sen, Goo Zien and Tai Shu-shan 1979 *Theoretical Method of the Ligand Field Theory* (Peking: Science Press)
[22] Kibler M 1968 *J. Molec. Spectrosc.* **26** 111
Kibler M 1969 *Int. J. Quantum Chem.* **3** 795
Kibler M 1969 *C.R. Acad. Sci. (Paris), Ser. B* **268** 1221
Kibler M R 1977 Group theory around ligand field theory *Group Theoretical Methods in Physics* eds R T Sharp and B Kolman (New York: Academic Press)
Kibler M R 1979 Finite symmetry adaptation in spectroscopy *Recent Advances in Group Theory and Their Application to Spectroscopy* ed J C Donini (New York: Plenum Press)
[23] Kibler M and Daoud M 1992 Symmetry adaptation and two-photon spectroscopy of ions in molecular or solid-state finite symmetry *Symmetry Methods in Physics* eds Yu F Smirnov and R M Asherova (Obninsk: Russian Federation Ministry of Atomic Energy – Institute of Physics and Power Engineering)
[24] Campigotto C and Smirnov Yu F 1991 *Helv. Phys. Acta* **64** 48
Drăgănescu Gh E, Campigotto C and Kibler M 1992 *Phys. Lett.* A **170** 339
Kibler M and Campigotto C 1993 *Int. J. Quantum Chem.* **45** 209
Kibler M and Campigotto C 1993 *Phys. Lett.* A **181** 1
Campigotto C, Smirnov Yu F and Enikeev S G 1995 *J. Comp. Appl. Math.* **57** 87
Smirnov Yu F and Campigotto C 2004 *J. Comp. Appl. Math.* **164–165** 643
[25] Smirnov Yu F and Kibler M R 1993 Some aspects of q-boson calculus *Symmetries in Science VI: From the Rotation Group to Quantum Algebras* ed B Gruber (New York: Plenum Press)
Kibler M, Campigotto C and Smirnov Yu F 1994 Recursion relations for Clebsch-Gordan coefficients of $U_q(su_2)$ and $U_q(su_{1,1})$ *Symmetry Methods in Physics* eds A N Sissakian, G S Pogosyan and S I Vinitsky (Dubna: Joint Institute for Nuclear Research)
Kibler M R, Asherova R M and Smirnov Yu F 1995 Some aspects of q- and qp-boson calculus *Symmetries in Science VIII* ed B Gruber (New York: Plenum Press)
[26] Kibler M and Smirnov Yu F 1995 *Int. J. Quantum Chem.* **53** 495
[27] Bonatsos D, Drenska S B, Raychev P P, Roussev R P and Smirnov Yu F 1991 *J. Phys. G: Nucl. Part. Phys.* **17** L67
Raychev P P, Roussev R P and Smirnov Yu F 1990 *J. Phys. G: Nucl. Part. Phys.* **16** L137
Bonatsos D, Raychev P P, Roussev R P and Smirnov Yu F 1990 *Chem. Phys. Lett.* **175** 300
Zhilinskiĭ B I and Smirnov Yu F 1991 *Sov. J. Nucl. Phys.* **54** 10
[28] Georgieva A and Dankova Ts 1994 *J. Phys. A: Math. Gen.* **27** 1251
Georgieva A I, Goleminov J D, Ivanov M I and Geyer H B 1999 *J. Phys. A: Math. Gen.* **32** 2403
[29] Barbier R, Meyer J and Kibler M 1994 *J. Phys. G: Nucl. Part. Phys.* **20** L13
Barbier R, Meyer J and Kibler M 1995 *Int. J. Mod. Phys.* E **4** 385
Barbier R and Kibler M 1995 A system of interest in spectroscopy: The qp-rotor system *Finite Dimensional Integrable Systems* eds A N Sissakian and G S Pogosyan (Dubna: Joint Institute for Nuclear Research)
Barbier R and Kibler M 1995 On the use of quantum algebras in rotation-vibration spectroscopy *Modern Group Theoretical Methods in Physics* eds J Bertrand, M Flato, J-P Gazeau, D Sternheimer and M Irac-Astaud (Dordrecht: Kluwer)
Barbier R and Kibler M 1996 *Rept. Math. Phys.* **38** 221
[30] Neudatchin V G and Smirnov Yu F 1969 *Nucleon Clusters in Light Nuclei* (Moscow, Nauka)
Nemets O F, Neudatchin V G, Rudchik A T, Smirnov Yu F and Tchuvil'sky Yu M 1988 *Nucleon Clusters*

in Atomic Nuclei and Many-Nucleon Transfer Reactions (Kiev: Naukova Dumka)
[31] Asherova R M, Smirnov Yu F and Tolstoy V N 1973 *Theoret. Math. Phys.* **15** 107
Tolstoy V N 2006 *Phys. Atom. Nuclei* **69** 1058
[32] Smirnov Yu F and Tchuvil'sky Yu M 1984 *Phys. Lett.* B **134** 25
[33] Smirnov Yu F, Shirokov A M, Lurie Yu A and Zaytsev S A 1995 Harmonic oscillator representation in the theory of scattering and nuclear reactions *Harmonic Oscillators* eds D Han and K B Wolf (Washington: NASA)
Bang J M, Mazur A I, Shirokov A M, Smirnov Yu F and Zaytsev S A 2000 *Ann. Phys., NY* **280** 299
[34] Smirnov Yu F, Suslov S K and Shirokov A M 1984 *J. Phys. A: Math. Gen.* **17** 2157
Braun P A, Shirokov A M and Smirnov Yu F 1985 *Molec. Phys.* **56** 573
Shirokov A M and Smirnov Yu F 1991 *J. Phys. A: Math. Gen.* **24** 2961
Zaitsev S A, Smirnov Y F and Shirokov A M 1988 **Teoret. Mat. Fiz. 117** 1291
[35] Neudatchin V G, Popov Yu V and Smirnov Yu F 1999 *Uspekhi Fiz. Nauk* **169** 1111

Point and counterpoint between Mathematical Physics and Physical Mathematics

PGL Leach[1,1] **and MC Nucci**[2]

[1] Department of Mathematics and Statistics, University of Cyprus, CY 1678 Lefkosia, Cyprus
[2] Dipartimento di Matematica e Informatica, Università di Perugia, 06123 Perugia, Italy

E-mail: `leach@ucy.ac.cy; leachp@ukzn.ac.za; leach@math.aegean.gr; nucci@unipg.it`

Abstract. In recent years there has been a resurgence of interest in problems dating back for over half a century. In particular we refer to the questions of the consistency of quantisation and nonlinear canonical transformations and the quantisation of higher-order field theories. We present resolutions to these questions based upon considerations of symmetry. This enables one to examine these problems within the context of existing theory without the need to introduce new and exotic theories.

Keywords: Quantisation; Lie symmetry; Noether symmetry; reduction of order; $\mathcal{PT}$-symmetry
PACS numbers: 02.30.Hq; 02.30.Xx; 03.65.Fd

In the recent literature there has been a resurgence of interest in two questions relating to the quantisation of classical systems. By coincidence the two questions arose at roughly the same time almost sixty years ago. One question related to the interchange of the actions of making a canonical transformation and performing quantisation on an Hamiltonian. In a lengthy and detailed paper of 1951 van Hove [31] demonstrated that in general there could not be consistency under the interchange. In 1950 Pais and Uhlenbeck [27] introduced a fourth-order field-theoretic model which, when quantised, led to the presence of eigenstates of negative norm. In both cases there was a conflict between the expectation of consistency of description, which is what Physics demands, and the results which Mathematics produced.

It is a commonplace to describe Mathematics as the Queen and the Servant of Science, by which one generally thinks of Physics since it is the most mathematically developed of the sciences. The description is quite eloquent and yet one can wonder if its implication is generally appreciated. One can relate to the concept of 'Servant' in which role Mathematics provides the tools for the resolution of problems in Physics in that the Mathematics is subservient to the realistic demands of Physics. Independently of Science Mathematics can evolve along manifold paths into areas which are not obviously connected to the reality as observed in Nature. Nevertheless one must expect, nay demand, a mathematical description of a physical problem

[1] permanent address: School of Mathematical Sciences, University of KwaZulu-Natal, Private Bag X54001 Durban 4000, Republic of South Africa

to be consistent with the Physics as it is observed experimentally. No matter the method of description of a physical system the bottom line should always be the same.

Hamilton [13] developed his theory of Mechanics. Dirac [9] found that Hamilton's theory was most acceptable for his operator theory of Quantum Mechanics. van Hove demonstrated that the generality permitted in the canonical transformations of Hamilton's theory caused problems with quantisation[2]. In the last decade Calogero (with Graffi in one paper) [6, 7, 8] examined a number of nonlinear oscillators classically related by nonlinear canonical transformations and demonstrated some of the problems associated with a consistent quantisation. In an attempt to bring consistency between Mathematics and Physics Nucci *et al.* [24] proposed that the criterion for a valid quantisation procedure between two Hamiltonians related by a canonical transformation be the preservation of the Lie point symmetries of the corresponding Schrödinger equations and demonstrated this with several examples taken from the then current literature.

Hamilton's theory is essentially a first-order theory in that the canonical equations of motion derived from the Hamiltonian are of first order. A standard way to obtain the Hamiltonian is to make use of a Legendre transformation of a first-order Lagrangian. However, the Euler-Lagrange theory is not confined to first-order Lagrangians. Only a few years after Hamilton's presentation of his theory Ostrogradsky [26] proposed a procedure which would enable the construction of an Hamiltonian function from an higher-order Lagrangian. The field-theoretic model of Pais-Uhlenbeck is of the fourth order and has a second-order Lagrangian. They are, respectively,

$$\ddddot{z} + \left(\Omega_1^2 + \Omega_2^2\right)\ddot{z} + \Omega_1^2\Omega_2^2 z = 0 \tag{1}$$

and

$$L = \tfrac{1}{2}\gamma\left\{\ddot{z}^2 - \left(\Omega_1^2 + \Omega_2^2\right)\dot{z}^2 + \Omega_1^2\Omega_2^2 z^2\right\}. \tag{2}$$

In [17] and [18] the constraint method of Dirac was used to construct a Hamiltonian. The result was the same as if the method of Ostrogradsky had been employed.

The problem with the Hamiltonian

$$H = \tfrac{1}{2}\left\{\frac{p_y^2}{\gamma} + 2p_z y + \gamma\left(\Omega_1^2 + \Omega_2^2\right)y^2 - \gamma\Omega_1^2\Omega_2^2 z^2\right\} \tag{3}$$

obtained in this way was that it possessed 'ghosts' when considered from the point of view of Quantum Mechanics in the usual formulation of Dirac. Not surprisingly the existence of unwanted properties did not bring joy to physicists. The Mathematics may well have been impeccable, but the Physics was definitely peccable. Bender and Mannheim [4] sought to obviate the problem of ghosts by discarding the approach of Dirac of some eighty years ago and to approach the problem of a correct quantisation of the model of Pais-Uhlenbeck using the methods of the theory of $\mathcal{PT}$ symmetry rather than the Hermitian approach which underpins Dirac's route to quantisation[3]. Their analysis was successful in that ghosts no longer bedevilled the problem.

[2] This leaves aside the question of the correct quantisation of a function of the classical momentum and position [2]. Naturally the two questions are related, but for the nonce we concentrate upon the difficulties with any particular quantisation scheme.

[3] One should bear in mind that an essential feature of the Hermitian approach is that the eigenvalues of the operators are necessarily real. However, this does not mean that nonhermitian operators cannot have real eigenvalues and so be physically acceptable. A potentially interesting application has been recently proposed to deal with the almost eternally vexed question of the quantum mechanics of a system with dissipation such as

A critical point, which seems to have been generally overlooked in previous treatments of this problem, is the process whereby the fourth-order equation of Pais-Uhlenbeck and its second-order Lagrangian are reduced to an Hamiltonian. The method, as we mentioned above, is due to Ostrogradsky. However, it is not the only method available. Indeed one could opine that the method of Ostrogradsky is a quite complicated way to implement the methods of reduction of order. In [25] we used the method proposed by Nucci [21], which has seen some quite successful applications such as in [19], [22], [23], [10], [12], to reduce the fourth-order model of Pais-Uhlenbeck to obtain a first-order Lagrangian. The quantum mechanical treatment of the Hamiltonian derived from this Lagrangian saw no ghosts. We wrote the fourth-order equation, (1), as a system of four first-order equations and then recombined the variables to produce the Lagrangian

$$L = \tfrac{1}{2}\left(\dot{r}_1^2 + \dot{r}_2^2 - \Omega_1^2 r_1^2 - \Omega_2^2 r_2^2\right), \tag{4}$$

where

$$z = r_1 - r_2 \quad \text{and} \quad \ddot{z} = -\Omega_1^2 r_1 + \Omega_2^2 r_2. \tag{5}$$

It is obvious that the Hamiltonian is

$$H = \tfrac{1}{2}\left(p_1^2 + p_2^2 + \Omega_1^2 r_1^2 + \Omega_2^2 r_2^2\right) \tag{6}$$

and that it is quite free of ghosts in the quantum-mechanical context. We emphasise that there was no need to depart from the context of Dirac's formulation of quantum mechanics.

We have the situation in which two analyses of the same fourth-order equation with its associated second-order Lagrangian have led to two quite distinct quantum-mechanical treatments. Were the results of the treatments the same, one would not be perturbed since one would expect consistency from two different descriptions of the same physical system. However, there were some quite notable differences in results and it is incumbent upon anyone either to resolve the differences or to explain why they exist.

The method of Ostrogradsky as applied to the Lagrangian, (2), proceeds as follows according to the report in Whittaker [32]. Commencing with the Lagrangian, (2), we obtain the two momenta

$$\begin{aligned} p_1 &= -\gamma\left(\Omega_1^2 + \Omega_2^2\right)\dot{z} - \gamma\dddot{z} \\ p_2 &= \gamma\ddot{z} \end{aligned}$$

the damped linear oscillator [1] by making use of the fractional calculus introduced a quarter of a millennium ago by Euler subsequent to a proposal by l'Hôpital towards the end of the seventeenth century and building upon the treatment of Lagrangians and Hamiltonian functions for systems with cetain forms of dissipation by Riewe [29]. The operators are obviously nonhermitian, but the eigenvalues are certainly real. One recalls that Bateman [3] proposed a method – maybe better known to physicists through the works of Caldriola [5] and Kanai [14] – to obtain a Lagrangian and hence Hamiltonian for this problem. Subsequent quantisation led to results which were regarded as impossibly nonphysical. The approach in [1] still remains to be subjected to thorough examination by the scientific community, but it does provide a route which at least does not tread the doubtful path initiated by Bateman. Indeed Bateman's approach could be cited as an instance of Mathematics behaving as a Queen. Schuch [30] provided an alternate approach based on the Caldirola-Kanai formalism which was more acceptable physically. A key feature to his treatment was introduction of a nonunitary transformation which obviated the need to confront the nonphysical aspects of earlier treatments. Somehow it would seem that the avoidance of physically unacceptable conclusions necessitates a departure from Dirac's formulism when dealing with dissipation.

and the corresponding canonical coordinates are $q_1 = z$ and $q_2 = \dot{z}$. Using the prescription of Ostrogradsky as given by Whittaker for the number of variables in the present situation we obtain the Hamiltonian

$$H = -\tfrac{1}{2}\gamma\left\{\frac{p_2^2}{\gamma^2} - \left(\Omega_1^2 + \Omega_2^2\right) q_2^2 + \Omega_1^2\Omega_2^2 q_1^2\right\} + p_1 q_2. \tag{7}$$

One observes that the four coordinates introduced by Ostrogradsky involve the dependent variable and its derivatives according to a prescription. This is not greatly different from the method of reduction of order and yet one finds a rather dramatic difference in the outcome. For the benefit of the reader without access to either the original paper by Ostrogradsky or a succinct account such as presented by Whittaker we summarise Ostrogradsky's method for recasting a problem from the calculus of variations in one independent variable with an higher-order Lagrangian into Hamiltonian form.

Let $L\left(t, x, \dot{x}, \ddot{x}, \ldots, x^{(n)}\right)$, where x can be a multivariable and n a multi-index in a general treatment, but we confine our attention to a single dependent variable which may occur up to the nth derivative with respect to the independent variable, t, be a Lagrangian for which the Euler-Lagrange equation of the Calculus of Variations is

$$0 = \frac{\partial L}{\partial x} - \frac{\mathrm{d}}{\mathrm{d}t}\left(\frac{\partial L}{\partial \dot{x}}\right) + \ldots + (-1)^n \frac{\mathrm{d}^n}{\mathrm{d}t^n}\left(\frac{\partial L}{\partial x^{(n)}}\right). \tag{8}$$

To obtain an Hamiltonian representation a first-order Lagrangian is required. Ostrogradsky defines the momenta as thus.

$$\begin{aligned} p_1 &= \frac{\partial L}{\partial \dot{x}} - \frac{\mathrm{d}}{\mathrm{d}t}\left(\frac{\partial L}{\partial \ddot{x}}\right) + \ldots + (-1)^{n-1}\frac{\mathrm{d}^{n-1}}{\mathrm{d}t^{n-1}}\left(\frac{\partial L}{\partial x^{(n)}}\right) \\ p_2 &= \frac{\partial L}{\partial \ddot{x}} - \frac{\mathrm{d}}{\mathrm{d}t}\left(\frac{\partial L}{\partial \dddot{x}}\right) + \ldots + (-1)^{n-2}\frac{\mathrm{d}^{n-2}}{\mathrm{d}t^{n-2}}\left(\frac{\partial L}{\partial x^{(n)}}\right) \\ &\vdots \\ p_n &= \frac{\partial L}{\partial x^{(n)}} \end{aligned} \tag{9}$$

which ensures that all derivatives present in the Lagrangian are covered. The canonical coordinates are defined according to

$$q_1 = x,\ q_2 = \dot{x},\ \ldots,\ q_n = x^{(n-1)}. \tag{10}$$

The Hamiltonian function is then defined according to

$$H = -L + p_1 q_2 + p_2 q_3 + \ldots + p_{n-1} q_n + p_n x^{(n)}, \tag{11}$$

which reduces to the standard prescription when the Lagrangian is of the first order. When Hamilton's Principle is applied, we obtain the standard equations for an Hamiltonian describing a system of n degrees of freedom, namely

$$\frac{\mathrm{d}q_j}{\mathrm{d}t} = \frac{\partial H}{\partial p_j} \quad \text{and} \quad \frac{\mathrm{d}p_j}{\mathrm{d}t} = -\frac{\partial H}{\partial q_j}. \tag{12}$$

The formalism is fitting and yet we have found that the transition to quantum mechanics has led to a flawed description of the physical properties of the original system.

The first question to be resolved is at what point in this transition has there been a departure from a faithful description of the mechanics. Is it in the variation of the classical description implied by the application of the method of Ostrogradsky or does the problem arise afterwards when one attempts to do quantum mechanics? We apply Hamilton's equations of motion, (12), to the Hamiltonian, (7). We obtain

$$\begin{aligned} \dot{q}_1 &= q_2 & \dot{p}_1 &= \gamma\Omega_1\Omega_2 q_1 \\ \dot{q}_2 &= -\frac{p_2}{\gamma} & \dot{p}_2 &= -\gamma\left(\Omega_1^2+\Omega_2^2\right)q_2 - p_1. \end{aligned}$$

Recalling that $q_1 = z$ we keep differentiating the first equation and substituting from the others until we indeed return to (1).

This particular problem does not provide a counterexample to the validity of the method of Ostrogradsky in the context of classical mechanics. Consequently we infer that there is a problem in the transition to quantum mechanics. In a formal sense the quantisation of the Hamiltonian, (7), is not a difficult procedure. There is not even a case to be made for this method or that method of quantisation since there are no products of noncommuting operators. We can only conclude that the problem arises in the manner in which the method of Ostrogradsky affects the subsequent quantisation procedure. We can compare this method in a manner of speaking with the problems encountered with the quantisation of simple mechanical systems after they have been subjected to nonlinear canonical transformation. There has been more than a little literature on the subject over the last half century for which a relevant sampling is [31, 6, 7, 8, 24]. Of particular relevance to the present discussion is the proposal made in [24] in which the point was made, with supporting examples, that an essential feature of the quantisation procedure appeared to be the preservation of symmetry. This immediately begs the question of what kind of symmetry should be preserved. In this respect we are guided by a particular experience which is the relationship between the Noether point symmetries of the classical Lagrangian and the Lie point symmetries of the corresponding Schrödinger equation [16, 2, 15, 20]. Apart from the generic Lie point symmetries, *i.e.* those corresponding to the solutions of the linear Schrödinger equation and its very linearity, a one-to-one correspondence is observed between the two classes of symmetry. When that nexus has been broken, the results of the quantisation have been embarrassing [8].

It seems that the method of Ostrogradsky has something in common with the application of nonlinear canonical transformations to Hamiltonian systems and their subsequent quantisation. That has been demonstrated to cause a difficulty. This is not a good thing. It is ever the desire of a describer of physical processes to provide a consistent description, no matter the viewpoint. Following our success [24] with the correct quantisation of a variety of nonlinear oscillators rendered nonlinear by means of nonlinear canonical transformations from the standard representation of a linear oscillator we propose in this paper a procedure to rectify the problems of the appearance of ghosts in the reduction and subsequent quantisation of higher-order field theories. The philosophy of the procedure is quite simple. Preserve the symmetries!

We consider the Lagrangian of equation (1) namely

$$L = \tfrac{1}{2}\left\{\ddot{z}^2 - \left(\Omega_1^2+\Omega_2^2\right)\dot{z}^2 + \Omega_1^2\Omega_2^2 z^2\right\} + \frac{\mathrm{d}}{\mathrm{d}t}F(t,z,\dot{z}) \qquad (13)$$

where F is the gauge function. Equation (1) admits a six-dimensional Lie point symmetry algebra generated by

$$\begin{array}{ll} \Gamma_1 = \partial_t & \Gamma_2 = z\partial_z \\ \Gamma_3 = \cos(\Omega_1 t)\partial_z & \Gamma_4 = -\sin(\Omega_1 t)\partial_z \\ \Gamma_5 = \cos(\Omega_2 t)\partial_z & \Gamma_6 = -\sin(\Omega_2 t)\partial_z, \end{array}$$

while the Lagrangian (13) admits five Noether point symmetries and consequently five first integrals, namely

$$\begin{array}{lll} \Gamma_1 & \Longrightarrow & I_1 = \frac{1}{2}(-\Omega_1^2\Omega_2^2 z^2 - \Omega_1^2\dot{z}^2 - \Omega_2^2\dot{z}^2 - 2\dot{z}\dddot{z} + \ddot{z}^2) \\ \Gamma_3 & \Longrightarrow & I_3 = (\Omega_2^2 z + \ddot{z})\sin(\Omega_1 t)\Omega_1 + (\Omega_2^2\dot{z} + \dddot{z})\cos(\Omega_1 t) \\ \Gamma_4 & \Longrightarrow & I_4 = (\Omega_2^2 z + \ddot{z})\cos(\Omega_1 t)\Omega_1 - (\Omega_2^2\dot{z} + \dddot{z})\sin(\Omega_1 t) \\ \Gamma_5 & \Longrightarrow & I_5 = (\Omega_1^2 z + \ddot{z})\sin(\Omega_2 t)\Omega_2 + (\Omega_1^2\dot{z} + \dddot{z})\cos(\Omega_2 t) \\ \Gamma_6 & \Longrightarrow & I_6 = (\Omega_1^2 z + \ddot{z})\cos(\Omega_2 t)\Omega_2 - (\Omega_1^2\dot{z} + \dddot{z})\sin(\Omega_2 t). \end{array}$$

En passent we remark that these first integrals, except I_1, could not be obtained if the gauge function F was taken to be a constant and yet for some strange reason there are writers who would have that so.

If we suitably combine the first integrals I_3, I_4, I_5 and I_6 as $\frac{1}{2}(I_3^2 + I_4^2 + I_5^2 + I_6^2)$, then the autonomous first integral,

$$I_{\rm aut} = \frac{1}{2}\left[(\Omega_2^2 z + \ddot{z})^2\Omega_1^2 + (\Omega_2^2\dot{z} + \dddot{z})^2 + (\Omega_1^2 z + \ddot{z})^2\Omega_2^2 + (\Omega_1^2\dot{z} + \dddot{z})^2\right], \tag{14}$$

is obtained. If we make the obvious transformations

$$q_1 = \ddot{z} + \Omega_2^2 z \qquad q_2 = \ddot{z} + \Omega_1^2 z \tag{15}$$

$$p_1 = \dddot{z} + \Omega_2^2\dot{z} \qquad p_2 = \dddot{z} + \Omega_1^2\dot{z}, \tag{16}$$

from (14) we obtain the Hamiltonian

$$H \equiv I_{\rm aut} = \frac{1}{2}[p_1^2 + p_2^2 + \Omega_1^2 q_1^2 + \Omega_2^2 q_2^2] \tag{17}$$

and the corresponding canonical equations are

$$\dot{q}_1 = p_1, \quad \dot{q}_2 = p_2, \quad \dot{p}_1 = -\Omega_1^2 q_1, \quad \dot{p}_2 = -\Omega_2^2 q_2. \tag{18}$$

This is the appropriate Hamiltonian[4] for the quantization of the fourth-order field-theoretic model of Pais-Uhlenbeck, (1).

This shows that Noether's Theorem does provide a guide for the transition from classical to quantum mechanics and that it has a definite advantage over Ostrogradsky's method when it comes to to the nonexistence of ghosts and the necessity to exorcise them by means of the development of new theories. Indeed Ostrogradsky in his original paper[5] was not so much interested in Mechanics since in his setting the kinetic energy came to be negative, something

[4] We have put the parameter $\gamma = 1$.

[5] We wonder how many of the authors who have cited this work have actually "wasted their time" and read it. We did [28].

that Whittaker corrected in his book. We recall our remarks above regarding Mathematics and Physics!

The application of the Legendre transformation to the Hamiltonian (17) gives the Lagrangian

$$L = \frac{1}{2}[\dot{q}_1^2 + \dot{q}_2^2 - (\Omega_1^2 q_1^2 + \Omega_2^2 q_2^2)] \tag{19}$$

and the corresponding Euler-Lagrange equations

$$\ddot{q}_1 = -\Omega_1^2 q_1, \qquad \ddot{q}_2 = -\Omega_2^2 q_2 \tag{20}$$

which admit a seven-dimensional algebra of Lie point symmetries[6] generated by the following operators

$$\begin{array}{llll} \Gamma_1 = \partial_t, & \Gamma_2 = q_1\partial_{q_1}, & \Gamma_3 = \cos(\Omega_1 t)\partial_{q_1}, & \Gamma_4 = -\sin(\Omega_1 t)\partial_{q_1}, \\ & \Gamma_5 = q_2\partial_{q_2}, & \Gamma_6 = \cos(\Omega_2 t)\partial_{q_2}, & \Gamma_7 = -\sin(\Omega_2 t)\partial_{q_2}. \end{array} \tag{21}$$

We note that the Lagrangian, (19), possesses *five* Noether point symmetries which is in curious agreement with the number of Noether point symmetries admitted by (13).

We started with the observation that Mathematics and Physics are not the same. One can do things mathematically which make no sense physically. In the application of Mathematics to the resolution of problems in Physics one should always bear in mind that in this situation Mathematics is the humble and obedient servant of Science.

Acknowledgements
PGLL thanks DICSE, University of the Aegean, and the Department of Mathematics and Statistics, University of Cyprus, for the provision of facilities whilst this work was undertaken. This work was funded in part by the Research Promotion Foundation of Cyprus, Grant Number ΠΡΟΣΕΛΚΥΣΗ ΠΡΟΕΜ/0308/02. The University of KwaZulu-Natal and the National Research Foundation of South Africa are thanked for their support. The opinions expressed in this paper should not be construed as being those of the two latter institutions.

References

[1] Al-Okour Amin and Ajlouni Abdul-Wali 2009 Schrödinger equation of dissipative finite potential barrier using fractional calculus (presentation to the *International Conference on Mathematics and Information Security*, Sohag University, Sohag, Egypt, 13-15 November, 2009)
[2] Andriopoulos K and Leach PGL 2005 Some group theoretical aspects of nonlinear quantal oscillators *J. Nonlin. Math. Phys.* **12, S-1** 32–43
[3] Bateman H 1931 On dissipative systems and related variational principles *Phys. Rev.* **38** 815–819
[4] Bender CM and Mannheim PD 2008 Giving up the ghost *J. Phys.* A: *Math. Theor.* **41** 304018 (7pp)
[5] Caldriola P 1941 Forze non conservative nella meccanica quantistica *Il Nuovo Cim.* **18** 393–400
[6] Calogero F and Graffi S 2003 On the quantisation of a nonlinear Hamiltonian oscillator *Phys. Lett.* A **313** 356–362
[7] Calogero F 2003 On the quantisation of two other nonlinear harmonic oscillators *Phys. Lett.* A **319** 240–245
[8] Calogero F (2004) On the quantisation of yet another two nonlinear harmonic oscillators *J. Nonlin. Math. Phys.* **11** 1-6
[9] Dirac PAM 1932 *The Principles of Quantum Mechanics* (Oxford at the Clarendon Press)
[10] Edwards M and Nucci MC 2006 Application of Lie group analysis to a core group model for sexually transmitted diseases *J. Nonlin. Math. Phys.* **13** 211–230

[6] The minimal number for a system of two autonomous linear second-order ordinary differential equations [11].

[11] Gorringe VM and Leach PGL 1988 Lie point symmetries for systems of second-order linear ordinary differential equations *Quæst. Math.* **11** 95–117
[12] Gradassi A and Nucci MC 2007 Hidden linearity in systems for competition with evolution in ecology and finance *J. Math. Anal. Applic.* **333** 274–294
[13] Hamilton William Rowan 1834 On a general method in dynamics *Phil. Trans. Roy. Soc.* **Part II** 247–308
[14] Kanai E 1948 On the quantization of dissipative systems *Prog. Theor. Phys.* **3** 440–442
[15] Leach PGL 2005 The solution of some quantum nonlinear oscillators with the common symmetry group $SL(2, R)$ *J. Phys.* A: *Math. Gen.* **38** 1543–1552
[16] Lemmer RL and Leach PGL 1999/1420 A classical viewpoint on quantum chaos *Arab J. Math. Sci.* **5** 1–17
[17] Mannheim Phillip D and Davidson Aharon 2005 Dirac quantization of the Pais-Uhlenbeck fourth order oscillator *Phys. Rev.* A **71** 042110
[18] Mannheim PD 2007 Solution to the ghost problem in fourth-order derivative theories *Found. Phys.* **37** 532–571
[19] Marcelli M and Nucci MC 2003 Lie point symmetries and first integrals: the Kowalevsky top *J. Math. Phys.* **44** 2111–2132
[20] Naicker V, Andriopoulos K and Leach PGL 2005 Symmetry reduction of some nonlinear Hamilton-Jacobi-Bellman equations arising in the mathematics of finance *J. Nonlin. Math. Phys.* **12** 264–274
[21] Nucci MC 1996 The complete Kepler group can be derived by Lie group analysis *J. Math. Phys.* **37** 1772–1775
[22] Nucci MC 2004 Calogero's "goldfish" is indeed a school of free particles *J. Phys.* A: *Math. Gen.* **37** 11391–11400
[23] Nucci MC 2005 Jacobi's three-body system moves like a free particle *J. Nonlin. Math. Phys.* **12-suppl.1** 499-506
[24] Nucci MC, Leach PGL and Andriopoulos K 2006 Lie symmetries, quantisation and c-isochronous nonlinear oscillators *J. Math. Anal. Applic.* **319** 357–368
[25] Nucci MC and Leach PGL 2010 An algebraic approach to laying a ghost to rest *Phys. Scripta* (to appear)
[26] Ostrogradsky 1850 Mémoires sur les équations différentielles relatives au problème des isopérimètres *Mém. Acad. Imp. Sci. St. Pétersbourg* **VI** 385–517
[27] Pais A and Uhlenbeck GE 1950 On field theories with nonlocalized Action *Phys. Rev.* **79** 145–165
[28] Pizzigoni G 2009 The method of Ostrogradsky: from Lagrangian to Hamiltonian equations, Undergraduate Thesis in Mathematics, under the supervision of Professor MC Nucci [in Italian]
[29] Riewe F 1996 Nonconservative Lagrangian and Hamiltonian mechanics *Phys. Rev.* E **53** 1890-1899
[30] Schuch D 1997 Nonunitary connection between explicitly time-dependent and nonlinear approaches for the description of dissipative quantum systems *Phys. Rev.* A **55** 935-940
[31] van Hove L 1951 Sur certaines représentations unitaires d'un groupe infini de transformations *Mem. Acad. Roy. Belg. Cl. Sci.* **26** 1–102
[32] Whittaker ET 1944 *A Treatise on the Analytical Dynamics of Particles and Rigid Bodies* (New York: Dover)

A new model for the structure of spacetime: physical applications

Miguel Lorente
University of Oviedo (Spain), Department of Physics
University of Comillas (Madrid), Cátedra CTR
E-mail: lorentemiguel@uniovi.es

Abstract. From the properties of material beings, an infinite number of networks can be constructed, which are considered the arena where the elementary particles are emerging. For this arena we present three particular models described by simple graphs. These graphs are constructed by vertices and edges (interactions) which are nowhere in space and time. i) The n-dimensional hypercubic lattice, where each being is interacting with $2n$ different ones, giving rise to n-dimensional spacetime, where straight lines, orthogonal lines, orthogonal coordinates and metric distance can be defined intrinsically. ii) The n-simplicial lattice evolving with discrete time according to Pachner moves. iii) The planar graph with negative discrete curvature. Given a planar graph derived from hyperbolic tessellation by omitting the embedding space, we can define discrete curvature by combinatorial properties of the underlying discrete hyperboloid made up of vertices and edges. At the end some comparison will be made between our model and some current models on the structure of spacetime: spin networks (Penrose), spin foams (Rovelli et al.), causal sets (Sorkin et al.), quantum causal histories (Markopoulou).

1. An ontological interpretation of the structure of spacetime

Epistemological presuppositions: we can consider three levels of human knowledge in the comprehension of the physical world.

- Level 1: Physical magnitudes, such as distance, time interval, mass, event, force and so on, that are given by our sensations and perceptions.
- Level 2: Theoretical models, that are the generalization of metrical properties given by measurements and numerical relations among them.
- Level 3: Fundamental concepts, representing the ontological properties of the physical world given by our consciousness in an attempt to know the reality.

There must be some connections between the three levels. In Quantum Mechanics the theoretical models of microphysics in level 2 are related to observable magnitudes in level 1 by correspondence laws.

If we accept level 3 should be connected to level 2 an immediate question is to ask the justification of the rules governing the construction of theoretical systems.

For instancee, the unification of Quantum Mechanics and the theory of general Relativity should be made in level 2 where they belong to, but the underlying ontological background should be taken from level 3.

The ontological interpretation: Is it possible to make some Ansatz about the nature of these fundamental objects in level 3? If we take the extension as the first property of matter, as Descartes has claimed, space and time should be considered necessary at the beginning of a fundamental theory. We prefer the point of view that the most essential property of material objects is the capacity of producing effects in other objects, which was identified by Leibniz with the concept of force [4].

This interpretation of matter has been confirmed by modern philosophers who tried to explain cosmic beings in terms of two metaphysical principles.

> "The essence of human being consists on realizing him-self in conscious actions, namely, on realizing being-in-himself, and being-for-himself... The essence of material being consists on realizing-it-self externally, therefore it is necessary a different principle of being-in-himself. If the principle of being-in-himself is called principle of conscience the principle of acting out there and being-in-others is called principle of matter" [3].

> "Matter is the constituting system of material beings in their two aspects: to be real by it-self and to be a system of potentialities. Potentiality is the capacity of producing actions and, some times, is the capacity of determining the structure of reality" [15].

2. A relational model for the discrete spacetime

We take a network of material beings acting among themselves as the ontological background of the theoretical model in level 2 for the relational character of spacetime. This ontological background consists of a finite number of material beings (beings-in-others) endowed with the principle of causality.

A very suitable language to describe the network of causal effects among the material beings (hylions) is the concept of graphs.

Graph: $\Gamma = \{V, E\}$ V vertices, E edges. Every edge is incident with two vertices.

Geometrical objects in graphs (by combinatorial rules): path, circuit, length, distance, triangle, plane.

Embedding of a graph Γ into a manifold S. A drawing of Γ on S is a map of vertices and edges into points and curves of S which preserves the relations between edges and vertices.

We call it an embedding of Γ into S if all the regions of the drawing are homeomorphic to the plane E^2.

Important remark! The embedding is an artifice to use our intuition for a better understanding of the geometrical properties of the graph. The real networks of material beings and their interactions can be abstracted from this embedding. They are nowhere, as Penrose says.

From the properties of material beings, an infinite number of possible networks can be constructed, which are considered the arena where the elementary particles are moving. For this arena we present three particular models described by simple graphs:

2.1. The causal lattice [9]

In the ontological level 3 the lattice is made out of material beings (hylions) acting among themselves

- Two interacting hylions

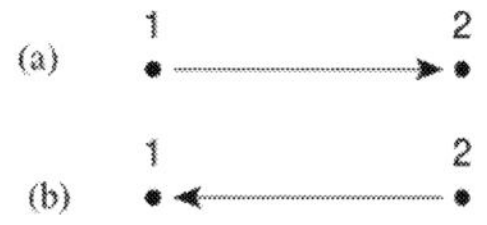

Action of 1 in 2 is a necessary condition for the action of 2 in 1.
Action of 2 in 1 is a necessary condition for a new action of 1 in 2.

- Chain of mutual interacting hylions in the relation 1 to 2

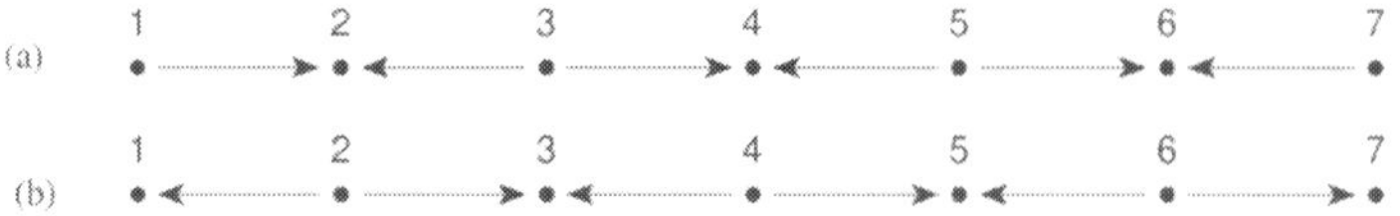

We postulate that the actions of (a) are necessary condition for the actions of (b) and actions of (b) are necessary for the new actions of (a).

- Network of mutual interacting hylions in the relation 1 to 4

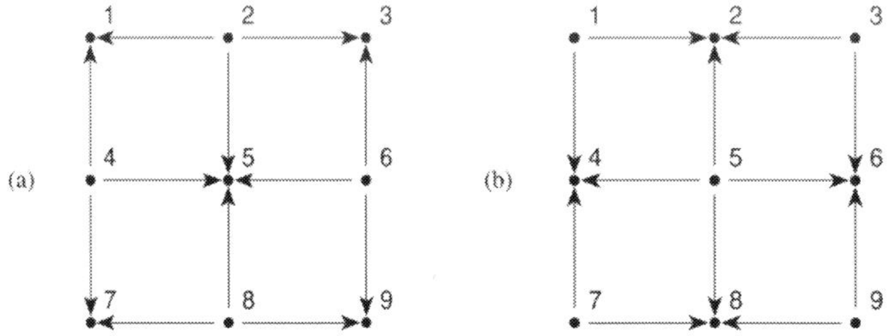

The actions of (a) are necessary conditions for the actions of (b)

The actions of (b) are necessary for new actions of (a)

This ontological model in level 3 can be considered an interpretation of the relational theory of spacetime in level 2

2.1.1. Consequences: interpretation of space Because relational theories use only objects and relations among them we use graphs that are composed of vertices and edges. Take a graph corresponding to the set of relations in the causal lattice:

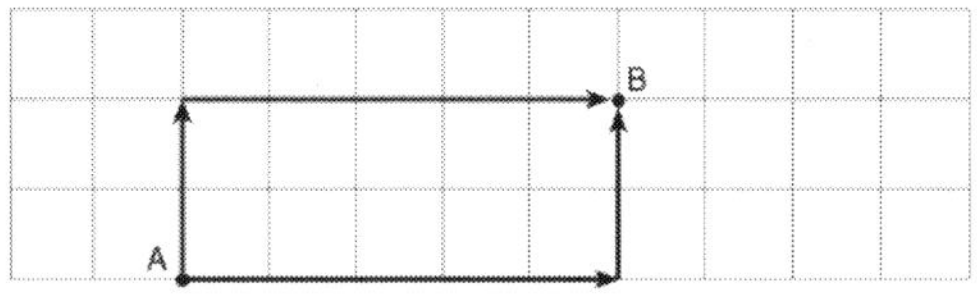

The logical structure is given in level 2 but it corresponds to some geometrical properties of physical space in level 1.

We take, for simplicity, a set of interacting points in relation 1 to 4.

We can define:

Path, length, principal straight line, orthogonal and parallel lines, Cartesian coordinates, Euclidean space.

This structure can be easily generalized to n-dim. Euclidean space by some network where each point is connected with no more and no less than $2n$ neighbour points.

2.1.2. Consequences: interpretation of time In this case the graph corresponding to causal lattices is oriented: one vertex (the cause) precedes logically the other vertex (the effect)

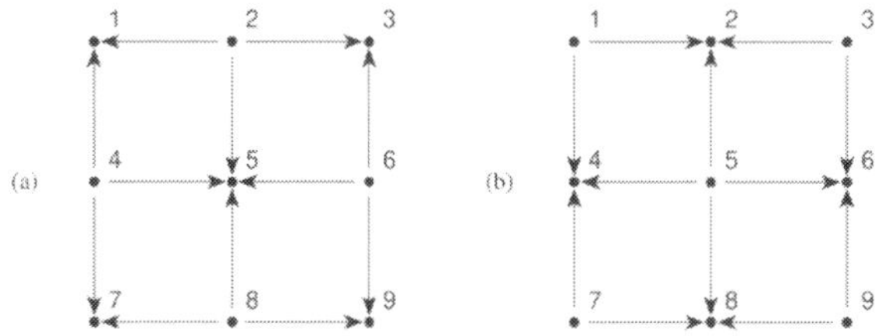

In the graph (a) the vertices $2, 4, 6, 8$ are prior to vertices $1, 3, 5, 7, 9$.
In the graph (b) the vertices $1, 3, 5, 7, 9$ are prior to vertices $2, 4, 6, 8$.
We have a series of actions in level 2 that correspond logically to successive instances of time.

2.2. The hyperbolic lattice [8]
Tessellation: a covering of 2-dim surface repeating the same figure without overlapping or gaps

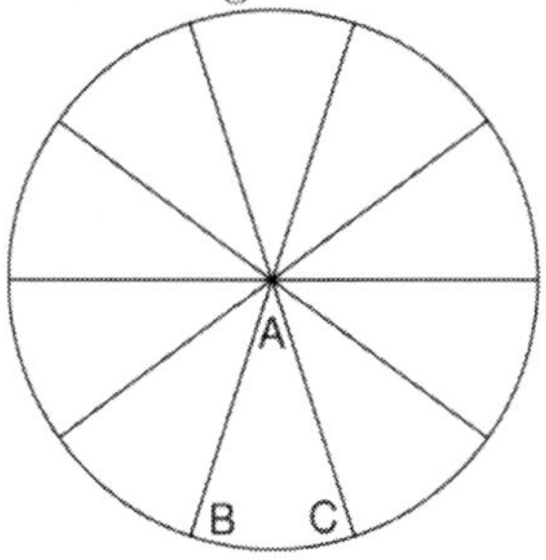

Figure 1. Tessellation of S^2 (in stereographic projection) by reflecting in the sides of the spherical triangle $T(2, 2, 5) \simeq \{A, B, C\}$

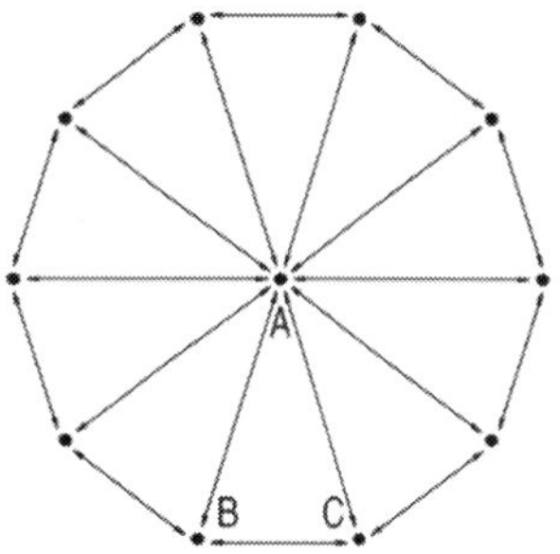

Figure 2. Graph obtained from spherical tessellation of figure 1.

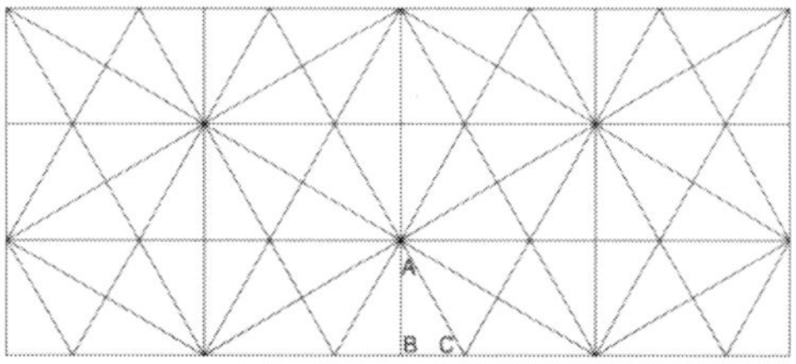

Figure 3. Tessellation of E^2 generated by reflecting in the sides of a Euclidean triangle $T(2, 3, 6) \simeq \{A, B, C\}$

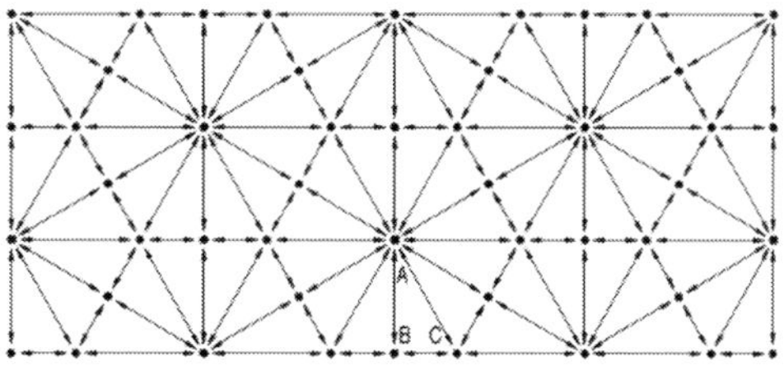

Figure 4. Graph obtained from Euclidean tessellation of figure 3.

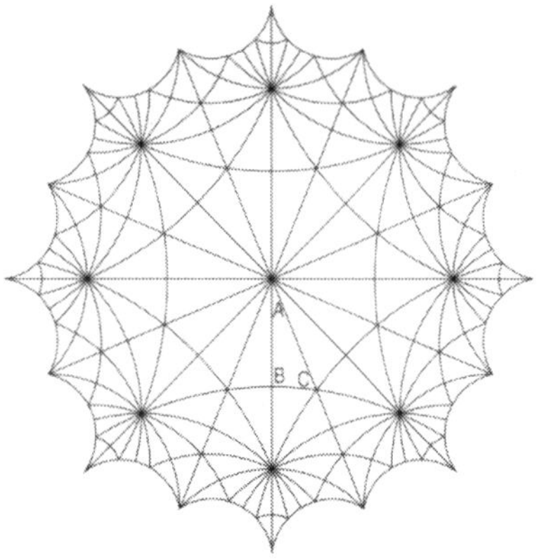

Figure 5. Tessellation of H^2 generated by reflecting in the sides of the hyperbolic triangle $T(2,3,8) \simeq \{A, B, C\}$

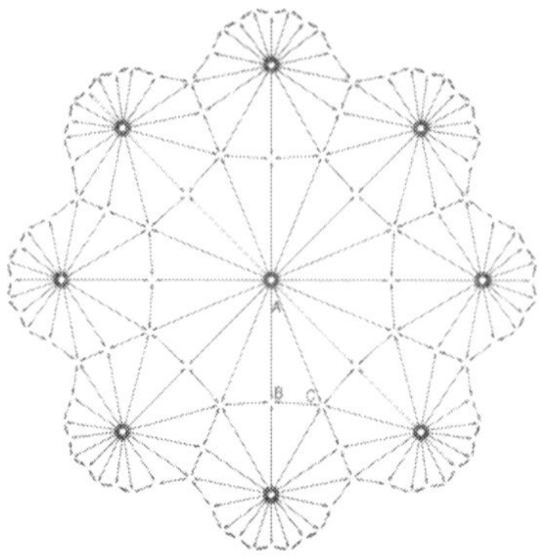

Figure 6. Graph obtained from hyperbolic tessellation of figure 5.

2.2.1. Gauss curvature of continuous tessellations: Two dimensional tessellations in X ($= S^2$, E^2 or H^2) are generated by 2-simplex (triangle) reflection group. In order to calculate the Gaussian curvature, we review some geometrical properties of geodesic triangles.

In S^2 the geodesic triangle (that is, triangle whose sides are arcs of geodesics) are spherical. Given a spherical triangle $T(x, y, z)$ with:

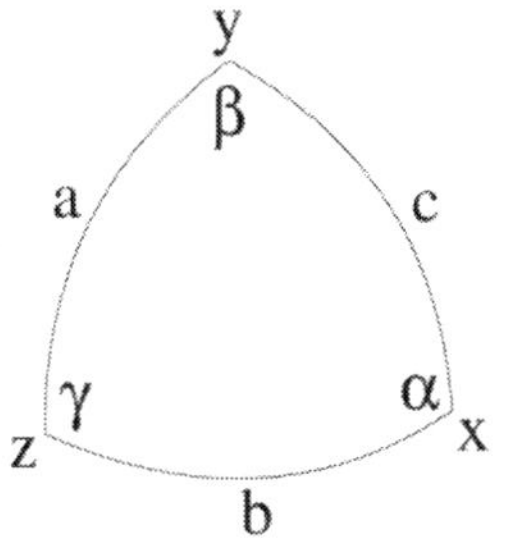

The excess of the interior angles of a spherical triangle is:

$$\epsilon = \alpha + \beta + \gamma - \pi \ .$$

It can be proved that this excess is always positive.

The area of the triangle $T(x, y, z)$ is

$$\text{Area}\{T(x, y, z)\} = \alpha + \beta + \gamma - \pi = \epsilon \ .$$

In E^2 we have Euclidean triangles $T(x, y, z)$

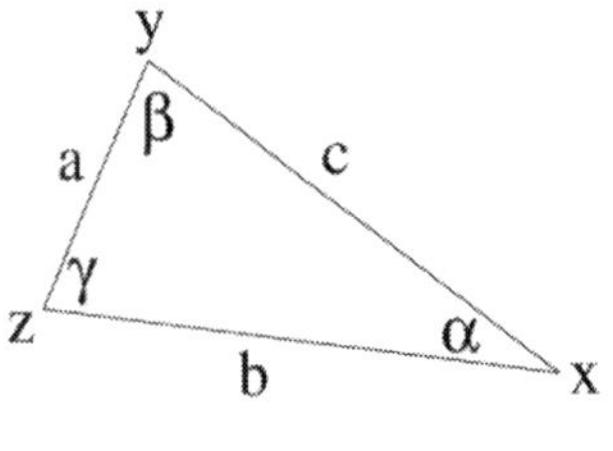

The excess of the interior angles of an Euclidean triangle is

$$\epsilon = \alpha + \beta + \gamma - \pi = 0 \, .$$

In H^2 the geodesic triangles are hyperbolic. For the hyperbolic triangle $T(x, y, z)$ we have

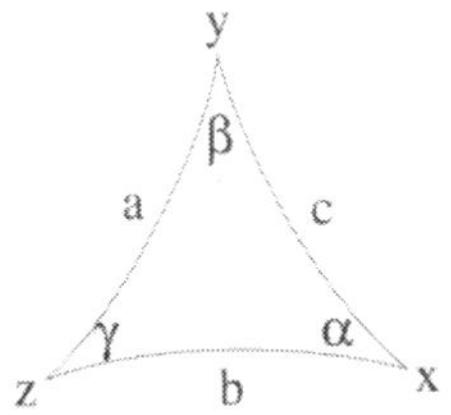

In S^2 the excess and area of the spherical triangle are, in units of π remember: $\alpha = \frac{\pi}{l}, \beta = \frac{\pi}{m}, \gamma = \frac{\pi}{n}$

$$\frac{\epsilon}{\pi} = \frac{1}{l} + \frac{1}{m} + \frac{1}{n} - 1 \tag{1}$$

$$\frac{A}{\pi} = \frac{1}{l} + \frac{1}{m} + \frac{1}{n} - 1 \, . \tag{2}$$

In E^2 the excess of the Euclidean triangles is $\epsilon = 0$.

In H^2 the corresponding excess and area of the hyperbolic triangles in units of π are $\left(\text{remember}: \alpha = \frac{\pi}{l}, \beta = \frac{\pi}{m}, \gamma = \frac{\pi}{n}\right)$:

$$\frac{\epsilon}{\pi} = \frac{1}{l} + \frac{1}{m} + \frac{1}{n} - 1$$

$$\frac{A}{\pi} = 1 - \left(\frac{1}{l} + \frac{1}{m} + \frac{1}{n}\right) \, . \tag{3}$$

We can now apply these results to the curvature of the surfaces corresponding to the 2-dimensional regular tessellations (spherical, Euclidean or hyperbolic). According to the Gauss-Bonet theorem the excess angle of some geodesic triangle T is equal to the integral of the Gaussian curvature over T

$$\epsilon = \alpha + \beta + \gamma - \pi = \iint\limits_T K \, d\sigma$$

where $d\sigma$ is the area element. If K=const.

$$K = \frac{\epsilon}{A} \, .$$

Applying this formula to the above results, we have:

$K = 1$ for spherical geodesic triangles

$K = 0$ for Euclidean triangles

$K = -1$ for hyperbolic geodesic triangle

We can use another interpretation of Gaussian curvature in terms of parallel transport. If $\Delta\varphi$ is the change of angle in the paraller transport of a vector along a curve C the trace of which is the boundary of a region R, containing the point p, then

$$\Delta\varphi = \iint\limits_R K \, d\sigma \, .$$

Since $\Delta\varphi$ does not depend on the choice of C (but it depends on the enclosed area $A(R)$)

$$\lim_{R\to p}\frac{\Delta\varphi}{A(R)} = K(p) \ .$$

This formula gives a method to calculate the curvature at a point in terms of the area and the parallel transport along the border.

2.2.2. Curvature on planar graphs Given a planar graph corresponding to figures 2, 4 and 6 where two adjacent vertices have always the same adjacent third vertex (different from the first two) one can define the excess of this triad of vertices as the quantity, in analogy with (1),

$$\delta = \frac{1}{l} + \frac{1}{m} + \frac{1}{n} - 1$$

where $2l, 2m, 2n$ are the number of edges incident in each of the three vertices, which correspond to $2l$–valued, $2m$–valued or $2n$–valued vertices, respectively. For instance, in figure 2, A, B, C are 10–valued, 4–valued, 4–valued vertices, respectively; in figure 4, A, B, C are 12–valued, 4–valued, 6–valued vertices, respectively; and in figure 6, A, B, C are 16–valued, 4–valued, 6–valued vertices, respectively.

If we define the spherical, Euclidean or hyperbolic graph, that is obtained from a spherical, Euclidean or hyperbolic tessellation respectively, we can check

$\delta > 0$, for a spherical graph (figure 2),

$\delta = 0$, for an Euclidean graph (figure 4),

$\delta < 0$, for an hyperbolic graph (figure 6).

In a similar way, we can define the areas and curvature of a triad in a graph, such that they become the standard quantities when the graph is embedded in a continuous manifold.

Therefore, we define the area of the triad $T(l, m, n)$ in a spherical graph, in analogy with (2),

$$\sigma(T) = \frac{1}{l} + \frac{1}{m} + \frac{1}{n} - 1 \ .$$

In fact, the quotients $\frac{1}{l}, \frac{1}{m}, \frac{1}{n}$ correspond to angles, and the sum $\frac{1}{l} + \frac{1}{m} + \frac{1}{n} - 1$ corresponds to an area. We define the area of the triad $T(l, m, n)$ in an hyperbolic graph, in analogy with (3)

$$\sigma(T) = 1 - \left(\frac{1}{l} + \frac{1}{m} + \frac{1}{n}\right) \ .$$

Similarly, we define the curvature of a triad $T(l, m, n)$ in a planar graph

$$K(T) = \frac{\delta}{\sigma} = \begin{cases} 1, & \text{for a spherical graph} \\ 0, & \text{for an Euclidean graph} \\ -1, & \text{for an hyperbolic graph} \end{cases}$$

an expression that can be considered the discrete version of the Gauss-Bonnet theorem. As in the continuous case, the curvature of a graph at a vertex can be calculated as the parallel transport of a path surrounding the m-valued vertex divided by the area of the triads embraced by the path. Obviously

$$K(P) = \frac{2m\delta}{2mA} = K(T) \ .$$

2.3. The evolution of simplicial lattice [11] [2]
Let $\mathcal{S}$ be a 2-dim network of N nodes with three edges arranged in the form of an hexagonal tessellation.

We associate a vector space $\mathcal{H}_n(n = 1, 2, \dots N)$ to each element of $\mathcal{S}$. The state space $\mathcal{H}$ is the direct product of all the constituents. Local dynamics on $\mathcal{H}$ are created by the following generators that replace some pieces of $\mathcal{S}$ with the same borders by new ones.

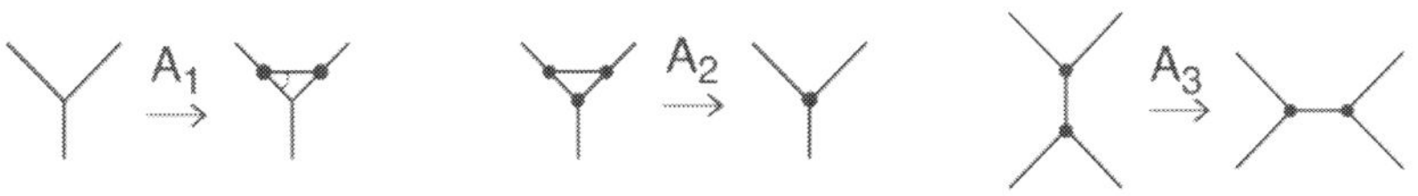

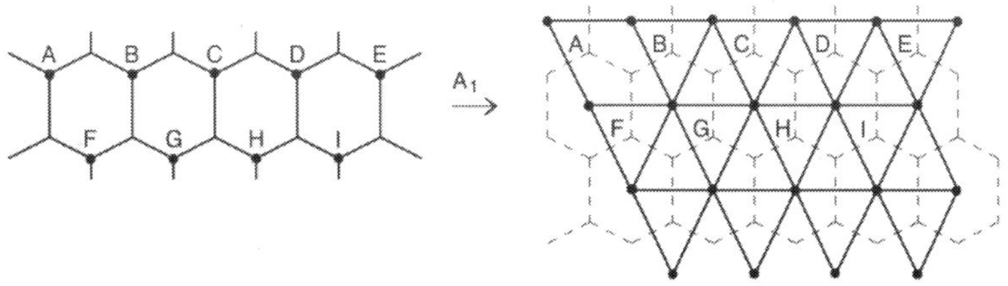

The evolution of the system $\mathcal{S}$ to $\mathcal{S}'$ is realized by Pachner moves with the use of combinatorial rules

$$\mathcal{S} \to \mathcal{S}' \to \mathcal{S}'' = \mathcal{S} \to \mathcal{S}'$$

Our model is not contained (embedded) in a continuous manifold, therefore the measure of length and time intervals is reduced to the problem of counting

3. R. Penrose's spin networks [14]

Discrete magnitudes in QM give rise to continuous spacetime

Total angular momentum $\vec{J} = \vec{L} + \vec{S}$

$\vec{n}$-units large spin component

$\vec{n}$-units with large orbital component

Two $\vec{n}$-units interactions

$\to$ angle

$\to$ displacements

Emergent spacetime from directions and positions

Analogies with our model

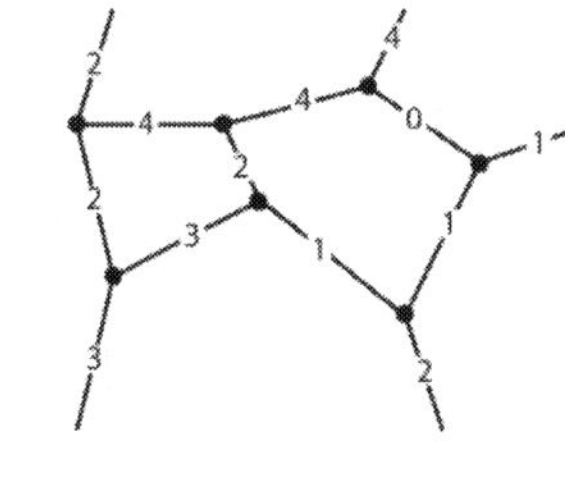

- "An object is thus located either in a certain direction or in a certain position in terms of its relations with other objects. One does not really need a space to begin with"
- The concept of Euclidean geometry emerges from the interactions of units among themselves.

Differences in comparison with our model

- For the construction of discrete magnitudes one needs a background space out of which the geometrical space emerges as a network of relations among $\vec{n}$-units.
- The elements of the spin networks are physical entities (elementary particles) with discrete magnitudes not evolving in time.

4. R. Sorkin's causal sets [1]

- Following the philosophical ideas of Sakata and Taketani causal sets can be considered "substances" of the real world, different from the pure observations which an operationalist philosopher could only admit.
- Underlying continuous spacetime there is a discrete causal set (a locally finite ordered set of events connected by the principle of causality)
- The structure of spacetime is nothing more than the set of relations coming from interactions among physical events.

Analogies with our model

- Causal sets are based in some ontological reality (substance) consisting of cause, effect and its production by the cause.
- Three levels of knowledge: phenomena, physical theory, substance.

Differences in comparison with our model

- Use of embedding space, necessary to calculate the metric.
- Omission of Hilbert space in order to introduce laws of quantum probability in causal interactions

5. F. Markopoulou's quantum causal histories [12]

- Causal set = partially ordered set locally finite with preceding relation.
- Quantum causal set = attach an Hilbert space to each event of a causal set.
- Quantum causal histories = the evolution of a quantum causal set is implemented by unitary operators between Hilbert spaces.
- The set of all causal relations among fundamental objects can be taken as the ontological background for the relational theories of spacetime.

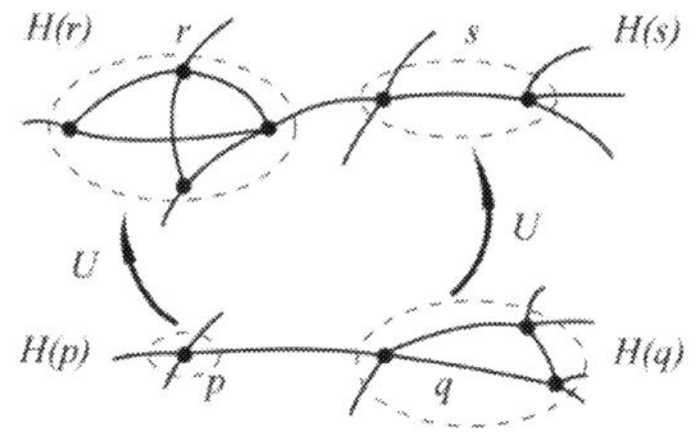

Analogies with our model

- Causal sets belong to the ontological level.
- Quantum operators introduce laws of probability.
- Evolution of a spin network is given by combinatorial rules.
- The last model is not embedded in a continuous manifold therefore the measure of distance and time intervals is reduced to the process of counting.

Differences in comparison with our model

- Lack of substantive character of fundamental objects.

6. C. Rovelli and L. Smolin's spin foam [13]

- A spin foam is a labelled 2.complex whose faces are labelled by representations of some group G, the edges by the intertwiners in the group, and the vertices carry the evolution amplitudes.
- A spin network is a graph whose edges are labelled by representations of the group G and its nodes are labelled by the intertwiners. We regard spin networks as the "spacelike slices" of a spin foam: If we take a spacelike cut through a spin foam, we obtain a graph; its edges are cuts through the spin foam faces, and so we label them with the same representations. Its nodes are cuts of the spin foam edges and so we label them with intertwiners.

Analogies with our model

- Abstract spin foam non embedded in some preexisting continuous manifold.
- Transitions from an initial spin network S, to another spin network by a set of combinatorial rules.
- The spin network diagonalizes the quantum area and volume operators whose spectrum was discovered to be discrete.

Differences in comparison with our model

- There is no ontological background for spin foams.

In the preceding sections we have presented a physical model for the structure of spacetime, the novelty of which lays in the origin of the model, that is to say, the model is based on some ontological properties of material beings. In other words, using the terminology of section 1, we start from level 3, from where we go to level 2; from this level we then go to level 1. In most of the physical models one starts from observations (level1) from which one derives general principles (level 2) that have to be accepted like axioms.

A physical consequence of our model is that space and time are discrete. It means that the laws of physics are written with the help of finite differences, that go to the continuous limit when the elementary units go to zero. We will describe with this method the most fundamental equations in Quantum Mechanics: the harmonic oscillator, the hydrogen atom, the relativistic wave equation and the covariant field equations. The solutions of the difference equations are the functions of discrete variables whose critical points are given by the eigenvalues of the discrete operators of the system and could be an experimental test of the model.

7. The quantum harmonic oscillator of discrete variable [7]

We start from the orthogonal polynomials of a discrete variable, the Kravchuk polynomials $K_n^{(p)}(x)$ and the corresponding normalized Kravchuk functions

$$K_n^{(p)}(x) = d_n^{-1}\sqrt{\rho(x)}\, k_n^{(p)}(x),$$

where $d_n^2 = \frac{N!}{n!(N-n)!}(pq)^n$ is a normalization constant, $\rho(x) = \frac{N!p^x q^{N-x}}{x!(N-x)!}(pq)^n$ is the weight function, with $p > 0, \quad q > 0, \quad p + q = 1, \quad x = 0, 1, \dots N + 1.$

The Kravchuk functions satisfy the orthonormality condition

$$\sum_{x=0}^{N} K_n^{(p)}(x) K_{n'}^{(p)}(x) = \delta_{nn'},$$

Journal of Physics: Conference Series **237** (2010) 012017 doi:10.1088/1742-6596/237/1/012017

and the following difference and recurrence equations:

$$\sqrt{pq(N-x)(x+1)}K_n^{(p)}(x+1)+ \\ +\sqrt{pq(N-x+1)x}K_n^{(p)}(x-1)+[x(p-q)-Np+n]\,K_n^{(p)}(x)=0,$$

$$\sqrt{pq(N-n)(x+1)}K_{n+1}^{(p)}(x)+ \\ \sqrt{pq(N-n+1)n}K_{n-1}^{(p)}(x)+[n(q-p)+Np-x]\,K_n^{(p)}(x)=0\,.$$

From the properties of the Kravchuk polynomials we can construct raising and lowering operators for the Kravchuk functions

$$L^{+}(x,n)K_n^{(p)}(x)=\sqrt{pq(N-n)(n+1)}K_{n+1}^{(p)}(x),$$

$$L^{-}(x,n)K_n^{(p)}(x)=\sqrt{pq(N-n+1)n}K_{n-1}^{(p)}(x).$$

In order to give a physical interpretation of the difference equation and the raising and lowering operators for theKravchuk functions we take the limit when N goes to infinity and the discrete variable x becomes continuous s .

First of all, we take the limit of Kravchuk functions. We write

$$K_n^{(p)}(x)\xrightarrow[n\to\infty]{}\left\{\frac{1}{2^n n!}\right\}^{1/2}\left\{\frac{1}{\sqrt{2\pi Npq}}e^{-s^2}\right\}^{1/2}H_n(s)=\psi_n(s),$$

where the last braket becomes the weight for the Hermite functions and the functions $\psi_n(s)$ are the solution of the continuous harmonic oscillator. We have

$$\frac{1}{\sqrt{Npq}}\,L^{+}(x,n)K_n^{(p)}(x)\xrightarrow[N\to\infty]{}\frac{1}{\sqrt{2}}\left\{s-\frac{d}{ds}\right\}\psi_n(s)=\sqrt{n+1}\,\psi_n(s),$$

$$\frac{1}{\sqrt{Npq}}\,L^{-}(x,n)\,K_n^{(p)}(x)\xrightarrow[N\to\infty]{}\frac{1}{\sqrt{2}}\left\{s+\frac{d}{ds}\right\}\psi_n(s)=\sqrt{n}\,\psi_n(s).$$

Therefore the raising and lowering operators for the Kravchuk functions become, in the limit, creation and annihilation operators for the normalized Hermite functions.

From the properties of spherical harmonics we obtain

$$A^{+}Y_{jm}=\sqrt{\frac{(j+m)(j-m+1)}{2j}}\,Y_{j,m-1}=\frac{1}{\sqrt{2j}}J_{-}Y_{jm}$$

$$A^{-}Y_{jm}=\sqrt{\frac{(j-m)(j+m+1)}{2j}}\,Y_{j,m+1}=\frac{1}{\sqrt{2j}}J_{+}Y_{jm}$$

where J_{+}, J_{-} are the generators of so(3) algebra.

From the commutation relations of these generators we have

$$(AA^{+}-A^{+}A)Y_{jm}=\frac{m}{j}\,Y_{jm}=\frac{1}{2j}2J_zY_{jm}=\left(1-\frac{n}{j}\right)Y_{jm},$$

and then take the limit:

$$[A,A^{+}]\,Y_{jm}=\left(1-\frac{n}{j}\right)Y_{jm}\xrightarrow[j\to\infty]{}[a,a^{+}]\;\psi_n(s).$$

For the anticommutation relation we have:

$$\left(AA^{+}+A^{+}A\right)Y_{jm}=\frac{1}{j}\left(j(j+1)-m^{2}\right)Y_{jm}=\frac{1}{j}\left(\vec{J}^{2}-J_{z}^{2}\right)Y_{jm}\,,$$

and take the limit

$$\begin{aligned}&(AA^{+}+A^{+}A)Y_{jm}\\&=\left\{(2n+1)-\frac{n^{2}}{j}\right\}Y_{jm}\xrightarrow[j\to\infty]{}\left(aa^{+}+a^{+}a\right)\psi_{n}(s)=(2n+1)\psi_{n}(s)\,.\end{aligned}$$

This correspondence suggests that the operator algebra for the quantum harmonic oscillator on the lattice is expanded by the generators of the SO(3) groups.

The eigenvalues of the Hamilton operator on the lattice are connected with the index $m=j-n$ of the eigenvectors Y_{jm}. These eigenvalues are equally separated by $\hbar\omega$ but finite $(m=-j,\ldots+j)$. The eigenvalues of the position operator on the lattice are connected with the index $m'=j-x$ of $Y_{jm'}$. These eigenvalues are equally separated by $\sqrt{\frac{\hbar}{M\omega}}$ but finite $(m'=-j,\ldots+j)$. Therefore the Planck constant $\hbar$ plays a role with respect to the discrete space coordinate similar to the discrete energy eigenvalues.

8. Wave equation for the hydrogen atom with discrete variables [7]

Our model is based on the properties of generalized Laguerre polynomials as continuous limit of the Meixner polynomials of discrete variable.

We start from the generalized Laguerre functions

$$\psi_{n}^{\alpha}(s)=d_{n}^{-1}\sqrt{\rho_{1}(s)}\,L_{n}^{\alpha}(s)\,.$$

In the discrete case, we defined the normalized Meixner functions

$$M_{n}^{(\gamma,\mu)}(x)\equiv d_{n}^{-1}\sqrt{\rho_{1}(x)}\,m_{n}^{(\gamma,\mu)}(x)$$

where $m_{n}^{(\gamma,\mu)}(x)$ are the Meixner polynomials,

$$d_{n}^{2}=\frac{n!\Gamma(n+\gamma)}{\mu^{n}(1-\mu)^{\gamma}\Gamma(\gamma)},\quad\rho_{1}(x)=\frac{\mu^{x}\Gamma(x+\gamma+1)}{\Gamma(x+1)\Gamma(\gamma)},$$

and γ,μ are some constants $0<\mu<1,\quad\gamma>0$. The Meixner functions satisfy the orthonormality condition

$$\sum_{x=0}^{\infty}M_{n}^{(\gamma,\mu)}(x)\,M_{n'}^{(\gamma,\mu)}(x)\frac{1}{\mu(x+\gamma)}=\rho_{nn'}$$

and the following properties:

i) Difference equation

$$\begin{aligned}&\sqrt{\frac{\mu(x+\gamma)(x+1)(x+\gamma)}{x+\gamma+1}}M_{n}(x+1)\\&\qquad+\sqrt{\mu(x+\gamma)x}M_{n}(x-1)-\left[\mu(x+\gamma)+x-n(1-\mu)\right]M_{n}(x)=0\,.\end{aligned}$$

ii) Raising operator

$$L^{+}(x,n)\, M_n(x) = \sqrt{\mu(n+\gamma)(n+1)}\; M_{n+1}(x)\ .$$

iii) Lowering operator

$$L^{-}(x,n)\, M_n(x) = -\sqrt{\mu(n+\gamma-1)n}\; M_{n-1}(x)\ .$$

The commutation relations read

$$L^{+}(x,n-1)\, L^{-}(x,n) - L^{-}(x,n+1)\, L^{+}(x,n) = -\mu(2n+\gamma)\ .$$

In order to make connection between the Meixner functions of discrete variable and Laguerre functions of continuous variable we substitute $\gamma = \alpha+1$, $\mu = 1-h$, $x = \frac{s}{h}$ and then take the limit $h \to 0$.

In the limit we obtain

$$M_n^{(\gamma,\mu)}(x) \xrightarrow[\substack{h\to 0\\ x\to\infty\\ hx\to s}]{} \sqrt{\frac{n!}{\Gamma(n+\alpha+1)} e^{-s} s^{\alpha+1}}\; L_n^{\alpha}(s) = \psi_n^{\alpha}(s)\ .$$

In order to make application to the hydrogen atom we take the reduced radial equation

$$\frac{d^2u}{d\rho^2} + \left[\frac{\nu}{\rho} - \frac{1}{4} - \frac{l(l+1)}{\rho^2}\right] u(\rho) = 0$$

where

$$\rho \equiv \frac{\sqrt{8M\,|E|}}{\hbar} r \quad , \quad \nu = \frac{Ze^2}{\hbar}\sqrt{\frac{\mu}{2\,|E|}}\ .$$

The energy eigenvalues are given by

$$E_{\nu l} = -\frac{1}{2}\,\frac{M\left(Ze^2\right)^2}{\hbar^2}\,\frac{1}{\nu^2} \quad , \quad \nu = 1,2\ldots$$

For fixed ν we still have degeneracy for $l = 0, 1, \ldots \nu-1$

The corresponding eigenvectors are given by

$$\psi_{\nu l}(\rho) = \left\{\frac{(\nu-l-1)!}{(\nu+l)!}\right\}^{\frac{1}{2}} \rho^{l+1}\, e^{\frac{\rho}{2}}\, L_{\nu-l-1}^{2l+1}(\rho)\ .$$

From the connection between the Meixner and Laguerre functions given above, we can make the ansatz of a discrete model for the hydrogen atom where the reduced radial equation is substituted by the difference equation

The anticommutation relations, which is proportional to the Hamiltonian of the hydrogen atom (for the radial part), is substituted by the anticommutation relations. The commutation relations for the raising and lowering operators defining the Lie algebra of the SU(1,1) group are substituted by commutation relations of discrete type.

The expectation value of the discrete variable x with respect to the Meixner functions $M_n^{(\gamma,\mu)}(x)$ is

$$\langle x\rangle_{n\gamma} = \sum M_n^{(\gamma,\mu)}(x)\; x\; M_n^{(\gamma,\mu)}(x),$$

that can be calculated with the help of the recurrence relation.

9. Exact solutions for Dirac equation on the lattice [6]

Given a scalar function $\Phi(n_\mu) \equiv (n_\mu \varepsilon_\mu)$ defined in the grid points of a Minkowski lattice with elementary lengths ε_μ, and the difference operators

$$\delta_\mu^+ \equiv \frac{1}{\varepsilon_\mu}\Delta_\mu \prod_{\nu \neq \mu} \tilde{\Delta}_\nu, \quad \delta_\mu^- \equiv \frac{1}{\varepsilon_\mu}\nabla_\mu \prod_{\nu \neq \mu} \tilde{\nabla}_\nu$$

$$\eta^+ \equiv \prod_\mu \tilde{\Delta}_\mu, \quad \eta^- \equiv \prod_\mu \tilde{\nabla}_\mu, \quad \mu, \nu = 1, 2, 3, 4$$

where the forward and backward differences are defined as

$$\Delta f(x) \equiv f(x + \Delta x) - f(x) \quad , \quad \nabla f(x) \equiv f(x) - f(x - \Delta x) \,.$$

Similarly, we can define the forward and backward promediate operator

$$\widetilde{\Delta} f(x) \equiv \frac{1}{2}\{f(x + \Delta x) + f(x)\} \quad , \quad \widetilde{\nabla} f(x) \equiv \frac{1}{2}\{f(x - \Delta x) + f(x)\} \,.$$

The Klein-Gordon equation can be read off

$$\left(\delta_\mu^+ \delta^{\mu^-} - m_0^2 c^2 \eta^+ \eta^-\right) \Phi(n_\mu) = 0 \,.$$

Using the method of separation of variables, we obtain the exact solutions of this difference equation

$$f(n_\mu) = \prod_{\mu = 0}^{3} \left(\frac{1 - \frac{1}{2} i \varepsilon_\mu k_\mu}{1 + \frac{1}{2} i \varepsilon_\mu k_\mu}\right)^{n_\mu}$$

with k_μ being continuous variables satisfiying the dispersion relations $k_\mu k^\mu = m_0^2 c^2$.

Starting from the Hamiltonian

$$\begin{aligned} H = \varepsilon_1 \varepsilon_2 \varepsilon_3 \sum_{n_1 n_2 n_3 = 0}^{N-1} & \tilde{\Delta}_1 \tilde{\Delta}_2 \tilde{\Delta}_3 \psi^+(n) \times \\ \times & \left\{ \gamma_0 \gamma_1 \frac{1}{\varepsilon_1} \Delta_1 \tilde{\Delta}_2 \tilde{\Delta}_3 + \gamma_0 \gamma_2 \tilde{\Delta}_1 \frac{1}{\varepsilon_2} \Delta_2 \tilde{\Delta}_3 + \right. \\ + & \left. \gamma_0 \gamma_3 \tilde{\Delta}_1 \tilde{\Delta}_2 \frac{1}{\varepsilon_3} \Delta_3 + \gamma_0 \tilde{\Delta}_1 \tilde{\Delta}_2 \tilde{\Delta}_3 \right\} \psi(n) \end{aligned}$$

we obtain from the Hamilton equations of motion, the Dirac equation

$$\left(i \gamma_\mu \delta^{\mu+} - m_0 c \eta^+\right) \psi(n_\mu) = 0$$

and from this we recover the Klein-Gordon equation.

Our model for the fermion fields satisfies the following conditions:

i) the Hamiltonian is translational invariant

ii) the Hamiltonian is hermitian

iii) for $m_0 = 0$, the wave equation is invariant under global chiral transformations

iv) there is no fermion doubling

v) the Hamiltonian is non-local

Finally, coupling the vector field to the electro-magnetic vector potential we construct a gauge invariant vector current leading to the correct axial anomaly

10. Integral Lorentz transformations [10]

We start with the integral transformations of the complete Lorentz group generated by the Kac generators

$$S_1 = \begin{pmatrix} 1 & 0 & 0 & 0 \\ 0 & 0 & 1 & 0 \\ 0 & 1 & 0 & 0 \\ 0 & 0 & 0 & 1 \end{pmatrix}, \quad S_2 = \begin{pmatrix} 1 & 0 & 0 & 0 \\ 0 & 1 & 0 & 0 \\ 0 & 0 & 0 & 1 \\ 0 & 0 & 1 & 0 \end{pmatrix}$$
$$S_3 = \begin{pmatrix} 1 & 0 & 0 & 0 \\ 0 & 1 & 0 & 0 \\ 0 & 0 & 1 & 0 \\ 0 & 0 & 0 & -1 \end{pmatrix}, S_4 = \begin{pmatrix} 2 & 1 & 1 & 1 \\ -1 & 0 & -1 & -1 \\ -1 & -1 & 0 & -1 \\ -1 & -1 & -1 & 0 \end{pmatrix} .$$

Any integral matrix of the complete Lorentz group can be factorized

$$L = P_1^{\eta} P_2^{\theta} P_3^{i} S_4 \ldots S_4 P_1^{\delta} P_2^{\varepsilon} P_3^{\zeta} S_4 \left\{ S_1^{\alpha} S_2^{\beta} S_3^{\gamma} \right\}_{\text{perm.}}$$

where $P_1 = S_1 S_2 S_3 S_2 S_1, \quad P_2 = S_2 S_3 S_2, \, P_3 = S_3, \qquad \alpha, \beta, \ldots = 0, 1$.

11. Discrete differential forms [5]

Given a vector space V^n over $\mathcal{Z}$ we can define a linear function

$$f(u) \equiv \langle \omega, u \rangle \quad u \in V^n .$$

The forms ω constitute a linear vector space (dual space) $^*V^n$.

The basis e_β of V^n and ω^α of $^*V^n$ can be contracted

$$\langle \omega^\alpha, e_\beta \rangle \, \delta_\beta^\alpha .$$

If we take $\omega^\beta = \Delta x^\beta$ as coordinate basis for the linear forms we can construct discrete differential forms (a discrete version of the continuous differential forms)

A particular example of this discrete form is the total difference operator:

$$\Delta f(x, y) = \left(\frac{\Delta_x \widetilde{\Delta}_y f}{\Delta x} \right) \Delta x + \left(\frac{\widetilde{\Delta}_x \Delta_y f}{\Delta y} \right) \Delta y .$$

We can define the exterior product of two form σ and δ: $\rho \wedge \sigma = -\sigma \wedge \rho$.

With the help of this exterior product we can construct a second order discrete differential form or 2-form, namely

$$\rho \wedge \sigma = -\rho_\alpha \Delta x^\alpha \wedge \sigma_\alpha \Delta x^\beta = \frac{1}{2} \left(\rho_\alpha \sigma_\beta - \rho_\beta \sigma_\alpha \right) \Delta x^\alpha \wedge \Delta x^\beta \equiv \sigma_{\alpha\rho} \Delta x^\alpha \wedge \Delta x^\beta$$

where $\sigma_{\alpha\rho}$ is an antisymmetric tensor.

We give now some examples:

- Energy-momentum 1-form

$$\mathbf{P} = -E \Delta t + P_x \Delta x + P_y \Delta y + P_z \Delta z$$

where $(P_x, P_y, P_z, iE) \equiv P_n$ is the four-momentum.

- Vector potential 1-form

$$\mathbf{A} = A_\mu \Delta x^\mu = A_x \Delta x + A_y \Delta y + A_z \Delta z + A_t \Delta t$$

where $A_\mu = (A_x, A_y, A_z, A_t)$ is the four-potential.

- Faraday 2-form

$$\begin{aligned}\mathbf{F} &= E_x \Delta x \wedge \Delta t + E_y \Delta y \wedge \Delta t + E_z \Delta z \wedge \Delta t \\ &+ B_x \Delta y \wedge \Delta z + B_y \Delta z \wedge \Delta x + B_z \Delta x \wedge \Delta y = \frac{1}{2} F_{\mu\nu} \Delta x^\mu \wedge \Delta x^\nu\end{aligned}$$

with $(B_x, B_y, B_z) \equiv \vec{B}$ and $(E_x, E_y, E_z) \equiv \vec{E}$ the magnetic and electric field, respectively.

12. Exterior calculus [5]

Given a 1-form in a two-dimensional space

$$\omega = a(x, y)\Delta x + b(x, y)\Delta y$$

we can define the exterior difference, in the similar way as the exterior derivative, namely,

$$\Delta\omega \equiv \Delta a \wedge \Delta x + \Delta b \wedge \Delta y = \left(\frac{\Delta_x \widetilde{\Delta}_y b}{\Delta x} - \frac{\widetilde{\Delta}_x \Delta_y a}{\Delta y} \right) \Delta x \wedge \Delta y \, .$$

Given a 2-form in a 3-dimensional space,

$$\omega = a(x, y, z)\Delta y \wedge \Delta z + b(x, y, z)\Delta z \wedge \Delta x + c(x, y, z)\Delta x \wedge \Delta y$$

we can also define the exterior difference as:

$$\begin{aligned}\Delta\omega &= \Delta a \wedge \Delta y \wedge \Delta z + \Delta b \wedge \Delta z \wedge \Delta x + \Delta c \wedge \Delta x \wedge \Delta y \\ &= \left(\frac{\Delta_x \widetilde{\Delta}_y \widetilde{\Delta}_z a}{\Delta x} + \frac{\widetilde{\Delta}_x \Delta_y \widetilde{\Delta}_z b}{\Delta y} + \frac{\widetilde{\Delta}_x \widetilde{\Delta}_y \Delta_z c}{\Delta z} \right) \Delta x \wedge \Delta y \wedge \Delta z \, .\end{aligned}$$

Given a 3-form in a 4-dimensional space

$$\omega = a\Delta y \wedge \Delta z + \Delta t + b\Delta z \wedge \Delta t \wedge \Delta x + c\Delta t \wedge \Delta x \wedge \Delta y + d\Delta x \wedge \Delta y \wedge \Delta z$$

we can define an exterior difference as before:

$$\omega = \left(\frac{\Delta_x \widetilde{\Delta}_y \widetilde{\Delta}_z \widetilde{\Delta}_t a}{\Delta x} - \frac{\widetilde{\Delta}_x \Delta_y \widetilde{\Delta}_z \widetilde{\Delta}_t b}{\Delta y} + \frac{\widetilde{\Delta}_x \widetilde{\Delta}_y \Delta_z \widetilde{\Delta}_t c}{\Delta z} - \frac{\widetilde{\Delta}_x \widetilde{\Delta}_y \widetilde{\Delta}_z \Delta_t d}{\Delta t} \right) \Delta x \wedge \Delta y \wedge \Delta z \wedge \Delta t \, .$$

From the Faraday 2-form we write down one set of Maxwell difference equations

$$\Delta \mathbf{F} = \Delta \left(\Delta \mathbf{A} \right) = 0$$

$$\left(\frac{\Delta_x\tilde{\Delta}_y\tilde{\Delta}_z\tilde{\Delta}_t B_x}{\Delta x} + \frac{\tilde{\Delta}_x\Delta_y\tilde{\Delta}_z\tilde{\Delta}_t B_y}{\Delta y} + \frac{\tilde{\Delta}_x\tilde{\Delta}_y\Delta_z\tilde{\Delta}_t B_z}{\Delta z}\right)\Delta x\wedge\Delta y\wedge\Delta z$$
$$+\left(\frac{\tilde{\Delta}_x\tilde{\Delta}_y\tilde{\Delta}_z\Delta_t B_x}{\Delta t} + \frac{\tilde{\Delta}_x\Delta_y\tilde{\Delta}_z\tilde{\Delta}_t E_z}{\Delta y} - \frac{\tilde{\Delta}_x\tilde{\Delta}_y\Delta_z\tilde{\Delta}_t E_y}{\Delta z}\right)\Delta t\wedge\Delta y\wedge\Delta z$$
$$+\left(\frac{\tilde{\Delta}_x\tilde{\Delta}_y\tilde{\Delta}_z\Delta_t B_y}{\Delta t} + \frac{\tilde{\Delta}_x\tilde{\Delta}_y\Delta_z\tilde{\Delta}_t E_x}{\Delta z} - \frac{\Delta_x\tilde{\Delta}_y\tilde{\Delta}_z\tilde{\Delta}_t E_z}{\Delta x}\right)\Delta t\wedge\Delta z\wedge\Delta x$$
$$+\left(\frac{\tilde{\Delta}_x\tilde{\Delta}_y\tilde{\Delta}_z\Delta_t B_z}{\Delta t} + \frac{\Delta_x\tilde{\Delta}_y\tilde{\Delta}_z\tilde{\Delta}_t E_y}{\Delta x} - \frac{\tilde{\Delta}_x\Delta_y\tilde{\Delta}_z\tilde{\Delta}_t E_x}{\Delta y}\right)\Delta t\wedge\Delta x\wedge\Delta y\,.$$

From the vector potential 1-forms $\mathbf{A} = A_\mu \Delta x^\mu$ we can construct Faraday 2-form $\mathbf{F} = \Delta\mathbf{A}$ from which the second set of the Maxwell equations are derived

$$\Delta^*\mathbf{F} = \Delta^*\Delta\mathbf{A} = 4\pi^*\mathbf{J}\,.$$

Taking the dual of this expression we obtain the wave equation for the vector potential

$$\Box A_\mu = {}^*\Delta^*\Delta A_\mu = 4\pi J_\mu\,.$$

Acknowledgments
The author wishes to express his gratitude to Professors Michael Ramek and Dieter Schuch for the invitation to participate in the Symposium "Symmetries in Science XIV" and to publish his contribution in the "Journal of Physics Conference Series". Part of this work was partially presented (unpublished) in the workshop "Quantum Integrable Discrete Systems" in march 2009 at the Isaac Newton Institute, Cambridge.
He also thanks M.E.C. for financial support, grant FPA2006-09199.

References
[1] Bombelli L, Lee J, Meyer D and Sorkin R 1987 "Space-time as a causal set" *Phys. Rev. Lett.* **59** 521
[2] Borisov R and Gupta S 1999 "Propagating spin modes in Canonical Quantum Gravity" *Phys. Rev.* D **60** 024002 arXiv: gr-qc/9810024
[3] Coreth E 1964 *Metaphysik* (Innsbruck: Tyrolia Verlag)
[4] De Risi V 2007 *Geometry and Monadology* (Basel: Birkhäuser) 530
[5] Lorente M 1995 "Discrete Differencial Geometry and Lattice Field Theories" *VII Int Conf Symmetry Methods in Physics* (Dubna: Joint Institute for Nuclear Research)
[6] Lorente M 1999 "Exact solutions of the Dirac Equation on the Lattice" *Nucl. Phys. Proc. Suppl.* **73** 679
[7] Lorente M 2001 "Continuous vs. discrete models for the quantum harmonic oscillator and the hydrogen atom" *Phys. Lett.* A **285** 119 arXiv: quant-ph/0401087
[8] Lorente M 2004 "A discrete curvature on a planar graph" *Encuentros de Física Fundamental Alberto Galindo* (Madrid: UCM) 339 arXiv: gr-qc/0412094
[9] Lorente M 2008 *Pensamiento* **64** n 242 665
[10] Lorente M and Kramer P 1999 "Representation of the discrete inhomogneous Lorentz group and Dirac wave equation on the lattice" *J. Phys. A: Math. Gen.* **32** 2481 arXiv: hep-lat/0401019
[11] Markopoulou F "New directions in Background Independent Quantum Gravity" (to be published by Cambridge UP) arXiv: gr-qc/0703097
[12] Markopoulou F 2000 "Quantum causal histories" *Class Quant Grav* **17** 2059 arXiv: hep-th/9904009
[13] Pérez A 2003 "Spin Foam Models for Quantum Gravity" *Class. Quant. Grav.* **20** R 43 arXiv: gr-qc/0301113
[14] Penrose R 1971 "Angular momentum: an approach to combinatorial space-time" Bastin T (ed.) *Quantum Theory and Beyond* (Cambridge: CUP)
[15] Zubiri J 1996 *Space. Time. Matter* (Madrid: Alianza Universidad)

Journal of Physics: Conference Series **237** (2010) 012018 doi:10.1088/1742-6596/237/1/012018

Colour vision and computer-generated images

Michael Ramek

Institut für Physikalische und Theoretische Chemie, Graz University of Technology, 8010 Graz, Austria

michael.ramek@tugraz.at

Abstract. Colour vision deficiencies affect approximately 8% of the male and approximately 0.4% of the female population. In this work, it is demonstrated that computer generated images oftentimes pose unnecessary problems for colour deficient viewers. Three examples, the visualization of molecular structures, graphs of mathematical functions, and colour coded images from numerical data are used to identify problematic colour combinations: red/black, green/black, red/yellow, yellow/white, fuchsia/white, and aqua/white. Alternatives for these combinations are discussed.

1. Introduction

Human colour vision is made possible by three types of colour sensitive cones in the retina with absorption maxima at 558.4±5.2 nm, 530.8±3.5 nm, and 419.0±3.6 nm [1]. According to wave length range, these cones are commonly called L, M, and S or simply red, green, and blue. The red cones are modifications of the green ones, which differ only in 14 amino acids of a 364 amino acid protein [2]. Therefore, the red and the green receptors overlap in a significant spectral range. It is this overlap, which enables us to distinguish various shades of green, yellow, and red.

Changes in the spectral properties of these colour receptors may be caused by genetic reasons or induced by illness (either directly or as a medication side effect). While the former reason affects mostly males [3] at a rate of about 8% [4], the latter affects both sexes at equal rate. Minor changes often remain undetected for a long time, whereas severe cases ("colour blindness") are usually noticed by those affected. The first detailed scientific report of colour blindness is the one given by John Dalton [5]. Although based on a wrong assumption, namely coloured liquid in the eyeballs, Dalton's report made this kind of deficiency widely known. Moreover, those affected became aware that their situation is not due to an individual inability, but rather a condition that others are also subject to. The long-lasting impact of Dalton's report is manifested by the fact that, until today, colour blindness is called "daltonism" in a number of languages.

In today's world, computers are used for more and more purposes. Information retrieval via the world wide web or computer-based presentations have become routine tasks. Astonishingly often, colour deficient viewers miss part of the information on monitors, video walls, or projections: a person, e. g., who is unable to see red, can neither spot a red laser pointer on a screen nor read red letters on dark background [6]. Photographs of real life situations are usually less problematic than computer-generated images, which are the focus of this work. Based on a number of examples from the area of quantum chemistry, for which the author had data at hand, a few rules of thumb are developed to avoid such pitfalls.

2. Colour vision types

The generally accepted types of colour vision are listed in table 1 together with approximate frequency values. It has to be stressed that these frequencies are global average values, ethnic groups may have different values that depend on the specific circumstances and history.

Table 1. Colour vision types with approximate frequency values. With the exception of achromatopia, which is not sex-linked, these values refer to the male population. Different ethnic groups may have deviating frequency values.

Type	Cones Affected	Effect on Vision
normal (91%)	red, green, and blue function normally	normal trichromatic vision
deuteranomalia (5%)	green cones are red-sensitive	green is perceived as black, red and yellow appear identical, turquoise as blue
protanomalia (1%)	red cones are green-sensitive	green is perceived as a shade of yellow, red as black, purple as blue
protanopia (1%)	red cones are insensitive	red is perceived as black, green and yellow as identical, purple as blue
deuteranopia (1%)	green cones are insensitive	green is perceived as black, red and yellow appear identical, turquoise as blue
tritanopia (10^{-3}%)	blue cones are insensitive	blue is perceived as black, purple and red are identical, turquoise appears green
monochromatopia (10^{-4}%)	only blue cones function normally	red, yellow, and green are perceived as black
achromatopia (10^{-5}%)	all colour cones are insensitive, only rods (night vision) function normally	day-blindness

While the different types listed in table 1 may already appear confusingly complex at first, individual situations may even be more complex. Human vision is a brain skill that is developed step by step in the first few months of a child's life. During this process, the brains learns to optimally discern the incoming signals, and occasionally an individual pigmentation results in unique vision abilities [7].

Despite the broad range of individual cases, several types can be discussed together: *deuteranomalia* and *deuteranopia*, e.g., are different conditions that are caused by different pigmentation patterns of the retina cones, yet both can be summarized as a *deuteran* group, a significant number of people who are unable to see green. Similarly, *protanomalia* and *protanopia* constitute a *protan* group of people, who are unable to see red. Cases that involve blue occur at a much smaller frequency.

3. Computer monitors and software

In painting or printing, subtractive colour mixing takes place: a mixture of all colours yields some kind of black. Moreover, contextual influences, like reflectance differences or the type of illumination, may alter the visual impression. In contrast, computer or TV monitors are light sources with an additive mixing: all colours together yield white. Monitors emit light that consists of a few frequencies only.

Table 2. The colours defined for use in hypertext documents [8], with their red, green, and blue components.

colour name	R	G	B
black	0%	0%	0%
gray	50%	50%	50%
silver	75%	75%	75%
white	100%	100%	100%
maroon	50%	0%	0%
red	100%	0%	0%
green	0%	50%	0%
lime	0%	100%	0%
navy	0%	0%	50%
blue	0%	0%	100%
purple	50%	0%	50%
fuchsia	100%	0%	100%
olive	50%	50%	0%
yellow	100%	100%	0%
teal	0%	50%	50%
aqua	0%	100%	100%

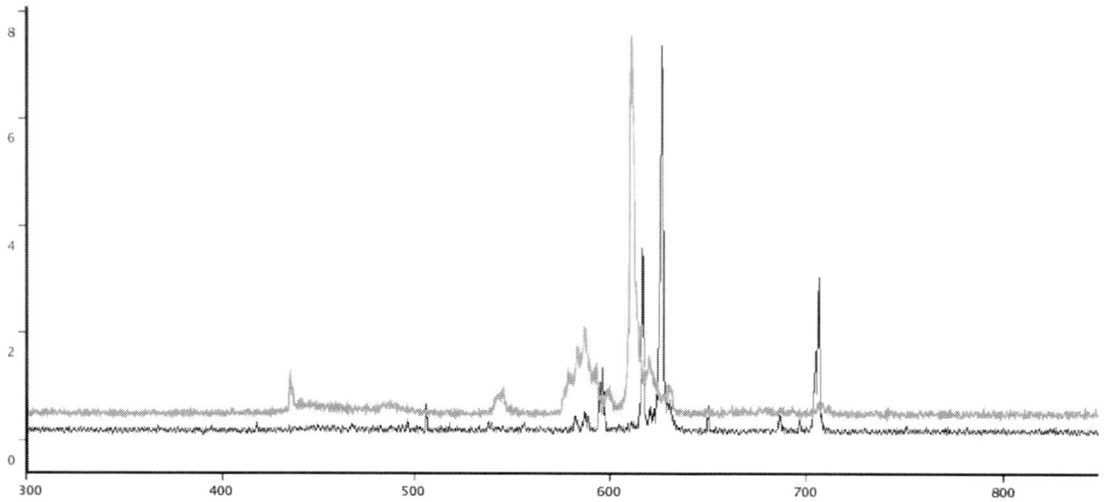

Figure 1. Emission spectra (intensity [arbitrary units] vs. wave length [nm]) of the R-channel of two RGB displays: a cathode ray tube monitor (black, lower graph) and a LCD monitor (gray, upper graph).

Figure 1 demonstrates this via the emission spectra of the red channel of two technically quite different commercial monitors. These few frequencies are specifically designed to exclusively excite the red, green, and blue receptors in the human eye. Contextual influences, which give important clues in the normal visual process, are not present in computer generated images, i.e., the light emitted by the three colour channels of the monitor is the only relevant information that reaches the observer's eye.

With these three channels, almost all colours can be generated that the human eye is capable of seeing. In practice, however, authors of software packages or software standards tend to use only one- or two-dimensional subspaces of this three-dimensional colour space. Table 2 lists the basic colours defined in the hypertext markup language standard [8] as an example. These colour names are also used in this work.

Since the majority of colour deficient viewers is either unable to see red or unable to see green, an easy method of predicting the effect of colour blindness would be to unplug the red or green channel (some older monitors indeed do have cables with three individual plugs). The same effect can be simulated by graphic programs that allow an intensity adjustment of the RGB channels. Reducing the red or green channel to zero intensity will change white to an intense colour (aqua or fuchsia in the notation of table 2), which looks strange to a person with normal trichromatic colour vision. The brain of a red blind person, however, that developed its visual skills under the absence of a red retinal pigment, will classify aqua a normal colour, often seen before. The essential point of this simple simulation is not the strange colour, to which white is transformed—the essential point is that any aqua-coloured display on a white background will be transformed to an aqua display on an aqua background that becomes invisible. A red blind person will, indeed, have difficulties seeing an aqua display on a white background, as it will have difficulties seeing red text on black background [6]. Hence, this simple simulation can be used as a rigorous test: If essential features get obscured by such a manipulation, chances are good that a non-negligible fraction of viewers will have difficulties.

As a general rule, all colours made up by only one non-zero component and all colours that consist of two 100% components are problematic. This includes the following:

a) Red/black and green/black: these combinations often occur when a black background is used. Figure 2 exemplifies this by means of a Rasmol [9] visualization of a molecular structure, using the default colour scheme of this program. In this colour scheme, some atoms remain invisible for *protan* or *deuteran* viewers, since for them these atoms are displayed in black—the same colour as the image background.

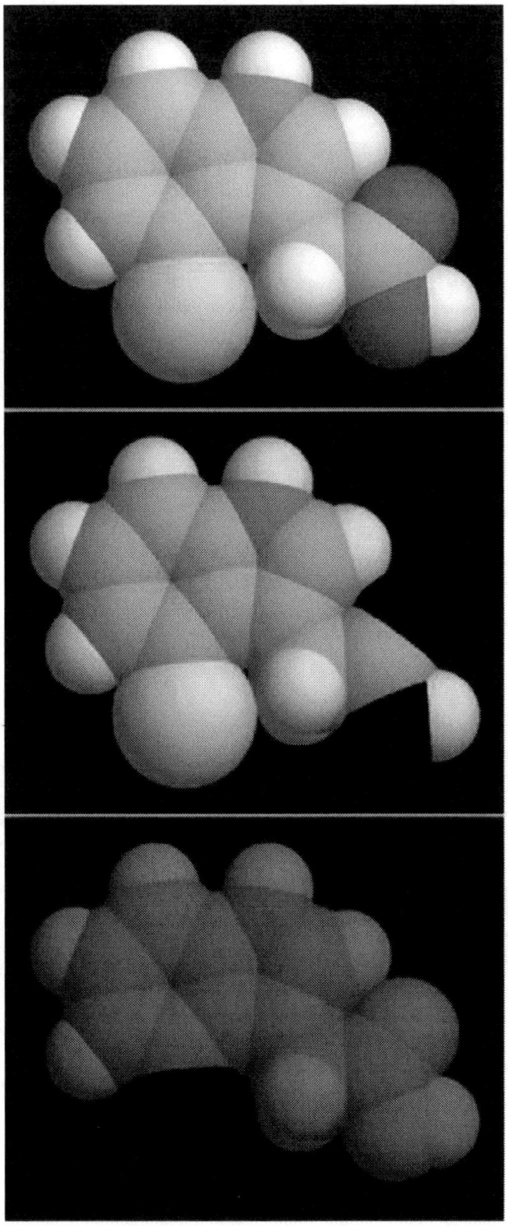

Figure 2. Visualization of the most stable 4-chloro-3-indole acetic acid conformer [10] by Rasmol [9] using the default color scheme of that program as seen by a person with normal color vision (top), by someone unable to see red (middle), and by someone unable to see green (bottom). Atoms depicted in these colors blend in with the black background.

b) Yellow/white, aqua/white, and fuchsia/white: these colour combinations often occur in graphs on white background. Figure 3 gives an example using the approximation of Slater-type orbitals by a number of Gaussian functions [11]. While the author has to admit that this figure uses intentionally chosen colours, figure 2 of [12] is another recent example, the authors of which certainly cannot be blamed for an intention to choose bad colours.

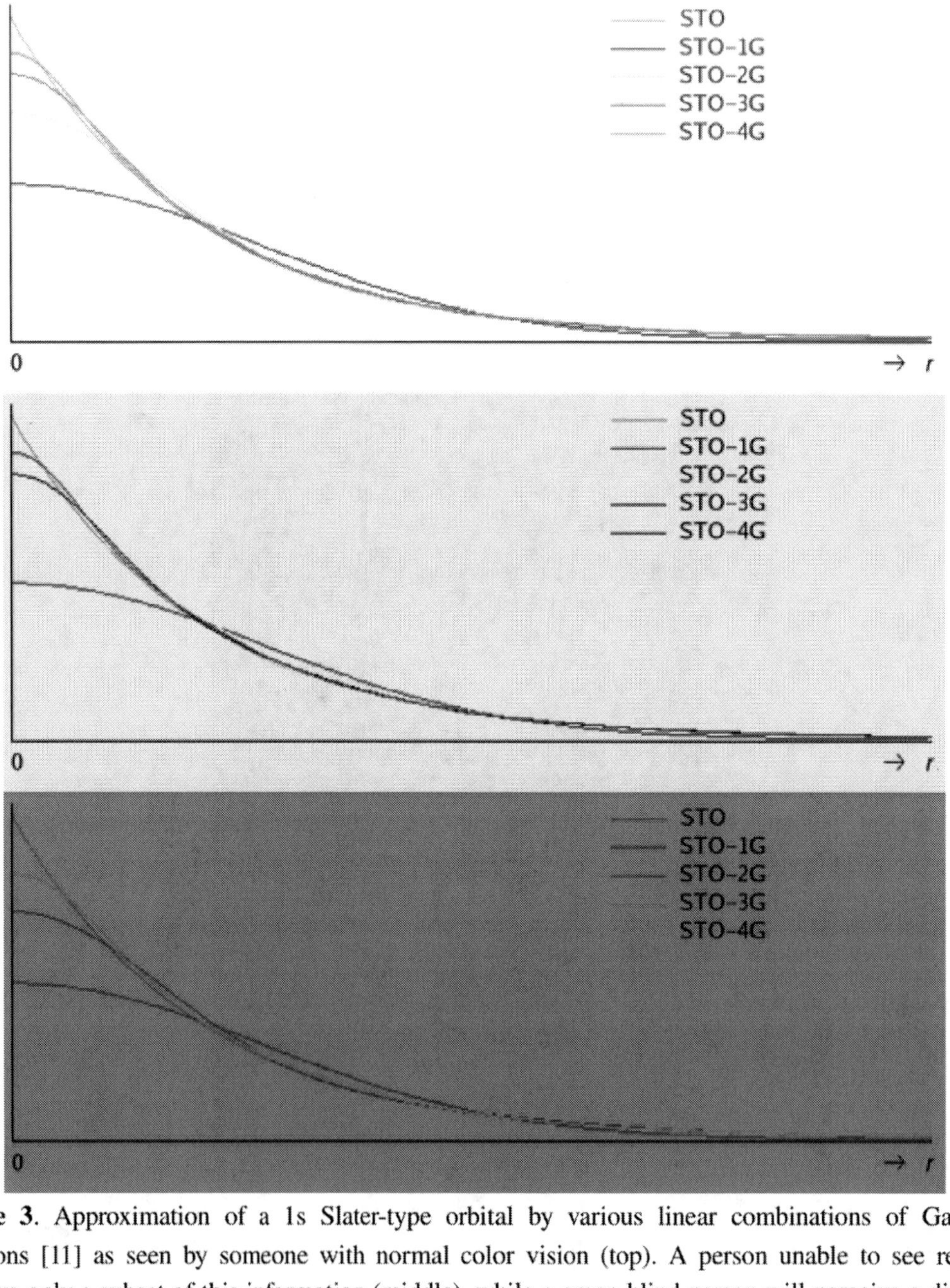

Figure 3. Approximation of a 1s Slater-type orbital by various linear combinations of Gaussian functions [11] as seen by someone with normal color vision (top). A person unable to see red can perceive only a subset of this information (middle), while a green-blind person will perceive a different subset (bottom).

c) Red/yellow: this colour combination often occurs in connection with rainbow-like colour coding, where red gradually turns into yellow and on into green and blue. Such colour coding is often used for maps, thermo imaging, or the visualization of electric properties. Figure 4 gives an example using the RHF potential energy surface of β-alanine [13] in the vicinity of an interesting conformation. Other recent examples are figures 1–3 of [14] as well as figures 2 and 8 of [15]. Figures 5 and 6 show the same energy surface as figure 4 without the red and the green component, respectively.

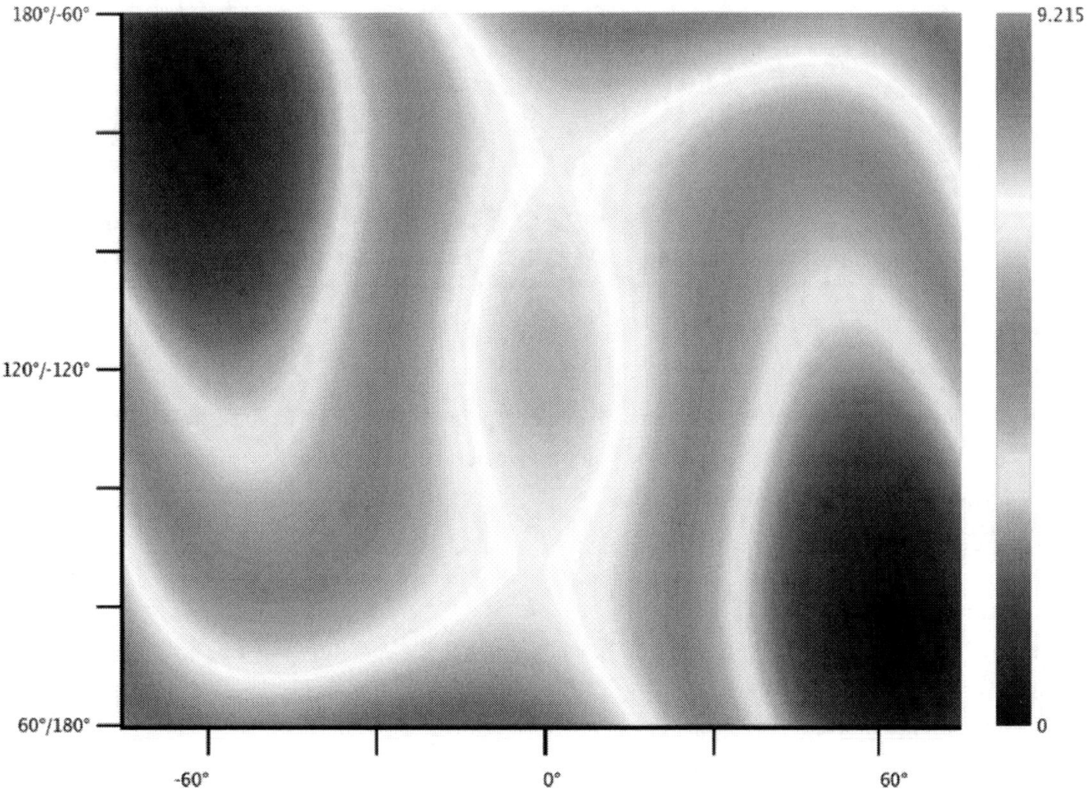

Figure 4. Relative energy (kcal/mole) of the vicinity of the two mirror-symmetrical conformers of neutral β-alanine, which are stabilized by an intramolecular hydrogen bond. The map shows RHF/4-31G energies[13] as a function of the amino group orientation (torsion angles H–N–C–C, ordinate) and the back bone orientation (torsion angle C–C–C–N, abscissa) in the colour coding indicated. The reaction does not pass the achiral structure of C_S symmetry, which corresponds to the center of the image; instead, two equivalent chiral reaction paths exist, both of which keep the hydrogen bond intact.

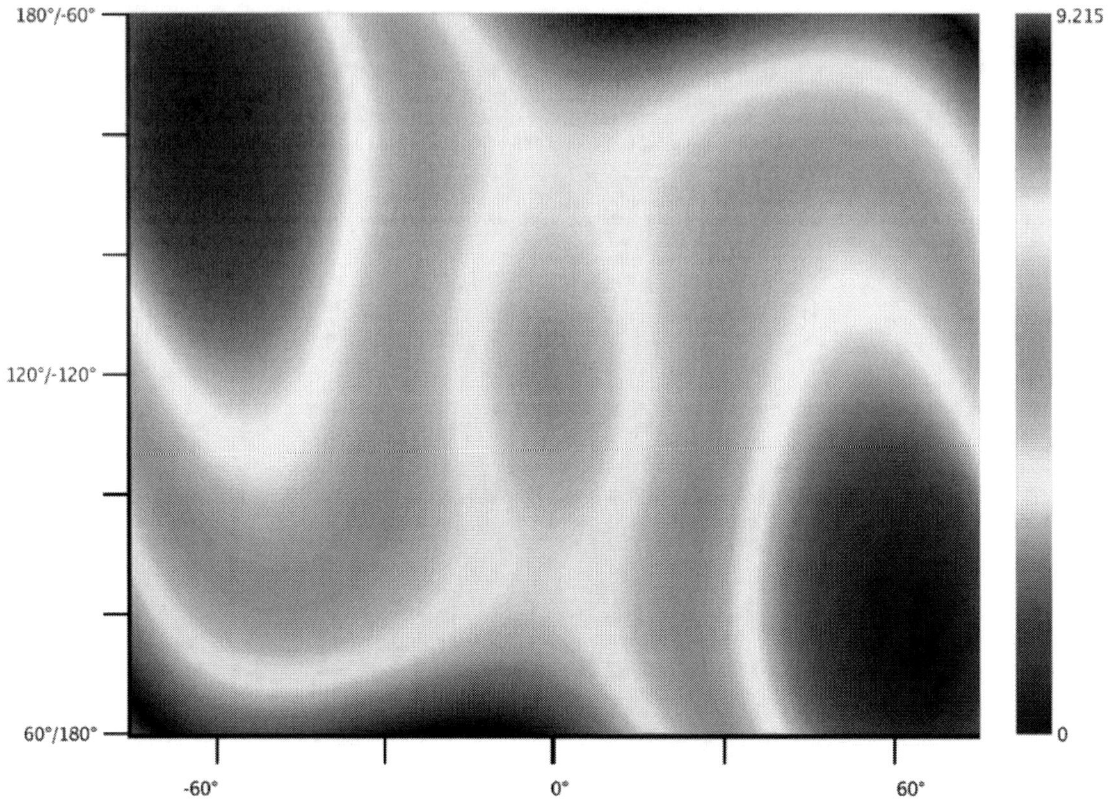

Figure 5. Figure 4 as seen by a red blind person. The point of achirality in the center of the surface is clearly visible, but a detailed knowledge of potential energy surface properties is necessary to recognize that it is a second order saddle point.

4. Alternatives

Most problematic colour combinations (red/black, green/black, red/yellow) can be avoided by changing black or yellow to some grey tone. A change of the background colour may already be sufficient: figure 7 shows the same molecule as figure 2, but with a grey background. This small change is enough to transmit significantly more information.

Colour combinations including yellow can be resolved by a transformation of yellow to off-white or grey. For RGB triples in the range [0, 255], such a transformation is very effectively performed by the following code segment, in which a reduction factor is determined from the absolute value of the R–G difference (which is zero for yellow or blue, and up to 255 for red or green tones); subsequently the R and G signals are reduced, whereas the B signal is increased if less than 255.

```
n = 7   # yellow → off-white ; use smaller values for yellow → light grey
f = max (0, abs(B–255) – abs(R–G ) ) / ( (2 × n + 1) × 255)
B = min ( 255 , B +  n × f × (R+G) )
R = (1 – f ) R
G = (1 – f ) G
```

Such a transformation is easy to incorporate into programs that generate RGB codes from numerical data as an alternative to a redefinition of the underlying grey scale to colour conversion. The effect can be seen in figure 8, which shows the same energy surface as figure 4, but with yellow converted to light grey (using the above code segment with n = 3). Figure 9 shows this transformed image without the red and the green component. Again, for colour deficient viewers significantly more information is transported by these figures.

The colours yellow, aqua, and fuchsia pose problems in connection with white background. For graphs of mathematical functions, data recordings, ... colour should therefore not be the only means to distinguish different entries. The use of different symbols (open and full circles, squares, triangles, ...) or line styles (solid, dotted, dashed, ...) in traditional black and white figures is an unproblematic way of presentation.

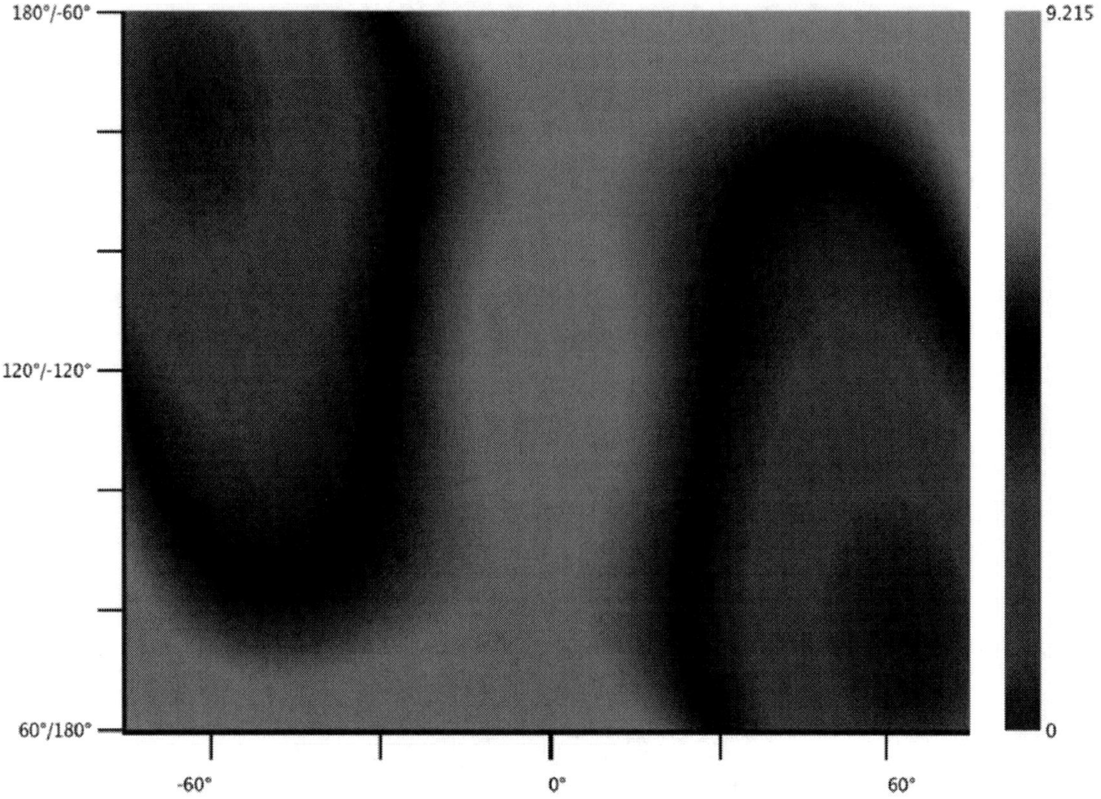

Figure 6. Figure 4 as seen by a green blind person. The second order saddle point in the center of the image is completely hidden in a broad area of uniform color.

Colour combinations that are easy to distinguish for most viewers and hence can be recommended, when strong colour contrast is desired, are: black/white, black/yellow, blue/white, blue/yellow. Pastel colours in which all RGB components are > 65% and different from each other are usually good choices for background colours in presentations, when visual clues to related or opposed topics or areas are intended.

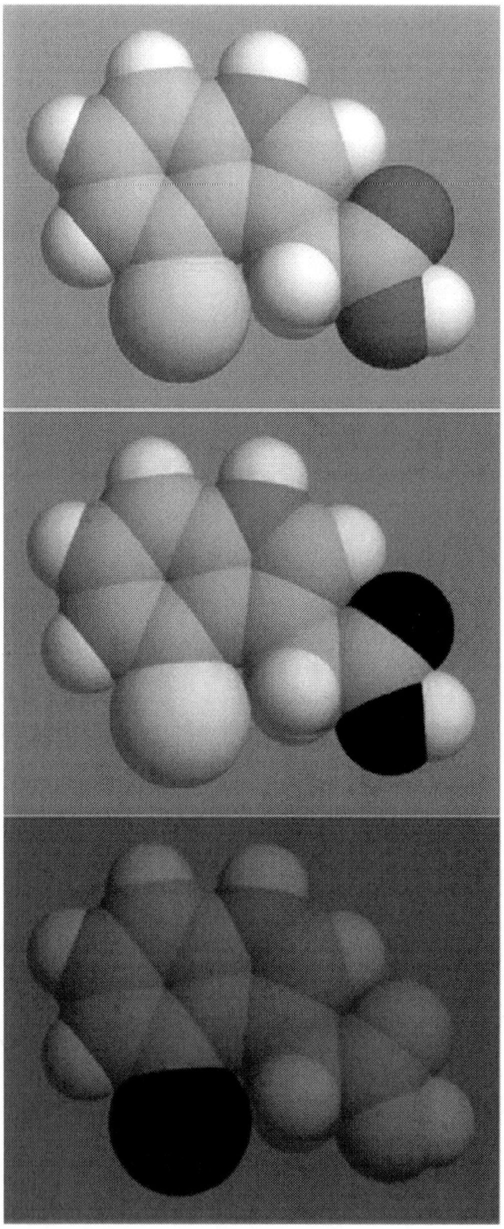

Figure 7. Figure 2 with a different background color. For people, who cannot see red or green, atoms in these colors still appear black, but the shape and structure of the molecule remains clearly recognizable.

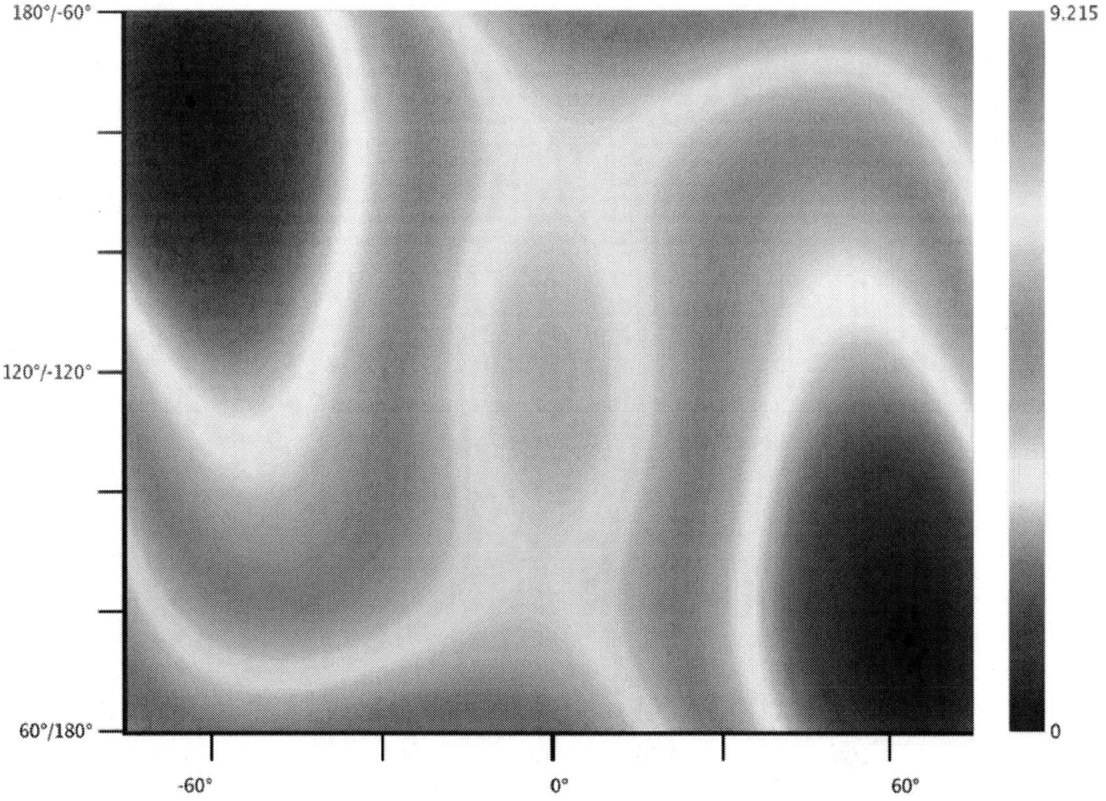

Figure 8. Figure 4 with yellow converted to grey (see text for details of this transformation).

Acknowledgement
The author is grateful to Drs. Karl Gatterer and Elmar C. Fuchs from the Graz University of Technology for recording the emission spectra in figure 1.

References

[1] Dartnall H J A, Bowmaker J K and Mollon J D 1983 *Proc. Roy. Soc. London. B, Biol. Sc.* **220** 115-30
[2] Natans L, Thomas D and Hogness D S 1986 *Science* **232** 193-202
[3] Bell J and Haldane J B S 1937 *Proc. Roy. Soc. London. B, Biol. Sc.* **123** 119-50
[4] Birch J 1993 *Diagnosis of defective colour vision* (Oxford: University Press) p 46
[5] Dalton J 1798 *Mem. Literary Philos. Soc. Manchester* **5** 28-45
[6] Romano P E 1998 *Opthalmol.* **105** 1797
[7] Jameson K A, Highnote S M and Wasserman L M 2001 *Psychosom. Bull. Rev.* **8** 244-61
[8] Ragett D, Le Hors A and Jacobs I 1999 HTML 4.01 Specification. *http://www.w3.org/TR/html4/*
[9] Sayle R and Milner-White E J 1995 *Trends in Biochem. Sci.* **20** 374-76
[10] Ramek M, Tomić S and Kojić-Prodić B 1996 *Int. J. Quant. Chem.* **60** 1727-33
[11] Hehre W J, Stewart R F and Pople J A 1969 *J. Chem. Phys.* **51** 2657-64

 IOP Publishing
doi:10.1088/1742-6596/237/1/012018

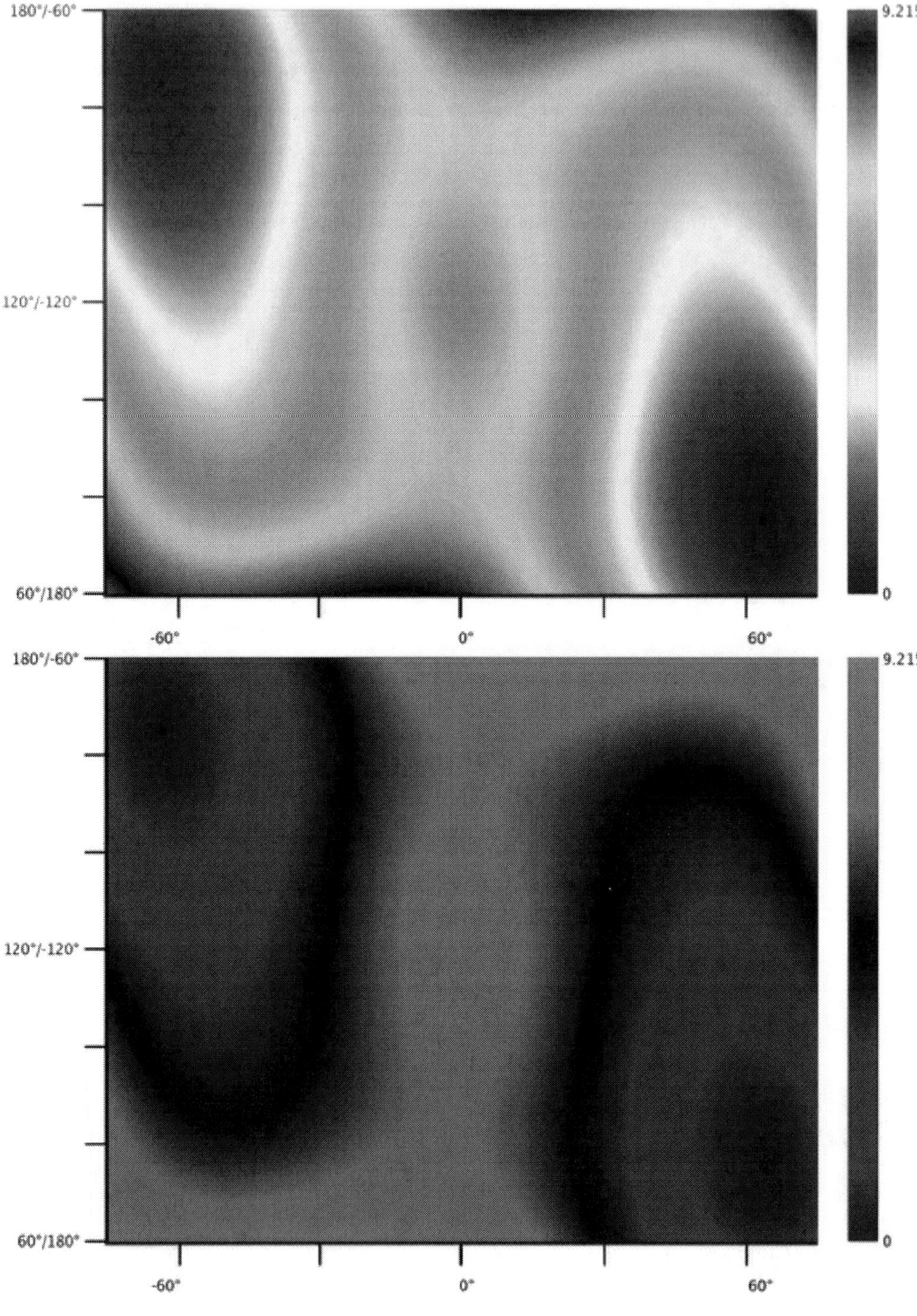

Figure 9. Figure 8 as seen by red- or green blind persons. In contrast to figures 5 and 6, the center of the image is clearly recognizable as a local maximum in this two-dimensional map in both cases.

[12] Lele B S, Murata H, Matyjaszewski K and Russell A J 2005 *Biomacromolecules* **6** 3380-7

[13] Ramek M 1990 *J. Mol. Struct (Theochem)* **208** 301-55

[14] Jones H G, Stoll M, Santos T, de Sousa C, Chaves M M and Grant O M 2002 *J. Exp. Botany* **53** 2249-60

[15] Aksimentiev A and Schulten K 2005 *Biophys. J.* **88** 3745-61

Journal of Physics: Conference Series **237** (2010) 012019 doi:10.1088/1742-6596/237/1/012019

Nonlinear Schrödinger-Poisson definition of fine-structure-constant's value $\sim 1/137$

Gilbert Reinisch
Université de Nice - Sophia Antipolis
CNRS, Observatoire de la Côte d'Azur
BP 4229
06304 - Nice Cédex 4 France

E-mail: Gilbert.Reinisch@oca.eu

Abstract. By numerically investigating the nonlinear Schrödinger-Poisson eigenstates of a condensed Bose gas of charged particles that is confined in a two-dimensional axisymmetric parabolic potential $\frac{1}{2}m_e\omega^2r^2$ (e.g. quantum-dot helium), it is shown that the probability amplitude between two nonlinear—and hence non-orthogonal—eigenstates displays an interference pattern scaled (within 0.03%) by $\sqrt{\pi\alpha} = 0.15141\ldots$ Since $\alpha = e^2/\hbar c \sim 1/137$ is the fine structure constant, this stunning result—indeed velocity of light c does not enter the present non-relativistic model—is tentatively explained by the existence of a "nonlinear" bound state of the trapped particle-particle interaction Coulomb field whose energy $\mathcal{E} = \frac{1}{2}\pi\,\hbar\omega$ defines the induced emission or absorption equilibrium processes between two appropriate chemical potentials. Besides, a non-decoherence quantum-classical transition with increasing nonlinearity is pointed out. As a possible experimental test for the present theory, the $0s^2 \rightarrow 1s^2$ nonlinear transition in an $\hbar\omega = 1.66\,\text{meV}$ GaAs quantum-dot helium is emphasized.

1. Introduction

The numerical value $\sim 1/137$ of the fine-structure-constant $\alpha = e^2/\hbar c$ has been a mystery ever since it was discovered [1]. "It's one of the *greatest* damn mysteries of physics: a *magic number* that comes to us with no understanding by man. We know what kind of a dance to do experimentally to measure this number very accurately but we don't know what kind of a dance to do on a computer to make this number come out—without putting it in secretly!", adds R. P. Feynman in his unique kidding style. The present paper suggests an original—and, remarkably enough: non-relativistic—scenario for such a "computer dance". It makes use of a Schrödinger-Poisson (SP) differential model for single-particle nonlinear eigenstates in a particular two-electron physical system, namely quantum-dot helium (QDH). However, we insist that the present SP description is also quite appropriate for any harmonically confined and condensed Bose gas of charged particles, the present QDH system being only a very simple special case with experimental interest.

Quantum-dots can be viewed as artificially structured, disk-like shaped atoms in heterojunctions or metal-oxide-semiconductor devices where few electrons are confined to a length comparable to the mesoscopic effective Bohr radius a_B ($a_\text{B} \sim 10^{-2}\,\mu$m in the case of GaAs) [2–4]. Since electron numbers N as low as one or two per dot have already been realized, QDH consisting of two electrons trapped in the two-dimensional (2D) axisymmetric harmonic potential $V(r) = \frac{1}{2}M\omega^2r^2$ (where $r^2 = x^2 + y^2$ and M is the effective electron mass) is actually considered as the simplest realistic model for an interacting quantum system [5–6]. As itself or amongst other such few-electron systems, it has been extensively studied in the relationship with the development of nanotechnologies [7–10]. Both its exact 2D-3D analytical [9] or 2D numerical [6] solutions are known.

Nonlinear eigenstates are as a rule not orthogonal [11]. The probability amplitude $<\Psi^i_{\rm NL}|\Psi^j_{\rm NL}>\neq 0$ of a couple of stationary (time-independent) nonlinear (hence the subscripts) normalized SP eigenstates $\Psi^{i,j}_{\rm NL}$ yields the occupation number $w^{i\,j}_{\rm NL}=w^{j\,i}_{\rm NL}=|<\Psi^i_{\rm NL}|\Psi^j_{\rm NL}>|^2\neq 0$ of eigenstate $\Psi^i_{\rm NL}$ when the system is assumed to be in $\Psi^j_{\rm NL}$, or reverse. Therefore the SP quantum system is in a *mixed state*: its appropriate description in terms of the density matrix ρ is

$$\rho=\frac{1}{1+w^{i\,j}_{\rm NL}}|\Psi^i_{\rm NL}><\Psi^i_{\rm NL}|+\frac{w^{i\,j}_{\rm NL}}{1+w^{i\,j}_{\rm NL}}|\Psi^j_{\rm NL}><\Psi^j_{\rm NL}|, \tag{1}$$

or reverse. The paradoxical property, compared with standard linear theory, is that although the system is in the mixed state described by Eq. (1) and *not* in a superposition state (indeed this later is *not* a solution of the SP differential system, due to nonlinearity), it nevertheless allows interference between the two eigenstates $|\Psi^{i,j}_{\rm NL}>$ to occur, according to the following theorem:

$$<\Psi^i_{\rm NL}|\Psi^j_{\rm NL}>=\frac{\Phi^i_{j,\,i}}{\mu_i-\mu_j}+\frac{\Phi^j_{i,\,j}}{\mu_j-\mu_i}. \tag{2}$$

The subscripts in $\Phi^i_{j,\,i}$ and $\Phi^j_{i,\,j}$ define the matrix elements—calculated by use of the SP nonlinear eigenstates $\Psi^{i,j}_{\rm NL}$—of the corresponding particle-particle interaction potential $\Phi^{i,j}$ while $\mu_{i,j}$ are those nonlinear eigenvalues (or chemical potentials) which respectively define $\Psi^{i,j}_{\rm NL}$. Eq. (2) is a direct consequence of the Hermiticity of the Laplacian operator in the Schrödinger equation of the parabolically-confined SP differential system [12]. The term on the rhs of Eq. (2) defines the probability amplitude for the system being in the nonlinear eigenstate $|\Psi^i_{\rm NL}>$ to be also in the nonlinear eigenstate $|\Psi^j_{\rm NL}>$, due to the interaction potential Φ^i defined by the probability density $|\Psi^i_{\rm NL}|^2$ through the Poisson equation, while the second term of Eq. (2) defines the reverse process, namely, the probability amplitude for the system being in $|\Psi^j_{\rm NL}>$ to be also in $|\Psi^i_{\rm NL}>$ as a consequence of the interaction potential Φ^j defined by $|\Psi^j_{\rm NL}|^2$. Eq. (2) is *exact*. Therefore no perturbative-like ordering in $\Phi^{i,j}$ is needed although, of course, a straightforward time-independent perturbation scheme that considers $\Phi^{i,j}$ as perturbative potential in the respective $\{i,j\}$ Schrödinger equation of the SP system (for instance when nonlinearity $\mathcal{N}$ is weak) allows one to recover it.

The two amplitudes in Eq. (2) interfere in the build-up of the probability amplitude $<\Psi^i_{\rm NL}|\Psi^j_{\rm NL}>$ because the two $i\leftrightarrow j$ processes are indistinghishable from the point of view of the (numerical) value of $<\Psi^i_{\rm NL}|\Psi^j_{\rm NL}>$. Therefore this later defines the *nonlinear quantum coherence* in our two-level system. The resulting interference pattern is achieved by the $\omega\to 0$ progressive "flattening" of the parabolic trap or, equivalently, by a progressive increase of the nonlinearity in the system as numerically shown by figure 3. The quantum coherence allows energy to be exchanged between the two energy levels or, more accurately, between the two nonlinear-eigenvalue chemical potentials. The present paper proposes that these energy exchanges be quantized in terms of the "nonlinear" (or "renormalized") non-integer bunch $\mathcal{E}=\frac{1}{2}\pi\hbar\omega$ of harmonic quanta $\hbar\omega$. It will be shown that the coupling probability per particle σ to the surrounding harmonic quanta—assumed to be in thermodynamical equilibrium between the emission and absorption processes—equals 1/137.17 in the present state of the numerical art. This result should be compared with the fine-structure-constant's value 1/137.04.

The explanation for such an amazing coincidence—recall that the whole SP model is non-relativistic—might be due to the fundamental equivalence between i) the classical non-relativistic description *à-la-Rutherford* of an elastic electron Coulomb scattering; and ii) its quantum electrodynamics (QED) analogue defined (in the lowest-order in α) by the probability amplitude related to the single-vertex electron-photon coupling process [13]. In the former classical case that corresponds to the Poisson definition of the Coulomb potential, there is no photon at all (and hence obviously no velocity of light c) while in the later QED case, the photon (and hence c) crucially mediates the interaction through the coupling constant α. Although c disappears at the end of the QED calculation in the non-relativistic limit, the present work shows that the nonlinear probability amplitude $< \Psi^i_{\rm NL}|\Psi^j_{\rm NL} >$ seems to keep a memory of this α (and hence c) dependence even in the non-relativistic limit that defines the SP differential system.

2. Nonlinear quantum-classical Schrödinger-Poisson transition

We start from the stationary Schrödinger equation with the classical potential $W(x,y)$ that defines the 2D orbital eigenstate wave function $\Psi(x,y)$ *per particle*, together with its corresponding energy eigenvalue μ:

$$-\frac{\hbar^2}{2M}\nabla^2\Psi(x,y) + W(x,y)\Psi(x,y) = \mu\Psi(x,y). \tag{3}$$

Then we assume that $W(x,y) \equiv V(x,y) + \Phi(x,y)$ is but the mere addition of both classical potentials *per particle*, namely the external parabolic confining potential $V(x,y) = V(r) = \frac{1}{2}M\omega^2 r^2$ and the mutual electrostatic repulsive interaction potential $\Phi(x,y)$ between the two particles which is defined by Poisson equation:

$$\nabla^2\Phi(x,y) = -4\pi\mathcal{E}\,|\Psi(x,y)|^2. \tag{4}$$

Since $|\Psi|^2 \propto [length]^{-2}$ and the two above potentials V and Φ are actually potential energies, we must introduce, for dimensional reasons related to Eq. (4), a characteristic energy $\mathcal{E}$ which is a typical measure of the SP nonlinearity and which will be specified as a functional of Ψ, Φ and μ later on. Note that the differential system of Eqs. (3–4) is only relevant for particles in the same orbital state Ψ: e.g. a Bose-Einstein condensate of charged particles. It also yields a non-relativistic Schrödinger description of a Cooper-like pair of opposite-spin electrons trapped in the same orbital bound state [11,14] whose physical meaning is transparent. Indeed, one electron, say electron a with orbital wave function Ψ_a, "feels" the repulsive electrostatic potential Φ_b that is being created by its fellow electron b with orbital wave function Ψ_b. This potential Φ_b is classically defined by the Poisson equation $\nabla^2\Phi_b \propto -|\Psi_b|^2$ while Ψ_a is solution of the one-particle Schrödinger equation including both classical potentials, namely the Coulomb interaction potential Φ_b and the external confining potential $V(r)$. The system is closed by the "orbital bosonic assumption" $\Psi_a \equiv \Psi_b$ which obviously agrees with the Pauli exclusion principle for two opposite-spin fermion electrons.

Due to axisymmetry, we have $\Psi(x,y) = \Psi(r,\phi) = \psi(r)\,e^{im\phi}$. The one-particle wavefunction Ψ is thus eigenstate of the angular-momentum operator $-i\hbar\partial/\partial\phi$ related to its eigenvalue $m\hbar$. Its radial part $\psi(r)$, which describes the 2D confinement of the Bose gas, is defined by the following dimensionless ordinary-differential system:

$$\ddot{u} + \frac{1}{X}\dot{u} + \left[C - \frac{X^2}{4}\right]u = 0;$$

$$\ddot{C} + \frac{1}{X}\dot{C} + \frac{4m^2}{X^4} = u^2, \tag{5a, b}$$

if $l_0 = \sqrt{\hbar/M\omega}$ is the characteristic harmonic length and if

$$X = \sqrt{2}\,\frac{r}{l_0} \quad ; \quad u = \frac{1}{\omega}\sqrt{\frac{2\pi\mathcal{E}}{M}}\,\psi \quad ; \quad C = \tilde{\mu} - \frac{m^2}{X^2} - \tilde{\Phi}. \tag{6a, b, c}$$

The dot stands for derivation with respect to the (dimensionless) radius X and the tilde superscript labels energy in units of $\hbar\omega$. The one-particle probability of presence $|\Psi|^2$ must be normalized to unity. Hence the corresponding norm $\mathcal{N}$ of the dimensionless solution u, which provides a first definition of $\mathcal{E}$, is given by

$$\int |\Psi|^2 \, d^2\mathbf{x} = 1 \quad \longrightarrow \quad \int_0^\infty u^2 \, X dX = \mathcal{N} = 2\tilde{\mathcal{E}}. \tag{7}$$

Eqs. (5) yield $u(r)$ and $C(r)$ as functionals of $\mathcal{N}$, once given the (actually spatial) "initial conditions" $u_0 = u(0)$, $\dot{u}_0 = \dot{u}(0)$, $C_0 = C(0)$ and $\dot{C}_0 = \dot{C}(0)$. Amongst them, u_0 and C_0 are left free and will be numerically chosen by dichotomy in order to yield regular bound-state eigenstates defined by $u(X) \to 0$ for $X \to \infty$. We consider in the present work and for the sake of simplicity the two first $m = 0$ zero-angular-momentum nonlinear eigenmodes per particle, namely $u_1(X)$ and $u_3(X)$, for which $\dot{C}_0 = \dot{u}_0 = 0$ (the nonlinear SP differential system works equally well with $m \neq 0$ eigenstates, as shown by figure 2). The subscripts (1) and (3) refer to the $u_0 \sim \sqrt{\mathcal{N}} \to 0$ linear limit of their energy eigenvalues [6]:

$$E^{\text{lin}}_{n,m} = (2n + |m| + 1)\,\hbar\omega. \tag{8}$$

Figure 1 displays the C_0 versus u_0 spiraling trajectories for u_1 (circles) and u_3 (stars). These trajectories are parametrized with respect to increasing values of the nonlinear parameter $\mathcal{N}$ defined by Eq. (7), i.e. with decreasing values of the trap harmonicity ω.

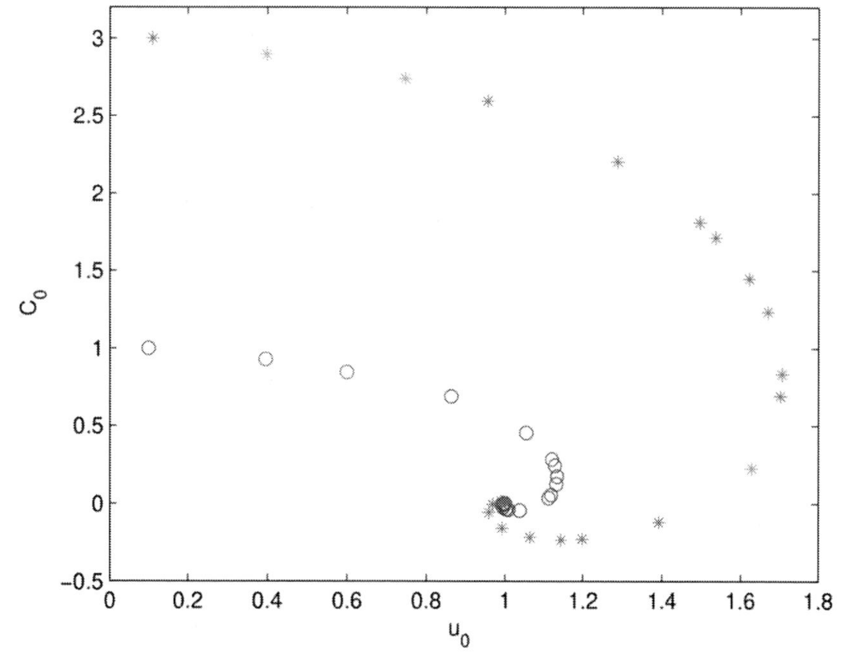

Figure 1. The convergence of the discrete SP nonlinear system towards the quasi-classical continuum Thomas-Fermi regime defined by the fixed point $u \equiv 1$ and $C_0 = \frac{1}{4}[X^2]_{X=0} = 0$ when $\mathcal{N}$ increases from $\sim 10^{-2}$ to $\sim 10^2$ in the C_0 vs u_0 boundary-condition phase space for the two first $m = 0$ nonlinear eigenmodes. This modes are defined by their corresponding linear quantum numbers, namely (cf. Eq. (8)) $n = 0$, $m = 0$ (u_1: circles) and $n = 1$, $m = 0$ (u_3: stars).

Indeed electron-electron Coulomb interaction becomes (relatively to quantum kinetic and external potential energy) more and more important when the two electrons are less and less confined (i.e. when the trap is "flattened"). Actually $\mathcal{N}$ varies in figure 1 from 10^{-2} ($u_0 \sim 0.1$) to 10^2

($u_0 \sim 1$) where one then reaches the quasi-classical asymptotic Thomas-Fermi regime. This regime is defined by neglecting the quantum kinetic Laplacian derivative terms in Eq. (5a), thus yielding $C(X) \sim X^2/4$ and hence $C_0 = C(0) = 0$, while $u(X) \equiv 1$ through Eq. (5b) with $m = 0$. Therefore the "initial conditions" for the two discrete modes u_1 and u_3 converge for $\mathcal{N} \to \infty$ towards the same Thomas-Fermi fixed point $\{u_0 = 1; C_0 = 0\}$ as evidenced by figure 1. Physically, this means that there is a continuous transition, through the increase of nonlinearity in the system (i.e. through $\omega \to 0$ flattening of the trap), from the $\mathcal{N} \leq \pi$ "pure" quantum regime towards the $\mathcal{N} \gg 1$ classical one where the dimensionless Schrödinger equation (5a) reduces to its last-bracket classical-energy term. As a consequence, the $\mathcal{N} \to \infty$ highly nonlinear case leads to the progressive merging of the two discrete energy levels corresponding to u_1 and u_3 into the single asymptotic one defined by the fixed point in figure 1. Quantum energy discreteness disappears, which is the hallmark of the classical regime.

3. Nonlinear eigenvalue: chemical potential and corresponding energy

The SP virial energy E per particle corresponding to the nonlinear eigenstate $u(X)$ is twice the quantum expectation value $\langle \frac{1}{2} M\omega^2 r^2 \rangle$ of the external parabolic potential energy [15]. On the other hand, the energy eigenvalue μ defined by Eq. (3) is that energy which is needed in order to add the second electron to the single-electron quantum-dot (Koopman's theorem [16]). Therefore it is indeed the chemical potential and can truly be regarded as the nonlinear eigenvalue of the SP differential system related to the corresponding nonlinear eigenstate Ψ (or u in the reduced units Eqs. (6)). Therefore we have:

$$\tilde{E} = \frac{1}{2\mathcal{N}} \int_0^\infty u^2\, X^3 dX = \frac{1}{2}(\tilde{E}^{\text{lin}} + \tilde{\mu}), \tag{9}$$

where E^{lin} is defined by Eq. (8). The nonlinear integrodifferential system Eqs. (5–9) is closed by the use of Eq. (6c) at $X = 0$,

$$\tilde{\mu} = C(0) + \tilde{\Phi}(0) = C_0 + \frac{e^2}{\hbar\omega} \int \frac{|\Psi|^2}{r} d^3\mathbf{x} = C_0 + \frac{\sqrt{2}\,k}{\mathcal{N}} \int_0^\infty u^2\, dX, \tag{10}$$

where we recall that C_0 is that "initial condition" which defines, in addition to u_0, the nonlinear eigenstate $u(X)$. The parameter $k = l_0/a^*$ is the usual dimensionless dot size that consists in measuring the characteristic harmonic length $l_0 = \sqrt{\hbar/M\omega}$ in units of the effective Bohr radius $a^* = \hbar^2/Me^2$ (ranging from $a^* = 67\,\text{nm}$ for InSb to $a^* = 9.8\,\text{nm}$ for GaAs).

Eq. (5–10) self-consistently define, for any given value of the trap harmonic frequency ω (or, equivalently, its reduced size k), the solution $u \equiv u_\omega(X)$, its norm $\mathcal{N} \equiv \mathcal{N}(\omega)$ as well as its corresponding energy $\tilde{E} \equiv \tilde{E}(\omega)$ per particle, together with the chemical potential $\tilde{\mu} \equiv \tilde{\mu}(\omega)$. In [6], for instance, where $\hbar\omega = 3.37\,\text{meV}$ for a GaAs parabolic quantum dot (M equals 0.067 electron mass while the charge is $1/\sqrt{12.4}$ electron charge), we have $a^* = 9.79\,\text{nm}$ and $l_0 = 18.5\,\text{nm}$. Hence $k = 1.89$. Then Eqs. (9–10) respectively yield $\mathcal{N} = 2.53$ and $\tilde{E} = \tilde{E}_{per\ particle} = 1.83$. Therefore $E_{\text{QDH}} = 2(1.83)\,\hbar\omega = 12.33\,\text{meV}$, which must be compared with the authors' value 12.28 meV. Similarly, in [10], $\hbar\omega = 2\,\text{meV}$ yields $k = 2.43$. Hence $\mathcal{N} = 2.97$ and $\tilde{E} = 2.00$, which yields $\tilde{\mu} = 2\tilde{E} - 1 = 3.00$ in accordance with Eq. (9), and therefore $\mu = 3.00\,\hbar\omega = 6.00\,\text{meV}$ which is in complete agreement with the Coulomb-interaction case ($d_1 = d_2 = \infty$) of that reference.

Therefore we recover an important property that has already been emphasized in the $N = 2$ Coulomb-trap case, both for highly-compressed giant-planet atomic helium [14] and for astrophysical H^- hydrogen ion [11], namely, that the SP nonlinear differential description yields surprisingly accurate values for the ground state energy when compared to the corresponding mean-field Hartree-Fock ones. Note that the use of nonlinear eigenstates and their related eigenvalues, together with Eqs. (7–9), yields in the quite opposite physical situation of a rotating condensed Bose gas of 10^5 neutral Rb atoms with hard-sphere particle-particle interactions the correct vortex thresholds, as compared with the corresponding experimental results [17].

A remarkable property of the SP nonlinear eigensolutions is their "universal" limit behavior for small $\mathcal{N}$, whatever the actual state's quantum numbers n and m in Eq. (8) are. Indeed we have

$$\tilde{E} \sim \tilde{E}^{\rm lin} + \frac{1}{4}\mathcal{N} \qquad \text{for} \qquad \mathcal{N} < 1 \quad \leftrightarrow \quad k < \frac{1}{2}, \tag{11}$$

(see figure 2). Therefore the characteristic energy $\mathcal{E} = \frac{1}{2}\mathcal{N}\hbar\omega$ (cf. Eq. (7)), which was introduced for dimensional reasons into the Poisson equation (4), is simply the smallest additional particle-particle interaction energy with respect to the linear regime that is due to nonlinearity. Indeed Eq. (11) yields for QDH's energy whose $\mathcal{N} \to 0$ non-interaction limit is $2E^{\rm lin}$

$$\lim_{\mathcal{N}\to 0} 2E = 2E^{\rm lin} + \mathcal{E}. \tag{12}$$

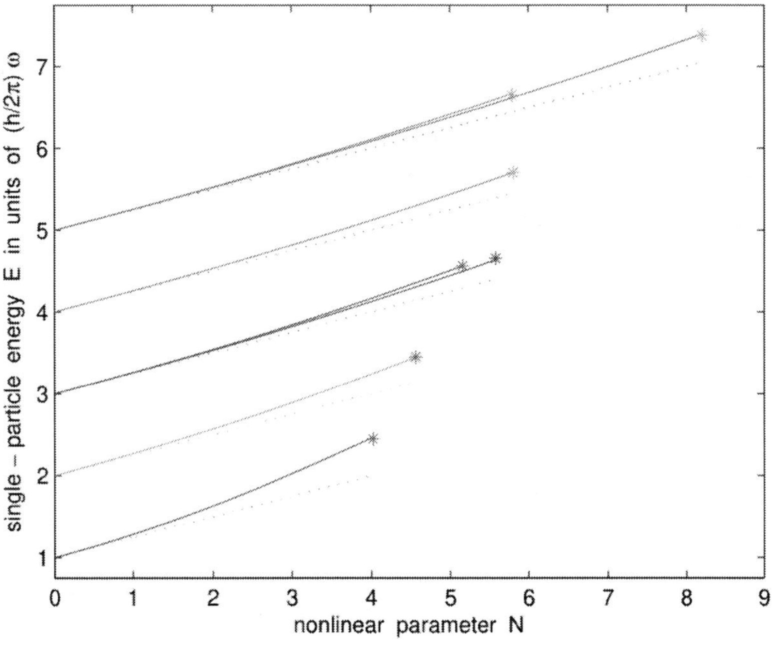

Figure 2. Illustration of Eq. (11) (dotted lines) for the $1 \leq \tilde{E}^{\rm lin} \leq 5$ nonlinear eigenmodes (as defined by their linear $\mathcal{N} \to 0$ quantum numbers n and m: see Eq. (8)). From bottom to top (a and b superscripts label linear degenerated states whose degeneracy is lifted by nonlinearity): $\{n = 0, m = 0\}$, $\{n = 0, m = 1\}$, $\{n = 1, m = 0\}^a$, $\{n = 0, m = 2\}^a$, $\{n = 1, m = 1\}$, $\{n = 2, m = 0\}^b$, $\{n = 1, m = 2\}^b$. The stars display the corresponding values of the first amplitude maxima reached at $X = 0$ by the nonlinear modes as $\mathcal{N}$ increases from its linear limit $\mathcal{N} \sim 0$ (see figure 1).

4. Nonlinear probability amplitude and renormalized state

Let us define the normalized probability amplitude

$$\langle u_1|u_3\rangle = \frac{1}{\sqrt{\mathcal{N}_1\mathcal{N}_3}}\int_0^\infty u_1 u_3\, X dX \tag{13}$$

of the two zero-angular-momentum eigenstates $u_1(X)$ and $u_3(X)$, together with their "matching condition"

$$k(\omega) = \frac{\mathcal{N}_1(\tilde{\mu}_1 - C_0^{(1)})}{\sqrt{2}\int_0^\infty u_1^2\, dX} = \frac{\mathcal{N}_3(\tilde{\mu}_3 - C_0^{(3)})}{\sqrt{2}\int_0^\infty u_3^2\, dX}. \tag{14}$$

Eq. (14) is the trivial condition that the trap profile defined by ω—or equivalently its dimensionless length $k = \sqrt{\hbar/M\omega}/(\hbar^2/Me^2)$—be identical for the two modes $u_{1,3}$ that enter the calculation of the Hilbertian scalar product (13): see Eq. (10). Let $\mathcal{P}_{1\,3} = \langle u_1|u_3\rangle^2$ (the eigenstates $u_{1,3}$ being real-valued, taking the modulus like for $w_{\rm NL}^{ij}$ in Eq. (1) is unnecessary). According to Eq. (1), we define the following respective population numbers of levels 1, 3:

$$n_1 = \frac{1}{1+\mathcal{P}_{1\,3}} \quad ; \quad n_3 = \frac{\mathcal{P}_{1\,3}}{1+\mathcal{P}_{1\,3}} = n_1\,\mathcal{P}_{1\,3}. \tag{15a, b}$$

Eq. (15b) can be loosely regarded as stating that the probability for the one-particle system to be in the nonlinear excited state $|u_3>$ is but its probability to be in the ground state $|u_1>$ multiplied by the mere nonlinear transition probability $\mathcal{P}_{1\,3}$ from $|u_1>$ to $|u_3>$. However we should keep in mind that the very concept of transition probability in a time-independent description of a stationary quantum system is questionable. Hence the concept of a mixed state like in Eq. (1) for the description of our nonlinear quantum system is much more appropriate.

Black-body thermodynamical equilibrium between the two levels is achieved if

$$n_1\, n^{\rm eq} = n_3\,(n^{\rm eq}+1), \tag{16}$$

where $n^{\rm eq}$ is that number of parabolic $\hbar\omega$ quanta which allows equal rates of stimulated emission and absorption processes between the two levels. Therefore Eqs. (15) yield

$$n^{\rm eq} = \frac{\mathcal{P}_{1\,3}}{1-\mathcal{P}_{1\,3}}. \tag{17}$$

Defining

$$\mathcal{N}_1^{\rm eff} = \mathcal{N}_1 + 2\,n^{\rm eq} \tag{18}$$

as the effective ground-state nonlinear parameter $\mathcal{N}_1$—or "renormalized nonlinear parameter" (renormalized actually by the equilibrium harmonic quanta which play here the role of QED vacuum photon fluctuations)—and scaling the probability amplitude (13–14) by use of $\gamma = \pi\alpha = \pi/137.036 = 2.2925\ldots 10^{-2}$, we obtain the remarkable interference pattern

$$< u_1|u_3 >= -\sqrt{\gamma}\,\sin\left(\frac{\mathcal{N}_1^{\rm eff}}{2}\right) + o(\gamma^{3/2}), \tag{19}$$

(see figure 3). The factor 2 in Eq. (18) originates from the corresponding renormalized nonlinear quantum defined by Eqs. (4) and (7),

$$\mathcal{E}^{\rm eff} = \frac{1}{2}\mathcal{N}_1^{\rm eff}\,\hbar\omega = \mathcal{E} + n^{\rm eq}\,\hbar\omega, \tag{20}$$

which now appears as "dressed" by the equilibrium harmonic quanta in much the same way as fermion lines are "dressed" by emission-reabsorption processes of one virtual photon in the QED calculation of second-order radiative corrections to Coulomb scattering [18].

Figure 3 also displays the following chemical-potential resonance at the maximum of $\mathcal{P}_{1\,3}$ that occurs at $\mathcal{N}_1^{\rm eff} \sim \pi$ (hence the subscripts),

$$\left[\tilde{\mu}_3 - \tilde{\mu}_1\right]_\pi \sim \tilde{\mathcal{E}}_\pi^{\rm eff}, \tag{21}$$

where $\tilde{\mathcal{E}}_\pi^{\rm eff} = \frac{1}{2}\mathcal{N}_1^{\rm eff} \sim \frac{\pi}{2}$ in accordance with Eqs. (19–20).

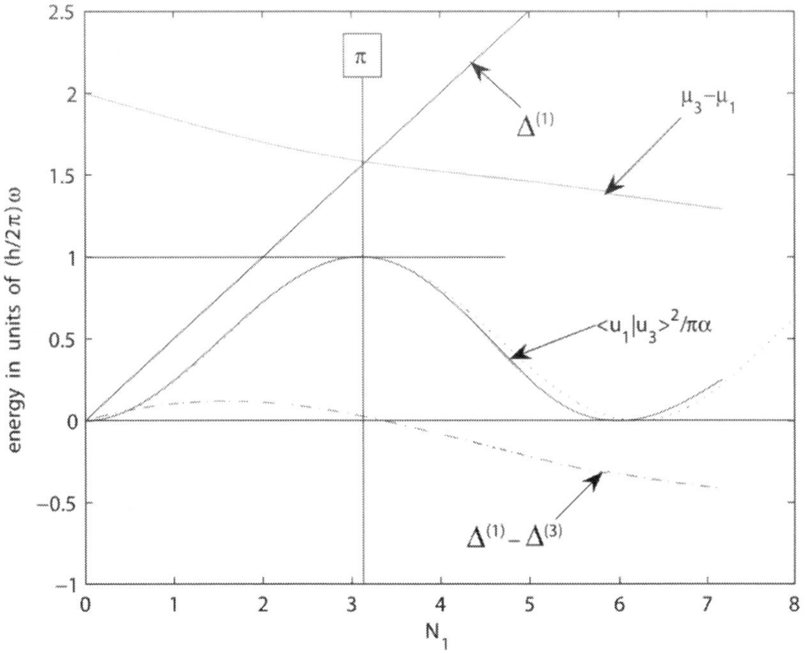

Figure 3. Eq. (19)'s square scalar product $\langle u_1|u_3\rangle^2/\pi\alpha$, as compared with its $\sin^2$ approximation (dotted line), together with the nonlinear resonance condition defined by Eq. (21) (in "tilde" units of $\hbar\omega$: intersection of the two upper plots; for some reasons, $\mathcal{E}$ is labelled Δ in the plot). Note the intersection of the dashed-dotted lower plot with the zero-energy value that yields $\tilde{\mathcal{E}}_1 \sim \tilde{\mathcal{E}}_3$, i.e. $\mathcal{N}_1 \sim \mathcal{N}_3$ at the resonance $\omega = \omega_\pi$ where $\mathcal{N}_1^{\rm eff} \sim \pi$.

This remarkable nonlinear resonance physically means that the quantum coherence maximum, reached at that particular value ω_π—or $\mathcal{N}_1^{\rm eff} = \pi$—of the trap parabolicity ω, is achieved when the renormalized quantum (20) equals the chemical-potential difference. It is thus tempting to extrapolate to the present case the standard picture of a discrete quantum jump, the energy quantum exchanged between the two corresponding "levels" (actually the two chemical potential nonlinear eigenvalues) being $\tilde{\mathcal{E}}_\pi^{\rm eff}$. From Eqs. (9–10), we numerically obtain $\hbar\omega_\pi \sim 0.14\,\epsilon$ where $\epsilon = Me^4/\hbar^2$ is the effective quantum-dot's atomic energy unit: $\epsilon = 11.86\,\mathrm{meV}$ for GaAs, thus yielding $\hbar\omega_\pi \sim 1.66\,\mathrm{meV}$ and $\mathcal{E}_\pi^{\rm eff} \sim 2.61\,\mathrm{meV}$ while $\epsilon = 27.21\,\mathrm{eV}$ if the dielectric constant of the bulk material equals unity, then yielding $\hbar\omega_\pi \sim 3.80\,\mathrm{eV}$ and hence $\mathcal{E}_\pi^{\rm eff} \sim 5.97\,\mathrm{eV}$. Therefore Eq. (21) provides parameters related to a possible experimental verification of the present SP theory. Note that $\mathcal{N}_1 \sim \mathcal{N}_3 \sim \pi$ at $\omega = \omega_\pi$ (dashed-dotted lower plot in figure 3). Therefore $\mathcal{E}_\pi^{\rm eff}$ is the common nonlinear quantum of the two eigenstates $u_{1,\,3}$ at their quantum-coherence resonance.

According to Eq. (7), Eq. (21) leads to the corresponding quantization rule for the total (two-particle) quantum-dot energy at $\omega \sim \omega_\pi$:

$$2(E_3 - E_1) \sim 2\hbar\omega_\pi + \mathcal{E}_\pi^{\rm eff} \sim \left(2 + \frac{\pi}{2}\right)\hbar\omega_\pi. \tag{22}$$

Eqs. (11), (21) and (22) show that the characteristic energy $\mathcal{E}$ which actually scales the electrostatic particle-particle interaction through the nonlinear differential Poisson equation (2) is in fact a true "nonlinear quantum". Indeed, on the one hand, it is the smallest particle-particle interaction energy that is present in the system at vanishing nonlinearity $\mathcal{N} \to 0$. On the other hand, the maximum of $\mathcal{P}_{13}$ "transition probability" between the two nonlinear eigenstates is achieved by the $\mathcal{E}_\pi^{\text{eff}}$ resonant quantum jump: i) either between the chemical-potential nonlinear eigenvalues μ_1 and μ_3; ii) or between both corresponding values $2E_{1,3}$ of the QDH energy, then with the addition of two standard "linear" radial quanta $\hbar\omega_\pi$.

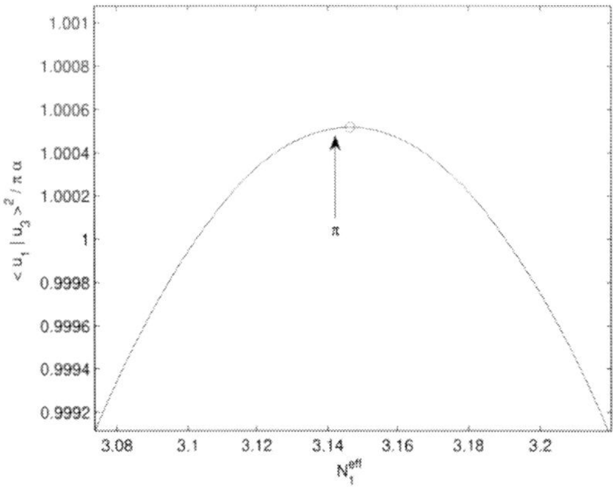

Figure 4. Numerical illustration of Eq. (23) which yields the fine-structure-constant's value within 0.1 % (the experimental value 1/137.036 would correspond in this figure to the small circle whose respective abcissa and ordinate are π and 1). The numerical calculations are done by use of standard tools [20] and consist in increasing nonlinearity through a three-loop iterative scheme from the $\mathcal{N} \ll 1$ linear regime. The two first loops define each eigenstate $u_{1,3}$ with a 10^{-8} accuracy at $X \sim 8$ (see figure 5) while the third one evaluates the matching condition Eq. (14) within 10^{-6} and then calculates the scalar product given by Eq. (13).

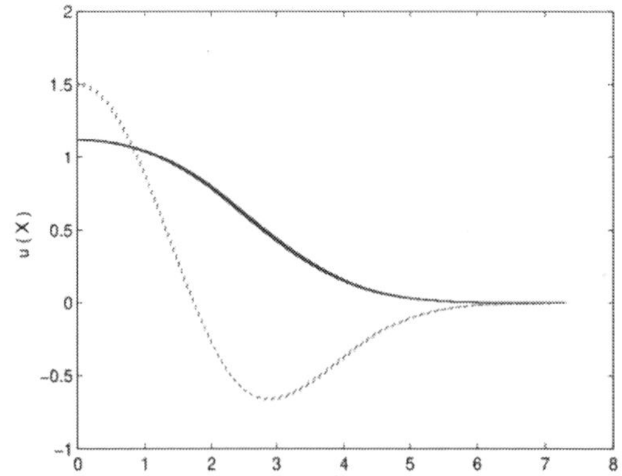

Figure 5. The nonlinear eigenstate profiles $u_1(X)$ and $u_3(X)$ of respective maximum amplitudes ~ 1 and ~ 1.5 corresponding to figure 4.

The nonlinear "quantum-transition" probability $\mathcal{P}_{13}$ takes into account both possible mechanisms related to Eq. (21) that interfere in accordance with Eq. (2), namely, $u_1 \to u_3$ ("absorption": $\uparrow$) or $u_3 \to u_1$ ("emission": $\downarrow$). Therefore *either* of these two processes has the probability

$\mathcal{P}_{\uparrow\downarrow} = \frac{1}{2}\mathcal{P}_{1\,3}$. At the nonlinear resonance defined by Eq. (21), the corresponding effective number of resonant harmonic quanta that enters the "transition" is $\tilde{\Delta}_{\pi}^{\rm eff} = \frac{1}{2}\mathcal{N}^{\rm max}$ by use of Eq. (20), where $\mathcal{N}^{\rm max} = \left[\mathcal{N}_1^{\rm eff}\right]_{\omega_\pi} \sim \pi$ is the abcissa of $\mathcal{P}_{1\,3} = [\mathcal{P}_{1\,3}]_{\rm max} = 1.0005\,\pi\,\alpha$ (see figure 4). Figure 5 displays the two corresponding nonlinear eigenstates.

Therefore the elementary probability σ per particle and per harmonic quantum for either "induced transition" process ("emission" or "absorption": like in QED, it does not make any difference which) is numerically given by

$$\sigma = \frac{[\mathcal{P}_{\uparrow\downarrow}]_{\rm max}}{\tilde{\Delta}_{\pi}^{\rm eff}} = \frac{\frac{1}{2}[\mathcal{P}_{1\,3}]_{\rm max}}{\frac{1}{2}\mathcal{N}^{\rm max}} = 0.3180\,\gamma = \frac{1}{137.17}\;. \tag{23}$$

This value of σ approaches the numerical value $1/137.036$ of Sommerfeld's fine structure constant α within $0.1\,\%$.

5. Conclusion

The present work emphasizes the physical properties of—and specifically the interference between—the SP nonlinear quantum *eigen*states as displayed by figure 5. It is useful to bear in mind the hierarchy of physical effects present in the system. Besides each particle's quantum kinetic effect defined by the Laplacian term in the Schrödinger equation (3) and its external parabolic confinement, the next significant effect is the particle-particle Coulomb interaction described by Poisson equation (4). Consequently, one may wish to define new quantum tools, namely, stationary eigenstates that take *all* these three dominant effects into account as already suggested by Cohen-Tannoudji [19]. The price to pay is high: abandoning the linearity of the quantum description. But it may be fruitful. New interference effects are indeed predicted that seem to be related to an "α-memory" in resonant energy-exchange nonlinear processes. If these later can be experimentally tested (we provide corresponding relevant parameters in the case of GaAs QDH), the crucial question raised by the present work—namely, which physical effect, if any, is described by the probability amplitude $\langle u_1|u_3\rangle$?— would be positively answered: $\langle u_1|u_3\rangle$ would then provide a nonlinear and, remarkably enough, non-relativistic definition of the numerical value of the QED coupling constant $\sqrt{\alpha}$.

At the present stage of the numerical investigations, we find $\sigma = 1/137.17$, see Eq. (23). It seems hard to ascribe all nonlinear quantum properties that yield this result to mere numerical coincidences. On the other hand, there is definitely no velocity of light c in the original differential system Eqs. (3–4) and its subsequent solution. Therefore, if numerical coincidences are discarded, α would then be a universal number, say of the Feigenbaum type in period-doubling transition to deterministic chaos, that would be related to some fundamental nonlinear quantum properties which the present particular model—quite fortunately actually—happened to hit. Then the numerical values of the elementary charge e, Planck's constant h and c would be related by this number.

This discovery would echo Feynman's concern for both further research concerning α's numerical value recalled in section 1 and fears about QED's mathematical self-consistency: "What *is* certain is that we do not have a good mathematical way to describe the theory of quantum electrodynamics..." [1]. By pointing out the specific role of *ab-initio* nonlinear quantum effects [11,14,17] as well as by using them in the present differential description of a 2D harmonically-trapped condensed Bose gas of charged particles, it is hoped that such a particular nonlinear quantum system might help progressing in the research trend advocated by Feynman [1], namely, "all good theoretical physicists should put this number [α] up on their wall and worry about it".

Acknowledgments

The author is deeply grateful to the late P. Valiron. He thanks J. Bec, V. Gudmundsson, P. Y. Longaretti, Ph. Nozières, Th. Passot, D. Schuch and X. Waintal for helpful discussions as well as useful critical comments. He acknowledges J. Bec's double-checking the numerical procedures and D. Kamm's expert technical assistance. Finally, he feels indebted to W. Kohn: his '96 UCSB "brown-bag seminars" initiated indeed the author's subsequent work concerning nonlinear eigenstates (-values) in N-particle interacting quantum systems.

References

[1] Feynman R P 1985 *QED: the strange theory of light and matter* Chap. 4 (Princeton: University Press)
[2] Sikorski C and Merkt U 1989 *Phys. Rev. Lett.* **62** 2164; Sikorski C and Merkt U 1990 *Phys. Rev. Lett.* **64** 3100
[3] Maksym P A and Chakraborty T 1900 *Phys. Rev. Lett.* **65** 108
[4] Ashoori R C, Stormer H L, Weiner J S, Pfeiffer L N, Baldwin K W and West K W 1993 Phys. Rev. Lett. **71** 613
[5] Merkt U, Huser J and Wagner M 1991 *Phys. Rev.* B **43** 7320
[6] Pfannkuche D, Gudmundsson V and Maksym P A 1993 *Phys. Rev.* B **47** 2244
[7] Maksym P A 1996 *Phys. Rev.* B **53** 10871
[8] Wagner M, Merkt U and Chaplik A V 1992 *Phys. Rev.* B **45** 1951
[9] Dineykhan M and Nazmitdinov R G 1997 *Phys. Rev.* B **55** 13707
[10] Hallam L D, Weis J and Maksym P A 1996 *Phys. Rev.* B **53** 1452
[11] Reinisch G 2004 *Phys. Rev.* A **70** 033613
[12] Bec J 2010 private communication
[13] see for instance Landau L and Lifchitz E 1989 *Électrodynamique quantique* 2nd Ed. (Moscow: MIR)
[14] Reinisch G, de Freitas Pacheco J and Valiron P 2001 *Phys. Rev.* A **63** 042505
[15] Park D 1990 *Classical Dynamics and its Quantum Analogues* 2nd Ed., Sect. 4.5 (Berlin: Springer)
[16] Szabo A and Ostlund N S 1996 *Modern Quantum Chemistry: Introduction to Advanced Electronic Structure Theory* (New York: Dover)
[17] Reinisch G 2007 *Phys. Rev. Lett.* **99** 120402
[18] Mandl F 1959 *Introduction to Quantum Field Theory* Chap. 17 (London: Interscience)
[19] Cohen-Tannoudji C, Dupont-Roc Jand Grynberg G 1988 *Processus d'intéraction entre photons et atomes* (Paris: InterEditions/Editions du CNRS)
[20] MathWorks 2004 *MATLAB* Version 7.0.1.24704 (R14), options = odeset('RelTol',1e-10) (MathWorks)

Journal of Physics: Conference Series **237** (2010) 012020 doi:10.1088/1742-6596/237/1/012020

Pythagorean quantization, action(s) and the arrow of time

Dieter Schuch

Institut für Theoretische Physik, Goethe-Universität Frankfurt am Main, Max-von-Laue-Str. 1, D-60438 Frankfurt am Main, Germany

E-mail: `schuch@em.uni-frankfurt.de`

Abstract. Searching for the first well-documented attempts of introducing some kind of "quantization" into the description of nature inevitably leads to the ancient Greeks, in particular Plato and Pythagoras. The question of finding the so-called Pythagorean triples, i.e., right-angled triangles with integer length of all three sides, is, surprisingly, connected with complex nonlinear Riccati equations that occur in time-dependent quantum mechanics. The complex Riccati equation together with the usual Newtonian equation of the system, leads to a dynamical invariant with the dimension of an action. The relation between this invariant and a conserved "angular momentum" for the motion in the complex plane will be determined. The "Pythagorean quantization" shows similarities with the quantum Hall effect and leads to an interpretation of Sommerfeld's fine structure constant that involves another quantum of action, the "least Coulombic action" e^2/c. Since natural evolution is characterized by irreversibility and dissipation, the question of how these aspects can be incorporated into a quantum mechanical description arises. Two effective approaches that also both possess a dynamical invariant (like the one mentioned above) will be discussed. One uses an explicitly time-dependent (linear) Hamiltonian, whereas the other leads to a nonlinear Schrödinger equation with complex logarithmic nonlinearity. Both approaches can be transformed into each other via a non-unitary transformation that involves Schrödinger's original definition of a (complex) action via the wave function.

1. Introduction

Looking for the first well-documented attempts of introducing some kind of "quantization" into the description of nature inevitably leads to the ancient Greeks, in particular, to Democritus, Plato and Pythagoras. The question of finding so-called Pythagorean triple, i.e. right(-angled) triangles where the length of all sides are integers, is surprisingly connected with complex nonlinear Riccati equations that occur in quantum mechanics. This Riccati equation, together with the Newtonian equation of motion for the respective system, leads to a dynamical invariant that has (almost) the dimension of an *action*. This is, insofar, remarkable since the physical quantity that is actually quantized in nature is *action* – Plancks constant (divided by 2π) $\hbar$ is a quantum of action. The often-quoted quantization of energy is only a consequence of multiplying this quantum of action by a constant with the dimension of a frequency. Another physical quantity with the dimension of *action* is *angular momentum* which, in quantum mechanics, is also quantized in terms of $\hbar$. But, like action, also angular momentum cannot be measured directly since it is a product of two measurable quantities (in this case, position and momentum).

In this context, it is interesting to state that even for systems where the Hamiltonian is no longer invariant in time (e.g., a harmonic oscillator (HO) with time-dependent frequency $\omega(t)$), or where not even a conventional Hamiltonian exists (e.g., for dissipative systems with (linear) velocity dependent friction forces), invariants still exist that have the dimension of action and are related to some kind of angular momentum.

In this paper, the following questions shall be essentially considered:

1. In the history of science and mathematics, what are the first know attempts at "quantization" and do they have any manifestation in modern (quantum) physics?

2. Apart from angular momentum, are there other quantities in physics that have the dimension of *action* and are they of any significant importance (e.g., as a constant of motion, etc.)?

3. Apart from $\hbar$, are there other quanta of action in physics? What is their physical meaning and where do they occur?

4. In cases where no conserved (physical) Hamiltonian exists (e.g., in dissipative systems), are there conserved action-type quantities and what is their meaning?

2. Pythagorean quantization and its manifestation in modern physics

The idea of dividing our world into the smallest components that are not further divisible is well known from the philosophy of Democritus (ca. 460 - ca. 371 BC). This led to the word *atom* that is still used to describe the building blocks of our physical world that, for some time, seemed further indivisible.

A similar idea was formulated in a more abstract way by another equally famous Greek philosopher living around the same time. In his work *Timaios*, Plato (428/27 - 348/47 BC) gives his view of how the world is built up in terms of right triangles. Werner Heisenberg, who was equally fascinated with and puzzled by this text, summarizes this idea in his book "Der Teil und das Ganze" [1] in the general sense as follows:"Matter is made up of right triangles which, after being paired to form isosceles triangles or squares, are just joined together to build the regular bodies of stereometry: cube, tetrahedron, octahedron and icosahedrons. These four solids are then supposed to be the basic units of the four elements earth, fire, air and water."

Even some 2000 years later this idea of Platonic polyhedral fascinated Johannes Kepler (1571-1630) so much that, in his quest for harmony in nature, he tried to explain the orbits of the planets in our solar system by fitting one polyhedron into another so that the radii of spheres enveloping this polyhedra would just represent the mean distance of the planets from the sun.

An aspect of quantization was brought into this picture by Titius of Wittenberg (1729-1796) and Johann Elert Bode (1747-1826) who proposed a series of numbers (integers!) that predicts the position of the planets (Titius–Bode law), showing similarities with Bohrs model of the atom. (Remarkably, a new formulation of this law has been found by Reinisch [2] using a nonlinear formulation of a (formal) macroscopic Schrödinger equation (SE) to describe the solar system. This nonlinear formalism is equivalent to the one that will be presented in section 2.1 in the context of time-dependent quantum mechanics.)

Kepler tried to connect the geometry of the planetary orbits and movement of the planets with some kind of imaginary sounds – the music of the spheres. This takes us back to the ancient Greeks and the right triangles.

The Greek philosopher Pythagoras lived around 570-500 BC. In todays society, even if (almost) nothing from our early mathematical education survived, most people can (hopefully) at least still recall the theorem named after him and might even be able to quote it as $a^2 + b^2 = c^2$ where a and b are the catheti and c the hypotenuse of a right triangle. Pythagoras and his disciples were well-known for their dogma "everything is number", with number meaning *integer*. They applied it to develop a musical scale (see Kepler's music of the spheres) and also to the right triangle. So, the Pythagorean triple are three integers denoting the lengths of the three sides of a right-angled triangle, thus fulfilling Pythagoras' theorem. The most common example is $(3, 4, 5)$ with $9 + 16 = 25$. But, asked for a few more examples of the kind, even people accustomed to mathematics have difficulties finding any – although an infinite number of triples exist! Moreover, there is even a rather simple rule to finding these triples. This rule was probably already known in Mesopotamia around 2000 BC, but at least Diophantus of Alexandria (around 250 AC) knew it [3]. [1]

That this rule is also somehow present in modern physics, particularly in quantum theory, will be discussed in the next sub-section.

2.1. Time-dependent quantum mechanics and dynamical invariants

In the following, it shall be shown how time-dependent quantum mechanics can be related to the "Pythagorean quantization" mentioned above. For this purpose, one-dimensional problems with exact analytic solutions in the form of Gaussian wave packets will be considered. This kind of solutions are obtained, e.g., for the free motion (potential $V(x) = 0$) and the harmonic oscillator (HO; $V = \frac{m}{2}\omega^2 x^2$) with constant frequency of oscillation, $\omega = \omega_0$, or time-dependent frequency, $\omega = \omega(t)$. In these cases, the solution of the time-dependent Schrödinger equation (SE)

$$i\hbar\frac{\partial}{\partial t}\Psi(x,t) \;=\; \left\{-\frac{\hbar^2}{2m}\frac{\partial^2}{\partial x^2} + \frac{m}{2}\omega^2 x^2\right\}\Psi(x,t) \tag{1}$$

can be written as

$$\Psi(x,t) \;=\; \Psi_{WP,L}(x,t) \;=\; N(t)\exp\left\{i\left[y(t)\tilde{x}^2 + \frac{<p>}{\hbar}\tilde{x} + K(t)\right]\right\} \tag{2}$$

with the shifted coordinate $\tilde{x} = x- < x >= x - \eta(t)$, where the mean value $< x >= \int_{-\infty}^{+\infty}\Psi^* x\Psi dx = \eta(t)$ corresponds to the classical trajectory, $< p >= m\dot{\eta}$ represents the classical momentum and the coefficient of the quadratic term in the exponent, $y(t) = y_R(t) + iy_I(t)$, is a complex function of time. The (possibly) time-dependent normalization factor $N(t)$ and the purely time-dependent function $K(t)$ in the exponent are not relevant to the following discussion in this section.

The parameters $y(t)$ and $\eta(t)$ have been introduced in the above definition of the Gaussian wave packet because the equations of motion for $\left(\frac{2\hbar}{m}y\right)$ and $\eta(t)$ will be the important ones in what follows. Inserting the wave packet (2) into the SE (1) yields these two equations, where the one for $\eta(t)$,

$$\ddot{\eta} + \omega^2\eta = 0 \;, \tag{3}$$

(overdots denote derivatives with respect to time; the case $V = 0$, in the following, is always obtained in the limit $\omega \to 0$), simply conveys that the maximum of the wave packet, located at $x =< x >= \eta(t)$, follows the classical trajectory. The equation for the quantity $y(t)$ has the form a of a (complex) nonlinear (NL) Riccati equation and describes the time-dependence of the wave packet width that is related with the position uncertainty. This can be seen from the fact that $y_I = \frac{1}{4<\tilde{x}^2>}$ with $< \tilde{x}^2 >=< x^2 > - < x >^2$ being the mean square deviation of position. Defining the complex quantity $\mathcal{C}(t) = \left(\frac{2\hbar}{m}y\right) = \mathcal{R} + i\mathcal{I}$, the Riccati equation can be written in the form

$$\frac{d}{dt}\mathcal{C} + \mathcal{C}^2 + \omega^2 \;=\; 0 \;. \tag{4}$$

This nonlinear complex equation will be the link to our "Pythagorean quantization" as well as to an action-type invariant.

[1] Fermat claimed to have found an elegant proof that for a, b and c being integers, the relation $a^n + b^n = c^n$ cannot be fulfilled for $n > 2$ but the margin of Diophantus' book was too small to write it down. This proof is still missing!

To show this, a new (real) variable $\alpha(t)$ is introduced via $\mathcal{I} = \left(\frac{2\hbar}{m} y_I\right) = \frac{1}{\alpha^2}$ (from which it follows that $\alpha = \left(\frac{2m<\tilde{x}^2(t)>}{\hbar}\right)^{1/2}$). Inserting this into the imaginary part of Eq. (4) allows one to determine $\mathcal{R}(t)$ as $\mathcal{R} = \frac{\dot{\alpha}}{\alpha}$, which, when inserted into the real part of (4) together with the above definition of $\mathcal{I}$, finally turns the complex Riccati equation into the real nonlinear so-called Ermakov equation for $\alpha(t)$

$$\ddot{\alpha} + \omega^2 \alpha = \frac{1}{\alpha^3} \ . \tag{5}$$

It had been shown by Ermakov [4] in 1880, i.e 35 years before quantum mechanics was formulated by Schrödinger and Heisenberg, that from the pair of equations (3) and (5), coupled via ω^2, by eliminating ω^2 from the equations, a dynamical invariant, the Ermakov-invariant

$$I_L = \frac{1}{2}\left[(\dot{\eta}\alpha - \eta\dot{\alpha})^2 + \left(\frac{\eta}{\alpha}\right)^2\right] = \text{const} \ . \tag{6}$$

can be obtained (this invariant had been rediscovered by several authors, also in a quantum mechanical context; see,e.g., [5, 6, 7]).

Two remarkable properties of this invariant are i) it is also a constant of motion for $\omega = \omega(t)$, i.e., in the case where the corresponding Hamiltonian does not have this property and ii) apart from a missing constant factor proportional to mass m, it has the dimension of an *action*, not of an energy. This missing factor m can be explained by the derivation of the invariant based on the *mathematical* properties of Eqs. (3) and (5). In a physical context, Eq. (3) would have to be replaced by the corresponding Newtonian equation of motion, i.e., by Eq. (3) multiplied by the missing factor m. Therefore, in the following we can assume that in a physical context the invariant has the dimension of an *action*. Furthermore, as will be shown below, an invariant of this type also exists for certain dissipative systems, i.e. systems for which a conventional Hamiltonian doe not even exist.

In order to make the connection with the Pythagorean triple and also to make the corresponding physical interpretation more transparent, the complex Riccati equation (4) shall be linearized using the ansatz

$$\mathcal{C} = \mathcal{R} + i\,\mathcal{I} = \frac{\dot{\lambda}}{\lambda} \ , \tag{7}$$

with complex $\lambda(t)$, leading to

$$\ddot{\lambda} + \omega^2 \lambda = 0 \ , \tag{8}$$

which has the form of the Newtonian equation (3) of the corresponding problem, but now for a complex variable.

First, a kind of geometric interpretation of the motion of λ in the complex plane shall be given. Expressed in Cartesian coordinates, λ can be written as $\lambda = u + iz$, or in polar coordinates as $\lambda = \alpha e^{i\varphi}$. Inserting the polar form into Eq. (7) leads to

$$\mathcal{C} = \frac{\dot{\alpha}}{\alpha} + i\,\dot{\varphi} \ , \tag{9}$$

where the real part is already identical to $\mathcal{R}$, as defined above.

The quantity α defined above in $\mathcal{I}$ as being proportional to the position uncertainty is identical to the absolute value of λ, if it can be shown that

$$\dot{\varphi} = \frac{1}{\alpha^2} \ . \tag{10}$$

This, however, can simply be proven by inserting real and imaginary parts of (9) into the imaginary part of the Riccati equation (4). Comparing relation (10), that can also be written in the form

$$\dot{z}u - \dot{u}z = \alpha^2 \dot{\varphi} = 1 \ , \tag{11}$$

with the motion of a particle under the influence of a central force in two-dimensional physical space, it shows that this relation corresponds to the "conservation of angular momentum", but here for the motion in the complex plane!

In the following, the case $V = 0$ shall be considered, turning the inhomogeneous Riccati equation (4) into the homogenous Bernoulli equation

$$\frac{d}{dt}\mathcal{C} + \mathcal{C}^2 = 0 \tag{12}$$

(in general, it is always possible to transform the inhomogeneous Riccati equation into a homogeneous Bernoulli equation if one particular solution of the Riccati equation is known).

From the definition of $\mathcal{R}$ and $\mathcal{I}$, it is obvious that

1) $\mathcal{R} = \frac{\dot{\alpha}}{\alpha}$ is related to the (relative) change in time of the *radius* of the motion of λ. In particular, for $\frac{\dot{\alpha}}{\alpha} = k =$ const., the radius changes exponentially according to $\alpha = \alpha_0 e^{kt}$, i.e., this radial change involves the number "e".

2) $\mathcal{I} = \frac{1}{\alpha^2} = \dot{\varphi}$ is related to the angular change of λ. In particular, for $\alpha =$ const., a constant angular velocity $\dot{\varphi} = \omega_0$ follows. The angular change involves the number "π".

The (negative) temporal changes of real and imaginary parts of $\mathcal{C}$ are given by

$$\begin{aligned} -\frac{d}{dt}\mathcal{R} &= \mathcal{R}^2 - \mathcal{I}^2 = \Re\{\mathcal{C}^2\} \\ -\frac{d}{dt}\mathcal{I} &= 2\,\mathcal{R}\mathcal{I} = \Im\{\mathcal{C}^2\} \ . \end{aligned} \tag{13}$$

Now, in the complex plane, the absolute value of $\mathcal{C}^2$, its real part and its imaginary part form a right triangle with the hypotenuse $\mid \mathcal{C}^2 \mid = \mathcal{R}^2 + \mathcal{I}^2$, i.e. the "length" of $\mathcal{C}^2$ and the two catheti $\Re\{\mathcal{C}^2\} = \mathcal{R}^2 - \mathcal{I}^2$ and $\Im\{\mathcal{C}^2\} = 2\mathcal{R}\mathcal{I}$.

For integers $\mathcal{R}$ and $\mathcal{I}$ (with $\mathcal{R} > \mathcal{I}$), this, however, supplies us with the desired rule for the construction of the Pythagorean triple since all possible triples can be obtained if $\mathcal{R}$ and $\mathcal{I}$ take all possible integer values. In this case, the three sides of the right triangle will obviously always also be multiples of integers. For our dynamical system, this also entails that the changes of our (integer) quantities $\mathcal{R}$ and $\mathcal{I}$, $\frac{d}{dt}\mathcal{R}$ and $\frac{d}{dt}\mathcal{I}$, will also be expressed in terms of integers, i.e., a "quantized" dynamics. This appears particularly interesting since complex Riccati equations and corresponding Ermakov systems occur in many different areas of physics, not only in time-dependent quantum mechanics, but also in time-independent QM, quantum optics, classical optics, Bose-Einstein condensates, systems in nonlinear dynamics, cosmological models and many more. So in all these systems, intrinsically, some kind of quantization is possible. (Relations to the complex quadratic family as it occurs in the theory of fractals are also obvious and will be discussed elsewhere.)

At this point, only one further aspect shall be mentioned. Considering the time-dependent SE in momentum space [8], similar Gaussian wave packet solutions can be obtained where the time-dependence of the momentum uncertainty is again determined by a complex Riccati equation, the complex variable now being just the inverse of the one in position space, i.e., $\mathcal{C}^{-1} = \frac{\lambda}{\dot{\lambda}}$, fulfilling,

$$\frac{d}{dt}(\mathcal{C}^{-1}) - \omega^2(\mathcal{C}^{-1})^2 - 1 = 0 \ . \tag{14}$$

In particular, for $V = 0$ (s. Bernoulli equation in position space), one obtains

$$\mathcal{C}^{-1} = t - i\,\alpha_0^2 \tag{15}$$

with $\alpha_0^2 = \frac{\hbar}{2m} < \tilde{x}^2 > (t = 0) = \text{const.}$, i.e., a kind of *complex time-variable*, where the imaginary part is related to the initial position uncertainty. In this context, it should be mentioned that the possibility of complexification of physical quantities (also in the context of dissipative systems or $\mathcal{PT}$-symmetry) has recently gained growing interest (some examples can also be found in other contributions to these proceedings; see also [9, 10, 11, 12, 13]).

A quantum phenomenon where the real part of $\mathcal{C}$, i.e. $\mathcal{R} = \frac{\dot{\alpha}}{\alpha}$, and its possible quantization play a role will be presented in section 3.

2.2. On the physical interpretation of the Ermakov invariant

It has been shown in section 2.1. that the position uncertainty can be expressed in terms of $\alpha(t)$. It is straightforward to show (see , e.g., [14]) that, similarly, the momentum uncertainty $< \tilde{p}^2 >=< p^2 > - < p >^2$ can also be expressed in terms of α, $\dot{\alpha}$ and $\dot{\varphi}$.

This finally allows one to write down the quantum mechanical contribution to the energy of the wave packet (2) (for the HO), which is identical to the ground state energy $\tilde{E} = \frac{\hbar}{2}\omega$, in the form [15]

$$\tilde{E} = \frac{< \tilde{p}^2 >}{2m} + \frac{m}{2}\omega^2 < \tilde{x}^2 > = \frac{\hbar}{4}\left\{\dot{\alpha}^2 + \alpha^2\dot{\varphi}^2 + \omega^2\alpha^2\right\}\,. \tag{16}$$

In search of a Hamiltonian formalism for the uncertainties similar to the classical Hamiltonian formalism for position and momentum, one can try using the difference $\tilde{\mathcal{L}} = \frac{<\tilde{p}^2>}{2m} - \frac{m}{2}\omega^2 < \tilde{x}^2 >$, expressed in terms of α, $\dot{\alpha}$ and $\dot{\varphi}$ and as a kind of Lagrangian to determine the corresponding canonical momenta, leading to

$$\frac{\partial\tilde{\mathcal{L}}}{\partial\dot{\alpha}} = p_\alpha = \frac{\hbar}{2}\dot{\alpha} \text{ and } \frac{\partial\tilde{\mathcal{L}}}{\partial\dot{\varphi}} = p_\varphi = \frac{\hbar}{2}\alpha^2\dot{\varphi} = \frac{\hbar}{2}\,! \tag{17}$$

So, the ground state energy $\tilde{E}$ can finally be written as a Hamiltonian in the form

$$\tilde{\mathcal{H}}(\alpha, p_\alpha, \varphi, p_\varphi) = \frac{p_\alpha^2}{\hbar} + \frac{p_\varphi^2}{\hbar\alpha^2} + \frac{\hbar}{4}\omega^2\alpha^2 \tag{18}$$

which yields the correct Hamiltonian equation of motion equivalent to the Ermakov equation (5) for $\alpha(t)$.

It shall only be mentioned here (for details see [16]) that the time-dependent Green function (also called Feynman kernel) for the time-propagation of the initial wave packet corresponding to (2) can be given in terms of u and z, the real and imaginary parts of $\lambda(t)$. Comparing the resulting time-dependent wave packet with the form used in section 2.1. shows that u and z can be expressed in terms of η, $\dot{\eta}$, α and $\dot{\alpha}$. In particular,

$$z = \frac{m}{\alpha_0 p_0}\eta(t) = \alpha\,\sin\varphi \tag{19}$$

can be verified.

Rewriting the Ermakov invariant in terms of z and $\dot{z}$, instead of η and $\dot{\eta}$, yields

$$I_L = \frac{1}{2}\left(\frac{\alpha_0 p_0}{m}\right)^2\left[\left(\dot{z}\alpha - z\dot{\alpha}\right)^2 + \left(\frac{z}{\alpha}\right)^2\right] = \text{const.}\quad . \tag{20}$$

Since $(\frac{z}{\alpha})^2 = \sin^2\varphi$, it is necessary for I_L to be constant that $(\dot{z}\alpha - z\dot{\alpha})^2 = (\frac{u}{\alpha})^2 = \cos^2\varphi$. Therefore, it follows that

$$u = \dot{z}\alpha^2 - z\dot{\alpha}\alpha = \left(\frac{m}{\alpha_0 p_0}\right)\alpha^2\left(\dot{\eta} - \frac{\dot{\alpha}}{\alpha}\eta\right) \tag{21}$$

and

$$I_L = \frac{1}{2}\left(\frac{p_0\alpha_0}{m}\right)^2 , \tag{22}$$

i.e., the Ermakov invariant corresponds (apart from the constant factor $\frac{1}{m}$) to the initial energy (for $\eta_0 = 0$) times the imaginary part of the "complex time" (15).

As we have seen before, there are two conserved quantities in the Ermakov systems; one is the Ermakov invariant (6), the other is given by the relation (11), $\dot{z}u - \dot{u}z = 1$, connected with the conservation of angular momentum in the complex plane.

Expressing now u, z, $\dot{u}$ and $\dot{z}$ according to Eqs. (19) and (21) in terms of η, $\dot{\eta}$, α and $\dot{\alpha}$ and using the Eqs. (3) and (5) for $\eta(t)$ and $\alpha(t)$, relation (11) can be rewritten in the form

$$\begin{aligned} \dot{z}u - \dot{u}z &= \left(\frac{m}{\alpha_0 p_0}\right)^2 \left[\dot{\eta}^2\alpha^2 - 2\eta\dot{\eta}\alpha\dot{\alpha} + \eta^2\left(\dot{\alpha}^2 + \frac{1}{\alpha^2}\right)\right] \\ &= 2\left(\frac{m}{\alpha_0 p_0}\right)^2 I_L , \end{aligned} \tag{23}$$

i.e., the action-type invariant I_L is (up to a constant factor) identical to the conserved "angular momentum" for the motion of $\lambda(t)$ in the complex plane.

3. Currents, (fractional) Quantum Hall Effect and more action(s)

The probability density $\varrho = \Psi^*\Psi$ corresponding to the solution of the time-dependent SE (1) obeys the continuity equation

$$\frac{\partial}{\partial t}\varrho + \frac{\partial}{\partial x}(\varrho\, v_-) = 0 \tag{24}$$

with the probability current density $j = \varrho v_-$ and the velocity field $v_- = \frac{\hbar}{2mi}\left(\frac{\frac{\partial}{\partial x}\Psi}{\Psi} - \frac{\frac{\partial}{\partial x}\Psi^*}{\Psi^*}\right)$. For Gaussian wave packets, this can be written as

$$v_- = \dot{\eta} + \frac{\dot{\alpha}}{\alpha}\,\tilde{x} \tag{25}$$

with the classical contribution $\dot{\eta} = < v >$.

For $\dot{\alpha} = 0$, i.e. constant wave packet width, there is no quantum mechanical contribution to the current. However, for $\dot{\alpha} \neq 0$, typical quantum effects like tunnelling or Quantum Hall Effect (QHE) may occur. [2]

In the QHE, the current j (here only one-dimension shall be considered)

$$j = \varrho\, v_- \propto \sigma\, E \tag{26}$$

[2] The contribution $\frac{\dot{\alpha}}{\alpha}\,\tilde{x}$ to the velocity should not be mistaken for the so-called "osmotic velocity" $u = \frac{\hbar}{2m}\frac{\frac{\partial}{\partial x}\varrho}{\varrho}$ applied in some alternative interpretations of quantum mechanics. In the case of our wave packet, the osmotic velocity would be $u = -\frac{1}{\alpha^2}\tilde{x} = -\dot{\varphi}\tilde{x}$ and therefore not be related to the radial change of λ, $\frac{\dot{\alpha}}{\alpha}$, but to the angular change $\dot{\varphi}$.

is proportional to the electric field strength E and the conductivity σ with

$$\sigma = n\, c \left(\frac{e^2}{c\hbar}\right) = n\, c\, \alpha_{Som} \tag{27}$$

where n = integer, c = velocity of light and $\alpha_{Som} = \left(\frac{e^2}{c\hbar}\right) \approx 1/137$ is Sommerfeld's fine structure constant (not to be confused with $\alpha(t)$ as defined in section 2!).

For fixed position of the system's centre of mass, $\dot{\eta} = 0$, the conductivity or current displays the following proportionality:

$$j \propto n \left(\frac{e^2}{c\hbar}\right) \propto \frac{\dot{\alpha}}{\alpha}\,, \tag{28}$$

therefore changes of j by integers n would correspond to integer changes of $\frac{\dot{\alpha}}{\alpha}$, a possibility discussed in section 2.1..

Apart from the QHE, there is also a fractional QHE observed where

$$j \propto \frac{n}{k} \left(\frac{e^2}{c\hbar}\right) \tag{29}$$

with n, k both being integers. There are different interpretations of this effect in the literature and not all possible combinations of n and k (being odd, even or both) have been observed. To my knowledge, none of the existing theories is able to explain all observed combinations of n and k or even predict yet unobserved ones. Therefore, there might still be room for some unknown in the theory of the fractional QHE.

One interpretation that had been given by the author and colleagues [17] some time ago assumes that the fractional QHE is a manifestation of a phenomenon where two effects are involved, each being quantized separately. Since it was already stated in the introduction that *action* is the quantity that is usually quantized in nature, we assumed that in the fractional QHE one of the effects is quantized in terms of $\hbar$, while the other is in terms of $\frac{e^2}{c}$, so the ration of both is just Sommerfelds fine structure constant $\alpha_{Som} = \frac{e^2}{c}/\hbar$.

Certainly the action $\hbar$ is well known in quantum mechanics; however, the action $\frac{e^2}{c}$ was also already mentioned by Einstein [18], Schrödinger [19] and Eddington [20], assuming its importance but not knowing its physical relevance. Our interpretation of the fractional QHE is, therefore, that this is the first observed physical effect that shows the relevance of a second quantum of action $\frac{e^2}{c}$.

In [17], we called $\frac{e^2}{c}$ the "quantum of electrostatic action". Why this name? For two electrons interacting via Coulomb's law, the electrostatic action involved is

$$\text{action} = \text{energy} \times \text{time} = \frac{e^2}{r} \times t\,. \tag{30}$$

Since the distance between the electrons, r, and their interaction time, t, can be related via the interaction velocity $v = r/t$, relation (30) can be written as

$$\text{action} = \frac{e^2}{v}\,. \tag{31}$$

This action becomes a minimum when the velocity attains its maximum value $v = c$ = velocity of light. So, the least Coulombic action is $\frac{e^2}{c}$.

But then the question arises, since nature is usually very economical, as expressed by the various extremal principles, e.g., in optics, mechanics, etc., why does it allow the luxury of two elementary quanta of action? Of course, no definite answer shall be given at this point; but the following remarks, based on the aforementioned shall be allowed. Planck's constant (divided by 2π), $\hbar$, is well known particularly in quantum mechanics of oscillating systems, where, e.g., the energy is quantized in terms of $\hbar\omega = \hbar\dot{\varphi}$, i.e., the frequency is related to some kind of angular velocity and thus to some *angular* aspect. On the other hand, it has just been shown that $\frac{e^2}{c}$ is somehow related with a physical situation when a *radial* distance plays the important role.

As radius and circumference of a circle cannot be expressed in terms of the same units ("quanta") since they are related via the irrational number π, maybe the "radial" *action* and the "angular" *action* also cannot be expressed in terms of the same fundamental units.

As seen in our discussion of integer value for $\frac{\dot{\alpha}}{\alpha}$ and $\dot{\varphi}$ related to $\left(\frac{2\hbar}{m}y\right) = \frac{\dot{\lambda}}{\lambda}$, also there, the radial aspect is for integer values of $\frac{\dot{\alpha}}{\alpha}$ connected with $e = 2.718...$, whereas the angular aspect $\dot{\varphi} = \frac{1}{\alpha^2}$ is connected with $\pi = 3.141...$. So, maybe $\frac{e^2}{c}$ and $\hbar$ just offer the possibility to quantize in a (complex ?) two-dimensional (phase ?) space radial changes as well as angular ones.

4. Irreversibility (the "arrow of time") and dissipation

In classical Hamiltonian mechanics, the time-evolution of a dynamical system can be described by canonical transformations that leave the algebraic structure based on the Poisson brackets unchanged and (usually) conserve the energy of the system. (For systems with explicitly time-dependent Hamiltonians, e.g. HOs with time-dependent frequency $\omega(t)$, the Hamiltonian is no longer a constant of motion, but the energy is still not lost due to a dissipative force.)

In quantum mechanics, the time-evolution of a corresponding quantum system can be described by unitary transformations which, essentially, represent a pure rotation of the wave function or state vector in an abstract Hilbert space with conserved norm, i.e., length of the vector. [3]

Considering natural evolution processes as observed in the macroscopic world around us, growth processes (in two or more dimensions) are usually also connected with some kind of radial changes. So the shell of a nautilus or the horns of a ram not only undergo angular changes when growing, but also radial ones leading to forms like (often logarithmic) spirals.

So, not only rotation but, also, radial expansion (or contraction) are essential elements of these kinds of irreversible evolutions thus offering also the possibility of defining something like a direction of time (e.g., larger radius = later in time). Another characteristic feature of macroscopic evolution processes is the loss of energy by transfer from the dynamical system to the environment due to dissipation, transforming (mechanical) energy into heat.

Irreversibility and dissipation are usually not elements contained in the description of a single system in terms of the conventional Hamiltonian formalism. In the following, it shall be shown how these essential elements of nature might be incorporated into this formalism (classical and quantum mechanical) and what price must be paid for this. It should also be mentioned here briefly that irreversibility and dissipation often occur simultaneously and are often also not independent of each other (see, e.g., fluctuation-dissipative theorems, etc.). However, this coupling is not necessarily always the case. There are, for instance, irreversible processes possible without any dissipation of energy and we will mention such cases subsequently.

4.1. Irreversible aspects in conventional quantum mechanics

One example of an irreversible behaviour of a quantum system is well known from the textbooks but not usually recognized as such.

[3] Due to the probabilistic interpretation of the wave function this normalizability condition has to be fulfilled, but still irreversible effects can occur that would cause a change of the length of the state vector if they would not be compensated by some kind of "renormalization"; for an example in conventional quantum mechanics, see section 4.1.

Considering the time-evolution of a dynamical system, by "reversible" we usually mean that we cannot distinguish between the motion of the system going forward or backward in time. This is reflected in the corresponding equation of motion where, under change of t into $-t$ (and consequently also changing the signs of velocities or momenta), the form of these equations does not change. This also applies to quantum mechanical equations where one must take into account that time-reversal also requires the change $i = \sqrt{-1}$ into $-i$ in these equations (which, in position space, automatically changes p into $-p$).

Taking a motion picture, e.g., of a classical HO, invariance under time-reversal implies that if a spectator watching the film closes the eyes in the same moment the film is stopped and ran backwards, on opening the eyes he/she will not be able to tell if the film is running forward or backward. The same would apply if one would show a film of the motion of the Gaussian wave packet (with constant width) that is usually given in the textbooks as solution of the SE for the HO.

However, if one would try the same with the very basic example of the Gaussian wave packet solution of the SE for the free motion, $V = 0$, the situation would change drastically. Due to the spreading of the wave packet width, it would immediately be possible to distinguish between the film running forward or backward in time; broadening width means forward, shrinking width means backward in time.

But how is this distinction possible although we know that the SE is invariant under time reversal? Obviously, the reason for this discrepancy is somehow connected with the temporal behaviour of the wave packet width, the typical quantum mechanical aspect of the system (as previously mentioned in section 3). Indeed, it is straightforward to show that for wave packets with time-dependent width, i.e., $\alpha(t) \neq$ cont., two *imaginary* contributions result when the kinetic energy operator is applied to the corresponding wave packet, $i\{-\frac{\hbar}{2}\frac{\dot{\alpha}\alpha}{+}m\frac{1}{\alpha^2}\left(\dot{\eta}+\frac{\dot{\alpha}}{\alpha}\tilde{x}\right)\tilde{x}\}$. Shouldn't these imaginary contributions cause problems with the physical interpretation of the kinetic energy? Obviously not since the mean value of the sum of these two terms vanishes. How about the unitarity of the time-evolution and the normalizability of the wave packet? The second of the two terms definitely would cause an exponential decay of the wave packet, but, the first term, which is just the negative mean value of the second, compensates for this. This first term, however, is essentially the time-dependent normalization factor of the wave packet. A wave packet with normalization "constant" would decay; this decay, however, is "renormalized" by the purely time-dependent imaginary term $-i\frac{\hbar}{2}\frac{\dot{\alpha}}{\alpha}$.

Another consequence of the irreversible spreading of the wave packet should also briefly be mentioned. Adding a real term, like a potential, to the SE does not change the corresponding continuity equation (24) for the probability density $\varrho = \Psi^*\Psi$. However, any additional imaginary term in the SE will also cause an additional term in the continuity equation (and vice versa, as will be shown below). In particular, the second of the above-mentioned terms occurs in the velocity field $v_- = \dot{\eta} + \frac{\dot{\alpha}}{\alpha}\tilde{x}$ and contains the contribution connected with the quantum aspects of the current (i.e., tunnelling, QHE, etc., see above). In contrast with the corresponding classical situation, for the quantum system now $\frac{\partial}{\partial x}v_- = \frac{\dot{\alpha}}{\alpha} \neq 0$ is valid for time-dependent wave packet width, just representing the first of the two above-mentioned imaginary terms. For classical Hamiltonian systems $\frac{\partial}{\partial x}v_-$ always vanishes! (For further details, see also [21]).

As this rather trivial textbook example already shows, imaginary contributions to the Hamiltonian can still lead to physically acceptable results, providing their mean value vanishes, thus avoiding imaginary contributions to the energy of the system and guaranteeing normalizability of the corresponding wave function. They introduce an element of irreversibility without changing the energy of the system, particularly without dissipation.

4.2. Conventional treatment of dissipation in quantum mechanics

Let us now, however, just turn to this other aspect that is usually not part of the conventional formalism, namely, dissipation. In an attempt to describe dissipative systems within the conventional formalism of classical Hamiltonian mechanics or quantum mechanics, one usually applies the so-called system-plus-reservoir approach, i.e., system of interest and interacting environment together are considered a closed Hamiltonian system. With the help of additional assumptions and approximations (projection techniques, averaging procedures, Markovian limits, etc.), one finally tries to obtain irreversible equations of motion for the relevant subsystem that also properly describe the energy loss, e.g., due to the action of some velocity dependent friction force.

A conventional way of treating dissipative quantum systems uses the (reduced) density matrix/operator approach. In this context, frequently-used models are i) the (linear) coupling of the system to a bath of many HOs (often attributed to Caldeira and Leggett [22, 23]) which, however, can violate the positivity requirement of the density operator [24]; ii) and the application of the so-called Linblad generators [25] (proposed also by Kossakowski et al. [26] at the same time), whose specific form guarantees positivity but the particular choice of the operators occurring therein is usually "guided by intuition" [27] since no obvious physical guideline exists.

From a practical viewpoint, both approaches have the disadvantage that, in numerical calculations, the computational effort scales in the most favourable case at least with N^2 if N is the size of the relevant Hilbert space. Furthermore, for a realistic description, usually a large number of environmental degrees of freedom must be considered, drastically increasing the computational effort and severely limiting the systems that can be treated. In contrast, methods based on pure state wave packets have the chance of scaling in the most favourable case linearly with N.

Therefore, in the following, modified SEs for "one-particle" wave functions/ wave packets shall be considered for the description of dissipative quantum systems. There also seems to be no conceptual objection against the use of pure states for the description of the system of interest since, due to the interaction with the environment, the system itself does not decay, it only changes its dynamical properties like momentum and energy. Therefore, a description of the system alone in terms of a wave functions/ wave packets should be acceptable. The degrees of freedom of the environment will not enter explicitly into our descriptions; only the effect of the environment on the system of interest will be taken into account by the modification of the SE.

4.3. Effective modified Schrödinger equations for the description of dissipative systems

Early attempts at finding classical Lagrangians/Hamiltonians that provide irreversible equations of motion including linear velocity dependent dissipative friction forces and allowing for usual canonical quantization to obtain the corresponding quantum mechanical versions are due to Caldirola [28] and Kanai[29]. They considered the Lagrangian

$$\hat{L}_{CK} = \left[\frac{m}{2}\dot{x}^2 - V(x)\right] e^{\gamma t} \tag{32}$$

yielding, via the Euler-Lagrange equation, the equation of motion

$$m\ddot{x} + m\gamma\dot{x} + \frac{\partial}{\partial x} V(x) = 0 \tag{33}$$

including the desired damping force, with friction constant γ. In the usual way, the canonical momentum $\hat{p}$ is obtained as

$$\frac{\partial}{\partial \dot{x}} \hat{L}_{CK} = m\dot{x}\, e^{\gamma t} = p\, e^{\gamma t} = \hat{p}\ . \tag{34}$$

Note that this canonical momentum $\hat{p}$ is different from the physical kinetic momentum $p = m\dot{x}$ and the transition from the physical variables (x, p) to the canonical variables $(\hat{x} = x, \hat{p} = e^{\gamma t}\, p)$ is given by a *non-canonical* transformation. It is straightforward to obtain the corresponding Hamiltonian $\hat{H}_{CK}(x, \hat{p})$,

$$\hat{H}_{CK} \;=\; e^{-\gamma t}\, \frac{\hat{p}^2}{2m} + e^{\gamma t}\; V\,(x) \tag{35}$$

which also supplies the correct equations of motion including the friction force.

Canonical quantization, i.e., replacing the *canonical* momentum $\hat{p}$ by the operator $\hat{p}_{op} = \frac{\hbar}{i}\frac{\partial}{\partial x}$ and applying the resulting (linear, but explicitly time-dependent) Hamiltonian $\hat{H}_{CK,op}$ to the *canonical* WF $\hat{\Psi}_{CK}\,(x,t)$, leads to the modified SE

$$\begin{aligned} i\hbar\frac{\partial}{\partial t}\;\hat{\Psi}_{CK}\,(x,t) \;&=\; \hat{H}_{CK,op}\;\Psi_{CK}\,(x,t) \\ &=\; \{e^{-\gamma t}\left(-\frac{\hbar^2}{2m}\frac{\partial^2}{\partial x^2}\right) + e^{\gamma t}\; V(x)\}\hat{\Psi}_{CK}\,(x,t)\;. \end{aligned} \tag{36}$$

Also for this SE in the cases considered in this work, i.e., $V = 0$ and HO, it can be shown that exact analytic solutions in the form of Gaussian wave packets exist. The maximum follows the trajectory, determined now by Eq. (33), whereas the wave packets width can again be expressed in terms of a quantity $\alpha(t)$ as $\alpha_{CK} = \left(\frac{2m<\tilde{x}^2(t)>_{CK}}{\hbar}\right)^{1/2}$, obeying the Ermakov-type equation

$$\ddot{\alpha}_{CK} + \gamma\dot{\alpha}_{CK} + \omega^2\alpha_{CK} = \frac{e^{-2\gamma t}}{\alpha_{CK}^3}\quad . \tag{37}$$

Equation (33) together with (37) leads, also in the dissipative case, to an exact Ermakov invariant [30] of the form

$$I_{CK} = \frac{1}{2}\left[e^{2\gamma t}\left(\dot{\eta}\;\alpha_{CK} - \eta\;\dot{\alpha}_{CK}\right)^2 + \left(\frac{\eta}{\alpha_{CK}}\right)^2\right] = \text{const.}\,. \tag{38}$$

However, apparently a problem arises when calculating the uncertainty product of position and *physical* momentum according to

$$U_{CK,phys} =<\tilde{x}^2>_{CK}<\tilde{p}^2>_{CK}=<\tilde{x}^2>_{CK}<\tilde{\hat{p}}^2>_{CK}\to e^{-2\gamma t} \tag{39}$$

where the brackets $< \dots >_{CK}$ denote mean values and the subscript CK indicates that they are calculated with the functions $\hat{\Psi}_{CK}\,(x,t)$ that are solutions of Eq. (36). In relation (39) $p \;=\; \hat{p}\; e^{-\gamma t}$ has been applied to replace the operator for the physical momentum p by the operator for the canonical momentum $\hat{p}$ where the uncertainty product of position and canonical momentum $\hat{p}$ always retains the minimum contribution $\hbar^2/4$. But relation (39) would lead to arbitrarily small values for $U_{CK,phys}$ (smaller than $\hbar^2/4$) and let it even finally vanish for increasing time t.

Therefore, this approach was criticised by several authors [31, 32, 33] and considered inappropriate for the quantum mechanical description of dissipative systems. Nevertheless, others still applied this approach or tried to justify or eliminate [34, 35] the problematic behaviour expressed by (39). The situation became even more puzzling after Yu and Sun [36, 37] showed that the CK-Hamiltonian *operator* can be derived from the conventional system-plus-reservoir approach. A solution to this paradoxical situation will be given below (see also [38]).

Other approaches try to avoid the problems arising from non-canonical transformations in classical mechanics and start already on the quantum mechanical level, trying to modify the

Hamiltonian operator by adding effective dissipative terms. The guiding line for the choice of the friction terms is Ehrenfest's theorem in the sense that the classical equation of motion including the damping term should, on an average, be valid, i.e.

$$\frac{d}{dt} < p > + \gamma < p > + < \frac{\partial}{\partial x} V > \; = \; 0 \tag{40}$$

which corresponds to the classical equation (33).

Assuming the modifided Hamiltonian to be of the form

$$H_{NL} = -\frac{\hbar^2}{2m} \frac{\partial^2}{\partial x^2} + V(x) + W \; = \; H_L + W \; , \tag{41}$$

applying Heisenberg's equation of motion and comparing it with Eq. (40) leads to

$$< \frac{\partial}{\partial x} W > \; = \; \gamma < p > \tag{42}$$

for the definition of W. Obviously, this definition is not without ambiguity as it only contains the derivative of W and, moreover, only the mean value of this quantity. Therefore, a variety of different modified Hamiltonians of this kind exist in the literature, usually leading to nonlinear (NL) SEs, at least due to the occurrence of mean values $< ... > = \int_{-\infty}^{+\infty} \Psi^* ... \Psi dx$ in the definition of W.

The most frequently quoted and applied one is Kostin's NLSE [39] with the friction term

$$W_K \; = \; \frac{\gamma}{2} \frac{\hbar}{i} \left(\ln \frac{\Psi_{NL}}{\Psi_{NL}^*} - < \ln \frac{\Psi_{NL}}{\Psi_{NL}^*} > \right) \; . \tag{43}$$

However, this approach also has several shortcomings. It allows all the solutions of the undamped HO (which are real functions with $\Psi = \Psi^*$) to also be solutions of the problem including damping. Additionally, the solutions of the damped HO contain the undamped frequency ω instead of the reduced frequency $\Omega = \left(\omega^2 - \frac{\gamma^2}{4} \right)^{1/2}$. These problems are related to a third one, namely, that W_K is a *real* quantity and could thus be considered an additional potential, therefore not contributing to the continuity equation for the density ϱ.

Thus the inconsistency occurs in that the equation of motion for the mean values contains the time-symmetry-breaking friction term $-\gamma < p >$, whereas the equation for the corresponding density ϱ is still the reversible continuity equation, which is unphysical since the system under consideration should display irreversible dynamics on all levels of description.

In order to avoid the conflicting aspects connected with the above-mentioned "derivation" of the "friction potentials" and to guarantee an irreversible time-evolution for the density ϱ with corresponding (non-hermitian) imaginary contributions in the respective modified SE, starting point of our approach is the addition of a time-symmetry-breaking diffusion term to the continuity equation, thus turning it into an irreversible Fokker–Planck-type equation (FPE), (more precisely, a Smoluchowski equation),

$$\frac{\partial}{\partial t} \varrho \; + \; \frac{\partial}{\partial x} (\varrho \, v_-) \; - \; D \frac{\partial^2}{\partial x^2} \varrho \; = \; 0 \; . \tag{44}$$

To avoid problems with particle/antiparticle mixing etc., this real FPE was supposed to be separable into a pair of c.c. equations containing only Ψ or Ψ^*, respectively (using an ansatz by Madelung [40] and Mrowka [41]). Due to the diffusion term, separability in general is not possible but can be achieved if an additional separation condition,

$$-D \frac{\frac{\partial^2}{\partial x^2} \varrho}{\varrho} \; = \; \gamma \left(\ln \varrho - < \ln \varrho > \right) \tag{45}$$

with D = diffusion coefficient (possibly time-dependent), is fulfilled.

This leads to a NLSE with *complex logarithmic* nonlinearity [42, 43, 44],

$$\begin{aligned} i\ \hbar\ \frac{\partial}{\partial t}\Psi_{NL}(x,t) &= \{-\frac{\hbar^2}{2m}\frac{\partial^2}{\partial x^2} + V(x) \\ &+ \gamma\frac{\hbar}{i}(\ln\Psi_{NL} - <\ln\Psi_{NL}>)\}\Psi_{NL}(x,t) \\ &= \{H_L + \tilde{W}\}\Psi_{NL}(x,t) = H_{NL}\ \Psi_{NL}(x,t)\ . \end{aligned} \tag{46}$$

Considering the complex nonlinearity $\tilde{W}$ in terms of real and imaginary parts,

$$\begin{aligned} \tilde{W} &= \tilde{W}_R + i\tilde{W}_I = \frac{\gamma}{2}\frac{\hbar}{i}\left(\ln\frac{\Psi_{NL}}{\Psi^*_{NL}} - <\ln\frac{\Psi_{NL}}{\Psi^*_{NL}}>\right) \\ &+ \frac{\gamma}{2}\frac{\hbar}{i}(\ln\varrho_{NL} - <\ln\varrho_{NL}>) \end{aligned} \tag{47}$$

it is obvious that the real part is identical to Kostins term and thus provides the dissipative friction term in the equation of motion for the mean values. On the other hand, the imaginary part, originating from the diffusion term in the FPE, introduces irreversibility [4] but, due to the property of vanishing mean value, causes no problems with imaginary energy contributions and also guarantees normalizability of the WF. In addition, the imaginary part also solves Kostin's other problems since the solutions of the undamped oscillator are no longer admitted and the oscillation frequency of the damped motion is now the correct reduced frequency $\Omega = \left(\omega^2 - \frac{\gamma^2}{4}\right)^{1/2}$.

Although the mean value of the energy is still given by the mean values of kinetic and potential energies (as the mean value of $\tilde{W}$ vanishes), this energy is no longer constant since it is calculated with the help of $\Psi_{NL}(x,t)$, the solution of the NLSE. In particular, the classical part of the energy decays exactly as expected from the classical equation of motion proportional to kinetic energy T_{cl}, i.e., $\frac{d}{dt}E_{cl} = -2\gamma T_{cl}$. The quantum mechanical contribution can display rather unexpected behaviour; for details see, e.g., [47].

For the cases considered in this paper ($V = 0$ and HO) also Gaussian wave packets solutions of the form (46) exist for the log NLSE. Inserting this wave packet into the log NLSE (46) yields a pair of equations of motion for the maximum,

$$\ddot{\eta} + \gamma\dot{\eta} + \omega^2\eta = 0, \tag{48}$$

and for the width, expressed in terms of the complex variable $\mathcal{C} = \left(\frac{2\hbar}{m}y\right) = \mathcal{R} + i\ \mathcal{I}$ as the complex Riccati equation

$$\frac{d}{dt}\mathcal{C} + \gamma\,\mathcal{C} + \mathcal{C}^2 + \omega^2 = 0\ . \tag{49}$$

Using the same definition for $\mathcal{I}$ as in the conservative case, i.e., $\mathcal{I} = \frac{1}{\alpha^2_{NL}} = \frac{\hbar}{2m<\tilde{x}^2>_{NL}}$, the real part $\mathcal{R}$, obtained by inserting $\mathcal{I}$ into the imaginary part of Eq. (49), is modified according to

$$\mathcal{R} = \frac{\dot{\alpha}_{NL}}{\alpha_{NL}} - \frac{\gamma}{2}\ . \tag{50}$$

[4] The formal similarity between our nonlinear imaginary term and an approach by Beretta introducing irreversibility and thermodynamic aspects into quantum mechanics without necessarily dissipative effects should be pointed out. See also Ref.[45-46] and Berettas contribution to these proceedings.

Inserting $\mathcal{I}$ and $\mathcal{R}$ into the real part of (49) yields, again, an Ermakov equation for $\alpha(t)$ and thus for the wave packet width,

$$\ddot{\alpha}_{NL} + \left(\omega^2 - \frac{\gamma^2}{4}\right)\alpha_{NL} = \frac{1}{\alpha_{NL}^3} \quad . \tag{51}$$

Also in the dissipative case where no conserved physical Hamiltonian exists that corresponds to the actual energy of the system, the pair of equations of motion for the wave packet maximum (corresponding to the classical trajectory) and the wave packet width (proportional to the position uncertainty) still provides an exact dynamical invariant of Ermakov type [48],

$$\begin{aligned} I_{NL} &= \frac{1}{2}\, e^{\gamma t}\Bigg[\left(\dot{\eta}\,\alpha_{NL} - \left(\dot{\alpha}_{NL} - \frac{\gamma}{2}\alpha_{NL}\right)\eta\right)^2 \\ &+ \left(\frac{\eta}{\alpha_{NL}}\right)^2\Bigg] = \text{const}\;.\,. \end{aligned} \tag{52}$$

Again, the complex nonlinear Riccati equation (49) can be linearized using the ansatz

$$\mathcal{C} = \left(\frac{2\hbar}{m}y\right) = \frac{\dot{\tilde{\lambda}}}{\tilde{\lambda}} = \frac{\dot{\lambda}}{\lambda} - \frac{\gamma}{2} = \left(\frac{\dot{\alpha}}{\alpha} - \frac{\gamma}{2}\right) + i\,\dot{\varphi} \tag{53}$$

with $\tilde{\lambda} = \lambda e^{-\gamma t/2} = \alpha_{NL} e^{-\gamma t/2 + i\varphi}$, i.e., the radius of $\tilde{\lambda}$ is exponentially shrinking, which is also expressed by the equation of motion for $\tilde{\lambda}$,

$$\ddot{\tilde{\lambda}} + \gamma\dot{\tilde{\lambda}} + \omega^2\tilde{\lambda} = 0\,, \tag{54}$$

which is just a complex Newtonian equation including the friction term.

Compared with the conservative HO, where $\lambda(t)$ moved on a circle with constant radius $\alpha = \alpha_0 = 1/\sqrt{\omega_0}$, for the damped HO, $\tilde{\lambda}(t)$ spirals towards the origin with the radius decreasing exponentially. (Again, as in the conservative case, it can be shown [49] that now the imaginary part of $\tilde{\lambda}(t)$ is directly proportional to the classical trajectory, i.e., $\tilde{z} = \frac{m}{\alpha_0 p_0}\eta(t)$.)

Expressed in terms of λ (i.e., compared to $\tilde{\lambda}$, in a system expanding exponentially), the linearized Riccati equation (49) takes the form

$$\ddot{\lambda} + \left(\omega^2 - \frac{\gamma^2}{4}\right)\lambda = 0\,, \tag{55}$$

i.e., like for an undamped HO with shifted frequency $\Omega = \left(\omega^2 - \frac{\gamma^2}{4}\right)^{1/2}$.

4.4. Nonunitary connection between the CK-Hamiltonian and the log NLSE

It can be shown that both approaches, the linear explicitly time-dependent one by CK and the NLSE with complex logarithmic term, are equivalent descriptions of the same physical situation. Only i) the CK-approach keeps the canonical Hamiltonian structure and the linearity of the quantized version, but looses the direct physical interpretation of the occurring quantities like operators and WFs; whereas ii) the log NLSE keeps the direct physical interpretation of all these quantities, but looses properties like the superposition principle (thus, solving actually *en passant* the problem of Schrödinger's cat for a non-isolated system).

The price one must pay in order to link these two approaches is that of a *non-unitary* transformation, based on the definition of action according to Schrödinger, used in his original derivation of the SE. In his first communication [50] Schrödinger starts from the Hamilton–Jacobi equation

$$\frac{\partial}{\partial t} S_C + H \; = \; 0 \tag{56}$$

with the action function S_c and the momentum $p_C \; = \; \frac{\hbar}{i}\frac{\partial}{\partial x} S_C$. He introduces the wave function $\Psi(x,t)$ through

$$S_C = \left(\frac{\hbar}{i}\right) \ln \Psi \tag{57}$$

(where in his original paper he omitted the imaginary unit $i = \sqrt{-1}$, which he corrected later on after he was forced to accept that Ψ and thus also S_c can be complex functions of space and time (therefore the subscript "c")). Via a variational ansatz Schrödinger finally arrived at the (linear) SE.

One can now reverse Schrödinger's procedure, starting with the log NLSE and, using the definition (57) of S_c, arriving at

$$\left(\frac{\partial}{\partial t} + \gamma\right) S_C + H \; = \; -\gamma < S_C > \; . \tag{58}$$

This is, of course, as little rigorous as Schrödinger's first attempt was; however, it follows his way of thinking of how to connect the classical Hamilton–Jacobi theory with a wave (mechanical) equation. The term $-\gamma < S_C >$ is necessary mainly for normalization purposes, as mentioned above, and shall be neglected for the moment. Multiplying the remaining lhs of Eq. (58) by $e^{\gamma t}$ and using the definitions

$$\hat{S}_C \; = \; e^{\gamma t} S_C \quad \text{and} \quad \hat{H} \; = \; e^{\gamma t} H \tag{59}$$

finally yields the transformed Hamilton–Jacobi equation

$$\frac{\partial}{\partial t} \hat{S}_C + \hat{H} \; = \; 0 \; . \tag{60}$$

From the definition of the action, it follows that the WF $\hat{\Psi}$ in the transformed (canonical) system is connected with the WF Ψ_{NL} in the physical system via the *non-unitary* transformation

$$\ln \hat{\Psi} \; = \; e^{\gamma t} \ln \Psi_{NL} \; . \tag{61}$$

Consequently, the complex momenta in the two systems are connected via

$$\hat{p} \; = \; \frac{\hbar}{i}\frac{\partial}{\partial x} \ln \hat{\Psi} \; = \; e^{\gamma t} \, \frac{\hbar}{i}\frac{\partial}{\partial x} \ln \Psi_{NL} \; = \; e^{\gamma t} \; p_C \; , \tag{62}$$

which is equivalent to the connection between canonical and kinetic momentum in the CK theory. The *non-canonical* connection between the classical variables (x, p) and $(\hat{x}, \hat{p})$ corresponds to the *non-unitary* transformation between Ψ_{NL} and $\hat{\Psi}$.

The transformed canonical Hamiltonian, the corresponding SE and the transformed WF are identical to those of the CK approach. This, however, explains the apparent violation of the uncertainty principle by the CK approach. The mean values of position and momentum uncertainty are first calculated with the operators corresponding to the canonical variables, $\hat{x}_{op} = x$ and $\hat{p}_{op} = \frac{\hbar}{i}\frac{\partial}{\partial x}$, using the canonical WF $\hat{\Psi}_{CK}(x,t)$. The apparent violation occurs when only the *canonical* momentum *operator* $\hat{p}_{op}$ is changed to the *physical* momentum *operator* p_{op} via $p_{op} = \hat{p}_{op} \; e^{-\gamma t}$,but then the mean values are *still* calculated, applying the *canonical WF* $\hat{\Psi}_{CK}(x,t)$, *instead* of the *physical WF* $\Psi_{NL}(x,t)$, since, according to Eqs. (61) and (62), a *change* of the *momentum* is also *connected* with a *change* of the definition of the *action* and thus a *corresponding (non-unitary) change* of the *WF*. Unfortunately, this is not so obvious at first sight since in the sets of physical and canonical variables, the position variable remains unchanged under the non-canonical transformation so both, the physical WF $\Psi_{NL}(x,t)$ and the canonical WF $\hat{\Psi}_{CK}(x,t)$, depend on the same variables, however having a different physical meaning (and different analytical forms as functions of x and t), in particular when mean values are calculated with them. Taking this into account, the violation of the uncertainty principle vanishes. That means, to be *consistent*, WFs *and* operators must be *transformed simultaneously*.

This particularly means that the canonical position uncertainty $<\tilde{x}^2>_{CK}$ and the physical position uncertainty $<\tilde{x}^2>_{NL}$ for the dissipative case, and hence α_{CK} and α_{NL}, are not identical but, using the non-unitary transformation (61), are related via

$$\alpha_{CK} = e^{-\gamma t/2}\alpha_{NL} \ . \tag{63}$$

Inserting this into Eq. (37) for α_{CK}, turns it exactly into Eq.(51) for α_{NL} and, hence, the Ermakov invariant I_{CK} is *identical* to I_{NL} in Eq. (52),

$$I_{CK} \equiv I_{NL} \ . \tag{64}$$

In an earlier work [51], it had been shown that the dissipative equation of motion (33) can also be obtained via the canonical formalism if a *different*, but also *non-canonical* transformation, is involved; in this case leading (at least for the damped HO) to a canonical Hamiltonian that is a constant of motion. This transformation uses an *expanding* coordinate system according to

$$\begin{aligned} Q &= e^{\gamma t/2}x \\ P &= e^{\gamma t/2}p + \frac{m}{2}\gamma\, e^{\gamma t/2}x \ , \end{aligned} \tag{65}$$

leading to the Hamiltonian

$$\hat{H}_{exp} = \frac{1}{2m}P^2 + \frac{m}{2}\left(\omega^2 - \frac{\gamma^2}{4}\right)Q^2 = \text{const.} \tag{66}$$

with the equation of motion

$$\ddot{Q} + \left(\omega^2 - \frac{\gamma^2}{4}\right)Q = 0 \tag{67}$$

which, after back transformation to the physical variables, yields Eq. (33). It should be mentioned that the variables (Q, P) of this expanding system and the variables $(\hat{x}, \hat{p})$ of the CK approach are connected via a *canonical* transformation,

$$\begin{aligned} Q &= e^{\gamma t/2}\hat{x} = e^{\gamma t/2}x \\ P &= e^{-\gamma t/2}\hat{p} + \frac{m}{2}\gamma e^{\gamma t/2}\hat{x} \ . \end{aligned} \tag{68}$$

Keeping in mind that the maximum of the wave packet solving the NLSE (46) obeys Eq. (48), which is identical to Eq. (33) for the classical trajectory, and applying the transformation (65), then Eq. (48) for $\eta(t)$ can be written in the form of Eq. (67). Except now, this equation and Eq. (51) for α_{NL} allow for rewriting the Ermakov invariant (52) in the form

$$I_{dissip} = \frac{1}{2}\left[\left(\dot{Q}\,\alpha_{NL} - Q\,\dot{\alpha}_{NL}\right)^2 + \left(\frac{Q}{\alpha_{NL}}\right)^2\right] = \text{const.} \ , \tag{69}$$

i.e., exactly in a form as in the conservative case (see. Eq. (6)).

It means, however, that this invariant is independent of ω and γ, i.e., it exists also for time-dependent parameters $\omega = \omega(t)$ and $\gamma = \gamma(t)$, in cases where definitely no conserved physical Hamiltonian function exists.

5. Conclusions

In the first part of this paper, it has been shown that the rule for finding the so-called Pythagorean triple, a rule that only involves integers and can therefore be considered as a kind of "quantization condition", is also found in different contexts in modern physics. In particular, all systems that can be described by complex nonlinear Riccati equations exhibit this property, e.g., time-dependent and time-independent [52] SEs. As an example, we considered the time-dependent SE and found that the formal similarity between the above-mentioned rule and the time-evolution of the quantum uncertainties, related to probability currents, showed the possibility for a quantization of the dynamical quantum properties (like they are already experimentally observed), e.g., in the QHE.

Linearization of the Riccati equation allowed for a geometric interpretation of the quantized dynamical properties in terms of angular and radial changes. It has also been shown that particularly the radial changes are related to an element of irreversibility (already in conventional quantum mechanics,) connected to imaginary contributions to the SE and thus additional terms (compared to the classical situation) to the continuity equation. Radial changes of a similar kind were also found in connection with dissipation, an aspect not usually directly included in conventional quantum mechanics.

The complex Riccati equation, together with the classical Newtonian equation of the system, allowed for the definition of a constant of motion, also in cases where the classical Hamiltonian of the system did not have this property. This Ermakov invariant has (apart from a constant factor $\frac{1}{m}$) the dimension of an action and can be identified with the conserved angular momentum for the motion in a complex plane. In our example, the value of this angular momentum is $\frac{\hbar}{2}$, usually attributed to the non-classical quantity "spin".

Quantization of currents connected to some quantized radial changes had already drawn our attention to the QHE. However, there is also a second element of quantization observable in the fractional QHE. The interpretation that this fraction might originate from two effects where each of them is quantized separately but only one in terms of $\hbar$, the other in terms of $\frac{e^2}{c}$, another quantum of action, the so-called *quantum of least Coulombic action*. The ratio of these two quanta actually yields Sommerfeld's fine structure constant. As reason for the occurrence of a second quantum of action in nature, it was speculated that both might be incommensurable in a sense like radius and circumference of a circle. One action, probably $\hbar$, should then quantize phenomena where angular aspects are relevant, whereas the other action, $\frac{e^2}{c}$, should quantize radial properties. These of course are, so far, only assumptions and further studies are necssary to clarify this; however, work in this direction is in progress (see also the contribution of Reinisch in these proceedings).

In the second part, systems were considered where no conserved physical Hamiltonian exists, in particular dissipative systems with linear velocity dependent friction forces. It had been shown that different approaches to incorporate these effects into the formalism of classical Hamiltonian mechanics as well as into quantum mechanics exist and what problems can rise in this context. An approach in the form of a NLSE with *complex logarithmic nonlinearity* was found that did not exhibit the problematic aspects of some other attempts, but combined their positive aspects. The imaginary part introduced an element of irreversibility but still kept real energy values and normalizable WFs while the real part introduced the irreversible friction term into the equation of motion for the classical aspect (mean value of position) and caused dissipation of the energy. Both terms however are, in this case, not independent of each other due to the coupling of amplitude and phase of the WF. Also in this case, an exact invariant of Ermakov-type with the same properties as in the conservative case could be found for our log NLSE and be compared with a corresponding approach by Caldirola and Kanai. An equivalence of these two approaches can be established but involves *non-canonical* transformations in the *classical* case and *non-unitary* transformations in the *quantum mechanical* case. This non-unitary transformation of the corresponding WFs is mainly a transformation of the (complex) action function, introduced by Schrödinger, when the description in terms of physical quantities obeying a NLSE shall be changed into a description in terms of canonical quantities obeying a linear SE. Again, in this case, the *action* (now complex) is the crucial quantity. Depending on the problem under investigation, one can choose the more convenient approach for the major calculations and still obtain the information contained in the other one via a non-unitary transformation. For an extension to a many-particle theory the linear CK-approach seems at first sight more appealing, but to obtain physically meaningful results, the necessary non-unitary transformation might in this case be more complicated. However, for a many-partile system where the corresponding wave function can be written as a product of one-particle wave functions, the logarithm allows for a convenient separation of the product into a sum, but terms that cannot be treated with a simple product ansatz might cause more problems. Work in this direction is in progress.

Although some answers to the questions posed in the introduction could be found, there are still sufficient unsolved problems, leaving plenty of room for further work and action!

Acknowledgment
The author wishes to thank Dr. Gilbert Reinisch for his very helpful and valuable comments.

References

[1] Heisenberg W 1973 *Der Teil und das Ganze* (München: Deutscher Taschenbuch Verlag) p 17
[2] Reinisch G 1998 *Astron. Astrophys.* **337** 299
[3] Courant R and Robbins H 1970 *Was ist Mathematik?* (Berlin: Springer Verlag) p 34
[4] Ermakov V P 1880 *Univ. Izv. Kiev* **20** 1
[5] Milne W E 1930 *Phys. Rev.* **35** 863
[6] Pinney E 1950 *Proc. Amer. Math. Soc.* **1** 681
[7] Lewis H R 1967 *Phys. Rev. Lett.* **18** 510
[8] Schuch D 2003 *IOP Conference Series* **173** *Proceedings of the 24th Intern. Colloquium on Group Theoretical Methods in Physics* Gazeau J P et al eds (Dordrecht: Kluwer Academic Publishers) p 741
[9] Dekker H 1975 *Z. Phys* B **21** 295
[10] Bender C M and Boettcher S 1998 *Phys. Rev. Lett.* **80** 5243
[11] Kaushal R S and Korsch H J 2000 *Phys. Lett.* A **276** 47
[12] Kaushal R S and Parthasarathi 2002 *J. Phys. A: Math. Gen.* **35** 8743
[13] Bender C M, Brody D C and Hook D 2008 *J. Phys. A: Math. Theor.* **41** 352003
[14] Schuch D 2005 *Phys. Lett.* A **338** 225
[15] Schuch D and Moshinsky M 2008 *SIGMA* **4** 056
[16] Schuch D and Moshinsky M 2006 *Phys. Rev.* A **73** 062111
[17] Wadlinger R L P, Hunter G, Kostro L and Schuch D 1990 *Phys. Essays* **3** 194
[18] Einstein A 1909 *Z. Phys.* **6** 185
[19] Schrödinger E 1922 *Z. Phys.* **12** 23
[20] Eddington A S 1935 *New Pathways in Science* (Cambridge: Cambridge University Press) p 234
[21] Schuch D and Moshinsky M 2005 *Rev. Mex. Fis.* **51** 516
[22] Caldeira A O and Leggett A J 1981 *Phys. Rev. Lett.* **46** 211
[23] Caldeira A O and Leggett A J 1983 *Ann. Phys. (N.Y.)* **149** 374; *ibid* 1983 **153** 445(E)
[24] Gao S 1997 *Phys. Rev. Lett.* **79** 3101
[25] Lindblad G 1976 *Commun. Math. Phys.* **48** 119
[26] Gorini V, Kossakowski A and Sudarshan E C G 1976 *J. Math. Phys.* **17** 821
[27] May V and Kühn O 2003 *Charge and Energy Transfer Dynamics in Molecular Systems* 2nd ed (Berlin: Wiley-VCH) chapter 3.12.3
[28] Caldirola P 1941 *Nuovo Cimento* **18** 393
[29] Kanai E 1948 *Progr. Theor. Phys.* **3** 440
[30] Nassar A B 1986 *J. Math. Phys.* **27** 755
[31] Brittin W E 1950 *Phys. Rev.* **77** 396
[32] Ray J R 1979 *Lett. Nuovo Cim.* **25** 47
[33] Greenberger D M 1979 *J. Math. Phys.* **20** 762
[34] Svin'in I R 1976 *Teor. Mat. Fiz.* **27** 270
[35] Messer J 1979 *Acta Phys. Austriaca* **50** 75

[36] Yu L H and Sun C P 1994 *Phys. Rev.* A **49** 592
[37] Sun C P and Yu L H 1995 *Phys. Rev.* A **51** 1845
[38] Schuch D 1997 *Phys. Rev.* A **55** 935
[39] Kostin M D 1972 *J. Chem. Phys.* **57** 3589
[40] Madelung E 1950 *Die Mathematischen Hilfsmittel des Physikers* (Berlin: Springer) p 432
[41] Mrowka B 1951 *Z. Phys.* **130** 164
[42] Schuch D Chung K-M and Hartmann H 1983 *J. Math. Phys.* **24** 1652
[43] Schuch D Chung K-M and Hartmann H 1983 *J. Math. Phys.* **25** 3086
[44] Schuch D and Chung K-M 1986 *Int. J. Quantum Chem.* **29** 1561
[45] Beretta G P 1987 *Found. Phys.* **17** 365
[46] Beretta G P 2006 *Phys.Rev.* E **73** 026113
[47] Schuch D 2002 *J. Phys. A: Math. Gen.* **35** 8615
[48] Nassar A B 1986 *J. Math. Phys.* **27** 2949
[49] Schuch D 2007 *J. Math. Phys.* **48** 122701
[50] Schrödinger E 1926 *Ann. d. Phys.* **79** 361
[51] Schuch D 1999 *Int. J. Quantum Chem.* **72** 537
[52] Reinisch G 1994 *Physica* A **206** 229

Lagrangians and Hamiltonians with friction

C E Smith

Center for Energy Systems Research, M.E. Department
Virginia Polytechnic Institute and State University, Blacksburg, VA, U.S.A.

E-mail: cesmith@vt.edu

Abstract. Methods are presented for extending a promising formalism that incorporates dissipative forces into Quantum Mechanics. Hamiltonians and Lagrangians are constructed that have additional force and frictional terms while still being consistent with the rules of the formalism. The simple systems of a damped particle in a force field and a quadratically damped oscillator are discussed.

1. Introduction

Not long after Quantum Mechanics was formulated, in the 1930's, questions arose as to how to account for the physical properties of entropy or dissipation in this new branch of physics. Schrödinger himself realized that the equation named after him was not adequate to the task, seeing that it was reversible.

Since that time, a particularly well-reasoned approach that puts dissipative forces into the Lagrange-Hamiltonian mechanics on which Quantum Mechanics is based has come from Schuch [1-3]. His formalism is based on the hypothesis is that dissipative systems can be formulated in a way such that they follow the same variational principles as conservative systems. At the heart of the analysis is the realization that the gap between conservative mechanics and analogous formalisms having friction must be bridged by *non-canonical* (classical) transformations and not just a changing of coordinate space via unitary (quantum mechanical) transformations. The resulting dissipative Lagrangians ($L's$) and Hamiltonians $(H's)$ have the following convenient properties. The variation of the Langrangian produces the correct equations of motion. The corresponding Hamiltonians represent the total energy for the system. The Hamiltonians are a constant of the motion. The Heisenberg uncertainty does not vanish with time. These properties support the idea that the formalism is indeed a generalization of the variational foundation that underlies reversible mechanics but with dissipative effects included. The list of systems that have been studied so far includes the damped harmonic oscillator [1]. The intent herein is to show that this formalism allows for other forms to have additional assortments of frictional forces and potentials. Other details are found in [4].

2. Results

The process of deriving the new examples is guided by the following assumptions. Note that the term "real space" here is synonymous with physical space.

1. All $L's$ and $H's$ must have units of energy
2. Transformed space (i.e. Q-space) has no dissipation or explicit time dependence

3. Transformed space is used to derive H from L
4. The variation of L, (i.e. $\delta L = 0$) in both real (i.e. q -space) and transformed space (Q -space) produce the correct equation of motion in real space.
5. As frictional and potential terms vanish, the correct simpler forms emerge.

Since there are an infinite number of frictional as well as conservative $L's$ and $H's$ that are associated with a given equation of motion [5, 6], the above assumptions greatly reduce the number of candidates. To obtain $L's$ and $H's$ that have the desired properties, the appropriate non-unitary transformations are needed to connect physical q -space, which has friction, to a transformed Q-space that does not have friction explicitly. As seen in the figure 1 below, the non-unitary nature of the transformation breaks the direct connection that the L in real space has with its respective H through the Legendre transform.

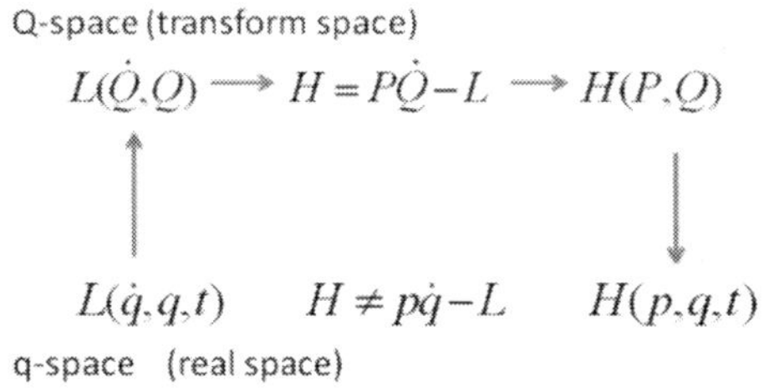

Figure 1.

We see also that momentum in physical space and transformed space is defined as $p = m\dot{q}$ and $P = \frac{\partial L}{\partial \dot{Q}} = m\dot{Q}$ respectively. The canonical definition $p = \frac{\partial L}{\partial \dot{q}}$ no longer holds for real space frictional systems, at least those with explicit time dependence. In addition to the guidelines above it is known as illustrated in [5] that the frictional $L's$ and $H's$ have "expansion" or integrating factors of the form:

$$X = \exp\left(- \int \frac{\partial G}{\partial \dot{q}} dt \right) \tag{1}$$

where G is the term on the right hand side of the generic equation of motion, $\ddot{q} = G(\dot{q}, q, t)$.

Simultaneous to determining the non-unitary transform, the appropriate L in transformed space, or $L(\dot{Q}, Q)$ must also be found. Each $L(\dot{Q}, Q)$ has a kinetic $\dot{Q}^2$ term added to a generalized potential function with the form, $V(Q, \xi_0, \xi_1, \ldots \phi_0, \phi_1 \ldots)$, with Q as the transformed position and the ξ_i and ϕ_i the relevant coefficients of friction and conservative potential. The generalized potential has no explicit time dependence. The basic approach for creating the generalized potential for a given equation of motion is to include any and all pertinent terms that have the units of energy. For example if gravitational potential g is present, a term of the form $2gQ$ is included. Viscous friction implies a term such as $\frac{\gamma^2}{4} Q^2$. For quadratic friction, which is proportional to the square of velocity, there evidently are no such potential terms that can be found.

As an example, the frictional L and H for the damped harmonic oscillator are shown below. The first equation is the non-unitary transformation of the coordinate space, the second is the L in the transformed space. Once these are known, the dissipative Lagranians and Hamiltonians follow in a straightforward way. It is seen that the variational criterion $\delta L = 0$ produces the expected equation of motion in both q -space and Q -space. All other rules of the formalism are satisfied and H is a constant of the motion. For the damped harmonic oscillator:

$$Q = q\exp\left(\frac{\gamma t}{2}\right) \tag{2}$$

$$L(\dot{Q},Q)=\frac{m}{2}\left(\dot{Q}^2-\left(\omega^2-\frac{\gamma^2}{4}\right)Q^2\right) \tag{3}$$

$$H(P,Q)=\frac{m}{2}\left(\frac{P^2}{m^2}+\left(\omega^2-\frac{\gamma^2}{4}\right)Q^2\right) \tag{4}$$

$$L(\dot{q},q,t)=\frac{m}{2}\left(\dot{q}^2+\gamma\dot{q}q+\frac{\gamma^2}{2}q^2-\omega^2q^2\right)\exp(\gamma t) \tag{5}$$

$$H(p,q,t)=\frac{m}{2}\left(\frac{p^2}{m^2}+\frac{\gamma}{m}pq+\omega^2q^2\right)\exp(\gamma t) \tag{6}$$

with (2) being the non-canonical connection between physical and transformed space and with $\exp(\gamma t)$ as the expansion factor.

The derivation of frictional $L's$ and $H's$ for a damped particle in a force field and a quadratically damped oscillator are now shown. The equations of motion for the two systems are:

$$\ddot{q}+\gamma\dot{q}+\alpha=0 \tag{7}$$

where α is a constant force field and

$$\ddot{q}+b\dot{q}^2+\omega^2q=0 \tag{8}$$

for the quadratically damped oscillator. Their corresponding q-transforms as well as $L(\dot{Q},Q)$ and $H(P,Q)$ are shown in table 1. The frictional $L's$ and $H's$ in physical space are shown in table 2.

Table 1.

Eqn of motion	*q − transform*	$L(\dot{Q},Q)$	$H(P,Q)$
$\ddot{q}+\gamma\dot{q}+\alpha=0$	$Q=\left(q+\frac{\alpha t}{\gamma}-2\frac{\alpha}{\gamma^2}\right)\exp\left(\frac{\gamma t}{2}\right)-2\frac{\alpha}{\gamma^2}\cosh\left(\frac{\gamma t}{2}\right)+4\frac{\alpha}{\gamma^2}$	$\frac{m}{2}\left(\dot{Q}^2+\frac{\gamma^2}{4}Q^2-2\alpha Q\right)$	$\frac{m}{2}\left(\frac{P^2}{m^2}-\frac{\gamma^2}{4}Q^2+2\alpha Q\right)$
$\ddot{q}+b\dot{q}^2+\omega^2q=0$	$Q=\frac{1}{b}(1-\exp(-bq))\exp(bq)$	$\frac{m}{2}(\dot{Q}^2-\frac{\omega^2}{b^2}(bQ+1)^2\ln(bQ+1)+\frac{\omega^2}{2b^2}(bQ+1)^2-\frac{\omega^2}{2b^2})$	$\frac{m}{2}(\frac{P^2}{m^2}+\frac{\omega^2}{b^2}(bQ+1)^2\ln(bQ+1)-\frac{\omega^2}{2b^2}(bQ+1)^2+\frac{\omega^2}{2b^2})$

Table 2.

Eqn of motion	$L(\dot{q},q,t))$	$H(p,q,t))$
$\ddot{q}+\gamma\dot{q}+\alpha=0$	$\frac{m}{2}((\dot{q}^2+\gamma\dot{q}q+\frac{\gamma^2}{2}q^2$ $-2\alpha q-\frac{\alpha}{\gamma}\dot{q}+2\frac{\alpha^2}{\gamma^2}-2\frac{\alpha^2}{\gamma}t+\alpha\dot{q}t$ $+\alpha\gamma qt+\frac{\alpha^2t^2}{2})\exp(\gamma t)$ $+\frac{\alpha}{\gamma}\dot{q}-3\frac{\alpha^2}{\gamma^2}+\frac{\alpha^2}{\gamma^2}\cosh(\gamma t))$	$\frac{m}{2}((\frac{p^2}{m^2}+\frac{\gamma}{m}pq$ $+\alpha q-\frac{\alpha}{m\gamma}p-2\frac{\alpha^2}{\gamma^2}+\frac{\alpha^2}{\gamma}t+\frac{\alpha}{m}pt)\exp(\gamma t)$ $+\alpha q+\frac{\alpha}{m\gamma}p+\frac{2\alpha^2}{\gamma^2}+\frac{\alpha^2t}{\gamma})$
$\ddot{q}+b\dot{q}^2+\omega^2q=0$	$\frac{m}{2}\left(\left(\dot{q}^2-\frac{\omega^2}{b}q+\frac{\omega^2}{2b^2}\right)\exp(2bq)-\frac{\omega^2}{2b^2}\right)$	$\frac{m}{2}\left(\left(\frac{p^2}{m^2}+\frac{\omega^2}{b}q-\frac{\omega^2}{2b^2}\right)\exp(2bq)+\frac{\omega^2}{2b^2}\right)$

What follows is an explanation in detail as to how the rules of the formalism and related guidelines were used to produce the above examples.

Example 1: The damped particle in a force field. The ansatz q-transform is taken to be that for the free damped particle as shown in [4] which is then combined with an as yet unspecified function of time:

$$Q = q\exp\left(\frac{\gamma}{2}t\right)+f(t). \tag{9}$$

The use of $f(t)$ is justified in [5]. The initial $L(\dot{Q},Q)$ is formed from the generalized potential terms that correspond to viscous friction and a constant field,

$$L(\dot{Q},Q)=\frac{m}{2}\left(\dot{Q}^2+\frac{\gamma^2}{4}Q^2-2\alpha Q\right). \tag{10}$$

The variation, $\delta L(\dot{Q},Q)=0$ is calculated in the same way as for a conservative system:

$$\delta L(\dot{Q},Q)=\frac{m}{2}\left(2\ddot{Q}-\frac{\gamma^2}{2}Q-2\alpha\right)=0. \tag{11}$$

Equation (9) is differentiated to find $\dot{Q}$ and $\ddot{Q}$ which are substituted into (11) to convert it to physical space:

$$(\ddot{q}+\gamma\dot{q})\exp\left(\frac{\gamma}{2}t\right)+\ddot{f}-\frac{\gamma^2}{4}f+\alpha=0. \tag{12}$$

The function $f(t)$ is determined by setting the differential equation in f in equation (12) to be equal to the term that is needed to complete the equation of motion:

$$\ddot{f}-\frac{\gamma^2}{4}f+\alpha=\alpha\exp\left(\frac{\gamma}{2}t\right). \tag{13}$$

The general solution for $f(t)$ is:

$$f(t)=c_1\exp\left(\frac{\gamma}{2}t\right)+c_2\exp\left(\frac{-\gamma}{2}t\right)+\frac{\alpha t}{\gamma}\exp\left(\frac{\gamma}{2}t\right)+4\left(\frac{\alpha}{\gamma^2}\right). \tag{14}$$

By using power series expansions of $c_1 \exp\left(\frac{\gamma}{2}t\right)$ and $c_2 \exp\left(\frac{-\gamma}{2}t\right)$, we can know how to set the two constants c_1 and c_2 so that the frictional L and H in q-space properly turn into the expected frictionless L and H as the viscous coefficient γ vanishes. We get

$$c_1 = -3\frac{\alpha}{\gamma^2}, \qquad c_2 = -\frac{\alpha}{\gamma^2}. \tag{15a, 15b}$$

Therefore $f(t)$ is:

$$f(t) = -3\frac{\alpha}{\gamma^2}\exp\left(\frac{\gamma}{2}t\right) - \frac{\alpha}{\gamma^2}\exp\left(\frac{-\gamma}{2}t\right) + \frac{\alpha t}{\gamma}\exp\left(\frac{\gamma}{2}t\right) + 4\left(\frac{\alpha}{\gamma^2}\right). \tag{16}$$

By adding $f(t)$ into equation (9) and using the result to convert the corresponding $L(\dot{Q},Q)$ and $H(P,Q)$ to real space, $L(\dot{q},q,t)$ and $H(p,q,t)$ can be determined.

Example 2: The quadratically damped oscillator. The ansatz q-transform is taken to be the free quadratically damped particle which is derived in [4]:

$$Q = \frac{1}{b}(1 - \exp(-bq))\exp(bq). \tag{17}$$

The initial $L(\dot{Q},Q)$ is taken to be:

$$L(\dot{Q},Q) = \frac{m}{2}(\dot{Q}^2). \tag{18}$$

There could also be a $\omega^2 Q^2$ term but it will be more convenient to leave this out. As a result:

$$\delta L(\dot{Q},Q) = \frac{m}{2}(2\ddot{Q}) = 0. \tag{19}$$

After converting the differential to physical space:

$$(\ddot{q} + b\dot{q}^2)\exp(bq) = 0. \tag{20}$$

It is clear that addition of $\omega^2 q\exp(bq)$ is needed to complete the equation of motion. Since there is no explicit time dependence the q-transform can be inverted so that the needed term as a function of Q is:

$$\omega^2 q \exp(bq) = \frac{\omega^2}{b}\ln(bQ+1)(bQ+1). \tag{21}$$

The right hands side of (21) can be integrated to find the needed term in $L(\dot{Q},Q)$ as seen as the complicated expression in table 1. By expanding $L(\dot{Q},Q)$ in power series it is seen that the $\omega^2 Q^2$ term is part of the series. After $L(\dot{Q},Q)$ is converted to real space using the q-transform, it is seen that the term $-\frac{\omega^2}{2b^2}$ is needed to allow $L(\dot{q},q,t)$ and $H(p,q,t)$ to correctly turn into simpler forms as ω or b vanish.

3. Discussion and conclusion

Additional combinations of friction and potential can be included into a variational formalism for mechanics that includes dissipative forces. The resulting Hamiltonians open up the possibility of modeling dissipative quantum systems. The example containing both a force field and viscous friction is noteworthy in that it can represent a steady state system as the terminal velocity of the particle in the field is reached. A frictional Hamiltonian like this may be useful, for instance, in the modeling phenomena such as lasers as is discussed in [7]. Since the dissipative terms of the Hamiltonians discussed herein are given as functions of velocity, they are in essence macroscopic and phenomenological in nature. Frictional forces of this type represent the average interaction of a larger environment on a particle or system of particles without all the microscopic details. Nevertheless, such simplifications, if justifiable, can be useful in capturing the larger aspects of a problem while reducing the calculational complexity.

References

[1] Schuch D 1990 *Int. J. Quant. Chem. Symp.* **24** 767
[2] Schuch D 1997 *Phys Rev. A* **55** 935
[3] Schuch D 1999 *Int. J. Quant. Chem.* **72** 537
[4] Smith C E 2006 *arXiv:physics*/0601133v1
[5] Yan C C 1978 *Am. J. Phys.* **46** 671
[6] Santilli R 1978 *Foundations of Theoretical Mechanics* (New York: Springer Verlag)
[7] Marcuse D 1980 *Principles of Quantum Electronics* (New York: Academic)

3D Quantum thermodynamic description of the non-equilibrium behavior of an unbounded system at an atomistic level

C E Smith[1], A Sciacovelli[2], M R von Spakovsky[1] and V Verda[2]

[1]Center for Energy Systems Research, M.E. Department
Virginia Polytechnic Institute and State University, Blacksburg, VA, U.S.A.
[2]Department of Energy Engineering,
Politecnico di Torino, c.so Duca degli Abruzzi 24, 10129 Torino, ITALY

E-mail: vonspako@vt.edu

Abstract. Quantum thermodynamics (QT) provides a general framework for the description of non-equilibrium phenomena at any level, particularly the atomistic one. This theory and its dynamical postulate are used here to model the storage of hydrogen on and in a carbon nanotube. The tube is placed at the center of a tank with a volume of 250 nm^3. The thermodynamic system of interest is the hydrogen, which is assumed isolated and having boundaries that coincide with the walls of the tank and the carbon nanotube. The hydrogen is initially prepared in a state far from stable equilibrium (i.e., with the hydrogen molecules probabilistically near one of the outer tank walls) after which the system is allowed to relax (evolve) to a state of stable equilibrium. To predict this evolution in state, the so-called energy eigenvalue problem, which entails a many-body problem that for dilute and moderately dense gases can be modeled using virial expansion theory, is first solved for the geometry involved. The energy eigenvalues and eigenstates of the system found are then used by the nonlinear Beretta equation of motion of QT to determine the evolution of the thermodynamic state of the system as well as the 3D spatial distributions of the hydrogen molecules in time. The simulation results provide a quantification of the entropy generated due to irreversibilities at an atomistic level and show in detail the trajectory of the thermodynamic state of the system as the hydrogen molecules, which are initially arranged to be far from the carbon nanotube, spread out in the system and eventually become probabilistically more concentrated near the carbon atoms, which make up the nanotube.

1. Introduction

To develop a complete *non-equilibrium* description at a nano-scale, quantum thermodynamics (QT) can be used since it extends the linear dynamics of quantum mechanics (QM) to the non-linear dynamics required to describe systems undergoing a *non-equilibrium* process [1-17]. Note that QT is a paradigm distinct from and at variance to statistical (not conventional) quantum mechanics (SQM). From SQM, a whole set of Markovian and non-Markovian quantum master equations (QMEs) for so-called "dissipative open systems" have been developed. Applications of these QMEs have been somewhat limited (e.g., to optics, quantum electronic devices, Brownian motion of single particles, absorption and emission spectra, and molecular wires [18-23]). Furthermore, the dissipative description of these "open systems" relies on a closed-system-plus-reservoir model for which the time evolution of state is governed by the standard Schrödinger and von Neumann equations. However, the underlying linear dynamics of these two equations is at variance with the dissipative (non-linear) nature of the QMEs derived from these equations (e.g., the well-known Kossakowski-Sudarshan-Gorini-Lindblad (KSGL) equation [24-27] or its equivalents). Irreversible relaxation to stable equilibrium is, thus, accomplished at the expense of introducing a paradox, i.e., the so-called Loschmidt paradox. In fact, common to all attempts at justifying the validity of these relaxations via QMEs is that they require additional, question-begging assumptions that are at odds with an overall unitary (i.e., reversible/linear) evolution [2]. In addition, although it is true that under appropriate conditions the response formalism of the QMEs can in certain cases be connected with the Onsager theory of irreversible processes [19], the perturbative theories governing these QMEs limit (as they do in the Onsager theory) their applicability to the near-equilibrium regime.

In contrast, QT and its equation of motion, the Beretta equation [2-11], circumvent these difficulties since the dissipations of the closed system minus the reservoir are assumed to be strictly endogenous to the system (contrary to the assumption made in SQM) and are fully captured by the nonlinear dynamics of the Beretta equation. QT, thus, is able to determine the *non-equilibrium* path, which the system takes and which once known, allows one to determine, for example, the *non-equilibrium* thermodynamic properties (e.g., the chemical potentials and affinities) of each species, the kinetics of any reactions present, and the entropy generated due to irreversibilities at each instance of time. The latter, in fact, represents the dissipation or destruction of the non-conserved system property, the available energy [28]. Furthermore, the path found is unique and fixed in an absolute sense at least to the extent that the relaxation times of QT are tied directly to experimental values and the time-energy Heisenberg uncertainty principle [16].

In this paper, QT is used to extend the work reported in previous papers [12-14] of modeling the non-equilibrium evolution in state of hydrogen storage on and in a carbon nanotube. The thermodynamic system of interest is the hydrogen, which is assumed isolated and contained within the walls of a 250 nm^3 tank at the center of which lies the carbon nanotube. The system boundaries coincide with the walls of the tank and the tube. To model this evolution, the so-called energy eigenvalue problem, which entails a many-body problem that for dilute and moderately dense gases can be modeled using virial expansion theory [29,30], is first solved in 3D to determine the so-called primitive-level energy eigenvalues and eigenstates of the system. This information is then used in a combinatorial problem to determine the system-level energy eigenvalues and eigenstates. These are subsequently used in the nonlinear Beretta equation of motion of QT [2-11] to determine the evolution (relaxation) of the thermodynamic state of the system to stable equilibrium, starting from some initially prepared state far from stable equilibrium (i.e., with the hydrogen molecules probabilistically near one of the outer tank walls). The spatial distributions of the hydrogen molecules in time are then determined using the solutions found from the spatial and temporal parts of the QT description. The results of our simulations provide a quantification of the entropy generated due to irreversibilities at an atomistic level and show in detail the trajectory of the state of the system as the hydrogen molecules, which are initially arranged to be far from the carbon nanotube, spread out in the tank and eventually become more concentrated near the carbon atoms that make up the nanotube.

2. Application of QT

2.1. System description

The thermodynamic system under consideration is that of 4 H_2 molecules contained in a 5 nm x 5 nm x 10 nm tank with a carbon nanotube at its center. The nanotube is constructed based on [31] and is a (15,15) armchair carbon nanotube. The tube consists of a total of 900 atoms. The radius and the length of the tube are 1.017 nm and 3.56 nm, respectively. The carbon nanotube depicted in figure 1 is located at the center of the tank so that the center of the tube is placed at coordinates (5.0, 2.5, 2.5). The red spheres that comprise the tube represent the carbon atoms. The bonds connecting these atoms are not represented. To simplify the model, it is assumed that the atoms and the structure of the nanotube remain stationary and do not vibrate.

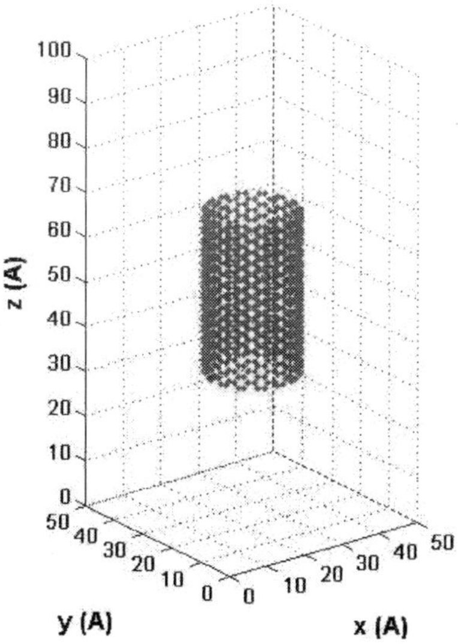

Figure 1. Schematic of the tank containing the carbon nanotube and the hydrogen gas.

2.2. Governing equations

Two systems of equations govern the evolution in time of the thermodynamic state and position of the H_2 molecules in the tank: one based on the Beretta equation of motion and the other on the energy eigenvalue problem. The former equation for a single constituent system is expressed in operator format by

$$\frac{d\hat{\rho}}{dt} = -\frac{i}{\hbar}\left[\hat{H}, \hat{\rho}\right] - \frac{1}{\tau}\hat{D} \tag{1}$$

while the latter is written as

$$\hat{H}u_i(\vec{x}_1, \vec{x}_2, ..., \vec{x}_n) = E_i u_i(\vec{x}_1, \vec{x}_2, ..., \vec{x}_n) \tag{2}$$

for $i = 1,2,...,L$. In Eq. (1), $\hat{H}$ is the Hamiltonian operator, τ a scalar time constant or functional, and $-\hat{D}/\tau$ the so-called dissipation term. The latter is a function of $\hat{\rho}$, $\ln\hat{\rho}$, and $\hat{H}$ and captures the nonlinear dynamics of an irreversible process by pulling the state operator or density matrix $\hat{\rho}$ in the direction of the projection of the gradient of the entropy functional $S = -kTr\,(\hat{\rho}\ln\hat{\rho})$ onto the hyper-plane of the constant energy (expectation value) of the system $E = Tr\left(\hat{\rho}\,\hat{H}\right)$.

Eq. (1) implies both the first and the second laws of thermodynamics [1]. Furthermore, τ has a lower bound and, thus, $\hat{D}$ an upper bound based on the time-energy Heisenberg uncertainty relation [16]. In addition, the state operator $\hat{\rho}$ is a linear, self-adjoint, non-negative definite, unit-trace operator (i.e., an operator whose diagonal elements sum to one) on Hilbert space $\mathcal{H}$. At a given instance of time, it is a representation of the state of a system which catalogues how the energy of the system is distributed amongst the various system energy eigenlevels. This state operator is based on a *homogeneous* ensemble of identical systems, identically prepared, i.e., on an *unambiguous* preparation [15], which contrasts with the state operator of SQM that is based on a *heterogeneous* ensemble of identical systems, not identically prepared, i.e., on an *ambiguous* preparation. The former leads to an unambiguous unification of thermodynamics and mechanics with a single kinematics and a single dynamics.

As to the system of Eqs. (2), E_i and u_i are the system-level energy eigenvalues and eigenfunctions, the $\vec{x}_k$ are the 3D particle position vectors, and L is the number of system-level energy eigenvalues. For the energies considered in this paper, only translational and rotational modes of energy storage and their degeneracies are relevant and, thus, included in the quantum mechanical description. These modes of storage are modelled and solved separately here: the translational as a set of two-body problems solved numerically using a finite element approach and the rotational analytically. The resulting expression for the energy eigenvalues and degeneracies for the latter are presented in the next section. For translational modes, the Hamiltonian operator is given by

$$\hat{H} = \sum_{k=1}^{n}\left(\frac{-\hbar^2}{2m}\nabla_k^2 + V_k(\vec{x}_1,\vec{x}_2,...,\vec{x}_n)\right) \tag{3}$$

and represents the kinetic and potential energies of the translational motions of the H_2 particles in the tank and the intermolecular interactions between H_2 particles and between the H_2 and C particles of the carbon nanotube.

2.2.1. The energy eigenvalue problem for rotation

Analytically, the rotational mode of energy storage for each H_2 molecule is modelled in its simplest form by assuming a rigid rotor model. The resulting eigenfunctions for the two degrees of rotational freedom are a product of an azimuthal quantum number (m) dependent exponential [32] and an angular momentum quantum number (l) dependent Legendre polynomial P_l^m [32] such that [29]

$$u_{l,m}(\theta,\phi) = \left[\frac{((2l+1)(l-|m|)!}{2(|m|+l)!}\right]^{1/2} P_l^m(\cos\theta)\frac{e^{im\phi}}{\sqrt{2\pi}}. \tag{4}$$

The corresponding rotational energy eigenvalue is expressed as [29]

$$\varepsilon_l = \frac{\hbar^2}{2I}l(l+1) \quad l \geq |m|, \quad m = 0,\pm1,\pm2,\ldots \tag{5}$$

where I is the moment of inertia equal to $m_{rm}r^2$, r the distance separating the hydrogen nuclei, and m_{rm} the reduced mass defined as

$$m_{rm} = \frac{m_1 m_2}{m_1 + m_2}. \tag{6}$$

To account for the degeneracies of the number of rotational orientations as well as of the spin that leads to ortho- and para-hydrogen, the following expressions are used for the rotational and the ground state spin degeneracies [29]:

$$g_l = 2l + 1 \tag{7}$$

$$g_{0_{sym}}^{(1,2)} = g_0(g_0 + 1)/2 \tag{8}$$

$$g_{0_{anti-sym}}^{(1,2)} = g_0(g_0 - 1)/2 \tag{9}$$

where g_0 is the ground-state spin degeneracy of a single nuclei equal to $2n+1$ and n is the quantum spin number, which for the ground state is 1/2. Eq. (8) represents the symmetric spin degeneracy which leads to ortho-hydrogen and Eq. (9) the anti-symmetric spin degeneracy which results in para-hydrogen.

2.2.2. The energy eigenvalue problem for translation

The computational difficulties of the eigenvalue problem, which represents a multi-body problem, quickly augment as the number of particles in the system increases. In fact, it may even become impossible to solve. To circumvent this, a set of 2-body problems [33], which define the motions and interactions between pairs of H_2 molecules (an H_2 and its nearest neighbour) and between each of these two H_2 molecules and the C atoms located on the nanotube, can be assumed. Such an approach has been used very successfully when using QT to derive stable equilibrium property relations such as, for example, the virial equation of state. For the H_2-C motions and interactions, which are of interest here, the 2-body problem is written as

$$\hat{H}_k u_j(\vec{x}_k) = \varepsilon_{k_j} u_j(\vec{x}_k), \quad j = 1,2,...,\mathcal{M} \tag{10}$$

for k = 1,2 (i.e., the 1st and 2nd H_2) and the translational Hamiltonian operator is given by

$$\hat{H}_k = \frac{-\hbar^2}{2m}\nabla_k^2 + \sum_l V_l(\vec{x}_k) \tag{11}$$

where the summation includes all of the H_2-C interactions. The potential functions $V_l(\vec{x}_k)$ are expressed by 6-12 Lennard-Jones potentials, i.e.,

$$V(\vec{x}_k) = 4\varepsilon\left[\left(\frac{\sigma}{r(\vec{x}_k)}\right)^{12} - \left(\frac{\sigma}{r(\vec{x}_k)}\right)^{6}\right] \tag{12}$$

where r is the distance between molecules/atoms. Values for the fitting parameters ε and σ specific to a particular interaction are given in table 1.

Table 1. Fitting parameters for the L-J potential.

Interaction	ε (J)	σ (nm)
H_2-H_2	5.24×10^{-22}	0.2915
H_2-C	4.502×10^{-22}	0.3137

In Eq. (10), each ε_{k_j} is a paired particle or so-called primitive-level energy eigenvalue related to the system-level energy eigenvalues E_{ki} by

$$E_{k_i} = \sum_{j=1}^{M} \nu_{ij}\varepsilon_{k_j} \tag{13}$$

where M is the total number of primitive energy eigenlevels utilized and ν_{ij} (the so-called occupation coefficient [33]) is the number of paired H_2-H_2 molecules for a given primitive energy eigenlevel j that occupies a given system energy eigenlevel i. If only a single pair of H_2 molecules were present in the tank (i.e., for N=1), all the occupation coefficients would be one. For the case of a closed, non-reacting system and $N \geq 2$, the number of system energy eigenlevels L is given by

$$L = (N + M - 1)!/(N!(M-1)!). \tag{14}$$

In general, the number of primitive energy eigenlevels is infinite for the translational Hamiltonian operator since this operator is unbounded. However, because of the computational burden of an infinite number of levels, a finite number $\mathcal{P}$ is assumed and of this finite number only a subset M is actually used, an assumption which under certain conditions is justified as described below.

For the paired H_2-H_2 motions, the 2-body problem is written as

$$\hat{H}_{1,2} u_j(\vec{x}_1,\vec{x}_2) = \varepsilon_{1,2_j} u_j(\vec{x}_1,\vec{x}_2),\ j=1,...,\mathcal{M} \tag{15}$$

and the Hamiltonian as

$$\hat{H}_{1,2} = \sum_{k=1}^{2}\frac{-\hbar^2}{2m}\nabla_k^2 + V(\vec{x}_1,\vec{x}_2). \tag{16}$$

This problem can equivalently be decomposed into a reduced mass (rm) and a center of mass (cm) problem [34] such that

$$\hat{H}_k u_j(\vec{x}_k) = \varepsilon_{k_j} u_j(\vec{x}_k), \quad j=1,2,...,\mathcal{M}. \tag{17}$$

In Eq. (17), $k = rm, cm$; and the Hamiltonian for both the reduced mass and center of mass problems is expressed by

$$\hat{H}_k = \frac{-\hbar^2}{2m_k}\nabla_k^2 + V(\vec{x}_k). \tag{18}$$

For the reduced mass problem, the potential function $V(\vec{x}_k)$ is given by the 6-12 Lennard-Jones potential (Eq. (12)), while for the center of mass, it is zero. Furthermore, the reduced mass is defined by Eq. (6) above, the center of mass by

$$m_{cm} = m_1 + m_2 \tag{19}$$

and the coordinates for each problem by

$$\vec{x}_{rm} = \vec{x}_2 - \vec{x}_1 \tag{20}$$

$$\vec{x}_{cm} = \frac{m_1\vec{x}_1 + m_2\vec{x}_2}{m_1 + m_2}. \tag{21}$$

Once the solutions to the set of 2-body problems listed above have been determined, they are assembled into a set of primitive-level energy eigenfunctions given by

$$u_p(\vec{x}) = u_{ijk}(\vec{x}_1, \vec{x}_2, \vec{x}_{rm}, \vec{x}_{cm}) = u_i(\vec{x}_1)u_j(\vec{x}_2)u_k(\vec{x}_{rm}) \quad i,j,k = 1,\ldots,\mathcal{M} \quad p = 1,\ldots,\mathcal{M}^3. \tag{22}$$

$\mathcal{M}$ represents the number of energy eigenvalue problems solved for each two-body problem. Neither the energy eigenfunctions for the center of mass nor for rotation are included in Eq. (22) because the energy eigenvalues for the former are two orders of magnitude below those of the other two-body eigenvalues and, thus, contribute little to each overall primitive-level eigenvalue and because for the latter the spatial and not the rotational orientation (a localized effect) of each hydrogen molecule in the tank is of interest here. However, as seen in Eq. (23), the rotational energies are included in the overall primitive-level energy eigenvalues since they are indeed significant, i.e.,

$$\varepsilon_p = \varepsilon_{ijkl} = \varepsilon_{1_i} + \varepsilon_{2_j} + \varepsilon_{rm\,k} + \varepsilon_{rot\,l} \quad i,j,k = 1,\ldots,\mathcal{M} \quad l = 1,\ldots,\mathcal{R} \quad p = 1,\ldots,\mathcal{P}. \tag{23}$$

Here $\mathcal{R}$ and $\mathcal{P}$ are the total number of possible degenerate rotational and primitive-level energy eigenvalues, respectively, that could be utilized. As explained below, the set of $\mathcal{P}$ primitive levels represents a much larger set than the set of M primitive levels actually utilized.

Now, when two or more H_2-H_2 pairs are present (i.e., when $N \geq 2$), a set of system-level energy eigenfunctions and eigenvalues must be determined based on the primitive-level ones utilized (i.e., the set of M levels). This is done by first determining the ν_{ip} coefficients in Eq. (24) via a combinatorial problem which fills an L x M matrix of possible occupation energies from which the system-level energy eigenvalues, E_i, are found using

$$E_i = \sum_{p=1}^{M} \nu_{ip}\varepsilon_p. \tag{24}$$

To find the system-level u_i, the primitive-level u_p are first projected from the decomposed space represented by $(\vec{x}_1, \vec{x}_2, \vec{x}_{rm})$ back onto 3D space represented by $(\vec{x} = \{x, y, z\})$ and then combined using

$$u_i(\vec{x}) = \prod_{p=1 \text{ and } u_p \neq 0}^{M} \left(u_p(\vec{x})\right)^{\nu_{ip}}. \tag{25}$$

This last result, Eq. (25), is a set of spatially dependent eigenfunctions, which are independent of time since the Hamiltonian operator and the boundary conditions (i.e., the tank walls and location of the carbon atoms) are fixed.

2.2.3. The equation of motion

To complete the thermodynamic description of the process, which the system of figure 1 undergoes, the solutions of the energy eigenvalue problem just described must be combined with those of the system of equations formed by the equation of motion, Eq. (1). In addition, since this is an initial-value problem, an initial condition for the system, i.e., an initial non-equilibrium state represented by an initial state operator or density matrix $\hat{\rho}$, must be determined. This can be done randomly or by using the procedure outlined in [35], which finds a set of partially canonical equilibrium states and then perturbs them to find a set of initial non-equilibrium states. Another approach and the one used here is to adjust the probabilities of the state matrix so that the likelihood of the hydrogen pairs at or near the boundary of the tank is very high.

Now, assuming that $\hat{\rho}$ is diagonal in the $\hat{H}$ representation, $\hat{H}$ and $\hat{\rho}$ commute and Eq. (1) reduces to

$$\frac{d\hat{\rho}}{dt} = -\frac{1}{\tau}\hat{D}. \tag{26}$$

Note that according to the Beretta equation, an initial $\hat{\rho}$, which is diagonal in the $\hat{H}$ representation, remains so at all times t. This special class of diagonal initial states is used here for simplicity and because in this case the eigenvalue ρ_i of $\hat{\rho}$ is readily interpreted as the probability of finding the system in the system-level energy eigenvalue E_i. Nonetheless, it should be emphasized that the Beretta equation yields well-defined evolutions for arbitrary non-equilibrium initial states $\hat{\rho}$ not necessarily diagonal in the $\hat{H}$ representation, i.e., when $\hat{H}$ and $\hat{\rho}$ do not commute.

Eq. (21) written out in terms of the eigenvalues ρ_i (probabilities) of $\hat{\rho}$ (for this particular case, the diagonal elements of $\hat{\rho}$) yields (for i=1,...,L)

$$\frac{d\rho_i}{dt} = -\frac{1}{\tau}\frac{\begin{vmatrix} \rho_i \ln \rho_i & \rho_i & E_i\rho_i \\ \sum \rho_i \ln \rho_i & 1 & \sum E_i\rho_i \\ \sum E_i\rho_i \ln\rho_i & \sum E_i\rho_i & \sum E_i^2\rho_i \end{vmatrix}}{\begin{vmatrix} 1 & \sum E_i\rho_i \\ \sum E_i\rho_i & \sum E_i^2\rho_i \end{vmatrix}}. \tag{27}$$

The dissipation term in Eq. (26) or (27) moves the system's entropy in the local direction of "steepest entropy ascent" [2] at constant energy $E = \sum_{i=1}^{L} \rho_i E_i$ and composition. This is expressed explicitly in terms of the time evolution of $S = -k\sum_{i=1}^{L} \rho_i \ln \rho_i$ by the entropy balance written as

$$\frac{dS}{dt} = \frac{k}{\tau}\frac{\begin{vmatrix} \sum \rho_i(\ln\rho_i)^2 & \sum \rho_i \ln\rho_i & \sum E_i\rho_i\ln\rho_i \\ \sum \rho_i \ln \rho_i & 1 & \sum E_i\rho_i \\ \sum E_i\rho_i \ln\rho_i & \sum E_i\rho_i & \sum E_i^2\rho_i \end{vmatrix}}{\begin{vmatrix} 1 & \sum E_i\rho_i \\ \sum E_i\rho_i & \sum E_i^2\rho_i \end{vmatrix}} \tag{28}$$

where the entropy generation term to the right of the equals is always non-decreasing.

2.2.4. Evolution in time of the thermodynamic state and position

Finally, with all the eigenvalues ρ_i of the state operator $\hat{\rho}$ known at each instance of time t and the system-level eigenfunctions u_i as a function of the spatial coordinates $\vec{x}$, the probability distribution function $\Phi(\vec{x},t)$ for the evolution in time and space of the state of the hydrogen particles in the tank is found from

$$\Phi(\vec{x},t) = \sum_{i=1}^{L} \rho_i(t)\left|u_i(\vec{x})\right|^2. \tag{29}$$

The first moment of this distribution function, used in the results presented below, is the mass density expressed as a function of time by

$$\rho(t) = \frac{\iiint \Phi \, m_{H2} n dV}{\iiint dV} \tag{30}$$

where m_{H2} is the mass of a hydrogen molecule and n the number of these molecules in the system.

3. Coarse graining the energy eigenlevels

3.1. Need for coarse graining

The complexity of solving the energy eigenvalue problem goes up markedly with the number of spatial dimensions considered (i.e., 1D, 2D, or 3D [12-14,35]). Consequently, for our 3D calculations, eigensolutions spanning only a limited energy range are calculated because our codes to date have only run on high-end PC workstations. Future work entails modifying and transporting the code to the massively parallel supercomputer, System X, available at Virginia Tech. For the results presented here, however, only two hundred eigensolutions ($\mathcal{M}$=200) are calculated for each translational two-body problem. Despite this limited number, the total number of possible primitive levels turns out to be $\mathcal{P}=\mathcal{M}^3\mathcal{R}= 1.2 \times 10^8$. If all of these are actually used so that $M=\mathcal{P}$, then for the case of N=2, the number of system levels L explodes to something on the order of 10^{15}. Obviously, for computational reasons, this number must be reduced to something more manageable, and this is done via the technique of coarse graining the energy eigenlevels.

3.2. Coarse graining

A coarse graining scheme is used here in which bundles of many system energy eigenlevel occupation trajectories are replaced with single representative trajectories. This is an improvement over our previous work [12], which also used methods that amounted to coarse graining in order to limit the number of levels and, thus, trajectories. However, the coarse graining employed in the present work is done in a more systematic way with the underlying justifications being more carefully considered.

In the context of our work, trajectories are traces that describe how the occupation probabilities of the energy eigenlevels of our system vary over time. They depict as a function of time the probability that a given energy eigenlevel will by occupied by the system. Figure 2 illustrates how the trajectories for energy eigenlevels that lie within a restricted energy range, i.e., within a so-called "bundle", tend to evolve similarly in time and might, therefore, be replaced by a single, judiciously chosen trajectory. The computational burden can thereby be reduced while retaining the major dynamical characteristics of the system

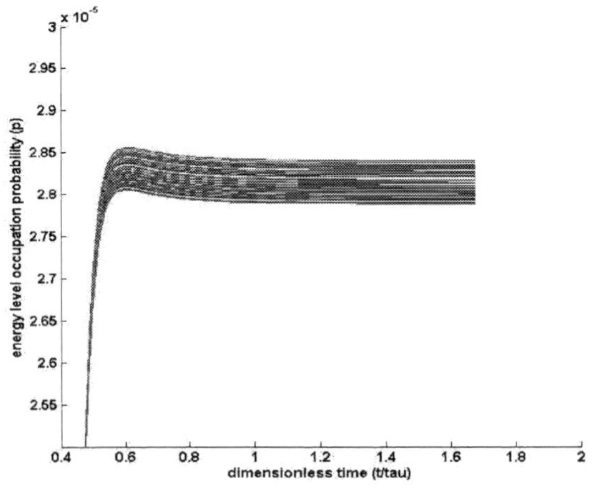

Figure 2. Solution of the Beretta equation for a set of energy eigenlevels over a limited energy range, i.e., for a so-called "bundle" of trajectories with energies close to one another and behaving similarly in time.

In order to help validate our coarse graining technique, we compare the results of coarse graining to the exact solutions for a given sample system. This is possible as long as the number of trajectories is not too large. The maximum number of trajectories that we have been able to solve to date on a high-end PC workstation has been on the order of a million. The exact results for a system having this many or fewer trajectories are, thus, compared to our coarse grained results. Provided the correlation between the coarse grained and the exact solution for this limiting case is good, a reasonable assumption is that the coarse grained results can be extrapolated to even higher numbers of levels for which exact solutions are not possible.

Figure 3 compares a set of trajectories from an exact solution with their coarse grained counterparts. In order to compare the fewer number of coarse grained trajectories with those for the exact solution, the number of trajectories for the latter is reduced by averaging their effects over the same energy ranges used to determine the bundles for the coarse grained trajectories. In figure 3, the averaged trajectories of the exact solution are steeper than those for the coarse grained solution, but the two sets of trajectories are seen to start at the same initial values and end at very near the same stable equilibrium values, thus, providing credence for the use of coarse graining as an approximate method. Of course, this leaves open the question of the proper scaling of the magnitude of the coarse grained solution since it necessarily underestimates that of the exact solution. This is addressed below.

As to determining the coarse graining itself, the technique used is to divide the entire energy range for the full number of energy eigenlevels into bundles having an equal number of levels. The level having the median index (quantum number) within a given bundle is chosen as representative of that bundle. The coarse grained levels, thus, reflect the overall density of levels for the larger system. Where the levels for the larger system are closely spaced or sparse, the coarse grained approximation is also closely spaced or sparse.

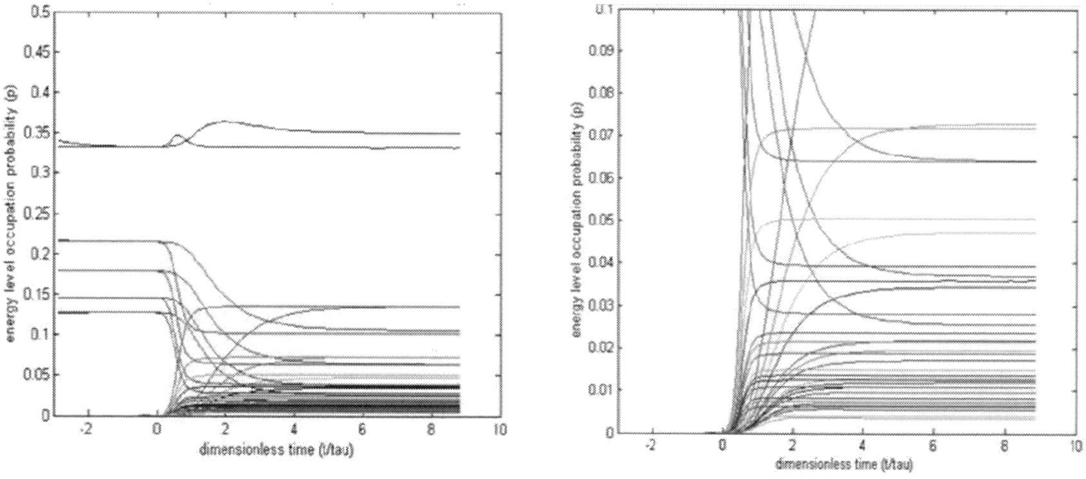

Figure 3. Coarse grained and exact trajectories are compared in a normal (figure on the left) and a magnified (figure on the right) view; paired trajectories (coarse and exact) have the same color with the steeper slope indicating the trajectory from the exact solution and the gentler the coarse grained one.

Now, the differences in trajectory slopes seen in figure 3 indicate that coarse graining introduces errors in magnitude that must be addressed if this technique is to be used to model larger systems. The number of trajectories affects, for example, the magnitude of the entropy of the system as well as the speed at which the system evolves towards stable equilibrium. Since the entropy for a given system energy is proportional to the natural log of the number of energy eigenlevels available to the system, the entropy of the coarse grained system will necessarily be lower than that for the larger (i.e., exact) system. In addition, systems accessing a higher density of energy eigenlevels also evolve more quickly. Thus, these effects must be taken into account by properly scaling the coarse grained results. If the scaling of the entropy and the time evolution of the coarse grained system is correctly done, the overall changes in entropy and the occupation probabilities with time can be made to closely represent the behavior of the larger system.

Figure 4 illustrates how coarse grained results are scaled to represent those of the larger system. In this figure, 20 energy eigenlevels representing the coarse graineed system are used to represent a system with 10,000 levels as illustrated by this diagram of entropy versus time. Proper scaling of the amplitude of the coarse grained curve allows one to match the full range of entropy for the larger system. Furthermore, by compressing the time axis, the slope of the trace is steepened and the rate of time evolution is made to correspond to the large system. Finally, a time shift is required to place the approximate curve on top of the curve for the larger system. Slight deviations are seen between the actual curve and the approximate one, but with these three simple scaling procedures a surprisingly close correspondence is achieved.

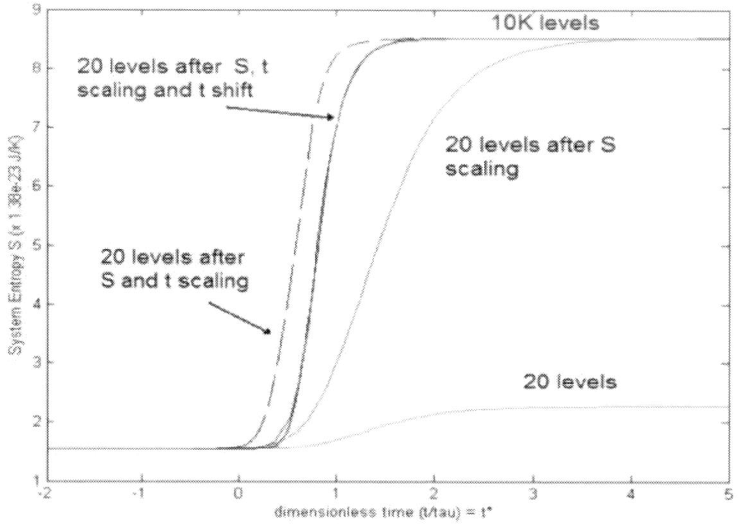

Figure 4. Scaling example of how a small number of coarse grained energy eigenlevels can closely approximate the collective behavior of many levels if proper scaling is used.

Figures 5 to 10 below provide relationships for the scaling factors for entropy, the time scale, and the time shift associated with the system depicted in figure 1. Two sets of scaling factors are needed: one for the primitive energy eigenlevels representing one hydrogen pair followed by the other for the system energy eigenlevels representing two hydrogen pairs. It is apparent from the primitive level data in figures 5 to7 that all three of scaling factors are proportional to the log of the number of

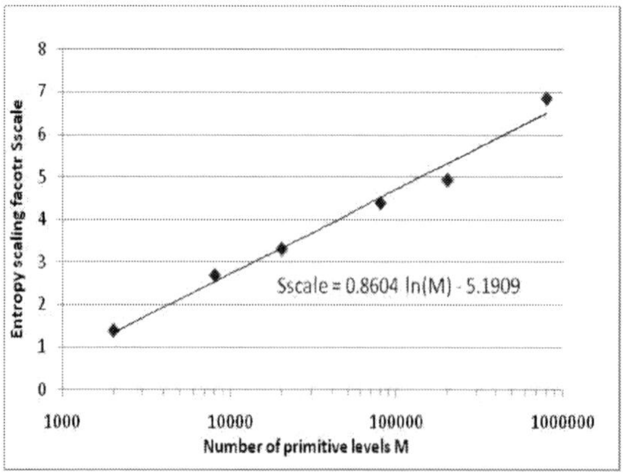

Figure 5. Entropy scale factor as a function of primitive levels; the scaling factor is based on 200 coarse grained levels.

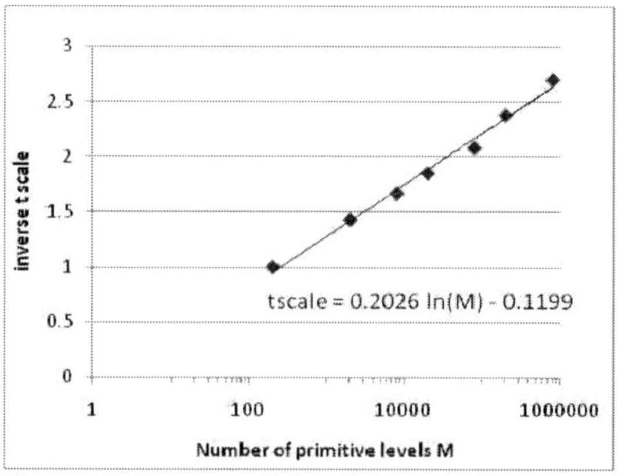

Figure 6. Inverse of the time scale factor as a function of primitive levels; the scaling is based on 200 coarse grained levels.

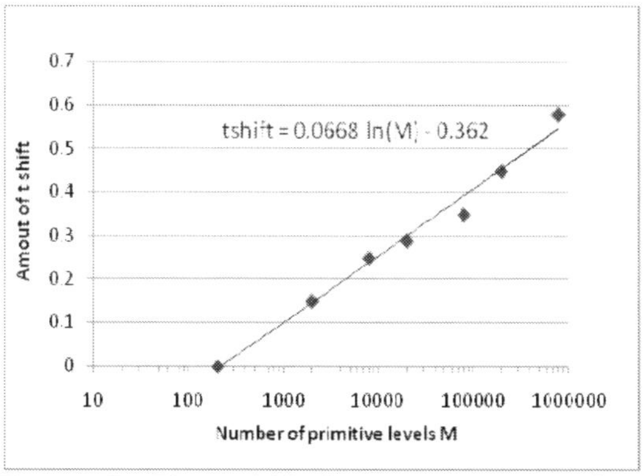

Figure 7. Time shift for 200 coarse grained levels as a function of the number of primitive energy eigenlevels.

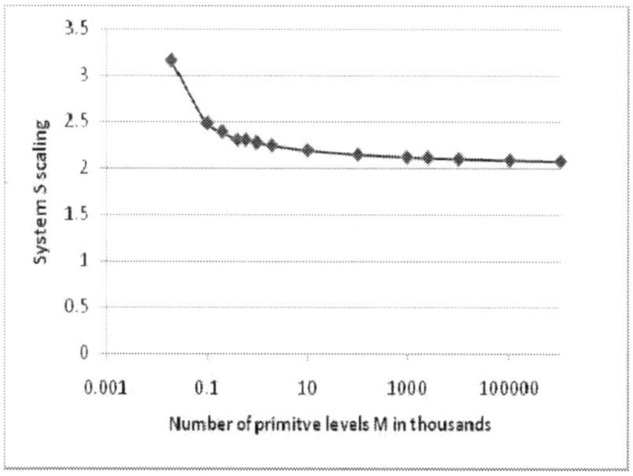

Figure 8. System-level entropy scaling factor as a function of the number of primitive levels for the case of N=2; the asymptotic limit is 2.

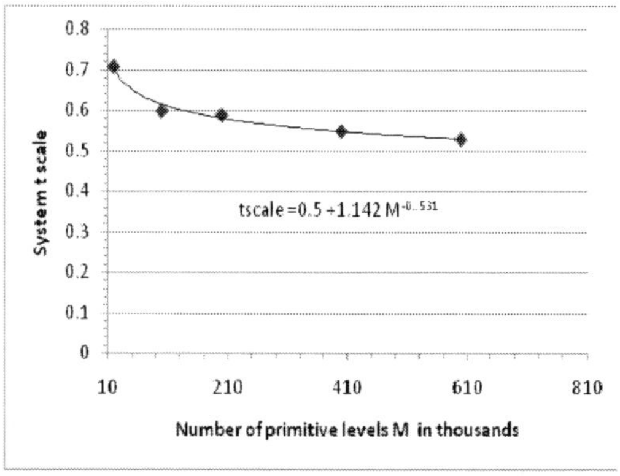

Figure 9. System-level time scaling factor as a function of the number of primitive levels for the case of N=2; the asymptotic limit is 0.5.

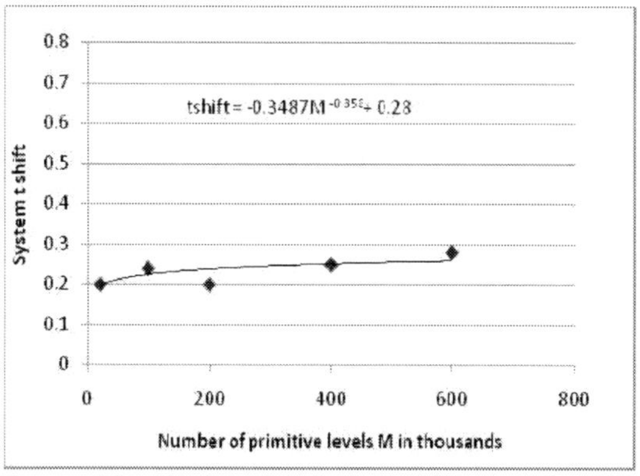

Figure 10. System-level time shift as a function of the number of primitive levels for the case of N=2; the asymptotic limit is 0.3.

primitive energy levels M that are utilized in the simulation. In contrast, the system-level scaling for two hydrogen pairs as shown in figures 8 to 10 is based on the expression

$$L = M(M+1)/2. \tag{31}$$

Since the number of system levels goes up geometrically instead of exponentially, the system-level scaling factors approach asymptotic values as M gets large.

4. Numerical approach

4.1. Numerical approach for solving the energy eigenvalue problems for translation

Each of the two-body eigenvalue problems for translation is numerically solved by the finite-element method (FEM). A proper weak formulation is used to apply FEM to this problem, namely,

$$\int \nabla u \cdot \nabla v \, d\vec{x} + \int V(\vec{x}) u \, v \, d\vec{x} = \epsilon \int u \, v \, d\vec{x} \tag{32}$$

where v is the test function and ϵ the eigenvalue considered. The solution u is approximated by second order Lagrange polynomials φ_j, i.e.,

$$u(\vec{x}) = \sum_{j=1}^{N_h} u_j \varphi_j(\vec{x}) \quad (33)$$

From Eqs. (32) and (33), the following generalized eigenvalue problem is deduced:

$$\hat{K}\vec{u} = \varepsilon \hat{M}\vec{u} \quad (34)$$

where $\vec{u} = \{u_j\}$, $\hat{K}$ is the stiffness matrix, and $\hat{M}$ is the mass matrix, the elements of which are given by

$$K_{ij} = \int \nabla\varphi_i \cdot \nabla\varphi_j d\vec{x} + \int V(\vec{x})\varphi_i \varphi_j d\vec{x} \quad (35)$$

$$M_{ij} = \int \varphi_i \varphi_j d\vec{x} \quad (36)$$

For the case of the center of mass *(cm)* problem, the potential is zero in which case the stiffness matrix becomes

$$K_{cm,ij} = \int \nabla\varphi_i \cdot \nabla\varphi_j d\vec{x} \quad (37)$$

Matrices $\hat{K}$ and $\hat{M}$ are properly modified to match the boundary conditions. The generalized eigenvalue problem is then solved by the Arnoldi algorithm [36,37] applied to a shifted and inverted matrix with restarts until the eigenvalues are found. Solutions are considered converged after the residuals are less than 1.0 x10^{-6}. The adopted grid consists of an unstructured mesh of 149,000 elements to ensure a grid independent solution.

4.2. Numerical approach for solving the Beretta equation of motion

A system of 20,100 first order ordinary differential equations (ODEs) are generated from the Beretta equation of motion and used to determine the evolution in state of the system in figure 1. This system of equations is solved numerically forward and backwards in time using a Runge-Kutta 4,5 (RKF45) method, starting from an arbitrarily chosen dimensionless time of $t^* = -2$. This Runge–Kutta–Fehlberg method [38-40] uses a fourth order approach together with a fifth order one by employing all of the points of the former plus one additional calculation required by a Runga-Kutta 5 method. The RKF45 method is, thus, able to estimate and control the error in the solution and determine an appropriate step size automatically. This makes the method efficient for ordinary problems of automated numerical integration of ODEs.

5. Results and discussion

Figures 11 to 14 show the evolution in time of the probability distribution function (Eq. (29)) for the

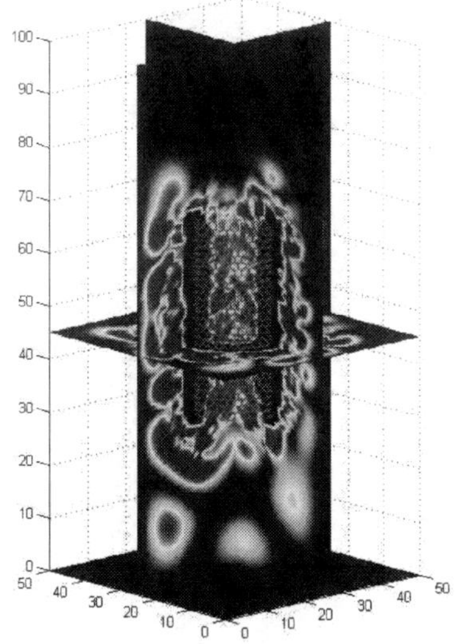

Figure 11. Plot of the probability density (probability distribution function) of particle position as a function of position at the initial non-equilibrium state at t^*=-2.0 (-9.5 unscaled) with N=2, M=200, and 20,100 system energy eigenlevels utilized.

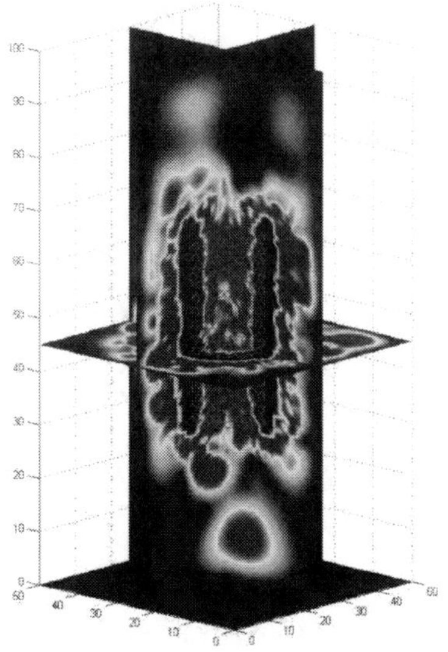

Figure 12. Plot of the probability density (probability distribution function) of particle position as a function of position at an intermediate non-equilibrium state at t^*=-0.224 (-2.0 unscaled) with N=2, M=200, and 20,100 system energy eigenlevels utilized.

system of figure 1 comprised of four hydrogen molecules (N=2) and the 900 carbon atoms that make up the carbon nanotube. At the initial (non-equilibrium) state at t^*=-2.0 (-9.5 unscaled), one can see in figure 11 that the highest values of the probability distribution function and, therefore, the highest densities (the dark red) of hydrogen occur in and around the tube as well as away from the tube and towards the bottom wall of the tank. In figure 12, this picture has evolved so that in this new non-equilibrium state at t^*=-0.224 (-2.0 unscaled), one can see a somewhat greater concentration of hydrogen inside the tube and nearer the outside tube walls. The concentration of hydrogen near the bottom of the tank has also increased as has the hydrogen in the upper part of the tank.

From this last state, the state of the system evolves to the state shown in figure 13. This new intermediate non-equilibrium state occurs at t^*=0.4894 (1.0 unscaled). As can be seen in this figure, the state of the system has evolved to the point that the value of the probability distribution function and, thus, the density of hydrogen has significantly decreased at the bottom of the tank and concentrated even more in and around the carbon nanotube.

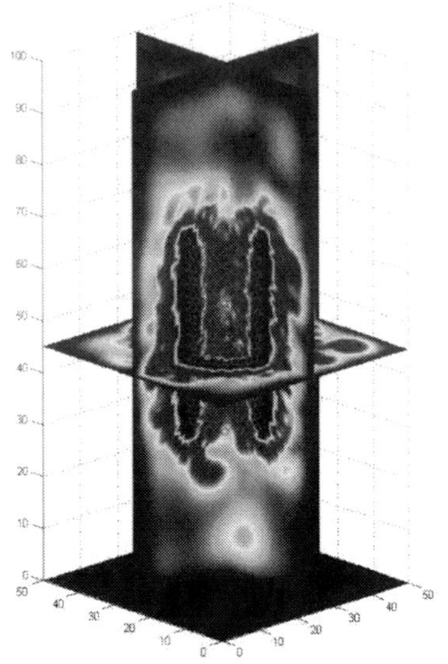

Figure 13. Plot of the probability density (probability distribution function) of particle position as a function of position at an intermediate non-equilibrium state at t^*=0.4894 (1.0 unscaled) with N=2, M=200, and 20,100 system energy eigenlevels utilized.

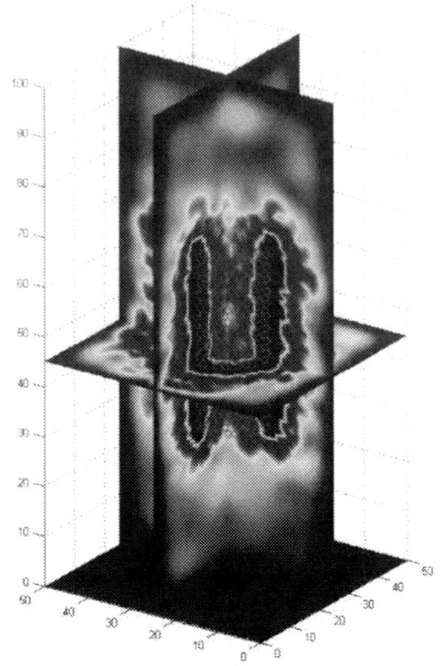

Figure 14. Plot of the probability density (probability distribution function) of particle position as a function of position at stable equilibrium at t^*=1.198 (4.0 unscaled) with N=2, M=200, and 20,100 system energy eigenlevels utilized.

Finally, figure 14 shows the final state of the system in stable equilibrium at t^*=1.198 (4.0 unscaled). As should be evident from the figure, the highest values of the probability distribution function and, thus, the hydrogen densities (the dark red) occur in the tube and close to the outer walls of the carbon nanotube. However, the hydrogen density inside the tube has decreased somewhat to 5.6 x 10^{-5} kg/m^3 from its previous value of 6.1 x 10^{-5} kg/m^3 at t^*=0.4894 (1.0 unscaled). This can be seen more clearly in figure 15, which shows the evolution of the hydrogen mass density as a function of time. Furthermore, the greater densities observed here in the interior of the tube are consistent with what has been observed both in a limited number of experiments and in, for example, the molecular dynamic (MD) simulations published in the literature [41]. Note also that the temperature at stable equilibrium is about 27.9 K which means that the hydrogen quantum molecular model used here is consistent with a gaseous state for the hydrogen since hydrogen has a triple point temperature of 14 K at 0.07 atm.

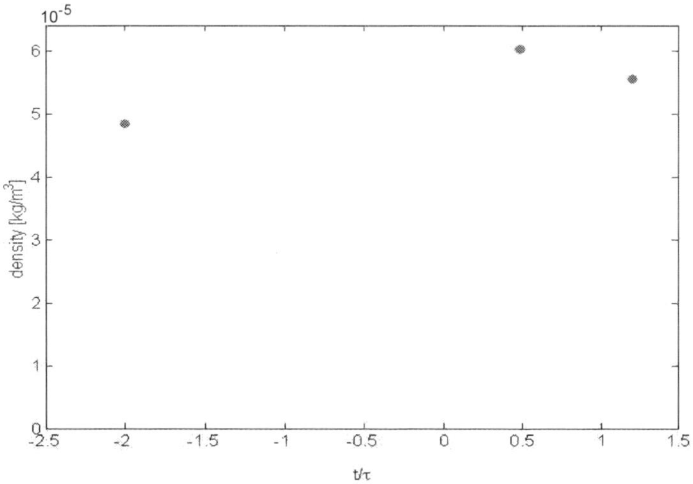

Figure 15. Plot of the mass density of hydrogen stored inside the carbon nanotube as a function of time.

Now, in order to assess the irreversibilities which occur during the thermodynamic process depicted in the previous figures, one can examine the rate of entropy generation as well as the change in the entropy due to irreversibilities. Both are shown in figure 16 and both results are scaled to represent the larger system with its 10^{15} system energy eigenlevels. The vast majority of the entropy creation and, thus, the increase in the entropy of the system occurs in a time interval of t^*=0.3 to t^*=0.6. This corresponds with the largest number of energy eigenlevels coming into play, taking up a share of the overall system's energy and becoming occupied as the state of the system moves towards the stable equilibrium (canonical) distribution (see figure 17). Note that the gaps in the energy eigenlevels seen in figure 17 are due to our current inability to calculate enough translational eigenvalues (M=200 currently) to fully span the energy range up to the lowest rotational energy eigenlevel.

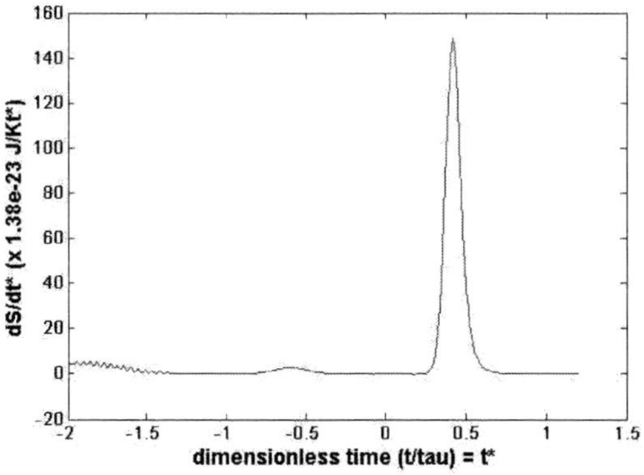

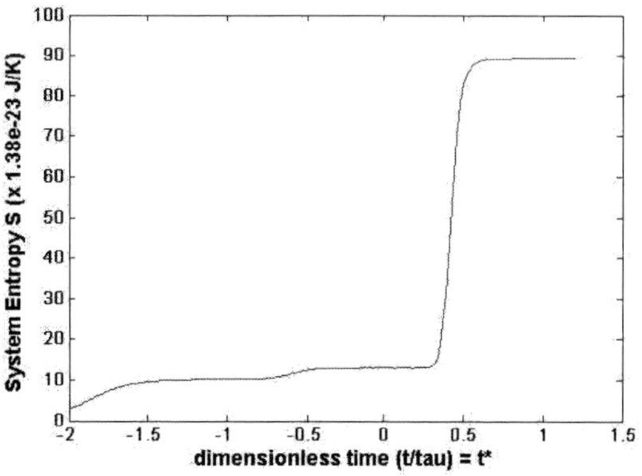

Figure 16. Evolution in time of the entropy generation rate (figure on the left) and the entropy (figure on the right); both have been scaled to represent the results of the larger system with 10^{15} system energy eigenlevels.

In order to view the evolution in state of the system in figure 1 in another more traditional light, the probability density distribution of energies is calculated for the non-equilibrium states through which the system passes as well as for the stable equilibrium state at which the system arrives. The

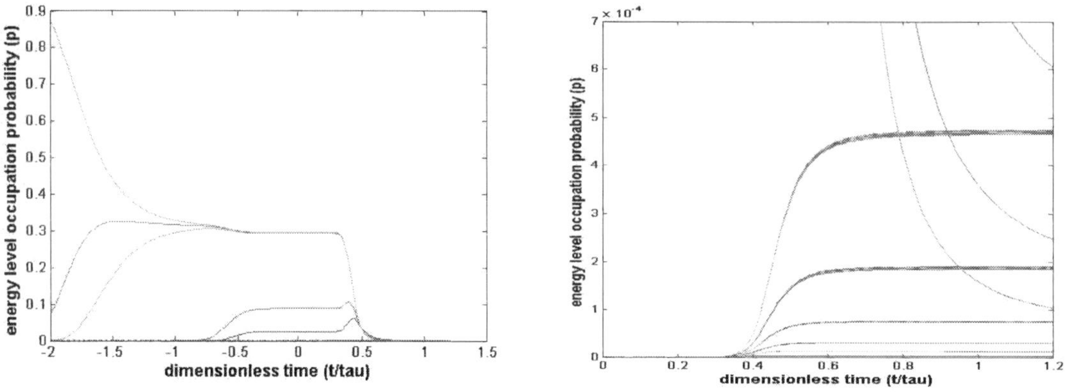

Figure 17. Normal (figure on the left) and magnified (figure on the right) views of the evolution in time of the 20,100 coarse grained energy eigenlevel trajectories; the trajectories represent simultaneous solutions to the Beretta equation.

results for the latter state and for two of the three non-equilibrium states are shown in figure 18. As can be seen, there is an evolution from a non-Maxwellian type of distribution at t^*= -2.0 and at t^*=0.224 to a Maxwellian type of distribution at stable equilibrium. This is what one would expect but what is significant here is that one does not have to guess at the distributions since they fall directly out of the physics/thermodynamics of the problem captured by both the Beretta equation of motion and the energy eigenvalue problem associated with the behavior of the hydrogen molecules relative to each other and to the carbon atoms in the tank. As in figure 17, the gaps in the distribution occur due to our current inability to calculate enough translational eigenvalues ($\mathcal{M}$=200 currently) to fully span the energy range up to the lowest rotational energy eigenlevel.

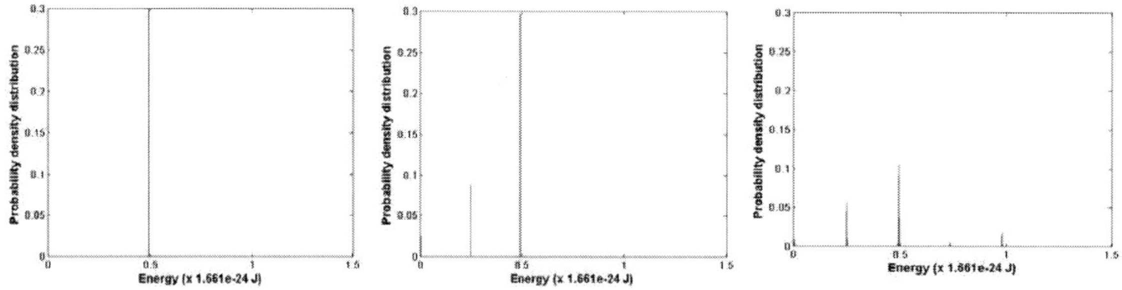

a) b) c)

Figure 18. Plot of the probability density distribution of energies at a) t^*= -2.0 (-9.5 unscaled), b) t^*=0.224 (-2.0 unscaled), and c) t^*=1.198 (4.0 unscaled) for N=2 and M varying between 200 and 20,100 coarse grained system levels.

Finally, to determine the minimum number of finite energy eigenlevels required for modeling the evolution of state of the system to an acceptable level of accuracy, a plot of the system energy E (an expectation value) versus the system entropy S (also an expectation value) is made for different numbers of eigenlevels including the so-called limit curve of an infinite number of levels. The results are shown in figure 19 along with the maximum thermodynamic temperature at which the accuracy for a given number of finite energy eigenlevels begins to decrease. As can be seen, with 20,100 coarse grained system energy eigenlevels utilized, the system can be modeled with good results for temperatures at stable equilibrium up to about 27.9 K. After that, the stable equilibrium curve for 20,100 levels begins to diverge significantly from the limit curve which is based on an infinite number of levels. To achieve higher temperatures, more system energy eigenlevels would have to be used. It should also be emphasized here that the thermodynamic temperature only has meaning at stable equilibrium and is not defined for non-equilibrium states. This, however, unlike in other approaches (e.g., MD), poses no problem for describing the thermodynamic evolution of state of the system since the equation of motion, the Beretta equation, does not require it.

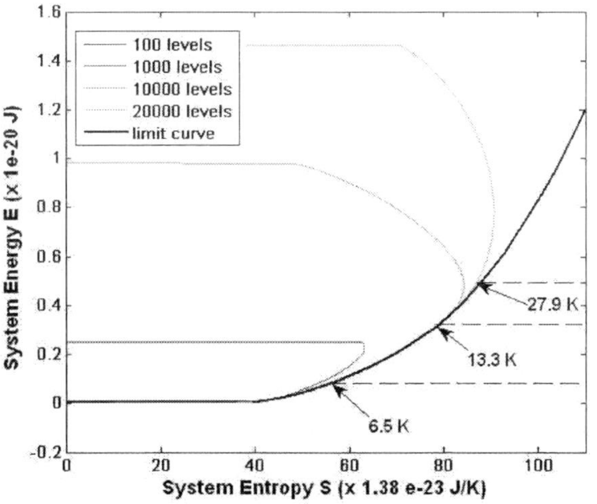

Figure 18. Expectation energy E versus the expectation entropy S for the hydrogen in the 3D tank with N=2 and M=200 and the number of coarse grained system energy eigenlevels varying between 100 and 20,000.

6. Conclusions

QT has been applied successfully here to modeling the 3D thermodynamic evolution in state and position of an atomistic or nanoscale system of hydrogen molecules contained in a tank at the center of which is a carbon nanotube. The thermodynamic description in time which results provides the exact path taken by the system in relaxing from some initial state far from stable equilibrium to a final state of stable equilibrium. The current improvement of our model to three dimensions holds the real prospect of being able to make future comparisons to the results of other well accepted simulation methods as well as experimental data. Such comparisons cannot help but be useful in making improvements to the present QT hydrogen storage model as well testing QT predictions with real world results.

Finally, the successful application of this new paradigm has been limited to date since any number of issues or problems remain to be resolved including how this paradigm can be applied to a much wider range of practical problems of interest (reacting systems, open systems, systems undergoing non-work interactions, high temperature systems, etc.). We are presently working on many of these issues.

Appendix

$\hat{D}$	dissipation operator
E	expectation energy
E_i	system-level energy eigenvalue
$\hat{H}$	Hamiltonian operator
$\hbar$	Planck's constant divided by 2π
k	Boltzmann's constant
L	number of system energy eigenlevels
M	number of primitive energy eigenlevels
m	mass
N	number of primitives

n number of particles or molecules
S expectation entropy
t time
t^* dimensionless time equal to t/τ
$u(\vec{x})$ energy eigenfunction
$V(\vec{x})$ interparticle potential function
$\vec{x}$ coordinate vector
x x-coordinate
y y-coordinate
z z-coordinate

Greek
ε fitting parameter for the L-J potential
ε_p primitive-level energy eigenvalue
$\Phi(\vec{x},t)$ probability distribution function of particle position in time
ν_{ip} occupation coefficients
$\hat{\rho}$ state operator or density matrix
σ fitting parameter for the L-J potential
τ scalar time constant or functional

References

[1] von Spakovsky M R 2008 *Proc. Am. Inst. Phys.*
[2] Beretta G P 2009 *Proc. Am. Inst. Phys.*
[3] Beretta G P 1981 *Dr. Sc. thesis*, MIT, Cambridge, MA
[4] Beretta G P, Gyftopoulos E. P, Park J L and Hatsopoulos G N 1984 *Il Nuovo Cimento* B **82** 169
[5] Beretta G P, Gyftopoulos E. P, Park J L and Hatsopoulos G N 1985 *Il Nuovo Cimento* B **87** 77
[6] Beretta G P 1984 *Proc. NATO Adv. Study Inst., NATO ASI Series* B*: Physics, Santa Fe, June 1984* **135** 193 (New York: Plenum)
[7] Beretta G P 2006 *Phys. Rev.* E **73** 026113
[8] Beretta G P 2005 *Modern Phys. Lett.* A **20** 977
[9] Beretta G P 2007 *Int J. Quantum Info.* **5** 249
[10] Gheorghiu-Svirschevski S 2001 *Phys Rev.* A **63** 022105 and *Addendum* **63** 054102
[11] Beretta G P 2008 *Int J. Quantum Info.* **11** 49
[12] Sciacovelli A, Smith C E., Verda V and von Spakovsky M R 2009, *Efficiency, Costs, Optimization, Simulation and Environmental Aspects of Energy Systems (ECOS'09)*, Parana, Brazil, ASME, Aug. 31 – Sept. 3
[13] Smith C E, Verda V and von Spakovsky M R 2008 *Efficiency, Costs, Optimization, Simulation and Environmental Aspects of Energy Systems (ECOS'08)*, Cracow, Poland, ASME, June 24-27
[14] von Spakovsky M R Smith C E, and Verda, V 2008 *ASME Int. Mech. Eng. Cong. and Expo. – IMECE'2008*, ASME Paper No. IMECE2008-67424, N.Y., N.Y., Oct.-Nov.
[15] Hatsopoulos G N and Gyftopoulos E P 1976 *Found. Phys.* **6** 15; Hatsopoulos G N and Gyftopoulos E P 1976 *Found. Phys.* **6** 127; Hatsopoulos G N and Gyftopoulos E P 1976 *Found. Phys.* **6** 439; Hatsopoulos G N and Gyftopoulos E P 1976 *Found. Phys.* **6** 561
[16] Beretta G P 2001 arXiv: quant-ph/0112046
[17] Beretta G P 2008 *Entropy* **10** 160
[18] Kleinekathöfer U 2009 *Springer Series in Chemical Physics* **93** DOI 10.1007/978-3-642-02306-
4-10 (Berlin: Springer-Verlag)
[19] Blum K 1996 *Density Matrix Theory and Applications, 2nd ed.* (New York: Plenum)
[20] Coffey W T, Kalmykov Y T, Titov S V and Mulligan BP 2007 *A Letters Journal Exploring the Frontiers of Physics, EPL* **77** 20011

[21] Yu T, Diosi L, Gisin N and Strunz W T 2000 *Phys. Lett.* A **265** 331
[22] Nakatani N and Ogawa T 2007 *Conference on Coherence and Quantum Optics (CQO)*, Rochester, New York, June 13
[23] Pomyalov A and Tannor D J 2005 *J. Chem. Phys.* **123** 204111
[24] Kossakowski A 1972 *Bull. Acad. Science and Math*. **20**, 1021
[25] Kossakowski A 1972 *Rep. Math. Phys.* **3** 247
[26] Ingarden R S and Kossakowski A 1975 *Annual Physics* **89** 451
[27] Lindblad G 1976 *Comm. Math. Phys*. **48** 119
[28] Gyftopolous E P and Beretta G P 2005 *Thermodynamics: Foundations and Applications,* 2nd ed.
(Mineola, NY: Dover)
[29] Tien C L and Lienhard J H 1979 *Statistical Thermodynamics, revised printing* (Washington,
DC: Hemisphere Pub.)
[30] Hill T L 1956 *Statistical Mechanics: Principles and Selected Applications* (New York: McGraw
Hill)
[31] Frey J T and Doren D J 2005 *TubeGen 3.3* (web-interface, http://turin.nss.udel.edu/research/ tubegenonline.html), Univ. of Delaware, Newark, DE
[32] Knuth E L 1966 *Introduction to Statistical Thermodynamics* (New York: McGraw Hill)
[33] Hatsopoulos G N and Gyftopoulos E P 1979 *Thermionic Energy Conversion – Vol. 2: Theory, Technology, and Application* (Cambridge, MA: MIT Press)
[34] Shankar R 1994 *Principles of Quantum Mechanics* (New York: Springer Science & Business Media, Inc.)
[35] Smith C and von Spakovsky M R 2007 *Int. Mech. Eng. Cong. and Expo. – IMECE'2007* Paper No. IMECE2007-42933 (New York)
[36] Arnoldi W E 1951 *Q. Applied Mathematics* **9** 17-29
[37] Quarteroni A, Sacco R and Saleri F 2000 *Numerical Mathematics* 215 – 218 (Springer)
[38] Fehlberg E 1969 "Low-order classical Runge-Kutta formulas with step size control and their application to some heat transfer problems," NASA Technical Report 315
[39] Fehlberg E 1970, *Computing (Arch. Elektron. Rechnen)* **6** 61–71
[40] Hairer E, Nørsett S and Wanner G 1993 *Solving Ordinary Differential Equations I: Nonstiff Problems*, 2nd ed. (Berlin: Springer-Verlag)
[41] Banerjee S and Puri K 2008 *Nanotechnology* **19** 155702 (6pp)

Marcos Moshinsky –teacher, colleague, scientist

Kurt Bernardo Wolf

Instituto de Ciencias Físicas
Universidad Nacional Autónoma de México
Av. Universidad s/n, Cuernavaca, Morelos 62210, México

bwolf@fis.unam.mx

Abstract. It is said that a man lives every time his name is pronounced, when his ideas are discussed, and when the result of his actions continues to benefit the community. In this regard, Marcos Moshinsky will be in the midst of our generation of students and colleagues as we often refer to his careful instruction and generous collaboration. And beyond, his name will be remembered for as long as his seminal works on quantum and nuclear physics, group theoretical methods, and mathematics, remain in the foundations of the edifice of science.

1. A child comes to Mexico, a student goes to Princeton, a scientist matures

Marcos Moshinsky was born in Kiev, Ukraine, in the spring of 1921 during very difficult times. His father, Israel Moshinsky, managed to escape with his family first to British Palestine, and then to Mexico, whose government was then fostering the influx of European migrants. As many refugees from a broken Europe, in time and with hard work he prospered and contributed generously to the community by funding the building of the Colegio Israelita in the 1940s, among other donations through the years.

Marcos was among the first students to graduate in Physics from the Facultad de Ciencias of the Universidad Nacional Autónoma de México (UNAM), which was then lodged at the School of Mining, a late eighteenth-century Borbonic building in downtown Mexico City. His B.Sc. thesis director was Dr. Manuel Sandoval Vallarta, who was also Professor at M.I.T. and had a prominent role in the study of cosmic rays. The young Marcos spent some months in Sandoval Vallarta's laboratory, housed on the School's rooftop, registering the crackling sound of a detector. He enjoyed telling us that he had become very excited when he noticed an almost-periodic increase in flux every 10 minutes or so. Could this be an important phenomenon at low latitudes? Was it some kind of a resonance? Then he would laugh: *No, it was the electric tram that passed by the street below*. Probably then he decided to be a theoretician.

Another anecdote that should not be lost concerns his doctoral studies in Princeton, recommended by Prof. Sandoval Vallarta. This was late 1947, when communism was starting to raise fear in the hearts of the U.S. administration. Marcos acted out the scene as told me of the occasion when he applied for a student visa at the embassy. *Born in Kiev? hmm...* Yes. *What do you want to study?* Nuclear physics. *...No!* shouted the clerk as if he were Senator Mc Carthy himself. Eventually, Prof. Sandoval Vallarta obtained for his student a diplomatic passport, and thus Marcos was able to reach

Princeton, to study under Eugene Wigner and contribute with him, and with Valentin Bargmann, several cornerstones of group theory and nuclear science.

By talent and circumstance, Marcos Moshinsky participated in the birth of many Mexican scientific institutions, such as the Sociedad Mexicana de Física, the Academia Mexicana de Ciencias, the Escuela Latino Americana de Física, the Revista Mexicana de Física, and the Centro Internacional de Ciencias (CIC), here in Cuernavaca, on which I will write some lines below. He was also a forceful figure behind the scenes, being widely known and respected by university rectors and government officials. Yet he always remained "a civilian," that is, he never had academic-administrative duties such a directorship (incongruously reserved by UNAM's statutes only to Mexicans by birth...), or other managerial tasks beyond the occasional organization of schools and conferences. Yet he became, surely in spite of himself, the popular icon of what a Mexican scientist looks like. The official school calendar issued by the Ministry of Public Education bears his photograph this year.

2. The teacher

As we started in the 1960s to learn physics at the Facultad de Ciencias of UNAM, in the new Ciudad Universitaria, we knew that during the senior year the course in Quantum Mechanics would be taught by Moshinsky. We trepidated as he entered the lecture room for the first time, but rather quickly realized that, even though he sometimes appeared formal and dour, his strictness resided in the treatment of this difficult subject in a precise and honest manner, and not at all in his person, who was very –very– human. I quote Alejandro Frank, one of his former students, who condenses Marcos' style of teaching: *Sometimes our questions were quite naïve, but Marcos pondered on them, and often he came with a different, deeper interpretation that rescued our query from total inanity.* And Marcos' view of his charges: *Scientific education is like a tennis match; at first, the student will return the ball in simple and predictable ways, but then you may have the pleasure of detecting talent when on occasion you see him answer with a strong hit in an unexpected direction.*

Many of us did our theses under his guidance, and now there are second and third generations of Moshinsky students; his grandstudents. Personally, I decided to continue my Ph.D. studies in Israel and do my postdoc in Sweden, but on returning to Mexico I came rather naturally to collaborate with Marcos on hidden symmetries and on canonical transformations. Then came years when our interests lost direct contact, but as with his other colleagues and former students, when a hard research problem appeared, we would discuss it with Marcos, who had the knack of asking the right questions and to find the essence of the problem in a simpler model –very often in the harmonic oscillator. As he used to say: *Two types of problems exist in quantum mechanics, those that you cannot solve and the harmonic oscillator. The trick is to push a problem from one category to the other.*

3. The role of Marcos Moshinsky toward the Centro Internacional de Ciencias –Cuernavaca

In 1980 Mexico had good fortune, with high oil prices and several intelligent people with scientific training in the Ministry of Public Education. It was the brainchild of Jorge Flores and Edmundo de Alba to set up in Cuernavaca a center for scientific meetings on the model of the International Centre for Theoretical Physics, then directed by Prof. Abdus Salam, which had established its home in Miramare, Trieste. Professor Moshinsky was offered the role of founding director; for his own reasons he declined, but kept close to the project during subsequent meetings with international functionaries, to which he invited me. It would be called CIFMO: Centro Internacional de Física y Matemáticas Orientadas (the term 'orientadas' –never fully defined– would be later changed to `aplicadas', which makes more sense). During the ninth group theory Coloquium (ICGTMP-1980), which Marcos Moshinsky and a group of us organized in Cocoyoc, I suggested to him that this would be the best forum to announce the International Center project. Rather wryly he told me: *be skeptical of such projects in this country, and take nothing for granted until it actually breathes.* And he was right, because the project was not put into the best hands, was deviated from its purpose and thus delayed; by the end of 1982 it fell prey to the economic crisis of that year and the change of administration. In 1984 the CIFMO project was revived, but in 1986 again it was nixed by superior ignorances.

Fed up with this dithering, among the original promoters of the project it was decided in December 1986 to set up the Cuernavaca Center as an Asociación Civil, independent of governmental patronage. Professor Moshinsky accepted to be President of the association and, in my view, he was the rock of its existence. During each of the following eight years CIFMA convened three or four national and international meetings on physics, mathematics, and several other related venues. For each event we requested and obtained funds from a variety of sources (UNAM, CONACyT, PEMEX, ICTP, IAEA, IUPAP, IAMP, NSF, and others). During that time, CIFMA was in two drawers of my desk at the UNAM campus in Cuernavaca, but we had the beautifully efficient help of Mrs. Petra Seligman, our indispensable and unforgettable administrator. We took care not to burden Marcos with micro-administration or crisis control; things were done as he would have approved, and his trust was always generous and firm.

During the summer of 1989, the Moshinsky family –Marcos and his sisters, Estela de Klip and Ruth de Rosengauss– gave the crucial support and momentum to turn CIFMA into the fully fledged Centro Internacional de Ciencias (as it is called since 1996), by offering the fund that their father Israel had bequeathed for the advancement of science in Mexico, as seed money to construct the excellent installations that CIC has now in Cuernavaca. I had first to find a good location –providentially provided by UNAM in commodatum at its joint campus with the Universidad Autónoma del Estado de Morelos (UAEM)– and then concurrent funds from UNAM and several private donors. Thus the building was inaugurated on May 25, 1996, with the presence of the rectors of UNAM, UAEM, and the state governor.

With the subsequent directors, Federico Sánchez, Thomas Seligman, and Marco José, CIC has focused its activities in *gatherings* rather than conferences. These gatherings convene small groups of scientists, nationals and visitors, active researchers and postdocs, for two or three weeks of close interaction; Marcos Moshinsky was often there and always a welcome guest.

4. The scientist

The contributions that Professor Moshinsky made to science are witness to his lasting presence among us and future generations. Many of them are well known to the participants of the Bregenz meetings on Symmetry in Science, which he rarely missed when convened by Bruno Gruber, and now by Dieter Schuch. A long time companion of Marcos Moshinsky, Professor Yuri F. Smirnov, is also remembered in these Proceedings; they formed a powerful team during the years they worked together on their book [1], a *tour de force* through the methods and results of group theory in n-body quantum systems. The last third of the book is devoted to the Dirac oscillator; it cannot be an accident that the Dirac equation, besides applying to the free particle, electromagnetism and the hydrogen atom, also describes systems with oscillator interparticle potentials. The expectation we all had was that this model would explain the hadron spectrum, which it did not; but to my biased understanding, Nature cannot let its elegance remain unrealized.

Foremost, Marcos Moshinsky was a researcher. Yet I would defer this fundamental aspect of his activity to the Laudatio written by Pavel Winternitz for the occasion of the Wigner Medal ceremony at the Group22 Colloquium, held in Hobart in July 1998 [2,3]. *I can think of only a few examples where the creation of a school of modern physics in a country can be attributed to one person. A person who has the necessary scientific talent, but also the correct personality to create a school. That is to devote time and attention to students, attract collaborators, influence authorities, generate ideas for many to work on, and in general, create a fruitful scientific and human atmosphere. Examples that come to mind are Enrico Fermi in Italy, Leopold Infeld in Poland, and Marcos Moshinsky in Mexico.*

During his long life, Marcos Moshinsky was awarded all Mexican scientific prizes and many international ones, including the 1988 Príncipe de Asturias prize by Spain, the 1991 Andrei Sakharov Medal for Human Rights, the 1997 UNESCO prize for Scientific Research, and the 1998 Wigner Medal. He was member by invitation to scientific academies of Argentina, Brazil, the E.U., U.S., TWAS, and the Vatican, and welcomed as Doctor Honoris Causa in several universities, including the Goethe University in Frankfurt. This recognition of the man means more than any count of articles, citations, H-numbers, or other numerical indicators for reports to which many of us have had to bow.

5. A pearl of wisdom

Certain events we lived through in our youth come back as blessed bits of wisdom. Years ago, as I was in Benares, it happened that I stumbled across a funeral procession and willingly joined the crowd on its way to the river Ganges. There were flowers, music and chanting; nobody seemed to be mourning at all. I asked for whom was this ceremony held; it was the matriarch, the mother, grandmother and great-grandmother of a large clan of the city's weavers. *One should not be saddened by the inevitable*, they told me, *and we are happy because we see the numerous progeny in which she continues to live*. We thus know that Marcos lives on in us.

References

[1] M. Moshinsky and Yu.F. Smirnov, *The Harmonic Oscillator in Modern Physics* (Hardwood Academic Publishers, 1996).

[2] *Group22. Proceedings of the XXII International Colloquium on Group Theoretical Methods in Physics*, Ed. by S.P. Corney, R. Delbourgo, and P.D. Jarvis (International Press Inc., Boston, 1999), pp. xvi-xxi.

[3] *Testimonios. Marcos Moshinsky: 80 Años de Vida y 60 Años de Trabajo Científico*, Ed. by A. Frank and K.B. Wolf (CIC-UNAM, México DF, 2001), pp. 160-165.

AUTHOR INDEX